Foraging for Survival

Foraging
for
Survival

Yearling Baboons in Africa

Stuart A. Altmann

The University of Chicago Press
Chicago and London

STUART ALTMANN is professor emeritus in the Department of Ecology and
Evolution, the Committee on Evolutionary Biology, and the Committee
on Human Nutrition and Nutritional Biology at the University of Chicago.
He is coauthor of *Baboon Ecology: African Field Research* and editor
of *Social Communication among Primates*.

The University of Chicago Press, Chicago 60637
The University of Chicago Press, Ltd., London
© 1998 by The University of Chicago
All rights reserved. Published 1998
Printed in the United States of America

07 06 05 04 03 02 01 00 99 98   1 2 3 4 5

ISBN: 0-226-01595-5 (cloth)

Library of Congress Cataloging-in-Publication Data

Altmann, Stuart A.
    Foraging for survival : yearling baboons in Africa / Stuart A. Altmann.
        p.   cm.
    Includes bibliographical references and index.
    ISBN 0-226-01595-5 (cloth : alk. paper).
    1. Baboons—Nutrition—Kenya—Amboseli National Park.
2. Baboons—Food—Kenya—Amboseli National Park.   I. Title.
    QL737.P93A58   1998
    599.8′65153—dc21   97-45726
                                                                    CIP

for Jeanne,

because

Feeding is such a universal and commonplace business
that we are inclined to forget its importance. The primary
driving force of all animals is the necessity of finding the
right type of food and enough of it.

—C. Elton, 1927

A defensible case can be made for the view that the main
driving force in the evolution of animals from the most
primitive forms onward has been the supply of nutrients
in the environment.

—H. N. Munro, 1986

Contents

Acknowledgments

Although my name is the only one on the title page of this book, let no one think that I carried out this project alone, or indeed that I could have. I am deeply grateful to many people and institutions whose contributions were essential.

Permission to carry out this project in Amboseli National Park was granted by the National Council for Science and Technology of the Republic of Kenya; by Perez Olindo, then Director of National Parks; and by Joseph Kioko, who at the time was Warden of Amboseli.

The research was supported by research grants MH 19617 from the National Institute of Mental Health and 15007 from the National Institute of Child Health and Development, and by the Abbott Fund of the University of Chicago.

The late Stephen Wagner collaborated with me on the optimal diet model that is at the core of this enterprise. Michael Altmann solved the critical problem of how to maximize rates in linear optimization (appendix 6) and wrote a computer program for locating references in text. Other computer programs for this project were written by Salvador Arias, William Bean, Sam Fenster, Simranjit Singh Galhotra, Rebecca McCauley, Douglas McIntosh, Jeff McIver, and Cathy Smithers.

Essential plant identifications were provided by Christine Kabuye and others at the East African (now Kenya) Herbarium. The staff of Kew Gardens, David Mitumu, and Robert Fadens provided additional help with problems in plant taxonomy.

For contributions to the long-term demographic data on Amboseli baboons, without which I could not have carried out the analysis of the fitness correlates of foraging, I am grateful to Susan Alberts, John Kamau, Barbara Kind, Bosco Kitibo, Janet Friedman Mann, Philip Muruthi, Raphael Supeet Mututua, Ronald Notë, Michael Pereira, David Post, Francis Saigilo, Amy Samuels, Sarah Sayialel, Carol Saunders, Jane Scott, Jennifer Shopland, Joan Silk, Sarah Sloane, Alberta Sluijter, Peter Stacey, David Stein, Karen Streier, David Takacs, and Jeff Walters.

Insightful comments on various drafts of this work were provided by Jeanne Altmann, Glenn Hausfater, David Post, and two anonymous reviewers.

I am indebted to many other people for assistance in various forms. This terse list cannot adequately express my appreciation for their help. Heartfelt thanks to Rachel Altmann, David Aulik, James Else, Sam Fenster, Nenetzin Gerald, Mimi Getz, Wendy Goldberg, Frederic Janzen, Carolyn Johnson, Christina and Joseph Kioko, David Klein, Hans Kummer and his family, Michael Liang, Joan Luft, Deborah Malamud, Peter McCullach, Jane Maienschein, Kenneth Nagy, Perez Olindo, John Roberts, Jane Scott, David Sherman, Irene Stadnyk, and Jeff Stelzner.

At the University of Chicago Press, the complex process of publishing was handled so professionally that a potentially traumatic experience was actually both pleasant and enlightening. The staff made the difficult seem easy, brought lucidity to the obscure, and handled everything with a sense of style and appropriateness. My thanks to David Aftandilian, Dennis Anderson, Alice Bennett, Teresa Biancheri, Liz Demeter, Russell Harper, Christie Henry, and JoAnn Kiser.

Above all, I want to thank Jeanne Altmann—for the many ways she contributed to every component of this project and for her unfailing personal support during its long gestation. To her, with all my heart and my deepest gratitude, this book is dedicated.

1

He who does not mind his belly will hardly mind anything else.
—Samuel Johnson

Introduction

This is the story of how eleven wild baboon infants fed themselves—some successfully and some not—at the time of weaning, when for the first time in their lives they could no longer depend entirely upon their mother's milk for sustenance (fig. 1.1). It is a story of adequate nourishment and a vigorous life for some and of hunger, malnutrition, and death for others. Yet all these infants lived in the same social group and moved with the other members of the group to sources of food, water, and shelter as they made their daily trek across the African savannah. The same foods were available to them all.

Baboons (*Papio*) are among the most widespread, abundant, and adaptable of the cercopithecine primates. They are found throughout most of the vast African continent.

> Although baboons are found in environments ranging all the way from moist, evergreen forest to semi-desert steppe, most of them live in savannah habitat, that is, in areas in which perennial grasses are the dominant ground cover, and in which trees, though of variable density, are never completely absent. . . . On the savannah, the success of baboons depends upon their ability to exploit a wide variety of plant and animal food sources, and to feed selectively on some of the most concentrated sources of nutrients in their environment. (S. Altmann and J. Altmann 1970, pp. 213–14)

That much was apparent from what was known by the end of the 1960s about baboon natural history. However, subsequent long-term studies of baboons in their natural habitat have revealed considerable variability in the biological fitness of baboons: in their ability to survive and reproduce under the conditions they encounter in the wild. This variability is apparent whether the animals are viewed at the level of the individual, family lineage, social groups, or whole populations: some survive and perpetuate their kind, others do not.

To what extent are such differences in biological fitness related to differences in foraging? At the level of individual differences, I address that question in this study of the feeding behavior of eleven yearling baboons. Doing so requires and entails answers to several other major questions about foraging and food selection. What is the array of foods that makes up a yearling baboon's

1

Fig. 1.1. An infant baboon being weaned. A. Rebuffed infant coos. B. Infant's attempt to get to mother's nipples is blocked. C. Infant grimaces and cackles. Photos by Jeanne Altmann.

diet? How much of each food does it eat? What amounts of nutrients and toxins is it getting from these foods? What about the composition of those potential foods that it discards or apparently ignores?

The foregoing questions are concerned with what the baboons eat and what these foods contain. Others focus on the consequences of the diets that they select. For example, can we say that one potential diet is in some sense the best diet for a young baboon, and if so, how close does each individual's diet come to this optimum? How different is the diet of one yearling from that of the other members of its cohort? What are the long-term consequences of these differences for survival and reproduction? What would be the consequences if the yearlings consumed a different array of foods, or the same foods in different proportions? These are the kinds of questions to which this volume is devoted.

Why Study Foraging?

The foraging behavior and diet selection of animals in their natural habitat have been the focus of many recent studies. In addition, numerous models of optimal foraging and diets have been developed, and in many cases these have been tested against what animals actually do when searching for food. Pyke, Pulliam, and Charnov (1977) and Stephens and Krebs (1986) provide excellent surveys of developments in this area.

Why such interest in what wild animals eat? The most obvious reason is that foods, or rather, particular nutrients obtained from them, are essential to running the machinery of the body. In addition, success at foraging affects an animal's chances of surviving and reproducing. For many animal populations, food is thought to be the primary limiting factor for population growth. In turn, demographic processes affect social behavior through their effects on group composition (S. Altmann and J. Altmann 1979). Potentially, feeding and social behavior also affect each other more directly. A working assumption in behavioral biology is that animal social behavior has evolved, and that much of this evolution reflects either effects of social interactions on foraging and feeding or constraints on social relationships that food imposes.

There are other reasons for this interest. Natural selection for effective foraging and feeding has affected almost every part of the body, not just the digestive system (S. Altmann 1984). An animal's success at foraging may depend on keen vision and a keen sense of smell to locate choice food, on lips, tongue, teeth, and in some cases hands or feet, for harvesting, hulling, and macerating the food, on an enzyme and gut system capable of extracting and absorbing usable compounds and detoxifying others, on sensitive ears to detect predators that approach while the animal is feeding and a good set of legs to outrun them, on display characters and weapons to defend the foraging area against encroaching competitors, on specializations that enable the animal to spend long periods foraging under adverse environmental conditions, and so forth. In the course of evolution, every organ system of animals probably has been altered by the exigencies of obtaining food. The orders of mammals primarily represent divergences along dietary lines.

Foraging also affects many of the relationships between animals and the other members of their ecosystems, particularly those they eat and those that eat them. "The whole of nature . . . is a conjugation of the verb to eat, in the active and the passive" (Inge 1927). Carnivores, herbivores, and plants have coevolved.

In sum, feeding and foraging have occupied a central role in the adaptations and evolution of animals. Just how much they contribute to differential survival and reproduction among members of any given population can only be determined empirically.

Intensive study of feeding behavior and dietary intake in wild animals, especially primates, may be of particular relevance to human biology. The diet of the protohominids must have been an important factor in human evolution (e.g., Isaac and Crader 1981; Johnston 1987; Jolly 1970; Mann 1981; Treviño 1991), and the study of dietary diversity in contemporary primates may help us to reconstruct the feeding niche of our ancestors (Harding and Teleki 1981).

Beyond that, studies of naturalistic diets in primates can be useful in research on human nutrition, in several ways. They may suggest dietary competences, such as the ability to detoxify certain naturally occurring toxins or to adapt to low levels of intake, that may not have been suspected in humans. They can also provide information on viable diets for animals that are used as models in biomedical research and can suggest criteria other than maximal growth rates for evaluating optimal diets. Perhaps most important, research on primate diets can serve as a testing ground for the development of research methods that may be directly applicable to studies of human nutrition. This last point can be illustrated by some of the techniques that were developed specifically for this project. They have enabled us to obtain records of dietary intake in free-living primates that, despite all the difficulties such field research involves, have a precision probably exceeding that of most human dietary intake studies, for which the error of estimation is unknown but is conservatively estimated to be 20% to 30% (Marr 1971; Stock and Wheeler 1972).

Criteria for Adaptiveness

In recent years, several attempts have been made to develop and test quantitative models of adaptive traits, including foraging and diet selection (Pyke, Pulliam, and Charnov 1977; Stephens and Krebs 1986). My approach to this problem (see also Post 1984) differs from the usual tack, in which a model of optimal foraging behavior or dietary intake is considered to be confirmed to the extent that observed foraging behavior and dietary intake conform to the predictions of the model. I refer to this as the descriptive approach in that the optimality model is taken to describe what organisms actually do.

With that procedure, a problem immediately arises if a deviation occurs: we cannot distinguish the shortcomings of the animals from the shortcomings of the model. That is, we do not know whether the animals are in fact optimal but our models fail to capture the optimal phenotype or whether our sample includes suboptimal individuals. Behind the descriptive approach lies the Panglossian assumption that natural selection has eliminated suboptimal variants, leaving only well-adapted organisms. To the contrary, the fitness of most organisms is zero: far more are born than breed. While some of this inequality is fortuitous, additive genetic variance and differential reproduction continue to characterize many natural populations. The discrepancy between the reproductive potential of populations and actual population growth was one of the cornerstones of Darwin's argument for natural selection. Dawkins (1982), Emlen (1987), and Van Valen (1960) discuss reasons why natural populations are unlikely to be at a fitness maximum.

An alternative approach, and the one I have adopted, is to consider optimality models as normative, not descriptive. Individuals differ in their fitness. At any one time in the history of a species or population, few if any individuals will be at the optimum point for any trait. In the normative approach, deviations of traits from the optimum that is specified by a model of adaptive traits are regarded not as tests of the model but as an indication of potential differences in fitness, "as grist for natural selection's mill, rather than as a cause for the model's rejection" (Post 1984). Such deviations are particularly likely in unstable habitats, such as the arid regions inhabited by most baboons, in which the optimal point may shift more rapidly than it can be tracked either by natural selection or by acquired characteristics such as learned behavior. The model itself would be tested by showing that those individuals whose traits are closer to the putative optimum have higher expected fitness as a consequence and, conversely, that those that deviate sufficiently from the optimum have predictable functional impairments, such as nutrient deficiencies, that lead to a reduction in fitness.

Consider the array of realizable diets, that is, those that an animal is capable of obtaining in a given habitat, and the corresponding array of actual or potential foraging patterns that would lead to these diets. (While a given diet may be obtained in more than one way, two foraging patterns will always be considered different if they lead to different expected diets, just as two theories would not be considered equivalent if they sometimes made different predictions.) Following Sih (1980b), I will use the term "adaptive" to describe any such realizable diet or foraging pattern that differs from a random selection of food in a direction that enhances fitness. In contrast, an "optimal" foraging pattern is one that maximizes fitness: it yields a diet that results in a fitness that could not be exceeded by consuming any other realizable diet. Thus, substantiating a claim that any animal's selective foraging behavior (or the resulting diet) is optimal requires at a minimum that we demonstrate a close approximation between what the animal eats and what it ought to eat. More generally, any claim that individual differences in selective foraging or diet reflect different degrees of adaptation is equivalent to a claim that individual differences in proximity to an optimal foraging pattern or diet contribute to differences in fitness.

Moreover, a mere correlation with survivorship or reproductive success would not explain why individuals having a particular trait are more successful, nor would it exclude the possibility that the trait is of no selective advantage whatever but is merely linked to or is a consequence of traits that are. What is required is some explanation for why animals that behave in an indicated manner have a better chance of surviving and reproducing, and why the most deviant ones fail: adaptation requires a mechanism. Such an analysis has rarely been attempted for the foraging behavior of any animal.

Let me reiterate the crux of my argument. Real animals are not ideal types. They vary in behavior, and they vary in fitness. Variance is not a nuisance; it is potentially the raw material of evolution. Beyond that, it provides us with some of our best opportunities to test our hypotheses about the consequences of deviations from putative optima. A normative model of optimal behavior should not be judged by the precision with which it predicts the behavior of individuals, but rather, (1) by how well distances from the putative optimum predict differences in fitness (correlation), (2) by how well the model explains the contributions of behavior to these differences (mechanism), and (3) by the extent to which the assumptions of the model approximate the real world (realism). From an evolutionary perspective, the primary task of a general theory of foraging is to specify the consequences of each pattern of foraging behavior that any given animal is competent to perform in the habitats in which it lives. In 1952 D'Arcy Thompson wrote that "sooner or later, nature does everything that is physically possible." In 1986 Waterlow replied, "Our problem is that what is possible may not be good enough."

Heritability

At this point a well-intentioned reader might assume that any difference in the baboons' diet or foraging behavior that enhances fitness must be the result of natural selection acting on heritable differences, but no such assumption is warranted. I do not know whether baboons differ in their diets as a result, in whole or in part, of genetic differences. Two individuals may differ in diet for many reasons. They may differ in what foods each has learned to use by trial and error or which each encounters by chance, in the contingencies of rein-forcement, or in what they have learned from their respective relatives or asso-ciates. Their diets may differ as a result of differences in size, strength, motor coordination, sensory thresholds, ability to tolerate prolonged and intense in-solation, or indeed virtually any other phenotypic differences, inherited or otherwise. In no case do we know the extent to which diet-relevant trait differ-ences in baboons are heritable, either genetically or "culturally," that is, through cross-generational transmission of learned information. The reason for this ignorance is that the requisite research, both genetic and ontogenetic, has not been done. Galef (1992) provides a penetrating review of purported evi-dence for cultural transmission in animals.

A pioneering study by Whitehead (1985) shows that, even with wild pri-mates, one may be able to differentiate aspects of diet selection that are the re-sult of trial-and-error learning from those that probably are based on social learning. The differences may hinge on major food classes. For example, ac-cording to Whitehead, howler monkeys (*Alouatta palliata*) learn the identity of selected leaves through a learning mechanism that requires a social context, but this is not true for the ingestion of fruits and fruit-like objects. Although it seems likely that learning of one sort or another is of great significance in the acquisition of food choice by anthropoid primates, for the present we must pro-ceed without knowing the sources of dietary differences.

Study Area and Population

This study was carried out on yellow baboons (*Papio cynocephalus*) in the for-mer Maasai-Amboseli Game Reserve of southern Kenya, part of which is now Amboseli National Park. Since 1963 the baboons of Amboseli have been stud-ied by a series of collaborating investigators in a long-term research program that has made this one of the most intensively studied populations of wild pri-mates. Most of the research to date has focused on members of three groups, Alto's Group, Hook's Group, and Lodge Group. In addition, repeated censuses and several shorter studies have been carried out on other groups in the area, some of whose ranges overlap those of our main study groups. All the re-searchers involved in the Amboseli baboon project have tried to minimize their influence on the baboons and their world. Over a period of years, the animals have become habituated to the continued presence of noninteractive human ob-servers. By the time of this study, most members of Alto's Group—the natal group of the subjects for this study—had had one or more humans accompa-nying their group most of the days of their lives.

The home range of Alto's Group was centered at about 2°40′ S, 37°10′ E, at an elevation of about 1,128 m (3,700 feet). (fig. 1.2.) Amboseli is a vast, semiarid plain, just north of Mount Kilimanjaro. Within the study area, the

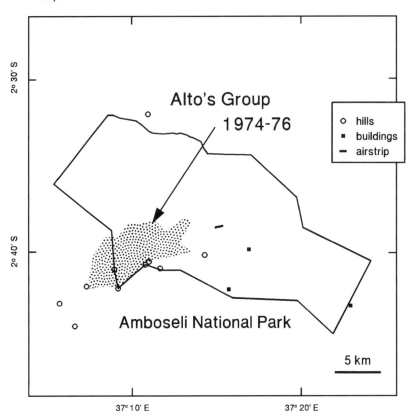

Fig. 1.2. Map of study area in Amboseli National Park, Kenya, showing home range of study group during 1974–76.

predominant habitat is savannah, that is, an area with some trees and in which grasses form the primary ground cover. Almost all the trees in the study area are acacias: the fever tree (*Acacia xanthophloea*) and the umbrella tree (*A. tortilis*).

Baboons apparently are obligate drinkers: their home ranges are restricted to the areas that lie within half a day's journey of water sources, and they drink almost daily. Here and there on the Amboseli plain, wherever surface depressions dip below the water table, natural springs have formed, fed by the underground runoff of rain and melted snow from Kilimanjaro. The resulting waterholes and swamps, along with their surrounding vegetation, make possible a spectacular assemblage of African wildlife. Except during the rainy season, drinking water in Amboseli is available only from these permanent waterholes and swamps.

The moist soil around these depressions supports a perennial green carpet of Bermuda grass (*Cynodon nlemfuensis*). Fever trees and their associated shrubs and forbs, such as *Withania somnifera, Trianthema ceratosepala, Azima tetracantha,* and *Salvadora persica,* also grow in these depressions. Baboons feed on various parts of each of these plants and can feed on first one then the other without moving very far. On higher ground umbrella trees grow, along with their understory burr grass, *Setaria verticillata.* Both provide food for the baboons.

On the plain beyond the woodland are edible halophytic grasses such as *Sporobolus consimilis* and *S. rangei* (called *S. kentrophyllus* in earlier Amboseli publications) and two associated dry-country sedges, *Cyperus obtusiflorus* and *C. bulbosa.* The corms of these grasses and sedges are important dry-season foods. (Corms are the subterranean growth centers from which leaves grow and to which the roots are attached. In some reports on baboon feeding, corms have been erroneously called "roots.") Various other plant assemblages can be distinguished within the baboons' home range, such as the band of regenerating fever trees and mats of tough *Psilolemma jaegeri* grass along the lake edge, the unique flora on the volcanic outcroppings, and localized dense growths of the shrub *Lycium "europaeum."* From each, baboons obtain food.

Further details about the Amboseli habitat are given by S. Altmann and J. Altmann (1970), Behrensmeyer and Boaz (1980), Behrensmeyer (1981), Post (1978), Western (1973a, 1973b), and Western and Van Praet (1973). Table 4.3 provides a list of the baboons' food taxa.

On any one day, the baboons' diurnal cycle of activity reflects their choice of routes from one plant association to another. Because their route varies from day to day and the available food in each plant assemblage changes through the year with the seasons, the mean annual diurnal pattern presented in chapter 6 represents a central tendency in a statistical sense, but it cannot accurately represent what happens on any one day. For baboons, the typical day does not exist. The ways baboons utilize the various parts of their home ranges have been studied by Post (1978, 1982) in Amboseli's yellow baboons and by Sigg and Stolba (1981) in hamadryas baboons (*Papio hamadryas*). In olive baboons (*P. anubis*), home range utilization has been studied by Barton, Whiten, Strum, Byrne, and Simpson (1992) and by Harding (1976).

Seasonality

To everything there is a season, and a time to every purpose under heaven.

—Ecclesiastes 3:1

The diet of baboons is not constant. It changes hour by hour as the animals move through their home range, passing from one plant zone to the next. On a longer time scale, the baboons' diet changes seasonally through each year, and even from year to year as the array of available foods shifts (Norton, Rhine, Wynn, and Wynn 1987; Rhine, Norton, Wynn, and Wynn 1989). In an arid, tropical area such as Amboseli, annual and long-term changes in the flora probably are dictated primarily by changing patterns of rainfall, whereas the baboons' diurnal cycle probably is largely constrained by light and by the spatial arrangement of their resources and hazards.

Temperature and Rain

Our twenty years of daily air temperature records, taken in and near Ol Tukai, Amboseli, during 1971–91, are summarized monthly in figure 1.3, along with values for the main study period, August 1975 through July 1976. Mean daily values were estimated from half the sum of the minimum and the maximum, as is done by the U.S. Weather Bureau. "The mean of the daily extremes is, as a rule, slightly too high, but it usually does not vary more than half a [Fahrenheit] degree from the true daily average" (Kincer 1941). In terms of human comfort, Amboseli air temperatures range from warm to hot at midday. Then, as is typical of arid regions, with their lack of an appreciable atmospheric greenhouse effect, the air cools quickly in the evening.

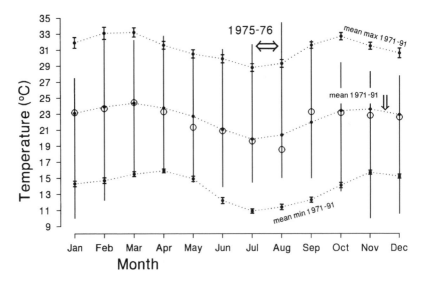

Fig. 1.3. Air temperature, Amboseli National Park. For each month, long-term values (1971–91, solid circles connected by dotted lines) are average daily minimum, maximum, and mean temperatures. Capped vertical bars are standard errors. For August 1975 through July 1976, values are monthly means (open circles) and ranges (vertical bars).

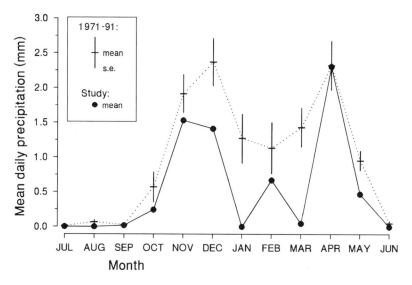

Fig. 1.4. Rainfall, Amboseli National Park. Monthly values for mean daily rainfall are given for the primary study period, July 1975 through June 1976 (solid circles) and long term (1971–91).

For rainfall, monthly values, based on our readings taken in and near Ol Tukai, Amboseli, during the same twenty-year period are shown in figure 1.4, along with data from the main study period. On average, two peaks in rainfall occur in Amboseli each year, one in March–April (the "long rains") and a second in November–December (the "short rains"). During the "inter-rain" period, January–February, scattered showers, sometimes quite heavy, are not uncommon. A long dry season extends from the end of the rains, typically in late May, until the rains begin again in late October or early November. This is the "typical" yearly cycle, but as is characteristic of arid regions, there is much year-to-year variability in the timing and intensity of rains.

In addition to those monthly and annual rainfall differences, a considerable spatial heterogeneity in precipitation exists in Amboseli. As the storm clouds from the Indian Ocean blow westward across the savannah, localized showers often wet some one part of a baboon group's home range, leaving the rest still swirling in the dust. My impression is that the eastern half of the 1974–76 range of Alto's Group, the area we refer to as the KB woodland, usually receives more rain than the western part, the KH woodland and the plains beyond that.

Rainfall during the main study period, August 1975 to July 1976 (fig. 1.4) was light even for this arid region. Total rainfall during the study year was 202 mm, which is 60% of the long-term average. At the end of the long dry season of 1975, strong rain began late, on November 17. Seven days later, the local Maasai left the area and moved to the north of Lake Amboseli, as is their custom during the rainy season. These rains were preceded earlier in the month by light, scattered showers, and showers continued into December. No rain fell in January 1976. During February there were several strong rains, but the total for that month was below average. Thereafter, except for scattered, light showers, there was no further rain until heavy rains began April 11. By that time, however, Amboseli was again very dry. A blinding dust storm—the worst I have experienced in Amboseli—occurred on April 6. At the time, articles in the Nairobi papers described drought conditions in most of Kenya. In total, the April rains were average for Amboseli despite their late start, but May rains were below average. On May 29, some of the baboons were again eating grass corms, their dry-season staple. The long dry season of 1976 can be taken to have started on June 1: after that date there were only a few light showers until we left Amboseli on July 28, 1976. During that year, mean temperatures (fig. 1.3) were above average for most months, though no temperature records were exceeded.

Phenology

Twice each year, the onset of the rains in Amboseli transforms the area. This effect is particularly marked in November because of the long dry period before it and the abruptness with which that rainy period usually begins. Overnight, the blowing dust becomes gooey mud. Within a few days, grasses begin sending up fresh, new leaves, and the predominant color of the landscape changes from beige to green. Numerous dormant insects emerge. Within a few weeks many plants are in bloom, and shortly thereafter their fruits are available.

Other plants have quite different cycles, particularly the trees and some of the large perennials whose deep roots may make them less dependent on the rains. (I do not know what triggers the blooming of the two species of acacia trees.) From the standpoint of a foraging baboon, a consequence of these phenological changes is that many foods are available for only part of each year. Furthermore, because the production of many foods is not synchronized, the array of available foods changes rapidly even within a season—indeed, from one fortnight to the next. Figure 1.5, based on my feeding records and phenology notes, summarizes many of the phenological events during the 1975–76 season.

Not surprisingly, this highly seasonal but asynchronous production of various foods is reflected in the baboons' diet. Challenges and benefits are entailed by these seasonal changes. For infants, the first year of life presents an ever-changing procession of new selections to be made among potential foods, of which a few are highly nutritious, some are edible but of little value, and others are toxic or otherwise hazardous to harvest. Some young baboons develop

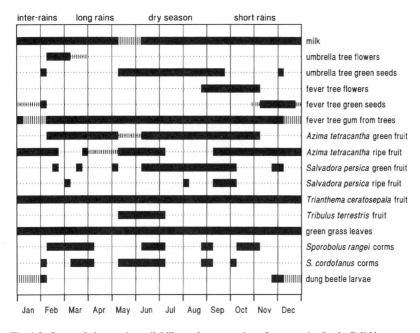

Fig. 1.5. Seasonal changes in availability and consumption of some major foods. Solid bars: periods of observed consumption (from table A8.2). Broken bars: periods in which lack of feeding records is assumed to be due to small sample sizes. Broken half-size bars: periods in which food was noted as available but was not consumed during feeding samples. Foods listed below milk include those from trees, from common bushes (*Azima tetracantha* and *Salvadora persica*), from two prostrate flowering plants (*Trianthema ceratosepala* and *Tribulus terrestris*), from corms of two favorite grasses in the genus *Sporobolus,* and from subterranean dung-beetle larvae.

skills for coping with these ephemeral menus, others do not; as I shall demonstrate, survivorship is a reflection of their diets during infancy.

The benefit of asynchronous food peaks for selective omnivores such as baboons is that some pairs of foods are sufficiently similar in composition to exhibit *partial substitutability* in the diet. For example, the green seeds of Amboseli's two main acacia trees are sufficiently similar in nutrient content that they are much more like each other than either kind is to, say, flowers. Because these seeds are available at different times of the year, the baboons have an abundant source of seeds over a greater part of the year than if the trees were in synchrony. Thus, asynchronous food productions tend to dampen seasonal fluctuations in food availability. Yet even with baboon foods grouped into major categories (flowers, seeds, and so on), seasonal shifts in diet occur from one fortnight to the next, as illustrated in figure 1.6.

The drawback to this asynchronous pattern of food production is that diet selection is vastly more complicated. Instead of facing, say, just two arrays of

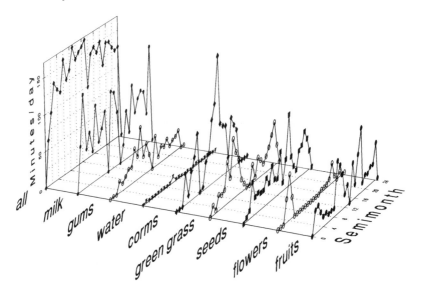

Fig. 1.6. Seasonal variation in time spent feeding on major food classes. Food classes as defined in table 4.4. "All" = the total of core and non-core foods and water. For graphic purposes, values for "all" have been plotted at half their actual values, and the beginning and end of each plot have been anchored at zero.

foods, one in dry seasons and one in the rains, baboons must cope with at least two dozen menus, each with a different selection that would provide an adequate or an optimal diet and a different array of hazards. In chapter 6, I compare the yearling baboons' mean annual diet with corresponding optimal diets. However, when dealing with individual differences (chapters 7 and 8), optimal diets for every individual are calculated for each fortnight.

Taxonomic Note

The genus *Papio* (setting aside drills and mandrills) comprises a cluster of phenotypically distinct, regional baboon morphs, some of which hybridize in nature or intergrade and which to varying degrees are geographically isolated from other baboons. These phenotypically distinct forms have some but not all of the attributes commonly used to define species. This has led to a confusing array of ways of classifying them. The proposed classifications of forms within the genus vary in which morphs they recognize as taxonomically distinct, in the taxon level (species, semispecies, subspecies, or superspecies) at which they designate each form, and in which are to be included in the genus or other collective taxa. Jolly (1993) summarizes the available evidence and literature on baboon systematics in an enlightening review on which this note is based.

As Jolly clearly indicates, no classification fully expresses what is now known about the complex relations among baboon morphs. Beyond that, some of the most crucial information is not yet available. Thus the choice of taxonomic levels for these forms must for the present be based largely on other grounds. In this monograph I refer to these phenotypically distinct baboon morphs as species (or semispecies), for example, *Papio hamadryas* for hamadryas baboons, *P. anubis* for anubis (olive) baboons, *P. cynocephalus* for yellow baboons, the subjects of my study, and so on. Jolly (1993) would designate them all as subspecies of *Papio hamadryas,* of which he lists ten, including *P. hamadryas ibeanus* for the yellow baboons of the Amboseli region.

Ironically, one of my reasons for preferring species designations for these forms is the same as Jolly's reason for preferring to call them subspecies. He claims, "Recognizing them as an equivalent number of full (phylogenetic) species would obscure the important attributes of subspecies." Consider, however, that many, perhaps most, laboratory scientists who work on primates have no direct interest in evolutionary processes and look upon them as irrelevant to their own research. They almost always ignore subspecies differences in their choice of animals for research and for their breeding colonies, and they almost never indicate the subspecies of their subjects in publications, if indeed they know them. Thus, classifying the major baboon morphs as subspecies would lead to conflated data on many distinctive attributes of these forms.

Furthermore, just how genetically isolated the various baboon forms are remains to be determined. A few areas of baboon hybridization have been located, but for none do we yet know how extensively and how far genes penetrate beyond these narrow hybrid zones. However, a short distance away, baboons on either side remain morphologically distinct, judging by field characteristics (personal observations).

Jolly further argues that species designations of baboon morphs would "conceal the overall closeness of their relationship and bias interpretations of their past and prediction about their future evolution." Unfortunately, subspecies designations would not solve that problem: "The typological division of species into subspecies does not give any information on the relation of populations to each other. Whether a subspecies is part of a cline or is isolated completely by geographic barriers is, however, of decisive influence on its evolutionary potential" (Mayr 1963, 366).

Habitat Changes, 1963–75

So far as we know, our 1963–64 study (S. Altmann and J. Altmann 1970) and the concurrent study of vervet monkeys by Thomas Struhsaker (1967a, 1967b, 1967c, 1967d, 1971) were the first systematic wildlife studies carried out in Amboseli. At that time the study area consisted primarily of an extensive fever

Fig. 1.7. An obligatory halophyte, *Suaeda monoica,* encroaching as fever trees die out. Successive photos of same area from same position. A. July 1964. B. August 1972. C. December 1981.

tree woodland, intermixed with areas of open grassland. In the decade that followed, major changes took place in the habitat, in particular, death of most of the fever trees and encroachment of halophytic plants, such as *Suaeda monoica* (fig. 1.7).

The death of Amboseli's fever trees has been attributed primarily to intrusion of salts into the root zone of the trees as a result of elevation of the water

C

Fig. 1.7. (*continued*)

table (Western and Van Praet 1973). Despite the proximity of Mount Kilimanjaro, Amboseli is remarkably flat, and most of the shallow depressions have no
external drainage. This leads to accumulation of salts in the soil. Such subsurface minerals tend to accumulate in a layer lying just above the water table,
carried downward by rainwater and moved upward by wicking action in the
capillary zone immediately above the subsurface water.

The elevation of Amboseli's water table began in the middle of the dry season of 1957, perhaps as a result of some underground disturbance (Smith
1986). Further increases during the 1960's, perhaps resulting in part from
higher than average rainfall, resulted in a rise of the water table by three to four
meters (Western 1973b). Subterranean salts, shoved up into the root zone of the
trees by the rising water table, probably destroyed the ability of the roots to
pump water. The fever trees died, typically from the top down. Because fever
trees occur on lower ground than umbrella trees and perhaps because of differences in salt tolerance, the former suffered a much higher mortality. Some fever
trees, lying in the lowest depressions, were drowned as the rising water table
enlarged the old waterholes and swamps and created new ones (fig. 1.8).

Young and Lindsay (1988) propose an alternative explanation for the die-off
of Amboseli's fever trees, namely, senescence in an even-aged population of
trees. That explanation may be correct, but it assumes low variance both in
potential life span and in dates of germination of the trees. No direct evidence
of an even-aged population is available, and the indirect evidence cited by
Young and Lindsay is equivocal. Their evidence for senescence rather than

Fig. 1.8. Fever trees drowning in rising water, in area that was formerly dry land. Lack of aquatic sedges indicates that the waterhole is new.

salinization as the primary cause of rapid mortality in Amboseli's fever trees is that "mature woodland has continued to die back in the central basin, while the water-table reportedly dropped in the 1970s and early 1980s (D. Western, pers. comm.) and was high again by 1984." In 1964 I began a series of samples of fever tree morbidity and mortality in the central basin of Amboseli, combined with observations on the growth of associated waterholes and of the obligatory halophyte *Suaeda monoica.* Although the water table in the area fluctuates seasonally and from year to year, it has not, at least as of 1996, receded to anything like the 1964 level, nor has the encroachment of *Suaeda monoica* abated.

If, as seems likely, a drop in the water table early in this century fairly quickly led to conditions favorable to fever tree growth (Western 1973b), the result would be fairly even-aged stands of trees. And if the rapid elevation in the water table about half a century later raised the salt layer of the soil into the root zone of the trees and thus fairly rapidly led to their demise (Western and Van Praet 1973), the result would be low variance in actual life span; but that hardly implicates senescence as a causal agent. Contrary evidence regarding senescence is that fever tree morbidity in Amboseli during the early 1970's was not restricted to the largest (presumably oldest) trees. Indeed, in samples of stands of fever trees, the percentage of trees with crown senescence—an early symptom of salt toxicity—was essentially independent of tree height and diameter at breast height (Western 1973b, table 12). Other possible sources of

Amboseli's habitat changes have been suggested but not documented. "The relative contributions of rising water table, climatic change, elephant damage, tourist effects, and normal senescence of the tree population remain a matter of considerable debate" (Behrensmeyer 1993).

Whatever the cause, the rate at which the fever tree woodland of Amboseli died off was astonishing. Western (1973b) estimated fever tree mortality from 1950 to 1967 by repeated counts of individual trees in the same thirty plots, each 0.18 km^2, that he established on government aerial survey photographs for 1950, 1961, and 1969. Mean tree counts were 49, 34, and 16 per plot in these three years, respectively (Western's table 11), indicating mortalities of 31% from 1950 to 1961 and of 53% from 1961 to 1967, or cumulatively, 67% from 1950 to 1967. In 1969 I took a sample of 782 standing fever trees that were green-barked, that is, that were alive or had very recently been so (S. Altmann 1969). (After the death of a fever tree, the green bark color probably persists only for a few months.) In that sample, 12% had recently died, as judged by a total lack of green leaves, and another 18% were judged to be moribund. Subsequent long-term sampling on marked, individual fever trees showed that the mortality rate of adult fever trees was 28% per annum in 1973–74 and 10% in 1974–75, with a projected total loss of 65% of all adult trees between 1973 and 1982 and the prospect of few juvenile trees to replace them (S. Altmann 1977). High mortality continued in the years after that, so that by the 1990's most of the trees in what had been one of East Africa's finest examples of fever tree woodland had died. Numbers of many other woodland plants, particularly those that flourish in the moist soil and filtered sunlight under fever trees, also declined. On the other hand, halophytes such as *Suaeda monoica* flourished. Furthermore, as swamps and waterholes grew larger, aquatic and swamp-edge vegetation became more abundant. The habitat was completely transformed.

Population Changes, 1963–75

The death of Amboseli's fever trees and the associated habitat changes profoundly affected populations of many animals. Although only one census of large mammals during the period 1963–75 has been published (Western 1980), my impression is that mammals that extensively use water-edge habitat, such as waterbuck and buffalo, increased in numbers during this period. On the other hand, populations of animals that depend on the acacia woodland declined. For example, a population of six groups of vervet monkeys (*Cercopithecus aethiops*) that live in the study area declined from an annual mean of 159.1 individuals in 1963–64 to 106 in 1971 (Struhsaker 1973). In February 1975 there were 105 individuals in six vervet groups in the same area, but that population figure is inflated by the abundance of infants at the time of that census; the juvenile and adult part of the vervet population declined from 103 individuals in 1971 to 83 in 1975 (Struhsaker 1976). High mortality continued to characterize this population in subsequent years (Cheney, Seyfarth, Andelman, and Lee 1988; Isbell 1990). Other woodland-dependent species of mammals probably also declined (Western 1973b).

The Amboseli baboon population decreased dramatically during this period of habitat changes. The decline between 1963–64 and 1969, from 2,589 baboons to 255, which occupy a larger area (described below), is equivalent to a mean loss (compounded continuously) of 46% of the population per year. Expressed in terms of density, the change is even greater, because in 1969 we censused over a larger area. From 1964 to 1969, the population density fell from 73 to 3.7 baboons per km^2, suggesting a loss of 95% of the population in about five years (J. Altmann, Hausfater, and S. A. Altmann 1985). At the time of my study of weanling feeding behavior, 1975–76, the population had stabilized, although at a much lower density than in 1963 (J. Altmann 1980). Group composition and demographic parameters for Amboseli baboons, including infant and juvenile mortality, are given by J. Altmann, S. Altmann, Hausfater, and McCuskey (1977), by J. Altmann, Hausfater, and S. Altmann (1985), and by J. Altmann (1980).

This dramatic decline in the number of baboons in Amboseli during our five-year absence, 1964–69, presented a puzzle. Exactly which of the many changes in the habitat had affected the baboons? David Western has suggested to us that the 1963–64 baboon population in our study area may have been augmented by migrants from the east of Enkongo Narok Swamp, where the decline of fever trees had begun earlier. That still leaves unanswered the question of the proximate cause of the population decline in either area.

Food as a Limiting Factor

I believe that food is the primary limiting factor for the baboon population of Amboseli and for many other primate populations. Let me give the grounds for that belief, as well as the reasons I cannot, at present, provide strong evidence for it.

How can one identify environmental factors that limit population size or test the hypothesis that a given component of the environment, such as food, is a limiting factor? One is tempted to rule out any environmental component that is not exhausted. For example, it has been said of howler monkeys that so long as the primary fruits that they eat in an area are abundant, their population cannot be food-limited (Coelho, Coelho, Bramblet, Bramblett, and Quick 1976; Coelho, Bramblett, and Quick 1977), or more generally, that herbivores are not ordinarily food-limited because they do not deplete the supply of green plants, though they are capable of doing so (Hairston, Smith, and Slobodkin 1960). This line of argument is not convincing. Food may be present, and even abundant, but insufficiently available for any of several reasons, including wide spacing, toxic secondary compounds, high lignin content, and hard shells—a condition referred to as *relative shortage* to distinguish it from the *absolute shortage* that arises when a resource is fully accessible but insufficient in quantity (Andrewartha 1971; Andrewartha and Birch 1954). Alternatively, foods that are abundant may be deficient in certain nutrients that are sufficiently abundant only in other foods that are sparse. Or obtaining some of the food may engender sufficient exposure to hazards, such as predation, dehydration, or energy depletion, as to make these foods unprofitable to harvest. Under any of these conditions, a population may be food-limited despite an apparently plentiful food supply (Murdoch 1966; Erlich and Birch 1967).

To say that a resource or hazard limits a population is to say that an increase in the availability of the resource (or a decrease in the hazard) would, within the range permitted by other limiting factors, produce an increase in the population, and conversely that a decrease in the resource (or an increase in the hazard) would result in a population decline. However, in a natural population, an observed *correlation* between, say, food supply and population density is insufficient to substantiate a claim—such as Cant (1980) has made for several primates, including the vervets and baboons of Amboseli—that the population is food-limited. The reason for this is that food abundance covaries with many other components of the environment: the same processes that have altered the food supply (primarily changes in groundwater, in the case of Amboseli) usually have also altered many other aspects of the environment. What is required is evidence that the putative limiting factor actually produces the changes in the population.

A particularly revealing way to isolate the effect of a putative limiting factor is to change its abundance artificially, thereby minimizing possible confounding effects of other, concurrent changes. For example, Miller, Watson, and Jenkins (1970) altered the abundance of heather, by either burning or fertilizing it, and showed how these changes affected local populations of red grouse (*Lagopus lagopus scoticus*) that feed on it. In a number of wild primates, artificial increases in the food supply have resulted in major increases in population size, through increases in both survival and fertility (Lyles and Dobson 1988), suggesting that many primate populations are indeed food-limited. Conversely, in two populations of Japanese macaques (*Macaca fuscata*), population declines occurred when artificial provisioning was reduced or eliminated (Mori 1979; Sugiyama and Ohsawa 1982).

For the baboons, vervets, and other animals of Amboseli, no such experimental evidence is available. In the absence of strong evidence, I believe that the combination of two lines of indirect evidence suggests food probably was the primary factor that led to the marked decline in the population of Amboseli baboons during the 1960's. The first of these is Cant's argument: the enormous changes in the Amboseli habitat during that period lead to a very marked reduction in many of the primary food plants of baboons.

Second, none of the concurrent changes in the habitat appear likely to have had a major impact on the baboon population. Greater predation seems highly unlikely, for by 1969 the baboons' archenemy, the leopard (S. Altmann and J. Altmann 1970), was completely missing from the study area: no leopard tracks were seen, none of the spectacular alarm calls that baboons give in response to these cats were heard during our 1969 study, and the game scouts of Amboseli reported they had not seen a leopard in the area for over a year. (Since 1969, leopards have been reintroduced into Amboseli.) Approximately 230 baboons that were trapped and removed from the area during 1965 by the Southwest Foundation (Maples, pers. comm.) represent about 10% of the 1964 population, whereas by 1969 about 95% of the population was gone.

Increased mortality from pathogens cannot be ruled out, and baboons are known to be susceptible to a wide variety of diseases (e.g., Basson and deVos 1971; Davies, Clausen, and Lund 1972; Kalter 1983; McConnell, Kim, Eugster, and Kalter 1968; Miller 1960; Vagtborg 1965), but superficially the animals showed no sign of any lethal epidemic disease except for some cases of hind-limb paralysis, apparently due to Coxsackie B2 and B4 (S. Altmann and J. Altmann 1970) and to one outbreak of a flu-like malady.

Could the reduction in shade with the death of many trees have resulted in greater heat stress or dehydration? We saw no symptoms of these. Water was, if anything, more abundant than before, because of the elevated water table, although during the drought of 1969 water was often inaccessible to the baboons owing to extreme competition from Maasai livestock.

One major possibility remained: that food was the primary factor in Amboseli's changing habitat leading to the great reduction in the population of baboons in the area. We have described the enormous reduction in the fever tree population during these years. These trees provide the baboons with several major foods: gum, flowers, pods, and seeds. Beyond that, other major changes occurred in the flora of Amboseli, including a reduction in various shrubs, forbs, and grasses that grow under or in close association with fever trees, and their replacement by halophytic plants of lower nutritional value (Western and Van Praet 1973). So the available evidence, however inadequate, strongly suggests that the baboons of Amboseli were primarily food-limited and that the rapid decrease in the Amboseli baboon population during the 1960's was primarily due to a shortage of food, which in turn was due to groundwater- and rainfall-induced changes in the flora of the region.

Disintegration of the last large groves of fever trees may have introduced a new limiting factor for baboons: a shortage of safe nighttime roosting sites. In more recent years, Amboseli baboons have taken to sleeping in umbrella trees (*Acacia tortilis*), whose low, rough-barked branches probably are not as safe from leopard attacks. One group, Hook's Group, occasionally sleeps in an isolated *Syzygium guineensis* tree.

If Amboseli baboons are primarily food-limited, then understanding the feeding behavior of infants at the time of weaning takes on a particular significance. J. Altmann, Hausfater, and S. Altmann (1985) review a growing body of evidence from several species of cercopithecine primates that rates of mortality for infants and juveniles are the demographic parameters that are most sensitive to changes in environmental conditions. Mammals may be particularly susceptible to nutritional stress after weaning, when they can no longer rely on their mothers' milk to make up for their inadequacies. Nutritional deficiencies during this period can result in the death of some severely malnourished infants. Even among those that survive, malnutrition can have marked effects on subsequent life expectancy, age at maturation, reproductive success, and other components of biological fitness.

Weanling mammals may not be the only animals for whom malnutrition during infancy is a primary cause of death. White (1978) argues that "for many if not most animals . . . the single most important factor limiting their abundance is a relative shortage of nitrogenous food for the very young."

Goals of the Research

By the time the present study was undertaken, the population of Alto's Group, on which most of our research has been carried out, had nearly stabilized (J. Altmann, Hausfater, and S. Altmann 1985), albeit at a level that made it susceptible to small-sample fluctuations. Although this transition precluded a direct study of the causes of the population decline, we were left with an unsolved and more general problem at the individual level. Under the new ecological conditions in Amboseli, could a weanling primate obtain a nourishing or even an adequate diet? Is selective foraging by baboons a major component of their adaptation to their habitat? Put the other way around, is much of the variability in baboon fitness attributable to individual differences in selective foraging, and are such differences particularly critical at the time of weaning?

These are the general issues I have addressed by posing five specific research questions: (1) What do weanling baboons eat, and what not? The answer to that question is the essential base for the rest. (2) What are the nutritional consequences of this selective foraging? That is, are the baboons meeting their nutritional requirements, and are they taking in excessive quantities of potentially toxic materials? (3) In view of the foods available to the baboons, what, optimally, should they eat? (4) Is their actual diet optimal? (5) Finally, do individuals whose diet comes closer to the optimum have higher expected fitness? Conversely, can one predict the nature of diet-induced malfunctions in individuals whose diets deviate markedly from the optimum?

An attempt to use a general model of optimal diets to relate selective foraging in baboons to the adequacy of their nutrition and to their biological fitness constitutes the more ambitious component of this project. Its more modest goal is to provide a quantitative description of key characteristics of that selective foraging. No field of science begins with its ultimate explanations: a priori it seems most unlikely that current attempts to formulate general quantitative models of food selection are correct. Under the circumstances, a solid empirical foundation for further development of hypotheses and models may in the long run be the more abiding part of the study. The need for such data is acute: many of the quantitative parameters of foraging that will be presented here have seldom if ever been estimated for any wild animals.

The second chapter ("Eclectic Omnivory") provides an introduction to common problems faced by foraging animals in their major dietary modes, particularly the selective omnivory that characterizes baboons. Detecting and measuring individual differences in such highly variable behavior require sensitive quantitative analysis; the methods I used in this study are presented primarily in chapter 3. A few special methods are presented in their respective chapters and appendices.

The first question above, about what baboons eat, is answered in chapter 4 ("Dietary Diversity"), which gives a general description of the feeding behavior and foods of weanling baboons. The third question, what baboons should eat, is the topic of chapter 5 ("Adequate and Optimal Diets"), and since the answer depends on the animals' nutritional requirements and the composition of their foods, that information is presented there and in appendix 1. In chapter 6 ("Real versus Ideal Diets"), I give quantitative data on weanling baboons' average intake of each food and nutrient (the second question) and relate these intakes to the specifications for adequate and optimal diets presented in chapter 5 (the fourth question). The individual subjects of this research—eleven yearling baboons of one cohort in one social group—are seldom mentioned by name in these chapters, in which average values are used to present a normative picture of foraging by yearling baboons. However, these eleven youngsters are the source of all the data presented there.

The transition from ensemble to solo contributions occurs in chapter 7. The fifth question above deals, at the individual level, with whether the baboons' diets were optimal and whether proximity of individuals to the optimum increases fitness. In "Individual Differences and Age Changes," I present evidence on several sources of age changes and individual differences in dietary intake—how often each food is eaten, for how long, and how rapidly—and show that the diets of individual yearlings differed, even those of the same social group, age, sex, and season of birth. Chapter 8 ("From Food to Fitness") focuses on answering the fifth question. In it I show how close the diets of individual baboons were to the diets that would be optimal for each of them and how well these individual differences predict fitness and fitness surrogates.

Chapter 9 ("Why Be Choosy?") provides further information on the consequences of selective foraging by baboons (the second question). In it I document the nutritional consequences of some of the baboons' fine-grain selectivity: of eating this species versus that, or this part of the plant versus another. Finally, in chapter 10 ("How to Be an Eclectic Omnivore") I discuss some of the characteristics of baboons that make them so successful, with an emphasis on those characteristics that affect foraging behavior or are affected by it.

2

Diversity rather than specialization is typical of primate diets.
—Robert S. O. Harding, 1981

Eclectic Omnivory

The Packaging Problem

For a foraging animal, the primary difficulty in diet selection is what I call the packaging problem. An individual cannot get at its vital resources without incurring some concomitant risk: "good" and "evil" always come packaged together. This is as true of foods as of other resources. The nutrients in foods are combined in different ways with a greater variety of other materials, some nutritious, some toxic, and some essentially neutral. Furthermore, by searching for foods and harvesting them, an animal often increases its exposure to several risks. Some of these are intrinsic to the foods or the food plants. These intrinsic hazards include toxic secondary compounds, thorns, siliceous material that wears down an animal's molars, and so forth. Other hazards are extrinsic, including food- and habitat-associated predators, food competitors, disease transmission, time-budget limitations, and the risks of dehydration or overheating in areas remote from water or shade. In these circumstances, the animal's task is to find a combination of foods that will satisfy its nutritional requirements and that can be searched for, harvested, and consumed without excessive risk.

Even if we focus solely on nutrients, two characteristics of potential foods aggravate the packaging problem for most animals. First, no single food provides an adequate diet, with the right proportions of all nutrients. No perfect food exists. Second, no nutrient occurs in isolation in the animal's environment, and in this nutrients differ markedly from foods. Although an animal requires specific *nutrients,* it can select and consume only *foods,* none of which can be eaten with impunity, none of which is nutritionally adequate, and for none of which the animal has any requirement whatever. The problem and the available solutions are at different levels of organization.

The baboons' solution to the packaging problem is eclectic omnivory: their foraging is extremely selective, yet their diet is highly diverse. The success of baboons depends on their ability to exploit a wide variety of plant and animal material in a complex habitat and, in the process, to feed selectively on some of the most concentrated sources of nutrients in their environment (S. Altmann

1974; S. Altmann and J. Altmann 1970). As we said in 1970, "Their lack of highly specialized anatomical adaptations, combined with their keen vision and their ability to climb, dig, pull, pluck, gnaw, and to move great distances, often far from trees, enable them to exploit a wide variety of foods, and thus to survive in a great diversity of habitats. Baboons are the most widespread and abundant nonhuman primates on the African continent" (144). As I document in later chapters, this pattern of diverse but highly selective diets, which is characteristic of adult baboons, begins at an early age. Solving the packaging problem is the challenge that every infant baboon faces and that it must resolve in order to become nutritionally independent of its mother. Those that fail to do so do not survive to adulthood.

The Threefold Symmetry

Certain properties of potential foods in any habitable region and their effects on foragers present problems for a foraging animal; others make possible the selection of diets that resolve these problems. As we shall see, these key properties form a threefold symmetry. They justify several aspects of the linear optimization methods that I used to find adequate and optimal diets (chapter 5), and at the same time they explain why an elegantly simple approach to optimal diets in the first models that were proposed—what I refer to as the "top-down" approach—cannot systematically do so.

The top-down approach is straightforward. One rank-orders the potential foods of a region by some scalable criterion, either of profitability (e.g., MacArthur and Pianka 1966; Emlen 1966) or (in inverse order) of the amounts of toxins and other deleterious components, as proposed by Stahl (1984). The well-adapted forager is then supposed to eat only items from the top of the list, and in various models, specific stopping rules have been proposed for finding the profitability threshold between acceptable and unacceptable foods. So, for example, Stahl (1984) writes, "If variables such as the availability, the cost of procurement, etc., could be held constant, the [food] categories at the top of [the rank-ordering in Stahl's] table 1 would have been selected by non-fire-using hominids over those at the bottom." MacArthur and Pianka put the principle succinctly: "The basic procedure for determining optimal utilization of time or energy budgets is very simple: an activity should be enlarged as long as the resulting gain in time [or energy] spent per unit food exceeds the loss."

The top-down method is both conceptually and computationally much simpler than the one I have used, and so, with progressive improvements in the data—say, sufficient to make finer distinctions within food classes or better information about toxin content and tolerances—if the top-down approach converged on adequate and optimal diets, it would be preferable. As I shall explain below, this is not the case, for a simple reason: plant foods do not have a scalable nutritional "value" independent of other foods in the diet (S. Altmann 1985; Rapport 1980).

Consider an herbivorous forager faced with an array of potential plant foods and attempting to solve its basic dietary problem, that of staying above its minimum for every nutrient and below its maximum for every toxin or other hazard. Among the various combinations of local foods that satisfy this criterion of adequacy, some may provide the forager with larger quantities of a fitness-limiting component, such as protein or energy, or result in less of some deleterious effect, such as those brought on by toxins. That is, some adequate diets are better than others. For such creatures, the task of finding an optimal or even an adequate diet is made difficult by two properties of all potential foods. First, for an herbivore, it is very unlikely that any food that is available in adequate quantities contains all the nutrients the animal requires and still less likely that any will contain them in the requisite proportions. For an herbivore, there are no perfect foods. All are nutritionally inadequate in one way or another (Pulliam 1975; Westoby 1978).

Second, no wild food, no matter how nutritious, can be eaten with impunity. All involve some form of risk, usually several. The result is what I referred to above as the packaging problem: in every potential food, the forager is faced with costs and benefits that are inextricably bound together. The animal's task is to get enough of the nutrients without incurring too high a cost.

Note that these two adverse characteristics of foods, incomplete nutrients and omnipresent risks, are in a sense mirror images of each other: not all nutrients are present in any food, and not all risks are absent.

The task of finding a good diet in the face of these two properties of all potential foods is made possible by two other mirror-image pairs of properties. First, a toxin or other hazardous component that occurs at dangerous levels in one potential food may occur at negligible levels or be entirely absent from others. That is, the potential foods of a region exhibit toxin complementarity.

Similarly for nutrients: they do not covary perfectly among the foods of any region, so that the nutrient inadequacies of potential foods are not identical. A nutrient that is absent or in short supply in one food may be adequately represented or even abundant in another. Indeed, some foods may be worthless or even deleterious except for some micronutrients, such as particular vitamins or minerals that they provide and that are not readily available elsewhere. That is, the potential plant foods of any habitable region, taken together, exhibit nutrient complementarity.

For example, judging by studies of cultivars (Davidson, Passmore, and Brock 1972), the seeds of legumes (pulses) typically are high in total protein but deficient in sulfur-containing amino acids (methionine and cysteine), whereas the seeds of grasses (grains) contain sulfurous amino acids but many are often deficient in lysine and tryptophane, which can in turn be obtained from many pulses. In many human cultures, combinations of pulses and grains, such as maize and beans, provide an amino acid balance that neither would supply alone. Although amino acid analyses have not yet been carried out on the pulses and grains baboons feed on, I would not be surprised if they exhibit a similar complementarity.

These two characteristics of the potential foods of a region, nutrient and toxin complementarity, are consequences of a more general property. Neither nutrients nor toxins occur with the same abundance or even have the same abundance rank in all foods, nor do all nutrients rank inversely with all hazards. Consequently, the "value" of a food is neither an independent property nor a scalable one. Its value depends on what other foods in the diet provide and fail to provide. As a result, no one-dimensional scaling of foods combined with a top-down rule will lead systematically to the set of adequate and optimal diets.

The question that must be asked of any potential food is not, What does it offer? but rather, How much does it offer of each nutrient that is otherwise in short supply in the other foods of the diet, and what respite does it offer from the hazards that other foods entail? The value of a food depends on how much it provides of each nutrient and hazard relative to what other potential foods provide (Rapport 1980), how much the animals require and can tolerate, and how their fitness is affected by their intake of each nutrient and hazard. Furthermore, the health impact of a given nutrient or toxin in a food depends not only on quantities of the same component in other foods but also on other components in the same foods: some nutrients interact with toxins and with other nutrients (Wise 1980), and some toxins no doubt affect each other. An animal's task is not so much selecting foods as selecting an ensemble of foods, that is, a diet.

Now to the third mirror-image pair. Of particular significance to a forager are cumulative hazards, that is, those for which the risk rate is an increasing function of the amount consumed (as with most toxins) or the duration or intensity of exposure (as with, say, dehydration). The cumulative risk entailed by feeding on any one food places an upper limit on the amount of that food that can be eaten in, say, a day. Furthermore, the very act of feeding by an herbivore induces some plants to increase their output of toxins within twenty-four hours (Haukioja and Niemala 1976; Ryan and Green 1974). At present, thousands of toxic secondary plant compounds have been identified, with an enormous variety of ill effects on animals (Harborne 1978; Montgomery 1978; Rosenthal and Janzen 1979). Every vital organ of an herbivore, and every essential metabolic pathway, seems to provide a potential site for a toxic compound. This great diversity among plant toxins probably reduces the chances that an omnivorous herbivore can develop an immunity to them. However—and this is the crucial point—the deleterious effects of many intake-limiting toxins do not add to the deleterious effects of others (Coon 1976). Consequently, an animal can with impunity eat two or more foods, each containing a different toxin, in a total amount that would be harmful or even lethal if any one of these foods were consumed in that quantity. Beyond that, such ameliorating effects of switching foods may be furthered by interaction effects: chemicals present in some foods may be antagonistic to toxins present in others (Coon 1976). For example, the toxic effects of oxalic acid in one food can be reduced by eating another food having abundant soluble calcium.

Conversely, the benefits that accrue from obtaining some nutrients from one food are not usually diminished by obtaining other nutrients from other foods, so long as those that react with each other—as, for example, essential amino acids do—are consumed before the first ones consumed are excreted or catabolized.

In short, the deleterious effects of many cumulative toxins are nonadditive, whereas for many nutrients benefits are nonsubtractive. This last pair of properties, combined with the complementary distributions of nutrients and toxins, provides possible solutions to the dual problems of the nutritional inadequacies and omnipresent hazards of wild foods: select an array of foods that in combination provide enough of every nutrient, without exceeding one's tolerance for any toxin or other deleterious food component. If among such diets one can find some that provide more of the fitness-limiting components, so much the better.

Not surprisingly, when foods are regarded as potentially complementary rather than as scalable on the basis of, say, energy or toxin concentration, quite different decisions for foragers result (Rapport 1980). The top-down approach served an extremely valuable heuristic function in the early development of optimal foraging models and may be relevant for some creatures, such as predators, that choose among food patches of the same food but of various sizes. However, for the problems facing many herbivores, more realistic formulations are needed (see also Schluter 1981). A common way of incorporating the effects of additional nutrients and hazards into optimality models is by adding constraint equations (see the survey by Stephens and Krebs 1986 for numerous examples).

Specialists versus Generalists

Versatility and flexibility in choice of foods is a great asset. In a changing environment, it is a much less Spartan solution than massive deaths resulting from natural selection against specialists whose food is on the decline. Omnivory allows a change in preferred basic food to occur within an organism's lifetime rather than over many generations. . . .

The problem is that experimentation with new foods can be dangerous, since such substances can be harmful. On the one hand, the omnivore should be familiar with and in touch with the various food sources in its environment; on the other hand, this involves risks, particularly needless risks, if there is already adequate familiar food. . . . The optimal solution to the omnivoral problem involves devoting quite a bit of brain circuitry to the food problem, and employing multiple mechanisms. Thus, we see instances of built-in programming, motivation through 'general experience,' more traditional learning, imprinting, social interactions, and culture or tradition all playing a role in food selection.

—P. Rozin, 1976

Old World monkeys (family Cercopithecidae) have diverged along dietary lines. Of the two subfamilies, members of one, the colobines or "leaf eaters," feed on large quantities of leaves and other material having a relatively low concentration of rapidly assimilable nutrients. The colobines have large, sacculated stomachs and an associated gut microflora that enables them to extract adequate nutrients from their bulky, leafy diet and to cope with toxic secondary compounds (Hladik 1978). But the extraction process is slow, and the leaf eaters spend much of their day "resting" while they digest their food (e.g., Milton 1980). As is often the case with folivorous mammals, their basal metabolic rate is low for mammals of their size (Cork 1994; McNab 1986).

The second subfamily, the cercopithecines or "cheek-pouched monkeys," includes baboons, which are the subjects of this study. Cercopithecine primates feed much more selectively and opportunely on less bulky but more nutritious foods than do colobines. They spend much time moving from one feeding site to the next, often over long distances, searching for food and then processing it by digging, hulling, and so forth. They are agile, alert, and active most of the day. With some exaggeration, we can say that colobines live by their stomachs while cercopithecines live by their wits.

For these two ways of making a living, the more conventional distinction is between dietary specialists and generalists. It is the distinction between animals with very limited diets, that is, that are stenophagous (in the extreme, monophagous) and those that eat a diversity of foods, that is, that are euryphagous (in the extreme, pantophagous or omnivorous). Of course, this dichotomy demarcates the extremes of a continuum along a gradient of dietary diversity, and only for convenience do we persist in assigning the animals to discrete dietary classes.

In an article on the feeding behavior of Amboseli baboons, Post (1982) wrote as follows:

As Schoener (1971, p. 384) has noted, the distinction between generalists and specialists can be drawn in several ways [e.g., dietary breadth or diversity (Morse, 1971), extent of behavioral flexibility during foraging (Homewood, 1978), and degree of dietary overlap with sympatric competitors (Struhsaker, 1978)]; baboons appear to be true generalists along virtually any axis. It is,

however, the extraordinary breadth of their diet that has occasioned the most
comment (see, e.g., S. Altmann and J. Altmann, 1970; DeVore and Hall, 1965;
Demment, 1983; Hamilton, Buskirk, and Buskirk, 1978b; Harding, 1976;
Rowell, 1966).

Dietary diversity is markedly affected by the spatial distribution of the plants
on which herbivores feed. As a result of localized conditions, primarily of soil,
moisture, and insolation, many of the terrestrial plants herbivores rely on have
a clumped or patchy distribution—that is, they are underdispersed relative to a
random (Poisson) distribution. In addition, the various tissues and organs of a
plant differ markedly in their nutritional value to an herbivore (Boyd and
Goodyear 1971; Cummins and Wuycheck 1971; Golley 1973; Milton and
Dintzig 1981), and in this respect herbivores face problems very different from
those of carnivores, whose diet is nutritionally far more homogeneous and
more like their own bodies in composition. Not only do the leaves of a plant
differ in chemical composition from the stems, and those in turn from the
blossoms, but new leaves differ markedly from old ones, and the leaves of one
species differ from those of another. Indeed, for a few plant species there is evidence that individual plants and even the branches of a single tree may differ
significantly in composition (Glander 1978, 1981, 1982). Beyond that, many of
the most nutritious components of plants, including the blossoms, fruits, and
new leaves, are among the scarcest parts (Bell 1970) and the most ephemeral
(Milton 1980). For all these reasons, most potential foods of an herbivore have
a distribution that is patchy, both in space and in time.

From the viewpoint of a foraging animal, the essence of a patch is that
within-patch distances are small but between-patch distances are large. This
basic attribute of patchy plants has a profound effect on herbivores. I suggest
that the degrees of dietary diversity in many herbivorous species result primarily from certain relationships between the sizes of the animals and the sizes of
their food patches. In particular, among animals that are very small and those
that are very large compared with the size of available food patches, the composition of their diets is relatively homogeneous over time compared with the
diets of animals of intermediate size. Why should this be so? Animals that are
very small feed closer to the level of localized biochemical homogeneity. For
such animals the environment is "fine grained" (Levins 1968), and most of their
life is spent in one patch or a few fairly uniform patches, within which nutritional variability is usually small enough to be ignored. In the extreme, patch
selection becomes synonymous with habitat selection. Perforce, the foraging of
such within-patch specialists is highly selective, and their diet is relatively low
in nutrient diversity.

The Virginia pitch-nodule moth (*Petrova wenzeli*) is typical of such animals. The larva, which is the only form that eats, feeds only on the tissues be-

tween the bark and the pith at one site on the twigs of just one individual of one species of pine tree (Miller and S. Altmann 1958). Other extreme examples are provided by the organ-specific parasites of endotherms. The diets of such animals probably vary relatively little from moment to moment and day to day, yet so long as they can synthesize all essential compounds not present in their food from those that are, such a monotonous diet is sufficient.

At the other end of the scale, animals that are very large relative to their available food patches also have diets of relatively homogeneous composition, but for a very different reason: by feeding relatively unselectively on large quantities of diverse food items, they rely on the mean quality of a statistical ensemble. Such animals may consume an extraordinary variety of foods—the diets of baleen whales doubtless hold the record for taxonomic diversity—but they probably are relatively unselective omnivores.

Thus, animals that are very large relative to patch size rely on large-sample statistics, and very small ones rely on extreme selectivity. Both experience the world as fine-grained.

For animals of intermediate size relative to their food patches, the world is inevitably experienced as heterogeneous (coarse-grained), and they have the worst of both worlds. Often no single food patch is both large enough and of adequate composition to sustain them, yet small-sample fluctuations (local differences) in community composition preclude their relying on the mean quality of an unselected diet. Such animals have two options: either specialize in a particular type of food that is readily available and master the art of finding and utilizing patches of it, or else become eclectics and search out concentrations of nutrients wherever they occur. Most mammals use the first strategy, and much of the evolution of the major groups of mammals represents divergences along specialized dietary lines. Conditions under which the second option, selective omnivory, is advantageous will be discussed below in the section on dietary diversity.

Thus we can distinguish four antipodal foraging categories: highly selective animals that are small relative to patch size and that feed on just one or a few patches of one food ("within-patch specialists"), large animals that feed relatively unselectively on many foods ("ensemble consumers" or "low-selectivity omnivores"), and animals of intermediate size relative to patch size that either specialize in locating foods of a few types ("multipatch specialists") or select among the best parts of a large number of food plants ("eclectics" or "high-selectivity omnivores").

These four foraging modes can be illustrated by the ways the exudates of an acacia tree are fed on by four African animals: an aphid, a kori bustard (a large, nearly flightless bird), a baboon, and an elephant. The aphid is a within-patch specialist. The secretions at one place on one branch of a single tree can provide more than enough nourishment to last it a lifetime. The bustard, too, makes extensive use of acacia exudate. No single source is big enough to sustain it, however, so it walks slowly around the base of the acacia, feeding on droplets and chunks of gum. After circling one tree it moves on to the next, always searching for new patches of gum and supplementing its diet with small animals, such as insects and frogs. When feeding on gum, the bustard is a between-patch specialist.

The baboon feeds on some of the gum the bustard misses or cannot reach. For the baboon as for the bustard, however, no patch of acacia exudate is large enough to sustain it; or put the other way around, the baboon is large relative to the size of the patches it feeds on, so it moves from patch to patch. Having eaten whatever acacia gum was readily available, the baboon turns to other foods. It eats the flowers of the tree when they are available, as well as the green pods. Later in the same day, the baboon finds many other sources of nutrients; it eats the corms and seeds of grass plants, the bulbs of lilies, the fruits of several shrubs, the new leaves of several forbs, and so forth—always searching out the most nutritious and palatable part of each plant—as well as eating a variety of small insects and even vertebrates on occasion. Baboons exemplify the extreme of selective omnivory.

Finally, along comes an elephant. It breaks off an entire branch of the acacia, chews it up and swallows it: flowers, pods, seeds, leaves, thorns, gum, wood, cambium . . . everything. Similarly, when feeding on grasses and sedges, the elephant swallows whole a grass plant that a baboon would have carefully dissected. Ensemble feeders must be sufficiently large relative to their available food sources, and those sources must be sufficiently abundant and nutritious, that small-sample fluctuations over time in the composition and amount of the diet are rarely large enough to be deleterious, as a result of either inadequate nutrients or excessive cumulative hazards.

Elephants and baleen whales are ensemble feeders not because they are large per se, but because they are large relative to the sizes of available, nutritionally distinct food patches. Earthworms illustrate the same strategy. So too do filter feeders, and these are often diminutive. Not surprisingly, however, the very largest herbivores in a taxon tend to be ensemble feeders, and many of the smallest are within-patch specialists. Jarman (1982) points out that "in a monophyletic radiation of many species there are typically a few large species, catholic in their diet and habitat preferences, several or many medium sized species, and many small species, often narrowly specialized in diet, feeding technique, and habitat," and this spectrum can be seen in many but not all groups of mammals. Here I make a further distinction between two quite different kinds of omnivores: highly selective, eclectic omnivores and relatively unselective ensemble feeders.

Colobine primates exemplify what I have called "multipatch specialists." Hladik (1978, 373, 386) has succinctly described this aspect of colobines: "The most specialized of the primates are the folivores. . . . With a sacculated stomach and its associated bacterial flora and fauna of flagellates (Kuhn 1964), the Colobinae show the best adaptation to leaf-eating among primates." In contrast, most cercopithecines, including baboons, are selective omnivores. Both strategies are viable ways for making a living, and in certain circumstances the two can coexist. In Kibale Forest, Uganda, where folivorous colobines and more omnivorous cercopithecines both occur, the specialized folivorous monkeys account for the largest part of the biomass (Struhsaker 1975).

After a masterful survey of the responses of mammalian herbivores to toxic secondary compounds, Freeland and Janzen (1974) spell out several expectations about the foraging behavior of those mammals that they refer to as "generalist herbivores."

> Herbivores are capable of detoxifying and eliminating secondary compounds. Limitations of these mechanisms force mammalian herbivores to consume a variety of plant foods at any one time, to treat new foods with caution, to ingest small amounts on the first encounter, and to sample food continuously. Selection of foods is based on learning in response to adverse internal psychological effects, and herbivores probably can predict these from the smell or taste of new foods. Herbivores prefer to eat familiar foods and can seek out and consume foods that rectify specific nutritional deficiencies induced by detoxification. They should prefer to feed on foods that contain small amounts of secondary compounds, and their body size and search strategies should be adapted to optimize the number of types of food that can be eaten and will be present in the future.

Now let us look more closely at the distinction between specialists and generalists. I indicated that having a diverse but selective diet can provide a solution to the dual problems of the nutritional inadequacies and omnipresent hazards of plant foods, but this is not the only way to cope with these problems. The specialist's solution is quite different from that of the eclectic: instead of avoiding the cumulative deleterious effects inherent in each food, as eclectics do, the specialist adapts to the peculiarities of a few of them. The adaptations of specialists to their limited array of foods are primarily physiological and anatomical, whereas those of eclectic omnivores are primarily behavioral (Janson and Boinski 1992). The specialist may cope with some nutrient inadequacies by catabolizing compounds that to the nonspecialist are indigestible, as is done by ruminants and colobines—or rather, by their gut microflora—in converting a large portion of the cellulose in their diet into metabolizable sugars. Alternatively, specialists may synthesize from available precursors some of the compounds that for other animals are required nutrients, as many animals do in synthesizing their own ascorbic acid, and as mammals other than felids do in converting beta carotene and other carotenoids into vitamin A (Gershoff et al. 1957; Rea and Drummond 1932). (Of course no such solution is possible for essential elements.) Further, in various ways, which have been little studied, many herbivores come to tolerate relatively large quantities of the toxins to which they are repeatedly exposed (Freeland and Janzen 1974; Fox and Morrow 1981) and develop mechanisms for minimizing the impact of other hazards associated with foraging.

Dietary specialization and omnivory are the extremes of a continuum, and most animals utilize a mixed strategy. They detoxify some compounds and avoid excess amounts of others. They synthesize some essential compounds, but for others they shift from one food to another to make up for the inadequacies of each.

Generalist and specialist herbivores inhabit the same areas; neither solution is unique for the peculiarities of any local flora. However, each of these basic strategies tends to become a commitment to a way of life. Given a small increase in, say, the toxicity of a food already at the animal's tolerance limits, a species of specialists may more readily develop a small change in its metabolic machinery for detoxification, whereas a selective omnivore may more readily switch to other foods.

Selection for Eclectic Omnivory

Under what conditions is it advantageous to be a selective omnivore? I have indicated two so far (see also Fox and Morrow 1981; Westoby 1978). First, given the commitment to a euryphagous diet, the six key properties of a community of foods that were presented above select for dietary diversity. So long as no abundant food is free of cumulative toxins, none can be used with impunity as the sole source of food. So long as the toxins differ from one food source to another, toxic effects can be minimized through dietary diversity. As Coon (1976) put it, "The wider the variety of food intake, the greater is the number of different chemical substances consumed and the less is the chance that any one chemical will reach a hazardous level in the diet." In addition, so long as the nutrients that are inadequate vary from food to food, a variety of foods is required to satisfy the organism's requirements. Thus, selective omnivory should be particularly advantageous in habitats with a shortage of foods that could provide both adequate quantities of nutrients and low toxicity. Candidates include habitats, such as those of low productivity, in which protecting their parts with toxins is less costly for plants than generating new ones. "When resources are limited, plants with inherently slow growth are favored over those with fast growth rates; slow rates in turn favor large investments in antiherbivore defenses" (Coley, Bryant, and Chapin 1985).

Second, selective omnivory is a solution to the problems presented by a patchy environment, one in which the size and spacing of food patches relative to the size and locomotor abilities of the animals is such that no single food patch, even if nutritionally adequate, is large enough to sustain the animals, yet the distances between the patches of any one type make it less profitable to specialize in one food type and not feed while going from patch to patch than to feed selectively on other foods during the interim.

Westoby (1978) pointed out two consequences of resource patchiness that lead to such differences in profitability: the cost of searching and the need for environmental sampling. Search costs can be biologically significant, as they are if the best foods occur only in small, highly dispersed patches, if searching brings with it a high risk of predation, if the environment is difficult (or energetically costly) to explore, or if, for purely statistical reasons, exploration is unlikely to turn up high quality foods in sufficient quantity that they can be relied upon. In those circumstances the organism should use less stringent rejection criteria when it encounters a food patch: search time can be reduced by taking more types of food. This is the strategy of the ensemble feeder, and in the extreme it means taking essentially all food items as they are encountered in the habitat. Conversely, high quality food is a rare commodity, and to succeed as a selective omnivore, an animal must be able to locomote efficiently over the distances necessary to get from one choice patch to the next. All else being equal, the more omnivorous of two closely related species should be the more mobile of the two.

Less stringent rejection criteria may serve another purpose: sampling the environment. Sampling behavior will be particularly important if the local habitat changes, either seasonally or long term, or if the animal moves from one habitat to another. A mechanism for sampling and testing potential new foods becomes mandatory once the diversity of foods in the diet of an omnivore exceeds its capacity to select the good food from the bad purely by "peripheral filtering," that is, by means of sense organs specifically tuned to the unique properties of good foods.

If, as I am suggesting, a broad but selective diet can provide a solution to the problems of patchily distributed foods, then we expect to find selective omnivores of several sizes in habitats that exhibit extensive hyperpatchiness, by which I mean that patches tend to be aggregated into larger patches, in which, as at the smaller levels, distances between patches tend to be large relative to those within each. So far as I know, the literature on the structure of plant communities is not yet adequate to identify these habitats.

In addition, though specialists are presumed to have the advantage of great efficiency in exploiting the few foods they rely on (e.g., Levins 1968; Slobodkin and Sanders 1969; Weis 1992 and references therein), no consistent differences in foraging efficiency have been demonstrated between generalists and specialists (Morse 1980; Fox and Morrow 1981). In contrast, generalists presumably have the advantage of flexibility: by not relying on any one food, they should be able to shift opportunely when some items in their diet become scarce and others more abundant.

If these assumptions about the relationship between omnivory and flexibility are correct, omnivory should be particularly advantageous in unstable habitats and when animals are colonizing new ones (cf. Fox and Morrow 1981). Baboons are a case in point: their ability to invade a wide variety of African habitats, many of which are characterized by marked seasonal changes, depends on their dietary flexibility.

Because of the ephemeral nature of several of the most nutritious plant parts, notably new leaves, flowers, fruits, and seeds (Milton 1980), the optimal diet shifts with time; perforce, so will the diet of a well-adapted eclectic omnivore. Such "temporal tracking" of seasonally available foods has been hypothesized to increase a selective animal's dietary diversity (Hamilton, Buskirk, and Buskirk 1978; Westoby 1978). This apparent relation between omnivory and dietary flexibility raises an interesting question: For selective omnivores such as baboons, are year-to-year fluctuations in diet (the amount of nonoverlap between the diets of adjacent years) greater in arid regions, despite their lower diversity of potential food species? This would not be surprising, because, as a rule of thumb, relative variance in yearly rainfall is higher in areas with lower annual mean rainfall.

I do not know of any given answer to this last question about year-to-year variability in primate diets. The number of multiyear feeding studies is quite small, so at present the data base probably is not adequate. However, Chapman and Chapman (1990) analyzed monthly variability in primate diets, using data from forty-six long-term studies of wild populations. All of these primates shifted their diet from month to month, sometimes appreciably. Indeed, when several studies were available on any one species, the authors' measure of temporal variability often differed severalfold between them. However, the Chapmans' measure of month-to-month diet variability did not differ significantly between families or putative diet types (frugivore, insectivore, etc.), nor was its logarithm significantly regressed on the logarithm of other independent variables, namely, habitat productivity (as judged by mean annual rainfall), seasonality of the habitat (judged by coefficient of variation of monthly rainfall), body size (in kilograms), population density, or number of potential sympatric primate competitors. Thus this study fails to confirm the hypothesized relationship between rainfall seasonality and feeding seasonality. However, the lack of any significant relationship of temporal diversity in diets with potential control variables suggests that the available methods may be of low sensitivity.

3

Though this be madness, yet there is method in 't.
—William Shakespeare, *Hamlet*, 2.2

Methods

The outstanding observational conditions during this study—close-range observations on habituated animals in relatively open terrain (fig. 3.1)—enabled me to obtain extensive quantitative samples of feeding behavior, despite the fact that the animals in my study were at the most difficult age to study. These direct observations provide quantitative data on diet of a kind that could not be obtained by analysis of feces or stomach contents, both of which have been used in studies of primate diets, for example, by Moreno-Black (1978) and by Jones (1970), respectively.

Yearling baboons are small and agile, they move rapidly, and they shift frequently from one activity to another. Even so, in the 18,460 feeding bouts in my samples I usually was able to identify the (plant) part eaten (205 cases), the species or other taxon of the food (218 cases), or both (15,786 cases, 85.5% of feeding bouts). I was not able to identify either the plant species or the part being eaten in 2,251 bouts (12.2%, table 4.1, items 246 to 250). These included 769 cases of material taken from mammalian dung, 664 cases of small objects—doubtless mostly acacia seeds—picked up from the ground, 277 cases of material from in, on, or under wood, probably consisting primarily of insects and fever tree sap, 12 cases of unidentified material picked from unidentified plants, and finally 529 unspecified cases (2.9%), which by my criteria should have been considered out-of-sight time.

The data were obtained during 19,979 minutes of in-sight sample time in 1,166 focal samples on eleven free-living weanling baboons (ages 30 to 70 weeks). These samples were taken throughout the year during twenty minutes out of each of the eleven hours from 0700 h to 1800 h. The samples of feeding behavior were combined with food samples and chemical analyses of nutrients and toxins so as to provide an overall picture of the foraging of weanling baboons and its nutritional consequences.

Subjects

The infants I studied are yellow baboons (*Papio cynocephalus*), and all were members of Alto's Group. All members of the group were individually

39

Fig. 3.1. Amboseli's short-grass savannahs and habituated baboons provide exceptional observation conditions. Photo by Jeanne Altmann.

recognized by their unique physical traits. During the study period, August 1975 to July 1976, I studied the feeding behavior and foods of every infant in the group between 30 and 70 weeks of age. At the onset of the study, I began with the four infants that were then in this age group. As the study progressed, new infants entered the sample population at the beginning of the week in which they reached 30 weeks of age, and infants were dropped at the end of the week in which they reached 70 weeks of age. An exception was made for female Eno, who was dropped from the sample schedule at 62 weeks of age in order to allow time to increase my sample size on the three youngest infants: Summer, Bristle, and Hans.

The main study of 1975–76 was preceded by a pilot project during July 1974 in which two yearlings were studied. With time to sample just two individuals in the pilot project, I thought that selecting two at random would be unlikely to produce a representative sample and would therefore not achieve the usual purpose of randomization. I therefore selected two yearlings, Dotty and Striper, who were as closely matched as possible, to see whether the sampling technique was sufficiently sensitive to distinguish between their food intakes. Because members of baboon groups remain together as they move from place to place, and because Dotty and Striper were members of the same social

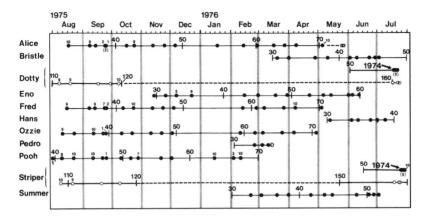

Fig. 3.2. Distribution of samples by individual, age, and date. Solid circles (•) represent days on which samples were taken on 30–70 week-baboons. Ten-week birthdays are indicated by numbered vertical bars. Open circles (○) represent days of follow-up samples, which are not analyzed here. On each sample day, 11 usable samples were taken, unless indicated otherwise by a small number above the circle. D = date of Pedro's death.

group, confounding effects that might result from differences in home range were precluded. The two of them were closely matched in age, and both were female. Their mothers were two of the three oldest females in the group and were adjacent in rank in the female dominance hierarchy. Despite this close matching of Dotty and Striper, the brief pilot study showed clear-cut differences between their food intakes and indicated that our quantitative sampling procedure would prove sensitive enough to detect individual differences in diet and presumably also in nutrient intake levels.

In all, eleven infants were included in the study. The dates and ages at which each infant was sampled are included in figure 3.2. Table 3.1 indicates the actual amount of sampling (minutes of in-sight sample time) by individual and age. The infants' birth dates and mothers are shown in table 3.2. Two of the sampled infants, Alice and Dotty, are sisters. None of the others are maternal siblings, though some could be paternal sibs. Other biographical information on most of the subjects has been published by J. Altmann (1980), who studied them during their first twelve months of life.

The Basic Formula

Nutritional requirements and toxin tolerances are usually expressed in terms of mass per unit time, for example, grams or milligrams per day. The intakes of nutrients, toxins, and other food components cannot be measured directly, but they can be broken down into estimable factors. For any food component i, the average amount ingested per unit time, say a day, can be calculated from component factors as follows:

$$(3.1) \quad \mu\left(\frac{\text{grams of } i}{\text{day}}\right)$$

$$= \sum_j \mu\left(\underbrace{\frac{\text{minutes of feeding on food } j}{\text{day}}}_{\substack{A_j \\ \text{Time Budget}}}\right) \mu\left(\underbrace{\frac{\text{units of food } j \text{ eaten}}{\text{minute of feeding on } j}}_{\substack{B_j \\ \text{Unit Intake Rate}}}\right)$$

$$\times \mu\left(\underbrace{\frac{\text{grams of food } j}{\text{unit of food } j}}_{\substack{C_j \\ \text{Unit Mass}}}\right) \mu\left(\underbrace{\frac{\text{grams of } i}{\text{gram of food } j}}_{\substack{D_{ij} \\ \text{Composition}}}\right)$$

where $\mu(x)$ indicates the mean of x and the summation is taken over all foods $(j = 1, 2, \ldots)$ in the diet. Although the components estimated by equation (3.1) are expressed in units of mass, the formula applies more generally, mutatis mutandis, to diet components that are measured in other units, such as energy (expressed in joules), vitamin A (international units), foraging time (minutes), or indeed, any "component" that is a linear function of the mass of each food consumed.

Much of this book is based on what was learned about foraging behavior from separate analyses of those four main factors of dietary intake and their combinations. Factor A_j was obtained from feeding bout samples. Factor B_j was obtained from samples of feeding rates. Factor C_j was obtained by weighing and counting representative food samples in the field. Finally, factor D_{ij} was obtained from chemical analyses. Of course these factors are random variables, not constants, and mean values were used, as described below. Means for factors A_j and D_{ij} could not be estimated directly but were estimated from component subfactors, as will be described. Beyond that, for some foods, additional partitioning of factors was required. For example, the mean rate of acacia seed consumption (seeds per minute) was obtained from the product of two subfactors, mean number of seeds consumed per pod and mean number of pods harvested per minute, each of which required a separate sample. Details of the way the mean of each of these factors was estimated are given below.

The method of factors presents one with the usual choice of whether to use pooled means or means of means, a choice that ideally depends on knowledge of the homogeneity of the subpopulations in the sample. I pooled data over the entire period of the sample, or partitioned them only by age or individual, and can therefore say nothing about the magnitude of day-to-day fluctuations in diet. Stacey (1986), on the other hand, estimated dietary intake for each day in his sample, then averaged the figures to get a mean daily intake. In chapters 4, 6, and 7, means are estimated from data pooled from all sampled infants, unless otherwise indicated. That is, I treated all observed feeding bouts as samples from a common distribution of bouts. In chapter 8, however, I shall show that individuals differ in some characteristics of feeding. For these characteristics, pooled means do not precisely characterize either the average of each individual's behavior, for which the individual means are needed, or the average behavior of individuals, for which the mean of individual means is preferable. I shall return to this problem in chapter 5.

Samples and Parameter Estimates[1]

Estimation of Time Budgets

The temporal components of this study are the hours of the day during which the baboons sometimes fed. Although the baboons often were still in their sleeping trees during the first sample of the day (0740 h to 0800 h), either asleep or relatively inactive, they sometimes fed at that hour. By 1800 h, the scheduled termination time of the last sample, the animals were usually near or under, and sometimes moving up into their sleeping trees again. At that time the feeding rate of the infants had dropped nearly to zero, and as revealed by samples that inadvertently began late and therefore ended after 1800 h, the feeding rate continued to drop rapidly thereafter. I cannot rule out the possibility of occasional nocturnal feeding on gum in the sleeping trees, all of which were *Acacia xanthophloea,* or of seasonal feeding on blossoms and pods of these trees. In addition, infants probably nurse at night. We have no information on how much milk they get then, or even how long they are on the nipple, though for most of the night they huddle within their mother's ventral flexure. Several all-night observations have convinced us that during most of the year no appreciable amount of any other type of feeding occurs at night. I therefore take the eleven hours from 0700 h to 1800 h to be the hours of potential feeding. However, my estimates of milk intake (chapter 7) take into account milk consumed throughout the twenty-four hours of the day.

As will be explained below, an important consequence of periods during which study animals are out of sight is that the mean food-specific time budgets (the A_j factors in equation 3.1) cannot be estimated directly from the observed proportion of time spent on each food. Instead, they were obtained as follows. For each food *j*,

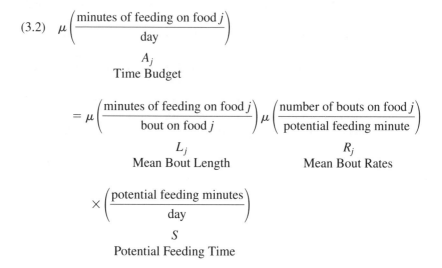

$$(3.2) \quad \mu \left(\frac{\text{minutes of feeding on food } j}{\text{day}} \right)$$

$$A_j$$
Time Budget

$$= \mu \left(\frac{\text{minutes of feeding on food } j}{\text{bout on food } j} \right) \mu \left(\frac{\text{number of bouts on food } j}{\text{potential feeding minute}} \right)$$

$$L_j \qquad\qquad\qquad R_j$$
Mean Bout Length   Mean Bout Rates

$$\times \left(\frac{\text{potential feeding minutes}}{\text{day}} \right)$$

$$S$$
Potential Feeding Time

where, as before, $\mu(x)$ indicates the mean of x (pooled mean or mean of means, as appropriate). Periods of nonfeeding were treated in the same manner. In equation (3.2), the product $L_j R_j$ is the fraction of potential feeding time spent feeding on the *j*th food. Multiplying by S converts this proportion into minutes per day. Factor S (potential feeding time per day) is the 660 minutes between 0700 h and 1800 h. Factor L_j (mean bout duration) and R_j (mean bout rate) were estimated from focal animal samples (J. Altmann 1974), using product-limit estimates of bout durations, as will be described later in this chapter.

Subfactors

Putting together equations (3.1) and (3.2), we see that the mean daily consumption in grams of any food component *i* is estimated from $\Sigma_j L_j R_j S B_j C_j D_{ij}$. I note here the interpretation of several other products:

Fraction of potential feeding time spent eating food *j*: $L_j R_j$;
Minutes of feeding on food *j* per day ("food-specific time budget"): $A_j = L_j R_j S$;
Units of food *j* consumed per day: $A_j B_j$;
Grams of food *j* consumed per day: $A_j B_j C_j$;
Grams of food *j* per bout: $L_j B_j C_j$;
Grams of food *j* consumed per minute of feeding on it: $B_j C_j$;
Grams of component *i* per minute of eating food *j* (e.g., "yield rate of nutrient"): $B_j C_j D_{ij}$;

Grams of component i per unit of food j: C_jD_{ij};
Daily intake of component i contributed by food j: $A_jB_jC_jD_{ij}$.

A note of caution must be added about this method of factors. For each factor in a product, its mean may be taken over all feeding bouts, over all bouts by a particular individual, or whatever, depending on the task at hand. In each case, it is a population mean. However, such a product of factor means is equivalent to the population mean of the factor products only if (as I have assumed) the factors are independent. The reason is that the product of the population means of random variables is not in general equivalent to the population mean of the individual products. So, for example, if X and Y are random variables—such as the measures in equation (3.2)—with means $\mu(X)$ and $\mu(Y)$ respectively, then $\mu(XY) = \mu(X)\mu(Y) + \text{cov}(XY)$, and clearly, the product of the means equals the mean of the products only if the covariance is zero (Welsh, Peterson, and S. Altmann 1988). The covariances between the various factors in equations (3.1) and (3.2) are at present unknown. So when, as in chapter 6, I discuss the diet of an "average" yearling baboon, this can be interpreted literally to mean not only an individual whose diet is characterized by the mean values of the factors in equation (3.1), but also one for whom these factors are independent of each other.

Standard Errors

For bout lengths (factor L_j in equation 3.2), standard errors of means were calculated for the product-limit estimates by the BMDP survival analysis program (Benedetti and Yuen 1977). For bout rates (bouts per sample minute) and feeding rates (food units per minute) the usual maximum likelihood mean rate estimates for Poisson processes were calculated, that is, from the number of events (n) divided by the amount of sample time on that food (t), each pooled over all appropriate samples, and the standard error of each such rate estimate was calculated from $n^{1/2}t^{-1}$ (e.g., Cox and Lewis 1966, 31; Johnson and Kotz 1969).

In equation (3.1) and (3.2), and at several other places in this book, estimates of the means of several random variables were combined by addition, by multiplication, or by various combinations thereof. Throughout this book, the standard error of a product of several estimated, mutually independent means $x_1, x_2, \ldots, x_n$ was obtained from

(3.3)
$$\left(\frac{S_{x_1\cdots x_n}}{x_1 x_2 \ldots x_n}\right)^2 = \left(\frac{S_{x_1}}{x_1}\right)^2 + \left(\frac{S_{x_2}}{x_2}\right)^2 + \ldots + \left(\frac{S_{x_n}}{x_n}\right)^2$$

where S_{x_j} is the standard error of x_j (Arkin and Colton 1955). This estimate is reasonably good if all (or all save one) of the coefficients of variation $S_{x_j}/x_j)^2$ are small. Otherwise it provides an underestimate of the exact variances, as pointed out by Goodman (1960), who gives a formula for the exact variance. The estimation also assumes that the sampling errors of the estimates are independent. Yule (1953, 196) has described sampling procedures that lead to independence. In equation (3.1), the food samples used to estimate mean compositions (factor D_{ij}) and mean unit mass (factor C_j) probably are independent of the feeding rate samples (factor B_j) and the time-budget samples (factor A_j), since these were obtained from different samples of the large population of food items in the baboons' home range. The estimates of mean feeding rates and of the two components of the time budgets, namely mean bout length and mean bout rate, were all obtained from the same feeding samples, and so their sampling errors may be correlated. In computing standard errors of products, no attempt was made to correct the estimates à la Goodman (1960) based on covariance.

For the sum of independent sample means, the standard error is the square root of the sum of their squared standard errors (Arkin and Colton 1955). The same comments about independence in my data apply here. The standard error of a constant times an estimated mean equals the constant times the standard error of the mean (Yule 1953, equation 4.5.b). This relationship was used, for example, when multiplying by factor S of equation (3.2). (Factor S, although literally an estimate, is treated here as a constant, since I have no estimate of its variance.)

At various places in this book, I use overlap in standard errors of estimated means as approximate indications of whether the estimates could have come from a common population. Ideally, instead of standard error intervals, one would like to obtain uncertainty intervals such that any two means are significantly different at a specified probability level if their uncertainty intervals do not overlap (Andrews, Snee, and Sarner 1980; Gabriel 1978); standard error intervals do not always correspond exactly with such uncertainty intervals. Similarly, I use standard error intervals to test the hypothesis that the mean has some specified value, such as that indicated by an optimal diet model. Suppose, as is often the case, that the sampling distribution of the mean is asymptotically normal, whether or not the population itself is. If the absolute value of the deviation between the sample mean and the specified value of the mean exceeds t_α (SE), where t_α is the two-tailed value of Student's t at $p = 1 - \alpha$, then the hypothesis will be rejected at the $p\%$ level and otherwise accepted. Because my means are often composites of other values (e.g., products of various means), I do not know how many degrees of freedom they involve. However, for degrees of freedom from 11 to infinity, $t_{.05}$, two-tailed, varies only from 2.20 to 1.96. I have therefore typically used twice the standard error for such tests, and looked skeptically at borderline cases.

Feeding Samples

The proportion of potential feeding time that the baboons spent feeding on each type of food was obtained from the product of factors L_j and R_j of equation (3.2). Estimates of L_j and R_j were obtained from extensive samples of feeding during 1974–76. On each sample day, just one individual was sampled, and samples were taken during every potential feeding hour of the day, from 0700 h to 1800 h. This procedure was followed not only because of day-to-day unpredictability in feeding time but also because of the possibility of dietary compensation during the latter part of the day. Such compensation may also take place over several days, as it sometimes does in humans (e.g., Taggart 1962), suggesting several consecutive days of samples per infant. Yet even one additional day on each individual before going on to the next seemed likely to result in phenophase differences between the first and last samples of a sample cycle and thereby seemed likely to reduce the validity of comparisons between individuals, which are of central importance to this study. Therefore, on each successive sample day another infant was sampled, until all that were in the sample population at that time were done, whereupon a new sample cycle was begun. The decision to rotate one-day samples is consistent with a recommendation by Thiemann and Kraemer (1984) that observations should be spread over the sampling period as uniformly as possible, especially for behaviors that are known to change over time.

The infants were done in the same order in each sample cycle, proceeding in age order from oldest to youngest; I relied on days off, addition of new infants, and deletion of old ones to avoid coincidence of the sample cycle with a possible environmental cycle. (The numbers of days per sample cycle were appropriately irregular: mean 19.4 days with a standard deviation of 10.6 days.)

All samples were taken on foot, except during heavy rain and during the pilot study, when the animals were observed from a vehicle. On foot, I was usually at quite close range, commonly within six meters of the subject. Binoculars were nonetheless useful for observing the details of selective feeding.

To avoid bias in starting and ending times of the samples, the following rules were followed. Scheduled starting time for all samples was 40 minutes 0 seconds after the hour, and except as indicated below, each lasted 20 minutes 0 seconds. (Because the scheduled starting time for my samples was 40 minutes after the hour, they emphasize the foods eaten in the latter part of each hour. Perhaps the samples should be regarded as representative of the hour intervals on which they center, that is, beginning 20 minutes after each hour.) If at the scheduled starting time I knew the whereabouts of the subject but it was momentarily out of sight, the sample was begun at the scheduled time. On the other hand, if the sample could not be taken, I continued searching until the subject was found, waited 15 seconds, then began the sample at the beginning of the next minute or half-minute, whichever came first. In total, 39% of all samples were started exactly at the scheduled time, and for the others, delay followed roughly a negative exponential distribution with a mean delay of 6.6 minutes. A negative exponential delay indicates independence of delayed start times from scheduled start times and implies the desired independence between delayed start times and ongoing activities.

Each sample record began with identification of the individual to be sampled and the time of day the sample began. (Watch times were checked almost daily against BBC time signals.) At that moment, the stopwatch was started. If a feeding bout, as defined below, was under way at the onset of the sample, this was indicated. Then for the next 20 minutes I recorded the onset and termination times of all feeding bouts, as well as the food being eaten, and the times of all out-of-sight periods (see below).

Timing Samples

All events were timed to the nearest second and recorded on a cassette tape. A digital electronic stopwatch, strapped to the flat side of an electret microphone, made a compact unit that I could hold in one hand, with all buttons readily available, leaving the other hand free for binoculars.

Because of the time required to identify and dictate the food being eaten in a bout and the times of bout beginning and ending, I found that the minimal practical length for rapidly occurring bouts was about 0.02–0.04 minute. Electronic event recorders and hand-held computers, which we have been using in our more recent studies in Amboseli, reduce this somewhat and give greater accuracy to records of brief events. A few foods, such as one acacia seed picked from the ground, were picked up singly and put into the mouth so quickly that feeding bouts on them could not be timed. In such cases the number of items was recorded, and each was given a nominal time of one second.

Sample Termination

All samples ended 20 minutes 0 seconds after they began, regardless of ongoing activity. The only exceptions were suckling bouts. These were always timed to their end, that is, until the nipple was no longer in the mouth. The exception was made because I expected suckling bouts in this age class to be long-lasting but fairly rare events, thus making it difficult to estimate the distribution of suckling bout durations if my sample included few suckling bouts and many of those were not observed to completion. Of course, this activity-dependent rule for sample continuation means that data obtained after 20 minutes should not be used to estimate the proportion of time spent in various activities ("time budgets") if the estimates are obtained directly from the fraction of actual sample time devoted to each activity. However, as will be explained below, the "censoring" of activity bouts that results from periods when the focal individual is out of sight produces a systematic bias in the observed activity distribution, so that such direct estimates cannot be used. The method I used to obtain activity-specific time budgets in these circumstances allows the use of data on the termination times of those bouts that are ongoing at the scheduled time of sample termination.

Out-of-Sight Records

If the subject went out of sight during a sample, I recorded the time at which it disappeared and (unless the sample ended first) the time at which I regained sight of it. If a feeding bout was under way at either of those times, that fact was noted. In total, the animals were out of sight for only 9.24% of scheduled sample time in this study, even by my strict criteria (below) and despite the fact that infants of this age probably are the most difficult class to sample. For the purpose of this study, an animal was considered to be in sight only if I could see whether and what it was eating. (I did not, however, exclude cases in which I had an unobstructed view of the animal's hands and mouth but still was not sure what it was eating, e.g., when the baboons plucked objects, such as acacia seeds, from elephant dung.) Otherwise it was considered out of sight, no matter how good my view in other respects. The special statistical treatment that was used to handle biasing effects of out-of-sight time will be presented below (see "Survival Analysis of Bouts").

Definition of Feeding Bouts

> Counting is easy—it is knowing when to start and when to stop that is difficult.
> —Nancy Howell, 1990

The food baboons eat is not like a strand of spaghetti being sucked continuously into the mouth. Instead, they consume discrete items: leaves, berries, corms, gum droplets, and so forth. The actual ingestion of each such item—its passage through the portals of the mouth into the oral cavity—takes but an instant. Nonetheless, there is a sense in which over some period a baboon can be said to be eating a particular food, for between bites the animal is processing the next food item, digging it up, plucking it, breaking off inedible portions, or whatever. Beyond that, a baboon repeatedly ingests the same food type, bite after bite, and while such volleys result in part from the patchy nature of the food, there are times when a given baboon's single-minded persistence with one food means it bypasses other foods that other members of the same group are eating and indeed that the animal itself will consume shortly thereafter. All such periods of feeding on one food will be referred to as "feeding bouts" or "meals." A practical reason for demarcating feeding bouts rather than just counting food items is this: there were many situations in which I knew very well what an animal was eating even though I did not have the near perfect view required to count every food item that entered the animal's mouth. Rather than discard all such data, I estimated the total intake of each food from the product of its intake rate and the amount of time devoted to it, as will be described below.

Because feeding bouts were a basic unit in my samples of feeding behavior, special attention was paid to their definition, that is, to criteria for judging when a feeding bout begins and when it ends. The choice of criteria affects important characteristics of the sample. For instance, in the extreme, one might partition the entire day into feeding bouts, each beginning whenever the animal switches to a new food, but that procedure would mean that some feeding bouts would include long periods during which the subject was not involved in feeding. The resulting inhomogeneity in feeding rates would lead to appreciable error when calculating the daily intake rate on each food from the product of items per food-bout minute times food-bout minutes per day.

In almost all circumstances, sustained contact with food plants (or animals) provided an adequate criterion for bridging the gap between contiguous bites in a volley of feeding on one type of food and for demarcating reasonably homogeneous periods. A feeding bout was considered to begin whenever the subject first put its mouth or hand in contact with any part of the food plant, such as when grasping a shrub to pull in fruit, but not including contact with the plant as a locomotor substrate, as when a grass plant is stepped on, a tree is climbed, or an infant clings to its mother. A feeding bout ended whenever the animal terminated contact with plants of that species, again excluding substrate contact. By this criterion, a feeding bout continued if, as rarely happened, food was carried away either in the hand or externally in the mouth (lips), and it terminated once the food material either was dropped or disappeared inside the oral cavity, thereby putting an end to contact as defined. So, for example, if an animal broke off a fruit-laden branch of a bush and carried it a few feet away before sitting down and plucking the berries, that entire process was considered part of a bout that began when the animal first made contact with the bush.

Contrary to the definition above, the entire sequence of chasing, repeated swatting at, and (if successful) capturing and eating of a grasshopper was treated as a bout of feeding activity, though by that definition the time from capturing the grasshopper to putting it in the mouth was on the order of a second—perhaps one or two more in the few cases in which the wings were first removed. These grasshopper chases will be discussed further in chapter 10 and appendix 3.

For some foods, contact with the plant was not always sustained between contiguous hand-to-mouth movements. Examples are grass blades, acacia seeds and blossoms picked up one by one from the ground, and berries of *Azima tetracantha,* a very thorny bush that the baboons do not touch if they can avoid doing so. For this reason, bouts were not terminated by brief contact pauses, two seconds or less. (If after two seconds the bout did not resume, those two seconds were not included in the bout.) On the other hand, a complete switch to a new food type always terminated the previous bout and initiated a new one, even if there was no break between the bouts. This was done to minimize statistical inhomogeneity.

Foods were considered to be of different types if known or suspected to be significantly different in nutritional composition. For this purpose I distinguished not only between species of plants and parts of plants, but also between degrees of fruit ripeness, between old leaves and young terminal growth, and so forth. On the very rare occasions when the animal retained contact with one food plant while beginning to feed on the next, of another species, overlapping bouts were recorded.

In short, a feeding bout begins whenever an animal first makes contact with a food plant other than as a locomotor substrate or, if already involved in a feeding bout, whenever it switches completely from one food to the next. It ends when the animal next loses nonsubstrate contact with plants of that species for at least two seconds or switches completely to another food. The same criteria were used for animal food and for suckling. When a yearling lost oral contact with its mother's nipple for two seconds, the suckling bout was considered to have terminated if it remained off for at least two seconds, but not when, within two seconds, the yearling switched from one nipple to the other.

The definition of a feeding bout given above served the purpose of this study. Bout end points could readily be recognized in the field, which is essential if consistent and replicable results are to be obtained. The definition is sufficiently inclusive that no feeding occurs outside feeding bouts, yet sufficiently exclusive as to preclude most other types of activity, thereby minimizing problems of statistical inhomogeneities. Substantial periods of nonfeeding activity got included as a result of the definition only on rare occasions, for example, when a yearling clutched a food plant in its teeth, then ran off to play; in future studies, some special rule might be invoked in such situations. With very rare exceptions—just 3 cases out of 18,460, all of nested bouts—feeding bouts as defined here were mutually exclusive: the baboons did not feed concurrently on more than one type of food, though of course by my definition of a feeding bout they could have more than one type of food at a time inside their mouths or cheek pouches.

Slightly different definitions of a feeding bout were used by Post et al. (Post, Hausfater, and McCuskey 1980; Post 1981) and by Shopland (S. Altmann and Shopland in prep.) in their studies of feeding by Amboseli baboons (see appendix 2). These alternative definitions doubtless produced some differences in our quantitative data; although the magnitude of these differences is unknown, I believe it is quite small.

Survival Analysis of Bouts

Survival analysis methods were used extensively in analyzing data on feeding bout durations. Because such methods are not yet common in animal behavior research and will be extensively relied on in what follows, a brief summary of the basic functions that are used is presented in appendix 3.

For reasons explained in appendix 3, I make extensive use of log-survivorship distributions. A survivorship distribution specifies the probability $S(t)$ that defined intervals (in this study, feeding bouts) will last at least t seconds. If $S(t)$ is a negative exponential distribution of t, so that log $S(t)$ is a linear function, then the termination rate is a constant, independent of how long the bout has been under way, as in a Poisson process. Consequently, a linear log-survivorship distribution is commonly the null hypothesis for interval distributions.

Although feeding bouts of some baboon foods, such as blade bases of *Sporobolus consimilis* (fig.3.3), grasshoppers (fig. A3.1), and leaves of *Azima tetracantha* (fig. 3.4), have log-survivorship distributions that are quite linear, others clearly deviate, for example, fever tree gum from trees (fig. 3.5), green leaves of grasses and sedges, and fruits (fig. 3.6). I tried to find a probability distribution that fit the observed feeding bout duration data, both to provide a compact summary of the results and to suggest suitable models for the underlying processes that generated the data. The Weibull distribution (Weibull 1951) was taken as the most likely candidate. The Weibull distribution applies to survival processes in which the failure rate—the probability per unit time

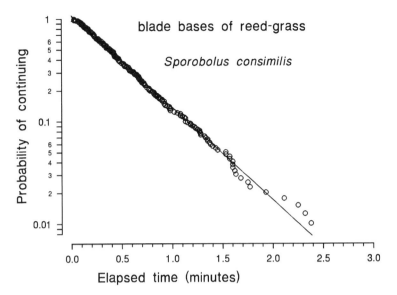

Fig. 3.3. A food with a linear log-survivorship distribution: blade bases (meristem) of reed grass, *Sporobolus consimilis,* category 191 of table 4.1. $R^2_{adj} = 0.99, p \leq 0.001, \log y = -0.86x.$ All log-survivorship graphs in this chapter show product-limit estimates, calculated via BMDP's survivorship program. Points are plotted wherever the survivorship program indicates a decrement in survivorship. In this and all other log-survivorship graphs in this chapter, probabilities below 0.008 (i.e., below 0.8%) are not printed but were included when calculating the regressions. Increasing fluctuations on the right sides of the graphs result from small-sample and integer-value fluctuations as the number of ongoing bouts approaches zero. All regressions are through $\log y = 0$ (hence through $y = 1$).

that an event under way after any specified amount of time will terminate in the next time unit—may increase with elapsed time, decrease, or remain constant, so long as it is a power function of elapsed time (see appendix 3). It thus subsumes a linear log-survivorship distribution as a special case. The Weibull distribution has two parameters, a scale parameter α and a shape parameter β. (A third parameter γ can be included if there is an initial period with no "deaths.") The method I used for estimating α and β is described in chapter 4, where I show that two-parameter Weibull distributions provide very good fits to the bout lengths of every primary food class in the baboons' diet.

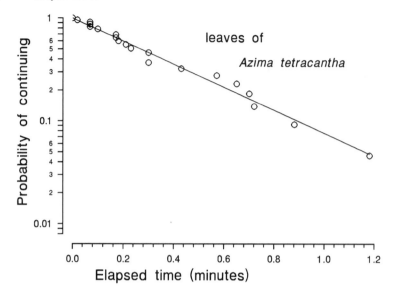

Fig. 3.4. A food with a linear log-survivorship distribution: *Azima tetracantha* leaves, category O of table 7.1. $R^2_{adj} = 0.99$, $p \leq 0.001$, $\log y = -1.12x$.

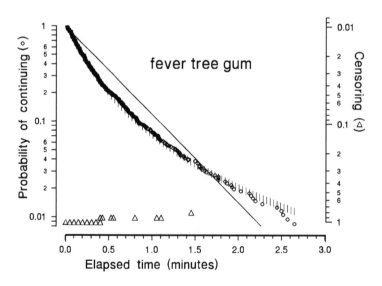

Fig. 3.5. A food with a concave log-survivorship distribution: fever tree gum, category M of table 7.1. Although the linear regression ($\log y = -2.1314x$) is highly significant ($p \leq 0.01$, $R^2_{adj} = 0.85$), a better fit can be obtained by a curvilinear function, such as that shown here by vertical bars (| | |) for $y = 0.2(x + 0.5)^{2.5}$ (fit by eye).

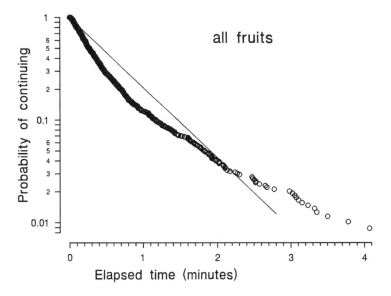

Fig. 3.6. A class of foods with a concave log-survivorship distribution: fruits (all species pooled), category FF of table 7.1. $R^2_{adj} = 0.83$, $p < 0.01$, $\log y = -0.6850x$.

Effects of Out-of-Sight Periods

Even in Amboseli, with its outstanding observational conditions, yearlings that were being sampled sometimes disappeared for short periods (mean 0.67 minute), usually as a result of intervening foliage. Other interruptions occurred at the beginnings and endings of the sample periods. The affected activity bouts are said to be "censored."

What is the magnitude of this censoring? For each of the 52 foods for which a separate product-limit estimate of the mean was calculated (as explained in the next section), table 4.1 shows the total number of bouts on that food and the number of complete bouts, that is, those seen from beginning to end. The difference between these two values is the number of bouts that were incomplete, because one or both ends were interrupted either by the beginning or ending of the scheduled 20-minute sample period or by the animal going out of sight (or reappearing) while the bout was under way. The values range from zero to 44% of bouts (for the semiripe fruit of a dense bush, *Salvadora persica*). For all foods pooled, 1,944 bouts (10.5%) were censored, or put more positively, 89.5% of all bouts were seen from beginning to end.

Another measure of the magnitude of censoring is the *censoring rate—* the number of incomplete bouts (of whatever cause) per minute of observed feeding. The censoring rate for all 277 foods pooled was 1,944 censored bouts in 8,621.56 minutes of in-sight feeding time, that is, an average of 0.225 censorings per minute of observed feeding, equivalent to one censoring every 4.43 minutes of feeding. For individual foods, the values range from zero or nearly so, for some foods eaten in open areas with no obstructing foliage, to about 1.2 censorings per minute of feeding, for leaves of *Salvadora persica.*

Closely related to the censoring rate of an activity is its *disappearance rate,* the rate (occurrences per unit time) at which an animal disappears from view while engaged in that activity. The disappearance rate provides a measure of observational conditions. I calculated the disappearance rate for feeding (all foods combined) by counting the number of feeding bouts that were right-censored because the animal disappeared from view (but not those that were right-censored by the ending of the sample period), then dividing that by the total number of minutes of focal sampling during which the focal individuals were feeding. The result is 612 disappearances in 8,621.56 minutes = 0.07 disappearances per minute while feeding, equivalent to one disappearance every 14.09 minutes of feeding. For all other activities combined (i.e., for nonfeeding periods), the disappearance rate was 2,704 disappearances in 11,357.18 minutes = 0.24 disappearances per minute—about three times higher than during feeding, perhaps because when feeding, the young monkeys are less mobile.

How does censoring affect the results? Of course it reduces the sample size, but does it introduce any systematic bias into the results? Perhaps the simplest assumption that one could make about these out-of-sight periods is that they begin and end at times that are independent both of the ongoing activity and of how long the interrupted activity has been under way. If so, the observed sample would provide an unbiased estimate of the proportion of time spent in each activity—the "time budget." Nonetheless, such samples would give systematically biased estimates of bout length distributions: the samples would be biased against long activity bouts, because the longer a bout lasts, the greater the probability that it will be interrupted by an out-of-sight period. Beyond that, even if out-of-sight periods are independent of how long the interrupted bouts have been under way, some activities may be more likely to be interrupted than others. If so, then even the time budget would be biased if it were based directly on the observed sample. The bias against long activity bouts that results from out-of-sight periods has been mentioned by several authors (e.g., Delius 1969; Slater 1974), whereas the potential bias in time budgets, when recognized, has usually been assumed not to occur.

In view of the above, I ask, Is a baboon more likely to go out of sight while feeding on some foods than on others? Of course, for foods on which an animal spends more time, more interruptions of observations are expected in total; I am in fact asking about *rates* of interruption, that is, the probability of

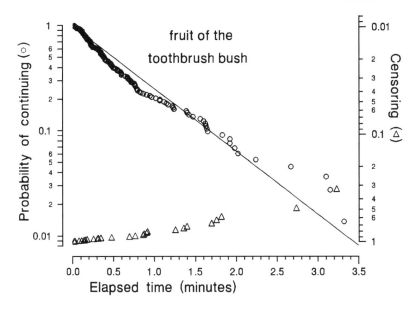

Fig. 3.7. Log-survivorship and out-of-sight censoring rate for fruit of the toothbrush bush, *Salvadora persica*, category G of table 7.1. Note the essentially linear (time-independent) censoring rate for this food. For fruit, $N = 148$, $R^2_{adj} = 0.95$, $p \leq 0.01$, $\log y = -0.5975x$. For censoring, $N = 30$, $R^2_{adj} = 0.94$, $p \leq 0.00005$, $\ln y = -0.2767x$.

In this and subsequent figures in this chapter, right censoring is plotted on an inverted *y*-axis for clarity. It is graphed as the probability that a feeding bout that has lasted this long is *still under observation* (the subject has not gone out of sight and the sample period has not ended at the indicated elapsed time since the bout began), whereas survivorship is the probability that a feeding bout is *still under way* (the animal is still feeding at the indicated time).

interruptions per unit time. Therefore, for a sample of six typical baboon foods, I computed the failure rate resulting from going out of sight—the "interruption rate"—as a function of "bout age," the time elapsed since a bout began. From that analysis (appendix 4, table 3.3, fig. 3.5, 3.7–3.11) I conclude that in my study, observations on behavior were interrupted at rates that depended on the behavior under way (the food being consumed), but that for each food, with perhaps two exceptions—fruits of *Azima tetracantha* and *Tribulus terrestris*—such interruptions were random. That is, for each food, interruptions occurred at a constant, food-specific rate that was independent of how long that bout of activity had been under way, so that for any food, the survival distribution of bouts interrupted at any particular bout age is the same as the conditional survival distribution of bouts that lasted beyond that bout age. These food-specific interruption rates mean that time budgets for individual foods cannot

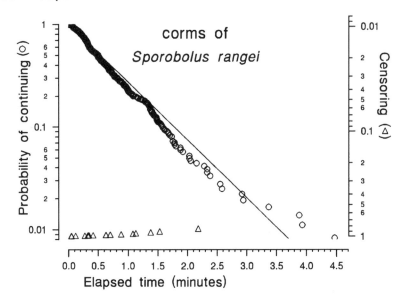

Fig. 3.8. Log-survivorship and out-of-sight censoring rate for *Sporobolus rangei* corms, category B of table 7.1. For corms, $N = 273$, $R^2_{adj} = 0.95$, $p \leq 0.005$, log $y = -0.5636x$. For censoring, $N = 15$, $R^2_{adj} = 0.95$, $p \leq 0.00005$, ln $y = -0.0683x$.

be estimated directly from the observed portions of the feeding samples. My solution to this problem follows.

Factor A: Food-Specific Time Budgets

I have demonstrated that foods differ in the rate at which baboons go out of sight while feeding on them. Consequently, the raw data on feeding time are systematically biased in two respects: (1) for each food, the observed bout durations are biased against long bouts (thus the sample means are also biased), and the magnitude of this bias depends not only on the frequency of long bouts but also on the food-specific censoring rate; (2) the "observed time budgets," that is, the proportion of in-sight time devoted to each food, are biased, and the magnitudes of these biases depend on the food-specific rates of out-of-sight periods.

The first bias was circumvented by using product-limit estimates of the mean bout durations as values of the L_j's of equation (3.2). Product-limit estimates (Kaplan and Meier 1958) of the mean and distribution of bout lengths were computed for each food (where the sample size was at least twenty) or aggregate food class. Again, whenever one food contributed more than 95% of

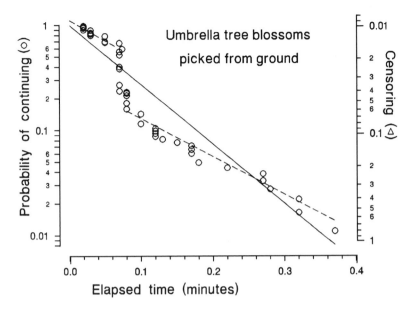

Fig. 3.9. Log-survivorship and out-of-sight censoring rate for umbrella tree blossoms picked up from the ground, in table 4.1. No censored bouts occurred in this sample. The data appear to be compounded of two bout length distributions, one consisting of short bouts that terminate by 0.05 minute or shortly thereafter, and another consisting of longer bouts that last beyond 0.08 minute and terminate at a constant rate thereafter. For each of these a dashed regression line is shown. For blossoms' overall regression, $N = 46$, $R^2_{adj} = 0.85$, $p \leq 0.001$, $\log y = -5.6474x$. For elapsed time of 0 to 0.05 minute, $N = 13$, $R^2_{adj} = 0.72$, $p \leq 0.0005$, $\ln y = -5.8423x$. For 0.08 minute onward, $N = 25$, $R^2_{adj} = 0.95$, $p \leq 0.0005$, $\ln y = 0.2965 - 8.2897x$. No right censorings occurred with this food.

the sample in an aggregate class and no other food in the class was observed twenty times or more, only the aggregate distribution was analyzed. Calculations were carried out by means of the survival analysis program of BMDP-77 (Benedetti and Yuen 1977), but without the reversal in coding described above for analysis of censoring.

In making these estimates I took bouts to be censored either if the animal went out of sight before the bout ended or if I terminated observations during a bout because the scheduled sample period ended. As before, only bouts with observed beginnings were used.

Next, the product-limit estimate of mean bout length L_j for each food j was multiplied by its estimated rate "per day," that is, per minute of potential foraging time in a day, factor R_j of equation (3.2). This product circumvents the second bias indicated above and provides time budget estimates as the fraction of potential foraging minutes in a day that the baboons devoted to feeding on

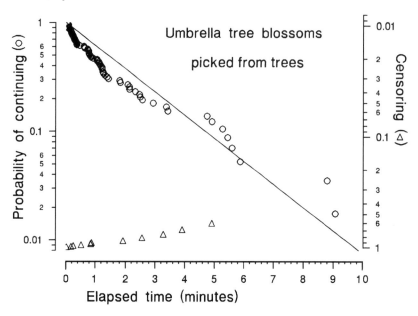

Fig. 3.10. Log-survivorship and out-of-sight censoring rate for umbrella tree blossoms picked from the tree, category 8 in table 4.1. $N = 76$, $R_{adj}^2 = 0.87$, $p \leq 0.005$, log $y = -0.2124x$. For censoring, $N = 11$, $R_{adj}^2 = 0.98$, $p \leq 0.0005$, ln $y = -0.0957x$.

the jth food. Finally, this fraction multiplied by the number of potential foraging minutes per day gives A_j, the number of minutes per day devoted to the jth food.

The numerator of my estimate of factor R_j is the number n_j of bouts on food j during an entire study or any designated subsample therein, for example, for a particular season, age class, or individual. For this purpose I used all bouts, whether or not they were censored.

One might argue that the inclusion in n_j not only of complete bouts (observed beginning and ending) but also of all bout fragments (one or both ends censored) would lead to an overestimate, since some left-censored fragments are continuations of bouts that were under way when an out-of-sight period began, and such bouts would thus be counted twice: my value for the numerator of R_j would be slightly too large. Only a very small proportion of the sampled bouts is involved here, namely, those that were left-censored and for which the preceding right-censored feeding bout was on the same food. By the same token, however, my estimate for L_j (mean bout length) is slightly too small, since I did not modify our estimate of L_j based on the information in such bouts, which could only have increased the estimate. I do not know whether

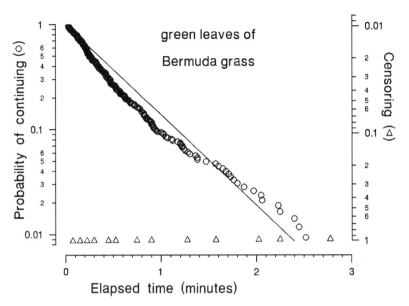

Fig. 3.11. Log-survivorship and out-of-sight censoring rate for green leaves of Bermuda grass, *Cynodon nlemfuensis,* category 90 in table 4.1. For leaves, $N = 278$, $R^2_{adj} = 0.93$, $p \leq 0.005$, $\log y = -0.8592x$. For censoring, $N = 23$, $R^2_{adj} = 0.97$, $p \leq 0.00005$, $\ln y = -0.1258x$.

these two sources of error cancel each other. Fortunately, the censoring rate was so low for almost all the foods in my samples (table 4.1) that the residual bias should be quite small.

What about the denominator of R_j? $\hat{L}_j$ is the product-limit estimate of the mean feeding bout length of the jth food, and n_j is the number of bouts of food j in the sample, as discussed above. Then $\hat{L}_j n_j / (\Sigma_{x = 0} \hat{L}_x n_x)$, where x indexes bouts on all foods and all other activities (all in-sight-not-feeding intervals), is an estimate of the proportion of sample time devoted to food j during $t = \Sigma_{x = 0} \hat{L}_x n_x$ total minutes of sampling on feeding and other activities. If the sample minutes are representative of the baboons' potential feeding time, this ratio also estimates the proportion of potential feeding time that they devoted to feeding on the jth food. I therefore used the *product-limit sample time t* as the denominator of factor R_j in equation (3.2). This method of estimating activity-specific time budgets is equivalent to assuming that the ratio of time spent on any two activities i and j is proportional to the ratio of $L_i n_i$ to $L_j n_j$, which seems reasonable.

Note that t, being based in part on product-limit estimates, may differ somewhat from actual in-sight sample time. (In my samples overall, the latter was

90% of the former.) If the latter has been used as the denominator of equation (3.2), the time budgets would not have summed to 100%. The *relative* values of the time budgets for feeding on each food or other activities are unaffected by this difference in scale.

In practice, I found it informative to break factor R_j into two subfactors, namely, the mean number of bouts of food j per minute of feeding and the mean fraction of the "day" spent feeding, that is, of minutes of feeding per minute of potential feeding time. For any food j, the first subfactor times L_j equals the fraction of feeding time devoted to that food. Further details are provided in chapter 4, under "Time Budgets of Foods."

For time spent on just one food, namely milk—or more exactly, time spent with the nipple in the mouth—activity bouts that were under way at the end of the twenty-minute samples were nonetheless timed until they ended. With my method of estimating time budgets, such bouts can be used in their entirety. Doing so improves the product-limit estimates of the mean durations of those activities (because more uncensored bouts are used) but does not change the number of bouts n_j, since only bouts whose onset occurred during the scheduled sample time were included. The result is to improve the estimate of L_jn_j for each such food, and thereby to improve the precision (decrease the variance) of the time-budget estimates.

Factors R_j are rates. Throughout this book, whenever rates of events were tested for equality, the method given in section 9.3 of Cox and Lewis (1966) was used. In brief, it is as follows. Suppose that in fixed time periods $t_1, \ldots, t_k$ the numbers of events observed in k independent Poisson processes are $n_1, \ldots, n_k$. Then the N events ($N = \Sigma n_i$) will be multinomially distributed among the k "cells," with probabilities t_i/T (where $T = \Sigma t_i$) and thus with expected values t_iN/T. We can therefore use a goodness-of-fit test by taking advantage of the fact that

$$(3.4) \qquad X^2 = \sum_{i=1}^{k} \frac{(n_i - t_iN/T)^2}{t_iN/T} = \frac{T}{N} \sum_{i=1}^{k} \frac{n^2_i}{t_i} - N$$

is asymptotically distributed as χ^2 with $k - 1$ degrees of freedom. (Notational differences aside, equation 3.4 is the same as equation 4 on p. 232 of Cox and Lewis 1966 except for a factor of k, which should be deleted from the right side of their equation.)

Finally, as indicated before, factor S of equation (3.2) is the 660 minutes of potential feeding time per day. That is the last component needed to calculate the time budgets, factor A_j of equation (3.2). It is possible that some food is eaten outside the sample period (0740–1800 h). Based on ad lib observations, I believe that the amounts the baboons consumed during other hours of the day were negligible. Nocturnal milk intake may be an exception; my method for estimating total daily milk intake is described in chapter 6.

Factor B: Unit Intake Rate

Data on feeding rates (factor B in equation 3.1) were obtained opportunistically, primarily during the food time-budget samples described above. The essential condition for a rate sample was that I have an unimpeded view of the animal's feeding activity so that a count can be made of the number of food units. Beginning and ending time of a rate sample did not always correspond with those of a bout sample, though I was sensitive to the fact that intake rate might change systematically during feeding bouts. (Extensive samples of food intake rates as a function of elapsed time in feeding bouts, obtained on Amboseli baboons by Jennifer Shopland [S. Altmann and Shopland in prep.], show no systematic change over time in unit intake rates, except for a very brief start-up interval.) Where food preparation time was required before consumption, for example, when digging up and tearing open grass corms, unit timing always included the preparation time. Some foods, such as berries and leaves, were consumed in natural units. For others, such as tree gum, an estimate was made of the amount consumed per bite. Based on these samples, pooled mean feeding rates were estimated.

Factor C (Unit Mass) and Factor D (Composition)

I tried to collect food samples that were as representative as possible of what the baboons were eating. All foods were collected during the same phenophase and the same hours of the day when they were eaten. All were collected within the home range of the study group, but not at patches that had been thoroughly "picked over" by the baboons. When gathering food samples I attempted to mimic any observed selectivity on the part of the baboons for plant part, size, degree of ripeness, or whatever.

Two samples of each collected food were obtained, a wet sample and a dry sample. The wet sample was preserved in a 3–5% solution of oxalic acid in UV-proof bottles. This was done to stabilize vitamins. For each such sample, about 20 g of material was put in about 200 g of preweighed oxalic acid solution immediately after being picked. Food items were counted as they were put in the container. The net mass of each sample was determined to 0.05 g. Very small items, such as the ovaries of *Trianthema ceratosepala*, were put directly into the solution. Larger items, such as the berries of *Azima tetracantha*, were first lanced or cut into pieces. (The *Azima* berry juice frosted a stainless steel knife blade!) The preserved samples were kept in a dark cabinet at our base camp until our next trip to Nairobi, then shipped by air freight to the nutritional biochemistry laboratory, WARF Institute, Madison, Wisconsin. All vitamin analyses were carried out at that laboratory from these wet specimens.

The oxalic acid method of field preservation was used on the advice of the WARF laboratory. The original source for this method (ICNND 1963), which I did not see until completion of my fieldwork, recommends adding alcoholic potassium hydroxide to prevent bacterial spoilage for specimens that cannot be refrigerated or frozen or ones that will be used for carotene determinations. In addition, I did not know at the time that folic acid may be unstable at low pH; hence values for folic acid reported here should be considered lower bounds of estimates of the true values.

For dry samples, food items were counted as they were collected, then about 20 g of material was placed in containers of known mass (Foil Pak no. 6, Champion Packages Co., Columbus, Ga.). (Particular care was taken with foods that were likely to desiccate rapidly after picking. As quickly as possible, they were put into plastic containers and processed.) The container plus contents was then weighed. Weighings were made on a triple beam balance, read to 0.05 g.

The net mass of all the food in the wet and dry samples combined divided by the number of food items gives the mean unit mass, factor C in equation (3.1).

The first three dry samples, which I collected during 1974, were desiccated in an "oven" (a heavy-gauge steel ammunition box kept warm by the coals of an outdoor fire). For most of the samples (1975–76), I replaced this cumbersome and variable method with fresh-frozen preparations: the Foil Paks were sealed, each was enclosed in a sealed polyethylene bag, and then as soon as possible after picking they were frozen in the freezer compartment of a small refrigerator. They were kept frozen, there and subsequently in the nutrition laboratory's deep freeze, until they were analyzed. I also made use of food samples taken in Amboseli by David Post during 1974–75 and by David Klein during 1975–76. Details of our samples are given in S. Altmann, Post, and Klein (1987).

Supplementary Data on Play

During all 1975–76 food time-budget samples, I recorded the start and stop times of all bouts of rough-and-tumble play in which the subject participated and the identity of its play partners. I used these data to analyze the consequences of individual differences in foraging (chapter 7).

Sources of Errors

In most studies of dietary intakes, little information is given on sources of errors and their potential contribution to variance in the estimates. (Exceptions include Beaton et al. 1979; Liu et al. 1978; Marr 1971; and Pekkarinen 1970.) My intention in writing this section is not to disarm my critics, who will no doubt find enough to quarrel with. Yet no one is more aware than I am of the numerous difficulties in carrying out a project such as this, and the ways they can affect the results. I hope that the following discussion of problems and potential sources of errors will be of use to those who carry out similar studies.

Sample Selection

For every sample involved in this study, one can ask whether the sample was taken by an unbiased method and whether it is large enough to avoid excessive small-sample fluctuations. Sharman and Dunbar (1982) present evidence that the study groups selected by primate fieldworkers are a biased sample of the available group sizes. All the animals in my study were members of one baboon group; I cannot claim that this group is "average," that is, that the mean diet of these animals is the same as the mean diet of other baboons, even other Amboseli baboons. I can only say that the group was selected independently of this study—indeed, several years before it was undertaken—and was, at the time of my study, the only group sufficiently accommodated to very close range observations. Within that group, I sampled all yearlings in the preselected age interval, 30–70 weeks of age. So, was this 30–70-week cohort representative even of other such cohorts in the same group during other years? Probably not entirely: the two yearlings in my brief pilot study, carried out a year earlier, a drought year, had a somewhat different diet, apparently because of phenological differences between years.

Beyond that, baboons living in different African habitats are faced with different selections of foods. If there is any generality to what I have done, it lies not in specifying, say, how many *Azima* berries any well-adapted baboon should eat, but rather, in providing and illustrating a procedure for deciding what diet would be optimal to a given animal, in view of the array of potential foods available to it, and of predicting the consequences of falling short of that optimum.

In some cases, the rationale for selecting a particular sampling method can be indicated. For example, by sampling each subject at virtually every hour of the day during which feeding occurs, I attempted to avoid a bias in favor of foods eaten more often during certain hours of the day. Starting samples at preselected times, to the second, eliminated activity-biased sampling periods. By repeatedly rotating among subjects, one day on each per rotation, I attempted to sample all subjects in each season and thereby to avoid confounding phenophase differences with individual differences. However, phenophase differences were not completely avoided in this way, because the infants were born at different times of the year, so at any given age they encountered different arrays of potential foods. Hans, for example, was the only subject to eat sticky-fruit (*Commicarpus pedunculosus*) at 30–40 weeks of age, but that was because he was the only one who was that age when these fruits were ripe.

In some situations, the validity of a sample must depend primarily on the judgment of an experienced observer rather than on a specifiable sampling scheme. When collecting food samples, I attempted to obtain representative samples of what I had seen the baboons eat, by trying to pick food items that had the same distributions of ripeness, ageing, position on the plant, phenophase, and so forth. I tasted almost all the foods that the baboons ate, and this sensitized me to some fine gradations in condition.

For practical reasons, my sampling was not distributed completely uniformly among the days of the year, so my samples are not perfectly representative of what the baboons ate in total through the year: foods eaten during phenophases during which I sampled more than average are overrepresented, and others are underrepresented. When required, an adjustment was made for seasonal differences in sample time.

Sample Sizes

I am acutely aware of how few individuals I sampled: eleven. Of these, six are female, and it is on these six that most of the fitness analysis is based, since only for females do we know reproductive success (chapter 7). Yet any attempt to increase the number of subjects without additional manpower would only have resulted in a reduced sample of the feeding behavior of each individual.

In future studies, the recording of core foods, identified in an earlier study such as this, could greatly reduce sampling effort. Indeed, one might search for a considerably smaller subset of the diet, say, ten foods, that would be nearly as effective at estimating total diet or predicting fitness as the larger set I recorded. Such methods would enable one to carry out feeding samples simultaneously with other sampling, for example, on social behavior. However, such reductions in sampling effort do not reduce the amount of sample time required per individual and so do not solve the sample size problem, except insofar as those researchers who are primarily engaged in other studies could contribute data from their subjects to a feeding study. Economical sampling methods are discussed further in chapter 7.

Within-Subject Variance

The diet of a baboon in Amboseli changes from hour to hour, from day to day, and from season to season. Even these changes are further compounded, in young animals, by secular changes in diet with age. All these sources of within-subject variance impede attempts at measuring differences between subjects.

At present the magnitude of within-subject variance in primate diets is largely unknown. Consequently, the minimal sample size needed for a particular purpose cannot be specified. Estimating day-to-day variance in a preliminary study could lead to a considerable reduction in sample time per individual, making possible corresponding increases in the number of sampled individuals. Excellent discussions of within-subject variability in human dietary studies have been presented by Waterlow (1986) and by Liu et al. (1978).

Between-Subject Differences

Each yearling's estimated mean intake of a given nutrient was obtained from the sum of several products (equation 3.1), each factor of which was based on its own sample. Similarly, food-specific time budgets and other "subfactors" described above were obtained from the products of the means of two or more random variables. I do not know of any discussion of how one statistically compares two or more such composite means. The method I used, overlapping standard error intervals (above and chapter 7), may not be the best way to proceed.

Computational Errors

I attempted to keep data handling and computational errors to a minimum, through vigilance, spot checks, cross-checks, and recalculations. Beaton et al. (1979) found a small contribution to variance attributable to hand-calculated values compared with those carried out on a mainframe computer. Even the latter are not infallible, however. For example, the PL-1 survival analysis program in BMDP, which was used extensively in this study, contained errors in its original version. Then, in a second version designed to correct them, other errors were introduced. I found out about this only after using these programs to carry out extensive computations, all of which had to be redone.

Computer hardware, too, can be a source of errors, as the recent Pentium fiasco illustrates. When I asked a computer repairman whether computers ever make mistakes, he replied, "How do you think I make a living?"

Choice of Computational Procedures

The actual and optimal diets presented herein are the end products of numerous calculations. For each, one can ask not only whether the computation was carried out correctly and whether the values used in the calculations were the appropriate ones to use, but also whether the calculation that was done is appropriate for the problem at hand. Some of these problems have already been discussed. For example, in equation (3.1), the mean intake of any nutrient or other food component was calculated from the product of several other means, one for each of several factors of the food component. Yet that method is valid only if the covariances among the factors are negligible (Welsh, Peterson, and S. Altmann 1988). At present, no estimates of covariances are available.

In a number of places, the mean values I used were pooled means rather than the averages of individual means. The latter might be preferable on statistical grounds when dealing with nonhomogeneous populations, but they were not always practicable. For example, my estimates of mean nutritive values for any food, say, *Azima tetracantha* berries, are pooled means: the berries were treated as if they were of uniform composition and were literally pooled for purposes of chemical analysis. A statistically preferable technique, but one not practicable in this study, would be to do separate assays on each of a random sample of berries, calculate the mean for each, and if these means differ significantly, average them. When presenting the mean diet of yearling baboons (table 6.1), I pooled data from all subjects. However, if individuals differ significantly in diet, the mean of individual means might be a preferable measure of the central tendency of the population. Some might advocate the use of modal rather than mean values.

One pooled mean that might better have been replaced by the mean of means is the unit intake rate for each food. With only a little more rate sampling than I did, a good estimate of these rates could be obtained for each subject, thus providing more information on individual differences in food intake. On the other hand, I see no practical solution to the problem that even such estimates would be pooled over all sample days on an individual: for animals that eat as many foods as baboons do, very few foods are eaten in sufficient quantities that one can get adequate estimates of daily mean values for intake rates of a given individual. One's decisions about how finely to partition a population for sampling purposes often are a compromise between research goals and feasibility, and the decision must often be made in the absence of adequate information about the major sources of inhomogeneity. Further comments on pooled means are given in chapter 6.

Diet Samples

In a study of human food intake on twenty-four-hour recall, Beaton et al. (1979) examined several potential sources of variance: effects of interviewer, training, changes in data coding rules, data handling, sex of subjects, and day of the week. Only the first two showed no significant effect. That study did not examine the amount of error in the twenty-four-hour recalls themselves, but Acheson et al. (1980) showed that their subjects underestimated their food intake over the previous twenty-four hours by 21% when responding to a printed questionnaire and by 33.6% when asked to write it on a blank sheet of paper—and this despite the fact that these subjects had been weighing and recording their own dietary intakes! Without such sensitization, twenty-four-hour recalls may be even worse (Morrison, Russell, and Stevenson, 1949). In this study I had the advantage of direct, close-range observation of consumption, but such observations are not immune to error. Direct tests of an observer's accuracy are not ordinarily possible in the field. Checks on observer agreement are, but they would have required a second trained observer and were not practicable in this study.

The convoluted method I used to estimate individual milk intakes from nursing time (chapter 6) is subject to error at every step. As in many other aspects of such a complex study, with many components, one hopes for a "cancellation of errors," that is, that the various positive and negative errors will behave something like Gaussian errors of measurement, with a zero mean. Yet no justification for this hope can be provided. Any zoo or laboratory with breeding mammals could, with some effort, greatly alleviate our present ignorance about mammalian milk intakes, by weighing infants before and after timed nursings or by using doubly labeled water to detect the amount of milk that mammals of various species consume. However, while such studies could provide estimates of age changes in mean milk intake of a given species, they would not satisfy my need to know the individual intakes of my subjects.

Nutrient Intake Estimates

How precisely can nutrient intakes be estimated in a field study? I estimated the intake of each nutrient from the mean values of five parameters for each food: nutrient composition, unit mass, unit intake rate, feeding bout length, and bout rate. For each of these estimated means except the first, I calculated the standard error, and from these I calculated the standard errors of the nutrient intakes. My method for doing so should be scrutinized by those more knowledgeable than I am about statistical matters.

In equation (3.1), the sigma indicates that the intake of any specified nutrient or other food component is the sum of its intakes from the animal's various foods; that is, I assumed there were no significant interaction effects on food absorption. However, several hundred interactions are now known to occur among food components (Freeland and Janzen 1974; Wise 1980). Whether any of these is of a magnitude to affect my results is unknown.

Biologists tend to regard chemistry as an exact science and to believe that errors made by professional analytic chemists are always negligible, but unfortunately that is not always the case (Eisenhart 1968). With only a few exceptions, I have no more than one chemical analysis of the nutrient composition of each of the baboons' foods and therefore have no way to know whether the results are even replicable. For one food, however, I attempted to check on consistency by having matched samples analyzed. Dry seeds of fever trees from one batch were divided into three parts; one was sent to the WARF Institute and the other two went to the Nutrition Laboratory of the University of Nairobi, along with other specimens. In each case a proximate analysis was requested; the laboratories were not notified that these were test samples. The differences between the largest and the smallest of the three values for each nutrient class were as follows: water 0%, minerals 0.2%, lipids 0.5%, protein 1.0%, fiber 8.1%, soluble carbohydrate 7.6%. That is, an appreciable discrepancy exists only for fiber (not surprising in view of the various ways used to analyze this component of food) and perforce also for soluble carbohydrate, since the latter is determined "by difference," that is, by subtracting the sum of all other values from 100%. The discrepancy was primarily between laboratories: results from the two samples analyzed in Kenya differ in fiber content by only 0.9%

Elsewhere (S. Altmann 1979) I describe some simple techniques that can be used in the field to check on the consistency and reliability of the laboratories to which food samples are sent, and the stability of field-preserved specimens.

The energy content of foods was estimated in a manner pioneered by Atwater, in which mean energy values per unit mass (Atwater factors) are assigned to lipids, proteins, and carbohydrates, respectively (Atwater and Bryant 1900b). The Atwater factors in the literature are based on human foods (Merrill and Watt 1973). I have modified these by assuming that much of the fiber is digestible by baboons, as indeed much of it is by humans (Ehle, Robertson, and Van Soest 1982), and by averaging the energy values of a set of least-processed plant foods (Merrill and Watt 1973, table 13), to approximate more closely the diets of wild baboons (see chapter 5). I have assumed that D-galactose and D-arabinose—the major sugars in fever tree gum (Hausfater and Bearce 1976)—are completely digestible by baboons. In each case, data on the digestibility of the baboons' wild foods would be preferable to my assumptions.

Nutrient Requirements

Present knowledge of nutrient requirements for mammals—even for humans, the mammal for which we have some of the best data—is unsatisfactory, a situation not likely to improve appreciably in the near future, since this type of research is not currently fashionable. The values I used and my rationale for doing so are given in appendix 1, but some of these values may be subject to appreciable errors.

Energy as the Objective

In chapter 5 I discuss the choice of objective functions for optimal diets. I used energy and protein maximization and feeding time minimization as objectives for the optimal diets I calculated. Yet I cannot provide direct evidence that any part of baboon life histories is limited by any of these three diet components. Conventional wisdom in ecology has it that energy is paramount. This perspective can be traced back to Boltzmann's concept that organisms are engaged in a struggle for entropy (Boltzmann 1886) and to Lotka's claim that "the laws which govern energy changes are laws governing evolution" (Lotka 1913). This conventional wisdom has rarely been checked, however. "In practice, . . . resource costs are usually expressed in energy terms. Yet the justification for this is usually a faint-hearted one: measuring energy is technically much easier than measuring most other resources" (Begon, Harper, and Townsend 1986).

Of course, biological fitness may be correlated with the energy and protein in the diet even if neither of these is causally responsible. Did two of my subjects, Pooh and Pedro, have poor diets because they were weak, or were they weak because they had poor diets? Both, I suspect. Whatever the ultimate causal relationships, the correlations are no less real and no less prognostic.

In chapter 5 I discuss the vexing question of maximizing daily intake (of energy, protein, or whatever) versus maximizing the intake *rate* per minute of feeding or foraging.

Plant Identifications

At the start of this study, the flora of Amboseli was poorly known, and no local collection of identified Amboseli plants was available. Taxonomic revision of several groups in the area was and still is under way, and this means one must also repeatedly revise the names given to foods. When taxa are lumped, this involved only an inconvenience, as when all the grasses that had been called either *Sporobolus robustus* or *S. consimilis* were put into the latter species. When taxa are split, it may or may not be possible to tell which species one's records apply to. Grasses from Amboseli that were initially identified as *Sporobolus marginatus* were recorded by me as either the large or the small form of the species, and when *marginatus* was split into *cordofanus* and *ioclados,* this size difference usually enabled me to know which species my earlier records applied to. Only well after my study was completed did I become aware that both *Commicarpus plumbagineus* and *C. pedunculosus* are present in Amboseli; the habits and habitat of the latter (prostrate on volcanic soils, away from trees) were used in retrospect to identify which of the two the baboons had eaten.

Identifying plants in the field, as they are being eaten, is quite different from working with herbarium specimens, and "field marks" often become more important than the distinguishing characteristics that systematic botanists use. For grasses, field identification is an acute problem: grass and sedge corms are fed on at times of the year when the seedheads may be completely absent, and the stubble left by the local grazers provides few clues. However, in some cases, these few may be distinctive. For example, the corms of the grass *Sporobolus rangei* can be distinguished from the corms of its intermixed "look-alike," the sedge *Cyperus obtusiflorus.* Those of the former are elongate, with a beige sheath; those of the latter are more bulbous and have a reddish saffron sheath. Leaves of both are convolute and of about the same length, but the *rangei* leaves are a more grayish green, those of *obtusiflorus* a slightly yellowish grass green. Alas, one is best at making such discriminations only at the end of one's study!

Finally, let me explain how plant species are referred to in this book. When an established vernacular name for a plant is available, I usually use it; otherwise I refer to a plant by its scientific name. So, for example, *Solanum incanum* is commonly referred to as "Sodom apple," and that name will be used here. I follow this practice because of my conviction that although to most readers the vernacular names of African plants are as unfamiliar as their scientific names, the former are easier to remember. A list of species the baboons fed on, with their scientific and vernacular names, is given in table 4.3. Some vernacular names I use were coined by those of us who do research in Amboseli and are familiar to us but have not yet become established elsewhere. Those names are given in quotation marks in table 4.3. A few vernacular names that I found in the literature are not well established even among systematic botanists and will not be used here.

Food names are even more specific, in that they require identification of the (plant) species, the part of it that is eaten, and its condition, such as its degree of ripeness. For brevity, I have devised a four-letter abbreviation for each baboon food, in which the first two letters indicate the species, the third indicates the part eaten, and the fourth indicates the condition. So, for example, "ATFR" indicates *Azima tetracantha* fruit, ripe. At some places in the text and particularly in tables, where a full verbal description of the food would be an impediment, I have used these four-letter abbreviations. Of course no reader can be expected to remember them. They are given in table 4.1. For the fifty-two core foods, an alphabetical list is given in table 6.2 and on the last page of this book.

Mechanical Failures

Repeated malfunctioning of my digital stopwatch led me to discard many samples in which the sum of periods of time-in and time-out did not add up to the total sample time. Similarly, several samples were lost because the tape recorder on which I dictated data malfunctioned. In our more recent studies in Amboseli, these problems have been minimized by recording data on portable computers and portable event recorders that produce computer-compatible records. The main advantage of these event recorders, however, is that they by-pass the error-prone and very time-consuming process of data encoding and entering.

At one stage I encountered several small but repeated discrepancies between my weighings of the preserved food samples and weighings by the American laboratory to which they were sent. We each checked what was weighed: sample + jar + lid + sealing tape + label, but not plastic bag. No discrepancy. Then we each recalibrated our scales. No discrepancy. Finally, I reweighed several samples just before shipping them to the lab, at the end of the month. They were lighter than before. Apparently, slow evaporative water loss was oc-curring at the imperfect seals in my jars. On the assumption that only water was lost, I made allowance for these shrinkages in calculating food compositions. "If the direct cost of making measurements is large, the indirect cost of making poor measurements must be huge" (Hunter 1980).

<big>4</big> Variety's the very spice of life,
That gives it all its flavour.
— William Cowper, "The Time-Piece," 1785

Dietary Diversity

W hat do yearling baboons eat, and how often is such food eaten? Once they have begun eating a particular food, how long do they persist? In total, how much time each day do they spend on each food? These questions will be answered in this chapter.

The quantitative data presented in this chapter all center on characteristics of feeding bouts. For feeding bouts, the relative frequency, duration, bout rate, and intake rate are more than descriptive parameters. The temporal distribution of feeding bouts and the frequency distribution of bout lengths may be controlled both by social influences such as displacement at food sites (Post 1978; Post, Hausfater, and McCuskey 1980) or the need to keep up with the group and by biotic factors such as the size and distribution of food patches (e.g., Hodges and Wolf 1981; Pyke, Pulliam, and Charnov 1977) or complementarity among the nutrients and toxins that foods contain. In turn, the length and rate of feeding bouts together determine the feeding "time budget" (the proportion of the day that animals spend feeding), and each of these factors may differ between individuals. For animals that are both time-limited and nutrient-limited— which I believe Amboseli baboons to be—the food-specific time budgets are critical. Food-specific time budgets, combined with feeding rates (food unit intake rates), determine the actual yield of each food. That is, the consumed amount of any food depends on how rapidly and for how long that food is eaten. While variations in feeding rates on a given food may primarily reflect differences in the food densities of local food patches, an upper limit on feeding rates is established by the animal's ability to process food. Consequently, individual differences in this characteristic of feeding bouts may be another source of food-dependent fitness variance.

What Do Weanling Baboons Eat?

One could describe the diversity of foods eaten by baboons in terms of either individual foods, food classes (primarily plant parts, such as leaves, flowers, and fruits), or species, and for each of these one could rank the baboons' dietary components in several ways: by the number of feeding bouts, by the mean duration of those bouts, by the product of the two (the amount of time spent feeding on each food), or by their nutritional contribution. Here I consider the frequencies with which the various dietary components were eaten during my samples on yearlings. Bout lengths and feeding time are considered in the rest of the chapter. The mass and nutritional contribution of individual foods are presented in chapter 6. In these chapters, data from all eleven yearlings are pooled. Individual differences are discussed in chapter 7.

In the field I recorded 903 categories of food during samples on baboons aged 30–70 weeks. By eliminating synonyms and combining categories that probably have little or no nutritional distinctiveness, I obtained a list of 277 foods, each known to be nutritionally distinct or suspected of being so (table 4.1). Each food is identified by the conjunction of its taxonomic group (species, when known), its food class (typically plant part, such as leaves or flowers), and its condition (e.g., degree of ripeness). A few of the food categories are artifacts of observational conditions, for example, pods of unknown ripeness from *Abutilon* versus unripe (green) pods from *Abutilon*. These artifacts are easy to recognize in table 4.1. Various other wild foods eaten by baboons in Amboseli are indicated in table 4.2, from observations made during several other projects.

In the following description of the baboons' diet, all identified foods and food species (table 4.1) that were eaten at least ten times (.054% of feeding bouts) in my 1974–76 samples on weaning baboons are mentioned. To help the reader appreciate the ephemeral nature of many components of the baboons' diet, I also include here a description of some of the major phenological changes in Amboseli.

Milk and Water

At 30 weeks of age, the yearling baboons suckled many times a day. At 70 weeks, when my sampling on them ended, they were nursing about a third as often. Weaning in baboons is a very gradual process. Consequently, the yearling could fall back on their mothers' milk to make up dietary inadequacies, particularly at the earliest ages I sampled. Over these 40 weeks, the baboons suckled from their mothers in 7.3% of feeding bouts. The baboons nursed throughout the year. Nursing is absent from my samples only in the second halves of May and July, when little sampling was done (183 minutes and 14 minutes, respectively). Age changes and individual differences in suckling will be discussed in chapter 6. For an analysis of earlier stages of development in Amboseli baboons, see J. Altmann (1980).

Baboons apparently are obligatory drinkers: their distribution in Africa seems to be restricted to areas that lie within cruising range of permanent drinking water. In Amboseli, baboons almost always drink at least once a day, more often when opportune (1.9% of feeding bouts). I have records of the weanling baboons drinking groundwater during every semimonth except the second halves of August and December, in each of which relatively little sampling was done.

Trees

The baboons of Amboseli live in open savannah habitat—that is, in an area in which perennial grasses form the primary ground cover and in which trees occur at low density. Judging by the number of plant species, the flora of Amboseli is impoverished. Virtually all the trees in the home range of the baboons are acacias of just two species: the fever tree, *Acacia xanthophloea* (18.4% of feeding bouts), and, on somewhat higher, drier ground, the umbrella tree, *A. tortilis* (5.0%). These two acacia trees provide the baboons with a wide variety of foods, particularly blossoms (0.46 and 1.6% of all feeding bouts, respectively), seeds (7.2 and 3.3%, respectively), and the gum of fever trees (10.4%).

Acacia Flowers

The small, spherical blossoms of both acacia species (fig. 4.1) are eaten in large quantities by baboons, who pick them from the tree or from the ground beneath it. When feeding in the trees, the baboons prefer the whitish blossoms to those that are gray-white (older?) and rarely eat the unopened flower buds. During 1976, the flowers of umbrella trees were available from January 6 to April 11 (Klein 1978) and were abundant from February 7 to 25 (February 11–29, according to Klein 1978). By March 24 almost all were dried up and gone. (One small, sick-looking umbrella tree bloomed out of season, starting October 18, 1975.) The blossoms of umbrella trees were eaten from the trees by the yearlings during February and the first half of March; they were picked from the ground during the second half of February and the first half of March.

Blossoms of the other acacia, fever trees, were fed on from the second half of August through the first half of November, with a peak during the second half of October. On October 9 I noted that flowers were superabundant and were beginning to drop. Klein (1978) gives October 15–31, 1976, as the period of peak availability. When the flowers of these trees are abundant, the fever tree woodland is filled with their sweet smell. Then the woodland hums with the sound of bees that are attracted to the acacia flowers, as are many other insects, including various moths and butterflies. J. Altmann (1980) and Klein (1978) have discussed the importance of these soft, nutritious blossoms as pri-

Fig. 4.1. Fever tree (*Acacia xanthophloea*). Flowers and flower buds.

mate "weaning foods." One very weak infant, Pedro, who could not climb a tree, survived perhaps longer than he otherwise would have by feeding on fallen umbrella tree blossoms, which began to fall during the last month of his life.

Acacia Pods and Seeds

Yearlings extract the green seeds from the umbrella tree pod; the pod itself is discarded. In early August of 1975, the green pods of umbrella trees were abundant. By the first week of September, few green ones were left. By September 20, virtually all were dry; although many were still hanging from the trees, they dislodged readily when the branches were shaken by the movements of the baboons in the trees. Green pods from the flower crop of February–March 1976 were first eaten May 4, though they were still not abundant ten days later. (Klein 1978 gives April 10 as the date of the first pods in his much smaller area of Amboseli.) Heavy feeding on seeds from green pods began June 22 for baboons (June 13 for vervets). A month later, green pods were superabundant above the highest level at which giraffes browse. The baboons' peak feeding rates on the green seeds were in late August.

On the fever trees, I saw the first light green, translucent pods from the August–November crop of flowers at the end of the first week of October 1975 (October 21–24; Klein 1978), when flowers were still very abundant. Green pods were available for the next few months; the peak of abundance was

December 7–11 (Klein 1978). By the end of January 1976 the pods were rapidly turning dry and brown, and the seeds were becoming hard (Klein 1978). By February 7, 1976, 90% of the pods in the fever trees had dried up. A month later, all were dry. Many of the dry pods remained hanging from the trees for several months thereafter; they were, for example, still abundant in some trees during the last week of June 1976.

Post (1982) has commented on the fact that the baboons' periods of peak feeding on *Acacia tortilis* flowers and "pods" (i.e., seeds) are more restricted than the period of availability: "This discrepancy may be due either to changes in the nutritional characteristics of these food items through time or to the highly patchy distribution of the *A. tortilis* trees."

The seeds of acacias are one of the most important components of the diet of baboons, not only in Amboseli but elsewhere; acacias of one species or another occur in almost every habitat occupied by baboons. As is typical of legumes, acacia seeds are highly nutritious. They are eaten both when young and green, and also later when dry and hard. The dry acacia seeds are either taken from the pods, which split open as they dry, or are picked up one by one from the ground. Because dry acacia seeds are available until they germinate during the rainy season, they provide a staple food for the baboons during a large part of the year (Post 1982). During the dry season, when the baboons are in the acacia woodland, we often hear the crackling sound made as they chew the hard, dry acacia seeds, and then we smell the characteristic mustard-garlic odor of the seeds. On one occasion I noted that Dotty, at age 32 months, could move a dry *tortilis* pod laterally through her lips without using her hands, stripping the seeds from the pod as she did so. None of the 30–70 week yearlings did this.

Dry acacia seeds are also picked from the dung of wild and domestic animals (0.12%). The boluses were torn open with the hands and teeth. Some of the highest acacia seed intake rates (up to 72 seeds per minute) occurred when baboons fed in a deserted Maasai village, picking seeds from the dung (primarily, I believe, of fat-tailed sheep), and in shallow depressions where gazelles had repeatedly defecated. Looking under dung and sorting through it are common activities of baboons (4.2%). Although I often could not tell what the items were that they picked and ate from it, observed items included not only acacia seeds but also insects. The same was true of the numerous small items that they picked up one by one from the ground (3.6%) or from logs (1.5%): many could not be identified, but those that could included acacia seeds and insects.

Feeding on old, dry acacia seeds has an unsuspected advantage. They probably contain lower concentrations of a toxin called "trypsin inhibitor," which is heat labile and even at room temperature degenerates completely in about a

Fig. 4.2. Impala gazelle eating pods dislodged from umbrella tree (*Acacia tortilis*) by baboons feeding in the tree.

month (Sohonie and Honawar 1955). See chapter 9 for further discussion of trypsin inhibitor.

Acacia seeds are eaten in much larger quantities during the relatively brief period when they are green. The green seeds are considerably easier to chew than the dry ones and are produced in great abundance, so that many can be picked from one place in the tree—which is fortunate, considering the sharp, hooked thorns of the umbrella acacia. When feeding on green umbrella tree pods, adults, perhaps only adult males, sometimes crush the entire pod in their molars, then spit out the fibrous residue. (According to Post 1982, the pod itself is sometimes consumed by adults.) However, adult females and immature baboons consume only the naked green seeds of acacias. To do so, they first remove the seeds from their pods and the seed coats from the seeds, as described just below. The rain of dislodged and discarded green pods sometimes attracts gazelles, such as impalas (fig. 4.2), who feed on them beneath a tree that baboons are in. Although such a concentration of potential prey might appeal to a predator, it would have a very difficult time approaching without being detected.

After half an hour or so of such harvesting, when the baboons have filled their cheek pouches to bulging, they move from the upper canopy of the umbrella tree to the deep shade of its lower branches. There they relax, sometimes grooming each other, as the green seeds are gradually brought back into the oral cavity, chewed, and swallowed. During this time the weanlings and perhaps also the adult females remove and discard the heavy seed coats, apparently by the joint action of lips, incisors, and tongue; the hands are not involved. As will be explained in chapter 9, this hulling of pods and seed coats from the seeds reduces the baboons' intake of toxins and increases the concentration of nutrients in their diet.

Acacia Gum

The viscous exudate ("gum") of fever trees is a baboon favorite and an important year-round food source, but that of umbrella trees is seldom eaten by baboons (0.01% in my samples on weanlings), perhaps because of differences in composition, abundance, and fluidity (Hausfater and Bearce 1976). I do not know whether the rate of exudate flow changes seasonally. The gum of fever trees is clear and colorless when first exuded, then ages through progressively darker amber tones until it eventually becomes brownish black and tarry hard; it is eaten at every stage.

The baboons search out droplets or larger chunks of fever tree gum, including accumulations at the base of the tree that are sometimes full of insect frass and quite fermented. Individual baboons are sometimes diverted from other activities when they catch sight of a concentration of gum in a fever tree. The youngsters are particularly agile at clambering about among the branches, eating one bit of gum after another, each usually just a droplet of about a twentieth of a gram.

The baboons sometimes make long treks to the dry lake edge, where young fever trees provide a particularly good source of gum. There they move along from tree to tree, scanning each and sometimes getting chunks weighing a gram or more.

Acacia Wood and Leaves

Weanling baboons often gnaw on the wood of acacias and on various other woody plants, twigs, and so on, that they pick up from the ground (1.0% in total). Most of the time, however, little or nothing is consumed. Perhaps by such gnawing these youngsters are relieving the irritation of "teething" (cf. chapter 7).

The leaves of fever trees were eaten only occasionally (0.14%). I have only one record of a weanling feeding on the leaves of umbrella trees, two records of them feeding on its exudates, and no record of them eating its thorns or the thorns of fever trees—rather surprising in view of how much time baboons spend in both these acacias harvesting other foods and of the importance of these foods in the diet of the local vervet monkeys (Klein 1978).

Grasses and Sedges

A large component of the baboons' diet is provided by grasses and sedges (25.7% of feeding bouts). Their leaves (14.1%), seeds and seedheads (0.49% combined), rhizomes (underground runners) and stolons (aboveground runners) (0.65% combined) were eaten primarily during early May and June, when they were new, low in fiber, and thus relatively tender. Baboons eat the green blades of several species of grasses and sedges. This means that such feeding is restricted to the rainy seasons, except on grasses such as *Cynodon nlemfuensis* that grow in or around the edges of permanent water sources and moist depressions. Although the baboons found some green grass blades to feed on throughout the year, a marked peak in such feeding occurred during the first half of December 1975 and the second half of April 1976.

Baboons feed on the corm of grasses and sedges (17.5%). Although corms are eaten at all times of the year, they are fed on most extensively during the dry season, when most preferred foods are not available. Thus grass corms are a major "fallback" food, probably the most important one for Amboseli baboons. They spend several hours of most days during the dry season moving to concentrated sources of favored corm species, then digging and pulling the plants from the ground. Dirt, roots, and sometimes an outer dry sheath are removed, and the corm is then eaten.

One of the baboons' favorite sources of corms is *Sporobolus rangei* (2.4% of feeding bouts), a short grass that grows abundantly but locally in dense clumps (tussocks) on the open plains, outside the woodland. Much of the baboons' day route and activity schedule in the dry season hinges on harvesting this plant. After manually digging up the edge of a *S. rangei* clump, the baboons use their hands or teeth to tear off a few plants and then to remove the roots and dry corm sheaths so that only the moister, inner blade bases remain; it is these that are eaten. Corms of *Sporobolus rangei* were fed on at peak frequencies during the first half of April and again during the first half of September, by which time the leaves of this grass had dried and turned beige. Other parts of *S. rangei* are seldom eaten (0.8%).

Cyperus obtusiflorus is a sedge that sometimes grows intermixed with *Sporobolus rangei,* though *obtusiflorus* is more abundant on the forest edge. Superficially *obtusiflorus* looks similar to *S. rangei,* suggesting convergent evolution, but the former can readily be distinguished by its seedhead, when present, and by its yellowish corm sheath. As with *S. rangei,* the corm of *obtusiflorus* is a dry season favorite of the baboons (0.15%), who readily switch back and forth between the two species. I recorded baboons eating the corms of this sedge during May and June 1976, but I believe the complete absence of records before then is due to my inability to recognize it earlier, during which times it would have contributed to my collective category "corms of unidentified grasses and sedges," and possibly to "corms of *Sporobolus rangei.*"

Sporobolus cordofanus (= *S. marginatus* in part) is a grass that is particularly abundant in open, flat areas in and near the acacia woodland. The soil around *S. cordofanus* is often soft, particularly after it has been trampled by ungulates, so that the corms are relatively easy for the baboons to harvest (0.95% of feeding bouts), and during the short period when the green leaves (0.24%) and closed seedheads (0.01%) are available, they too are eaten. For *S. cordofanus* corms, peak feeding frequencies occurred from late May to early June. On the edge of *S. cordofanus* areas, corms of the larger *S. ioclados* (= *S. marginatus* in part) are also eaten (0.03%).

Several other common grasses rank particularly high, judging by how often baboons ate them. *Cynodon nlemfuensis* (so-called Bermuda grass), which has been planted by humans in many parts of the world for lawn and pasture, occurs naturally in Amboseli, growing in thick mats in moist depressions. Because this grass grows around the edges of permanent waterholes and swamps, its green leaves are available to the baboons and are eaten by them at all times of the year. Baboons are fond of the young leaves (3.0%) but also eat several other parts of this plant, including corms (0.08%), seedheads (0.18%), and stolons (0.48%). The stolons are eaten during most months. Along with the seedheads, the stolons seem to show seasonal fluctuations in availability, even in the perennial moist areas, though my records on this are inadequate. The baboons' use of this plant fluctuates widely through the year, perhaps not so much because of varying nutritive value, but rather as a consequence of other factors, such as predator density or rainfall, that affect how much time the baboons spend in the vicinity of waterholes.

Setaria verticillata is a grass that grows abundantly under umbrella trees. Its fresh green leaves, especially the blade bases, are a baboon favorite when the new leaves sprout after the start of the rains (0.10%).

Psilolemma jaegeri is a very tough, prickly grass that covers large areas around the edge of Lake Amboseli. Young baboons sometimes ate the leaves, corms, and growing tips of stolons from the grass (0.07% total).

Sporobolus consimilis (= *S. robustus*) is the last of the frequently used grasses, and it is unique. The plants are tough and large, commonly over a meter high. They thrive in highly saline soils, so they form dense stands around the edges of pans and along the border of the dry bed of Lake Amboseli. The inner, white leaf bases at ground level on small plants and especially on lateral shoots (lateral meristem) of larger ones are eaten by the baboons, particularly when other foods are in short supply. The baboons tear off the shoots then shuck them with their teeth, thereby getting at the inner part of the leaf base. Rates of feeding on this material peaked in early March, and to a lesser extent in early May and at the end of the long dry season, in October and early November. Occasionally, baboons ate very young leaves of this grass (0.08%). I observed much more feeding by baboons on this tough grass during 1974–76 than during 1963–64, perhaps reflecting a decline in availability of preferred foods.

Other grasses used as food sources by Amboseli baboons include *Cynodon plectostachyus* (0.07%), *Sporobolus spicatus* (0.04%), *S. africanus* (0.03%, but see footnote to table 4.3), and *Dactyloctenium bogdanii* (0.04%). Post (1982), in what appears to be the only major discrepancy between the diet of the adult baboons in Amboseli that he studied and the weanlings I studied, lists *Sporobolus spicatus* as a major food species.

Bushes

Most of the bushes of the acacia woodland provide the baboons with some food. The berries of *Azima tetracantha* (3.0%), of *Salvadora persica* (1.5% of feeding bouts), and of *Lycium "europaeum"* (0.17%) are among their favorites, especially when ripe. Each of these shrubs also provided relatively small amounts of other foods, such as young leaves of *Salvadora* (1.1%), *Azima* (0.12%), and *Lycium* (0.61%), and the flowers of *Lycium* (0.17%).

The fruit of *Azima tetracantha,* a common, spiny evergreen shrub, is a year-round source of food for the baboons (cf. Post 1978; Klein 1978). They eat the fruit both green and ripe, although with a distinct preference for the latter, and particularly for the fully ripe fruit (translucent gray-white) instead of semiripe (opaque white). I occasionally noted that the seeds were extracted and consumed and that much of the rest of the fruit was discarded. Further observations on this are warranted. Ripe *Azima* fruits occurred in two peaks during the year, one centered on the first half of January 1976, the second on the end of May. A January but not a May peak of feeding on *Azima* fruit was noted in 1975 by Post (1978). Klein (1978) indicates peaks in mid-November 1975 and May–June 1976. Individual baboons varied in their tolerance for the unripe fruit.

The green fruit of *Salvadora persica* (toothbrush bush), a common, thornless evergreen shrub, is fed on by the baboons virtually throughout the year. However, their decided preference is for the ripe berries. Some ripe berries were found by the baboons during early March, but most of the fruit ripened during August to October, particularly in September, when the baboons often ate them in great quantities (SPFR in table A5.2). Many tiny new fruits were present during the first week of May. When ripe fruits of *Salvadora* are abundant, they attract many other animals, including vervet monkeys, impala gazelles, and numerous species of birds. The leaves of this shrub are present on the plant all through the year. They were nibbled by the baboons at various times of the year (SPLX in table A5.2), with feeding peaks during late February and early November. Perhaps at these times the new, red-tinged leaves, which the baboons prefer, were more abundant. Twigs of this bush were gnawed by the yearlings at various times, with an unexplained peak during early May.

The woody shrub *Lycium "europaeum"* (trumpet-flower bush) is locally concentrated, particularly near the base of volcanic outcroppings. Baboons eat the leaves, blossoms, and ripe fruit. These three foods are produced in that order after the onset of each rainy season. The leaves were eaten during much of the year, with peaks during December and late April (TFLX in table A5.2). On April 23, following the heavy rains of mid-April, new *Lycium* leaves were at peak abundance: all *Lycium* shrubs in a sample of 120 had new leaves (Klein 1978); no comparable December data were obtained. The short-lived blossoms were fed on during late April and early May, then again in early December.

Lycium is especially interesting because it occurred in greatest concentration well outside the fever tree woodland, immediately east of Naarbala Hill, on the periphery of the baboons' home range, and the animals made long treks to gorge on the fruit when it was ripe. The bright orange color of the ripe fruit suggested a large concentration of carotenoids, and indeed, chemical analysis revealed high levels of beta-carotene, a vitamin A precursor. The baboons ate so many of these fruits that their feces turned orange. I believe *Lycium* fruit would have been among the "core foods," which collectively account for 95% of feeding time, except that I did little sampling during its time of peak abundance, late December and early January. Another batch ripened during late April through May (cf. Klein 1978).

Suaeda monoica is a bush that has become much larger and more abundant in Amboseli as salts have accumulated in the upper soil zones (fig. 1.7). This species is well known as a halophyte (Waisel 1972) and in fact grows poorly in a salt-free medium (Shomer-Ilan, in Waisel 1972, 135–36). Sodium is most concentrated in the buds and young leaves of this plant (Ovadia, in Waisel 1972, 122)—exactly the parts eaten by the baboons (0.33%), which is surprising for animals living in a habitat that is rapidly becoming more halophytic and would thus seem to provide them with a more than adequate supply of salts.

Withania somnifera is a small shrub whose parchment-covered fruit is eaten both green and ripe, though the baboons have a decided preference for the latter (0.31%). The peak of feeding on the fruit was during late June and early July.

Ground Covers

The tiny ovaries of *Trianthema ceratosepala,* a procumbent, succulent perennial, were eaten by the baboons during every semimonth of the year (2.3%). Fluctuations in the use of this food probably were due not so much to changes in its abundance as to other factors affecting the baboons' movements through their home range and into the moist depressions where this plant grows in thick mats.

Tribulus terrestris (devil's thorn) is a procumbent annual herb that grows vigorously once heavy rains begin, and if the rains continue long enough these plants produce an abundance of small fruits, which are eaten by the baboons before the surrounding thorny calyx hardens. The fruits were fed on extensively from the first half of May (the peak time) until the end of June. Similarly, Klein (1978) gives May as the time of heaviest *Tribulus* fruiting in 1976; in May 1975 Post (1978) recorded a fruiting peak in February–March. By May 25, 1976, some were already dry enough to be rejected.

Several other baboon foods are available only after adequate rainfall. On the green plains, the flowers and perhaps other parts of *Rhamphicarpa montana* are eagerly eaten (0.24% of feeding bouts). During most years (e.g., 1974, 1975, 1976) the flowers of this inconspicuous little plant are uncommon, but when the rains are very abundant, as in 1963, some of the plains (e.g., east of Enkongo Narok Swamp) are blanketed by its pale mauve blossoms. *Commicarpus pedunculosus* is a prostrate, sprawling plant that grows on volcanic soils. Its tiny, sticky fruits, available for only a short while after the onset of the rains, were eaten in great numbers by the baboons (0.40%). The baboons fed on them from the second half of May to the first half of July 1976. They may also have been fed on during the short rains at the end of 1975, when I did not recognize them.

Scandents

Baboons occasionally eat the flowers, pods, and fruits of two capparaceous scandents that grow on acacia trees: *Capparis tomentosa* (0.28%) and *Maerua* (*angolensis?*) (0.14%). *C. tomentosa* blossoms were noted in October 1975. Ripe fruit was eaten by the baboons at least from mid-February through late March. By late May 1976, only empty shells were left on the plants.

Herbs

The weanling baboons occasionally ate various parts of several small herbaceous plants in the understory of the acacia woodland, including *Abutilon* sp. (*grandiflora?*) (0.05%), *Sericocomopsis hildebrandtii* (0.03%), *Chenopodium opulifolium* (0.02%), and the fruits of *Solanum dubium* (Synonym of *S. coagulans*), which grows on volcanic soils (0.11%).

Monocots

After the onset of the rains, baboons fed on the leaves, bulbs, and other parts of several species of lilies, *Chlorophytum* sp. nr. *bakeri* (0.11%) and the larger

Fig. 4.3. A lily (*Ornithogalum donaldsonii,* Liliaceae). Bulbs eaten by baboons.

Ornithogalum donaldsonii and *ecklonii* (0.05%, fig. 4.3), which grow very locally, on the open plains between the so-called KH and KB woodlands.

Aquatic Vegetation

A few plants of waterholes and swamps are occasionally eaten by baboons, in particular, *Ludwigia stolonifera* (0.06%) and the fleshy blade bases of several species of *Cyperus,* including *C. immensus* and *C. laevigatus* (0.03% total).

Fungi

Occasionally (0.05% of bouts) baboons find and eat mushrooms of several species (*Battarraea stevenii, Agaricus* sp., and perhaps others). So far as I know, all their other plant foods are angiosperms.

Animal Matter

Although baboons are primarily vegetarians, they sometimes eat animals and animal products (e.g., S. Altmann and J. Altmann 1970; Busse 1982; Dart

1953; Forthman Quick and Demment 1988; Hamilton and Harding 1973; Hausfater 1976; Kortlandt and Kooij 1963; Rhine, Norton, Wynn, Wynn, and Rhine 1986; Strum 1975). Baboons feed opportunistically on insects, particularly grasshoppers, at any time of the year, and except for mother's milk, grasshoppers were by far the commonest source of animal food for the weanlings in this study. Young baboons seem to be more agile than adults at catching grasshoppers and more willing to pursue them if they take flight, sometimes catching them in midair. With the onset of the short rains in November 1975, enormous numbers of insects emerged, and at that time the baboons' insect catching reached a marked peak. Seasonally, dung-beetle larvae were dug from the ground in places where they were abundant and the rain had softened the brown earth (0.62% of bouts). This digging is new to yearlings born since the last such beetle harvest and seems to require more strength than they have. The yearlings go to the diggings of first one adult, then another, watching them dig and picking up scraps. Once baboons have been weaned, the various grasshoppers, beetles, and other arthropods (1.5% total) eaten by baboons may be their primary source of vitamin B_{12}, which is almost completely absent from plant sources. A turban-shelled gastropod was the one other common source of invertebrate food eaten by the weanlings (0.07%).

The young baboons also ate the meat of several species of mammals, birds, and reptiles (0.07% total) and eggs of the latter two (0.02%) (table 4.1). As mentioned previously, they often removed acacia seeds from mammalian dung, but on a few occasions I was sure they ate dung itself—perhaps where acacia seeds were very concentrated. As they walked along, they often flipped over dried chunks of mammalian dung, even the large boluses of elephants, and sometimes picked up and ate small objects (seeds and insects) from beneath it.

Plants Not Eaten

In view of the baboons' eclectic feeding habits, the few commonly occurring large plants in their home range that they did not eat are of special interest. I know of four: *Volkensinia prostrata,* which we call "hairy-ball plant" because of the spheroidal bristle of sterile flowers around each inflorescence (0.06%, but see below), *Dicliptera albicaulis,* which our children nicknamed "butterfly bush" because of the many butterflies on it, suggesting a toxic or noxious secondary compound (Brower, Ryerson, Coppinger, and Glaser 1968), *Leucas stricta,* and *Solanum incanum* (Sodom apple). From all four species I have seen baboons pick off and eat small, unidentified objects, probably insects, but I have no record of their eating any part of these plants, though all of them are frequently encountered by baboons in the course of their daily movements.

Sodom apple fruit contains an alkaloid, solanine, and is more or less toxic, particularly when green (Watt and Breyer-Brandwijk 1962). It also contains a carcinogen, N-nitrosodimethylamine, that causes esophageal cancer (Leet 1973). I was therefore startled one afternoon by the following rapid sequence of events. First Ozzie, who was then fourteen months old, several times pulled at, once sniffed, and perhaps mouthed a Sodom apple still on the plant, but did not eat it. One minute later Alice (same age) held a Sodom apple in her mouth and nibbled on it but did not eat any of it. After that, her older sister, Dotty, nibbled on one briefly, then put it down. So did Striper, Hans, and Fanny (Ozzie's older sister). Janet gnawed on one for about two minutes but probably did not break the skin of the fruit before she finally discarded it. All these fruits were yellow (ripe). Perhaps this transient interest of young baboons in the fruit of *Solanum incanum* was a by-product of their response to *Solanum coagulans,* the fruit of which they eat despite its prickly calyx. In my focal samples on yearlings, *S. incanum* feeding was recorded twice. In one case Alice ate something, perhaps an insect, that she picked from a *S. incanum* plant. In the other, Bristle ate about one-quarter gram of a leaf that was from either *S. incanum* or *Withania somnifera.* Adult female Fem (not his mother) had been feeding on the leaf before he picked it up, but she discarded it.

The listing above of common Amboseli plants from which the young baboons in my study ate nothing can be compared with results obtained by Post (1982, table IIIB) during his 1974–75 study of adult baboons in the same group. He too includes *Solanum incanum, Volkensinia prostrata, Dicliptera albicaula,* and *Leucas stricta.* He includes six plants from which the youngsters in my study occasionally fed (table 4.1): *Maerua* sp. (for which I observed twenty-seven cases of feeding on blossoms, leaves, pods, and their enclosed fruits), *Kochia indica* (two cases in my study), *Sericocomopsis hildebrandtii* (six cases), *Psilolemma jaegeri* (twelve cases, probably all attributable to *Odysseus paucinerva:* see footnote to table 4.3), *Sporobolus africanus* (five cases, probably all attributable to *S. spicatus:* see footnote to table 4.3) and *Dactyloctenium bogdanii* (eight cases). Post also lists as not eaten *Trianthema triquetra* and *Aerva lanata,* which I did not recognize. Because of the likelihood that rarely eaten foods will be missed entirely in a given sample, these discrepancies are not surprising. In the case of *"Psilolemma jaegeri"* (probably *Odysseus paucinervus*), an extremely halophytic grass, we may have a new item in the diet, but most of this plant is far too tough ever to be a major food source for baboons.

Frequency Ranking of Foods, Food Taxa, and Plant Parts

Clearly, the 277 foods were not eaten with equal frequency. For each food, the observed number of feeding bouts in my samples on 30–70-week-old baboons is given in table 4.1. A few "favorite" foods were eaten very often. Below them, the frequency falls off rapidly. For foods, a graph of bout frequency versus fre-

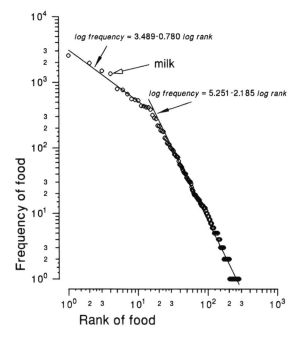

Fig. 4.4. Feeding bout frequencies (all samples pooled) of foods, rank-ordered. From data in table 4.1. For upper segment (ranks 1–13), regression $R^2_{adj} = 0.96$, $p \leq 0.001$; for lower segment (ranks 14–277), $R^2_{adj} = 0.99$, $p \leq 0.001$.

quency rank, each plotted on a log scale, is shown in figure 4.4. Except for the first thirteen foods, the relationship is markedly linear on a log-log scale and conforms to the equation: log food frequency = 5.251 − 2.185 log food rank, or equivalently, food frequency = 178237.9 food rank$^{-2.185}$. The exponent of food ranks in this regression is a measure of dietary diversity: the higher the exponent, the fewer foods contribute significantly to the diet, as judged by feeding frequency. For the thirteen highest-ranking foods, which deviate from the rest, the regression is log food frequency = 3.489 − 0.780 log food rank.

Alternatively, I have listed foods just by food type or class (e.g., plant part that was eaten: seeds, leaves, etc.). The data are tabulated in table 4.4 and graphed in figure 4.5. They conform fairly well to the equation, log class = 3.924 − 0.1517 class rank.

Cumulative frequency rankings for foods, food species, and food types are shown in figure 4.6. Clearly, while young baboons eat a great variety of foods, some foods are eaten far more often than others. Figure 4.4 spans more than three log cycles (orders of magnitude) between the rarest and the commonest foods. Of the weanlings' 277 foods, 184 were eaten fewer than eleven times,

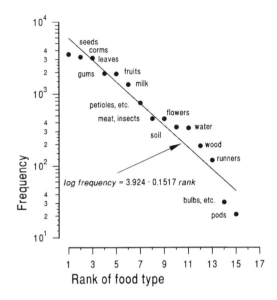

Fig. 4.5. Feeding bout frequencies of food types, rank-ordered. From data in table 4.1, classified as in table 4.4. Feeding bouts on foods of unknown type have been prorated among those of identified types. Regression $R^2_{adj} = 0.93, p \leq 0.001$.

that is, less than once per infant on the average. At the other extreme, their seven "favorites" accounted for over half their feeding bouts, and just 24% of the foods accounted for over 95% of the feeding bouts (fig. 4.6). Similarly for major plant parts (fig. 4.6 and table 4.4): more than half of the yearlings' feeding bouts involved just three plant parts: seeds, mostly of acacias, grasses, and sedges (20.2% of feeding bouts); corms of grasses and sedges (18.5%); and leaves, primarily of grasses and sedges (17.9%). In total, grasses/sedges and acacias accounted for 54% and 23%, respectively, of all feeding bouts. Finally, half the feeding bouts were accounted for by just three plant species (two grasses and one acacia), out of sixty-five species and other taxa of food that were observed in my samples on 30–70-week-old baboons (fig. 4.6 and table 4.3).

In short, even at this young age, the baboons of Amboseli depend primarily on acacia trees and grasses/sedges to sustain them. In this regard, the results of this study amply confirm our earlier impressions: "Plants other than acacias and grasses probably contributed less to the baboon's total diet, though some of these were seasonally very significant" (S. Altmann and J. Altmann 1970, 148). This dietary emphasis on grasses and acacias enables baboons to be among the most widespread mammals in Africa: they are successful at making a living

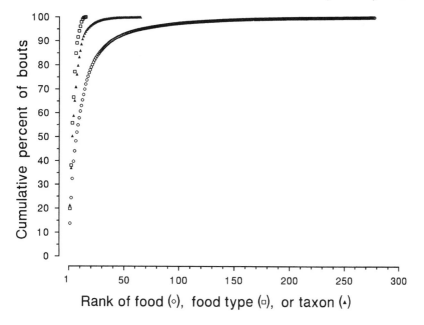

Fig. 4.6. Cumulative percentage of feeding bout frequencies for foods, food types, and food taxa. Data from table 4.1, food types classified as in table 4.4.

in acacia grassland, which is the most widespread habitat on the continent (Keay 1959).

So much, then, for the overall frequency with which each food was eaten. Of course, by pooling these data, which were gathered during every month of the year and every hour of the day, I obscure the marked seasonal and diurnal changes in the baboons' diets. Similarly, nothing has been said so far about age changes or individual differences in diet; these will be described in chapter 7.

Feeding Bout Rates

For each food, a pooled mean bout rate per minute of daytime—and thus per potential foraging minute, as in equation (3.2)—can be obtained by dividing that food's pooled bout frequency (table 4.1) by the product-limit sample time of 22,299.84 minutes, which is the sum of all food-specific times plus non-feeding time (see equation 4.2 below). These pooled bout rates are given in table 6.1. Because these bout rates are obtained from the bout frequencies by dividing all frequencies by the same number, the bout rates have the same rank order as the frequencies, which were discussed in the preceding section. As explained in chapter 3, these bout rates (factor R in equation 3.2) can be used in calculating the intake rates of nutrients and other food components. Feeding bout rates of individual baboons will be presented in chapter 8 and used to compute their component intake rates.

Any food with a rate greater than 0.1515 per 100 minutes in table 6.1 or a frequency in the whole study (table 4.1) of at least $34 = 22,299.84/660$ was eaten at least once a day, on average. The baboons averaged one bout of feeding on one of their 277 foods every 1.2 minutes (equivalently, 0.83 feeding bout per minute of daytime or 546 bouts per day). Clearly, decisions about what to eat, and how much, occur frequently. During the 277.7 minutes per day that the baboons fed (table 4.7), they averaged $546/277.7 = 1.97$ feeding bouts per minute of feeding. Conversely, the average feeding bout lasted approximately $1.97^{-1} = 0.51$ minute. Using the seasonally adjusted time budget of 240.5 minutes per day (table 6.5), these values become 2.27 feeding bouts per minute of feeding, with an average bout length of 0.44 minute.

When a baboon begins eating food of a particular kind, how long does it continue? In the aggregate, how much time is spent per day on each food? I begin with the first question.

Feeding Bout Durations

Some foods are harvested and eaten very quickly. Less than a second is required for a baboon to pick up an isolated dry acacia seed from the ground and toss the seed into its mouth. Most foods take considerably longer. I obtained data on the time it took yearling baboons to harvest and eat their various foods during focal samples by timing every observed feeding bout. However, as indicated in chapter 3, "censoring" of bout records, which occurred when the animal disappeared from sight so that the entire bout duration could not be timed, results in a bias against long bouts in the observed bout lengths. I therefore used product-limit estimates of mean bout lengths (appendix 3).

For each of the 277 foods eaten by weanling baboons (age 30–70 weeks), table 4.1, column f, gives the average "meal" length (mean feeding bout duration, p.l.e. = product-limit estimate) and its standard error, as calculated by the BMDP program (Benedetti and Yuen 1977). The rank order and frequency distributions of average feeding bout lengths for these 277 foods are shown in figures 4.7 and 4.8, respectively. The modal mean duration was between 12 and 24 seconds (fig. 4.8). The range of feeding bout lengths is surprisingly narrow compared with the variability of feeding frequencies that was depicted in figure 4.4. Of the 277 foods eaten by these weanlings, the average feeding bout

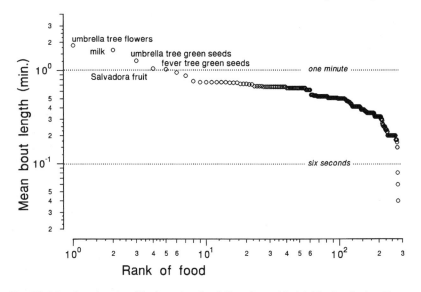

Fig. 4.7. Mean bout lengths of foods, rank-ordered. Data from table 4.1. The five foods with mean bout lengths of at least one minute are indentified.

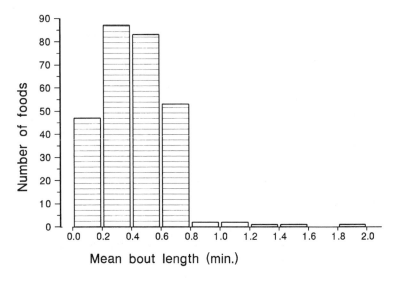

Fig. 4.8. Distribution of mean feeding bout lengths, 277 foods. Data from table 4.1.

for 258 of them (i.e., 93% of foods) lasted between 12 and 48 seconds inclusive, and for two-thirds of all foods the average feeding bout was within 18 to 42 seconds! The twenty foods with the longest mean bout lengths are listed in table 4.5. Umbrella tree blossoms, on which the baboons gorged seasonally, were fed on the longest (1.85 minutes on average) and were the only food fed on for longer than nursing bouts (1.65 minutes). The two acacia species, the toothbrush bush (*Salvadora persica*), and milk accounted for the foods with the seven longest mean bout lengths. The common denominators of all these foods on which the baboons fed nonstop for relatively long periods are that they are favorite foods of the baboons and occur in local patches large enough to sustain long bouts of feeding without pause.

So much for the bout lengths of individual foods. What about mean bout lengths if the baboons' foods are grouped into classes, that is, blossoms, leaves, and so forth? (For my third grouping, food species, mean bout lengths were not calculated because of the extreme heterogeneity of means for the various foods derived from any one plant.) All 277 foods were classified into the same set of sixteen exclusive and exhaustive classes used in the analysis of bout frequencies, and the mean bout duration for each class was estimated, using the same methods used for individual foods. The results are shown in table 4.4. These sixteen food classes were then rank-ordered by their mean bout lengths (product-limit estimates). Nursing bouts were by far the longest (1.65 minutes average). After that came pods (0.75 minute), blossoms (0.67 minute), and corms (0.53 minute). The remaining dozen categories ranked below these. From these data, one can easily discern the types of foods that took a long time to harvest: (1) small food items that are locally very abundant, so that they can be picked up one after the other in rapid succession, for example, blossoms on trees, seeds in mammalian dung, fruit on racemes; (2) foods that require extensive manipulation before being put in the mouth, for example, grass corms, seeds, or fruit in pods; and (3) "milk," for which the recorded bouts included not just suckling time but total continuous time on the nipple.

Duration Distributions

For each food the mean duration of feeding bouts is used in the next section to calculate the food-specific time budgets and in chapter 6 to calculate dietary intakes and the nutritional contribution of each food. Here I examine not just the mean bout duration for each food but the entire duration distribution and its behavioral correlates. As I shall presently show, virtually all baboon foods and food classes have Weibull-distributed feeding bout lengths. That is, the probability that a feeding bout on a given food will terminate after any elapsed time t in that bout is a characteristic power function of t.

The product-limit estimates for the log-survivorship distributions of feeding bout lengths of individual foods were divided by inspection into three categories: those that, small sample fluctuations aside, appear to be linear, or very nearly so; those that are concave upward; and a variety of more complex distributions. For the first category, their linear log-survivorship distributions indicate that for these foods, the probability of feeding termination in any small interval is independent of how long the feeding bout has been under way, as if the animal had no memory for when it began a feeding bout and merely had a constant termination rate that was characteristic of each food. Many baboon foods have log-survivorship distributions that are essentially linear, aside from small-sample and integer-value fluctuations (to which the right side of a distribution is progressively more susceptible); see, for example, figures 3.3 and 3.4.

For the second category, a concave log-survivorship distribution for a food (e.g., figs. 3.5 and 3.6) indicates that the probability of continuing during any small interval of time increases as the bout continues; that is, the probability of terminating decreases. This decrease can be seen graphically by plotting the failure rate (appendix 3), which decreases with time under this condition. I can think of several explanations for these concave log-survivorship distributions. First, the food may be "addictive" to the animals, in the sense that the longer they eat it, the less likely they are to quit. Second, the bout length distribution may reflect the distribution of patch sizes, with many small ones and few extraordinarily large ones. I do not have the data needed to decide between these alternatives. However, I note that several of the concave distributions are for foods that are high in sugars, such as various blossoms, fruits, and fever tree gum. Perhaps the reaction of weanling baboons to these foods is like that of human children to candy: the more they eat, the more they want.

A third possible explanation is this. Most of the concave distributions are not for single foods but for large aggregate categories, such as all pods, all fruits, all blossoms. Perhaps their concave distributions result from pooling several foods with different bout survivorship distributions. Mixed survivorship distributions are discussed by David and Moeschberger (1978), by Mann, Schafer, and Singpurwalla (1974), and by Gross and Clark (1975).

A fourth possibility is suggested by a study by Janson (1990) on foraging by brown capuchin monkeys, *Cebus apella.* When it is foraging on well-hidden or very tough substrates such as palm tree crowns ("long-search substrates"), the probability that a capuchin will be feeding increases as a function of time after an individual begins using such a substrate, perhaps because of the time needed to locate local concentrations of food (Janson 1990, fig. 5). Similarly for the baboons, suppose that, on making contact with a food plant and thus initiating a feeding bout by my criteria, the animals have an initial assessment period of variable length. A long feeding bout ensues only if the assessment proves positive, that is, if a large food patch is located. If little or no feeding took place during the initial assessment period, the results would be contrary to the bout-age-independent intake rates obtained for a sample (S. Altmann and Shopland in prep.) of six common foods eaten by Amboseli baboons, as described below. If feeding takes place at essentially the same rate as in the rest of the bout, then some other factor, such as those just described, may be the source of the concave feeding bout distributions.

The more complex distributions probably each have a unique explanation. Some appear to result from mixtures of two or more distributions. A particularly interesting one is that for umbrella tree blossoms picked from the ground (see fig. 3.9). At the time of year when the blossoms of an acacia are dropping, the baboons, walking across the savannah, encounter blossoms on the ground as soon as they get under the canopy of the tree. As they do so, they begin feeding. Some move toward the trunk, feeding as they go, then climb the tree to feed on the fresher and perhaps more abundant blossoms therein. The ascent of these baboons accounts for the sudden drop in the log-survivorship function at 0.07 to 0.10 minutes, about the time it takes them to move from the canopy edge to the trunk. The essentially uniform distribution beyond 0.10 minute results from those that remain on the ground.

The apparent break in the distribution of "suckling" bouts at about one minute (fig. 4.9) may likewise indicate a composite of two types of bouts: long suckling bouts, at the end of which the infants sometimes fell asleep with the nipple in their mouths (which by my criteria continues the bout), and short bouts, for infants that were rebuffed, or whose mothers were no longer lactating, or that held the nipple in their mouths only briefly as a "comfort reaction."

The preponderance of log-survivorship plots that appear to be either linear or fairly smooth and concave upward suggested that feeding bouts are Weibull-distributed; that is, that for most baboon foods, the decision of whether to terminate a feeding bout at any particular moment either is independent of how long the bout has been under way, leading to log-survivorship plots that are linear functions of elapsed time, or else is a negative power function of time elapsed since bout onset, leading to concave plots. Therefore, using the method described in appendix 3 (which also summarizes properties of the Weibull distribution), I estimated the Weibull parameters α and β for the primary food classes, including those of foods that appeared on inspection to have linear log-survivorship functions. Since such linear functions are special cases of the Weibull distribution, I was simultaneously testing for that possibility. Examples of the results are shown in figures 4.10–4.13. Estimates of the parameters α and β are given in table 4.6. For each food class in table 4.6, the standard error of β was estimated in the usual way for the slope of a linear regression,

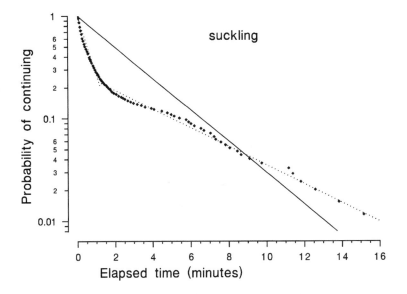

Fig. 4.9. Log-survivorship graph for suckling bouts. The data appear to be compounded of two bout-length distributions, one with a high rate of termination, lasting to about 1.2 minutes, and another thereafter, with a lower rate. For the overall linear regression (solid line), $R^2_{adj} = 0.59$, $p \leq 0.0005$, log $y = -0.3515x$. For the regression up to 1.2 minutes, $R^2_{adj} = 0.94, p \leq 0.0005$, log $y = -1.3310x$; thereafter, $R^2_{adj} = 0.99, p \leq 0.005$, log $y = 0.29 - 0.2097x$ (both regressions dotted). Because of the large size of this sample (1,350 bouts), only a selection of points was used for this graph.

and similarly for $\hat{a}$, the y-intercept value. (I do not know of a method for calculating the standard error of $\alpha = -\hat{a}/\beta$.) Table 4.6 also gives the correlation coefficients, which can be taken as approximate tests of goodness of fit.

The values are remarkably high: for only one food (no. 246, unidentified material picked from dung) is the correlation coefficient less than 0.90, and most are 0.97 or higher. This means that for all the primary food classes, the probability that a feeding bout will terminate after any specified amount of elapsed time is essentially a power function of time. Consequently, the distribution of bout lengths for each food can be well represented by its two Weibull parameters α and β, which are given in table 4.6. (See appendix 3 for some reservations about the accuracy of these parameter estimates.)

Because Weibull distributions closely fit the data for virtually all the foods tested, the Weibull parameters α and β completely characterize their bout length distributions (see appendix 3), so as a descriptive device they provide a remarkable economy. Beyond that, however, one wonders whether some

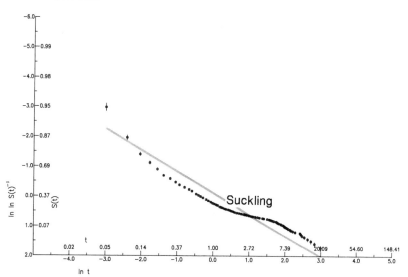

Fig. 4.10. Weibull graph for suckling. In this and all other Weibull graphs in this chapter: (1) The product-limit estimates of the mean and standard error are shown for each time interval; (2) time resolution of the data, graphing, and computing was 0.01 minute, and time, *t*, is in minutes; (3) whenever the BMDP survival analysis program gave more than one survival decrement value for a 0.01 minute interval, only the (higher) median survival value was used for that interval; (4) because of the regression program's limit of one hundred data points, only even-numbered intervals were used for distributions with 100–199 survival decrements, only every third for 200–299, and so forth; (5) the gray line indicates the linear regression of ln ln $S(t)^{-1}$ on ln *t* (the regression was not extended below 3 *s* (0.05 minute) because of the relative inaccuracy in timing very brief events and because of the Weibull transformation's exaggeration of deviations at small values of *t*); (6) on the *y* axis of these figures, increasingly negative values of ln ln $S(t)^{-1}$ are plotted progressively higher, so that higher survivorship values come out higher, as is conventional, and consequently the slopes of the plotted regression lines are opposite in sign from the α parameters in table 4.6; (7) the corresponding log-survivorship graphs will be linear (terminations will be time-independent) if the slope of the Weibull graph is -1, convex if it is steeper than that, and concave if it is shallower.

explanation can be provided for the fact that the probability of terminating feeding bouts is a power function of elapsed time, and whether the values of α and β can be given some interpretation. One way to approach this problem is to divide the foods and food classes into three categories, those in which the tendency to quit increases with elapsed time in the bout ($\beta < 1$), those in which it does not change appreciably with time ($\beta \approx 1$), and those in which the tendency decreases ($\beta > 1$). In table 4.6, asterisks indicate those foods for which the estimated value of the β parameter lies within one standard error of

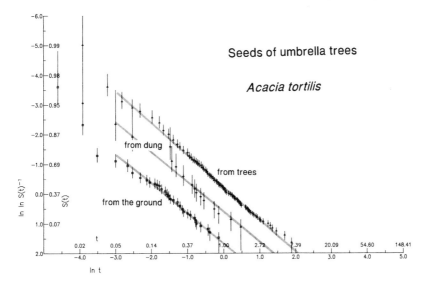

Fig. 4.11. Weibull graph for seeds from green umbrella tree pods (*Acacia tortilis*) picked from the tree, picked from mammalian dung, and gathered from the ground. See figure 4.10 for details of graphing.

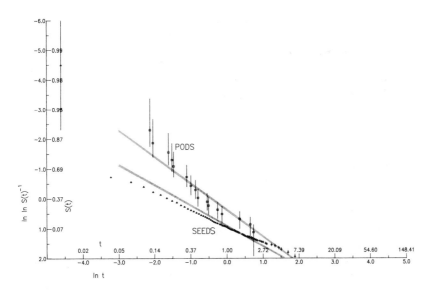

Fig. 4.12. Weibull graph for pods and seeds from all sources. See figure 4.10 for details of graphing.

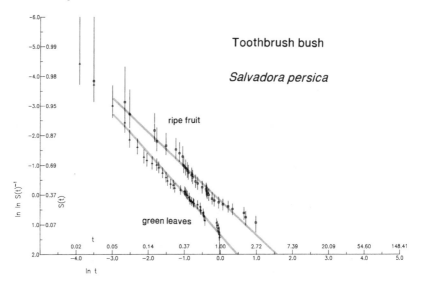

Fig. 4.13. Weibull graph for *Salvadora persica* ripe fruit and green leaves. See figure 4.10 for details of graphing.

$\beta = 1.000$. There are thirteen such foods among the eighty-five foods and food classes listed in table 4.6. A value of $\beta = 1$ indicates that the bout length distribution is consistent with the hypothesis that the baboons' feeding bout on these foods have an exponential survivorship distribution, that is, a linear log-survivorship distribution. This means that when a baboon is feeding on any of these foods, its tendency to stop feeding on that food is constant, independent of elapsed time in the bout: terminations are a Poisson process. Although constant for any given food, this tendency is food-specific, as indicated by the varied β parameter values.

I have not been able to discern any distinguishing characteristic of those foods with such time-independent stopping tendencies. However, two hypotheses come to mind. The elapsed time at which a feeding bout terminates may depend not so much on the yield rate of the patch that the baboon is in as on any of a wide variety of interrupting events: distractions from ongoing activities in the group, the approach of a competitor, the sound of a distant alarm bark, or whatever. So long as interruptions occur at random with respect to elapsed feeding time (albeit perhaps at food-specific rates), exponential survivorship distributions will result. An alternative is that the baboons leave a patch when their intake falls to a threshold level, as in most current models of patch utilization, and thus that variations in patch size result in the observed distributions of feeding bout lengths. I have no data either on patch size distributions or on the temporal distribution of potential interruptions.

Similarly, I have not discerned any common denominator for those foods for which β is greater than one, or for those for which it is less, though I expect yield rate to increase with bout age in the former and to decrease in the latter. This hypothesis can be tested. To date, bout-age-specific yield rates have been calculated for six foods commonly eaten by immature and adult Amboseli baboons in Alto's Group (S. Altmann and Shopland in prep.): fruit of *Withania somnifera,* of *Tribulus terrestris,* and of *Trianthema ceratosepala,* corms of *Sporobolus rangei,* corms of all other grass (but not sedge) species, and green grass leaves. These correspond with the following primary food classes in table 4.6: I ($\beta > 1$), TTFX.216 ($\beta \approx 1$), L ($\beta > 1$), B ($\beta > 1$), GRCX.117 ($\beta < 1$), and GRLU.121 ($\beta < 1$), respectively. For five of these six foods, the yearlings' tendency to quit feeding changed over time in the bout ($\beta \neq 1$); yet for each of these six foods, after an initial low-yield period of a few seconds, the baboons' food intake rate was essentially constant, independent of time, so my hypothesis is incorrect. The sixth food, the fruit of *Tribulus terrestris,* is consistent with the hypothesis (above) that foods with time-independent stopping tendencies will have intake rates that are essentially independent of time elapsed in the bout.

We are left, then, with a curious regularity: the probability that a feeding bout will terminate after any elapsed time t in that bout is a food-specific power function of $t,$ and for about one food out of seven the probability is a time-independent constant. The biological significance of this elegantly simple relationship between the probability that a baboon will stop feeding and how long it has been feeding is at present unknown.

Noteworthy is the result described above that for six of the most common foods in the baboons' diets (and thus foods for which we have large sample sizes), harvest rate is virtually independent of time elapsed in the bout. This nearly constant mean rate of intake results from the systematic way baboons feed on patches, foraging across them from one side to the other, rarely backtracking, then moving on to the next patch. Such time-independent intake is contrary to the assumption—widely accepted but seldom tested—that as a forager feeds on a patch, the resulting resource depletion leads to a continuously diminishing rate of return. Since 1976, when that assumption was presented by Charnov, Orians, and Hyatt, it has become commonplace and has been the basis for considerably modeling of strategies for optimal use of patchy resources. Yet depletion of resources in a patch need not depress harvesting rates (Stephens and Krebs 1986). Unfortunately, the requisite information to test the assumption, yield rates as a function of time in patches, is rarely published. Our results indicate that a diminishing intake rate cannot in general be assumed.

Time Budgets of Foods

For each food j of the 277 in my samples (table 4.1), its time budget A_j, in the sense of the average amount of time (minutes per day) spent feeding on it, was obtained from equation (3.2):

$$(3.2) \quad \mu \left(\frac{\text{minutes of feeding on food } j}{\text{day}} \right)$$

$$\underset{\text{Time Budget}}{A_j}$$

$$= \mu \left(\frac{\text{minutes of feeding on food } j}{\text{bout on food } j} \right) \mu \left(\frac{\text{number of bouts on food } j}{\text{potential feeding minutes}} \right)$$

$$\underset{\text{Mean Bout Length}}{L_j} \qquad\qquad \underset{\text{Mean Bout Rate}}{R_j}$$

$$\times \left(\frac{\text{potential feeding minutes}}{\text{day}} \right)$$

$$\underset{\text{Potent Feeding Time}}{S}$$

where, as before, $\mu(x)$ indicates the mean value of x.

It is useful to divide factor R_j into two subfactors, as follows:

$$R_j = \mu \left(\frac{\text{number of bouts on } j}{\text{minute of feeding}} \right) \mu \left(\frac{\text{minutes of feeding}}{\text{potential minute of feeding}} \right)$$

I made use of this factoring as follows. The first of these two subfactors multiplied by L_j equals the fraction of feeding time the baboons devoted to each food, for which values are given in table 4.1, column h. The first subfactor was estimated from $n_j / \Sigma_{x=1}^{277} n_x \hat{L}_x$, where n_j is the number of feeding bouts on food j in my samples, $\hat{L}_j$ is the product-limit estimate of food j's mean bout length, and the summation is taken over all 277 foods in my samples.

The second subfactor was estimated from $\Sigma_{x=1} n_x \hat{L}_x / \Sigma_{y=0} n_y \hat{L}_y$, where the extra product $n_0 L_0$ in the denominator is for the frequency and mean length (product-limit estimate) of the nonfeeding intervals. (I refer to the denominator of this fraction, $t = \Sigma_{y=0} n_y \hat{L}_y$, as the *product-limit sample time,* and as explained in chapter 3, it differs slightly from actual sample time. For my samples, product-limit sample time was 22,299.84 minutes.)

The standard error of each food's time budget was obtained from the time budget's components, as follows. For the product-limit estimate $\hat{L}_j$ of the mean bout length of any food *j*, standard errors, which are given in table 4.1, column f, were calculated by the BMDP program (see Benedetti and Yuen 1977 for method). For the estimated bout rate $\hat{R}_j$—which was obtained from the number n_j of observed feeding bouts on that food (table 4.1, column a), divided by the product-limit sample time *t*, above—the standard error was computed from $n_j^{1/2}/t$ and is given in table 6.1 for each of the "core foods" that together accounted for 95% of the baboons' feeding time. Finally, for each food's time budget, calculated from the product $A_j = L_j R_j S$, the standard error was calculated by equation (3.3). Time budgets and their standard errors are given in table 6.1 for the core foods.

Feeding Time Allocation

The time-budget values for all foods, food species, and plant parts (tables 4.1, 4.7, and 4.4, respectively) have been rank-ordered, then plotted cumulatively in figure 4.14. Again, the baboons' marked pattern of strong dietary preferences can be seen. Although the weanlings ate 277 foods in total, 95% of their

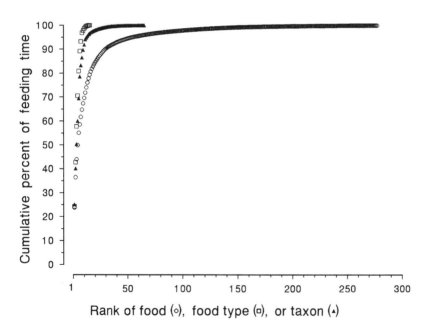

Fig. 4.14. Cumulative time budgets of foods, good types, and food taxa. Data from table 4.1. Feeding bouts on unidentified grasses and sedges have been prorated among the fifteen identified grasses and sedges. Bouts on unidentified acacias (seeds) have been prorated between the two acacias in the area. Bouts on foods of unknown taxa have been prorated among all known taxa.

observed feeding time was devoted to just fifty-four foods. Total feeding time, 277.74 minutes per day on average (table 4.7) is 42.1% of the 660 minutes per day of potential feeding time.

One might object that some of the apparent selectivity for grass corms results from the large amount of time the baboons spent on grasses whose species I could not at the time identify, so that "corms of grass or sedge, species unknown" is inflated by my ignorance. I therefore apportioned all feeding time on unknown grasses/sedges among the fifteen identified species of grasses and sedges, in proportion to their observed feeding time, and similarly for foods identified only as coming from acacias. The results, tabulated by species or other taxon in table 4.7, are even more striking. Setting aside the 5.3% of the time spent on foods of unknown taxa, these data indicate that the yearling baboons in my study spent 95% of their observed feeding time on milk plus just eleven plant species. The domination of the baboons' diet by acacias and grasses, a pattern revealed above in the analysis of bout frequencies, is borne out by these data on time budgets.

Finally, when we focus on food types, that is, at the part of the plant (or animal) that was eaten (table 4.4), the continued impact of nursing for baboons of this age becomes obvious: it accounted for a larger proportion of their observed feeding time (24%) than any other food category. If we leave out nursing, 95% of their remaining observed feeding time was devoted to just sixty-seven foods, thirteen food species, and eight plant parts.

During 1974–75, Post (1987) studied feeding behavior in three adult male baboons and two adult females, all from the same group as my subjects. (Indeed, during the later months of Post's study, his two females, Brush and Preg, were pregnant with two of my subjects, Bristle and Pedro, respectively.) Post noted, as I did, that feeding time was concentrated on a few items from a large repertoire of foods. In his study, over 60% of the animals' feeding time was spent on just five species; in my study the top five species accounted for 69% of the baboons' feeding time. In Post's study the two top-ranking species were, in order, fever trees and *Sporobolus rangei* grass; in my study these two species ranked the same way, and their time budgets were exceeded only by that for milk. Of the twenty-two major food species and parts fed on by the adults in Post's study (his table III), all but six are included in my table 4.7 of the top twenty-eight food species (judging by time budgets), and after eliminating taxonomic synonyms or not-distinguished species, the only food species on his list that I never recorded is *Cordia gharaf* (leaves).

In short, before baboons are 70 weeks old, they are eating essentially the same foods as adults. Indeed, when food classes of Amboseli baboons are ordered by feeding time, the ranking of adult foods (Post's fig. 3) and of yearling foods are nearly identical (fig. 4.15). Although my list includes several cate-

Yearlings
(Altmann)

milk
corms
seeds
leaves
fruits
gums
petioles, etc.
flowers, buds
other
prey
wood
water
runners
pods
bulbs, roots, etc.
soil

Adults
(Post)

corms
seeds
gum
leaves
fruits
flowers, buds
prey

Fig. 4.15. Food classes of yearlings (this study) and of adults (Post 1982), rank-ordered for each by feeding time, with the most time-consuming foods at the top. To make the lists comparable, I have combined Post's bud and flower classes, his blade and leaf classes, and his [seeds from] fresh and dry [acacia] pods.

gories that are not represented in Post's, for those food classes included in both lists the rank orders of all except gum are identical for yearlings and adults. If one compares Post's results and mine, adult baboons appear to spend more time feeding on fever tree gum than do yearlings. Perhaps adults are more skillful at feeding on gum because of greater strength or experience or a greater willingness to leave the others and forage by themselves (Post, pers. comm.). Alternatively, the difference in gum feeding between Post's study and mine may be due to year-to-year fluctuations in gum availability.

The similarities between the diets of adults and weanlings are not inconsistent with J. Altmann's 1980 discussion of "weaning foods" for baboons, that is, foods that are nutritious, abundant, and easy for young baboons to harvest and consume. Baboon infants are born throughout the year. What is not yet known is whether any advantage to certain birth dates accrues from greater abundance of such foods at the time the infant is becoming nutritionally independent.

The time-budget values given in the figures and tables in this chapter are based on all 277 foods that the yearlings ate during my samples. For a subset consisting of the 52 foods to which the baboons devoted the most time, I obtained more accurate time-budget estimates by adjusting for seasonal differences in food availability and sampling intensity. These seasonally adjusted feeding time budgets are fully presented in chapter 6 and appendix 8.

Relationships between Frequency, Duration, and Time Budget

I have now presented three quantitative characteristics of each food in the diet of weanling baboons: how often it was eaten, how long each such feeding bout (meal) lasted, and how much of the day the animals devoted to that food. In the literature on animal foraging, each of these three characteristics has been used to indicate the relative importance of the various foods that make up an animal's diet. Of course, their significance may hinge on features other than their nutritional value, and in the first section of this chapter I gave reasons why each of the three is central for studies of foraging ecology. Here I want to set aside these other factors and ask whether the relative feeding frequencies, meal durations, or feeding times can serve as indicators of relative nutritional importance. If, say, an animal feeds more often on leaves than on flowers, does this indicate leaves' greater nutritional value?

From a nutritional standpoint, food-specific time budgets have primacy, not because they are direct indicators of nutritive value but because, in conjunction with the mass intake rate of each food, they establish the amount of each food that is eaten (see equation 3.1), and thus the quantity of each nutrient and toxin consumed. The amount of time devoted to a given food can be estimated in several ways. I have used the product of frequency of use and average meal length. Should either of these two values be used alone, as they sometimes have been, to indicate a food's time budget and thus indirectly to measure the importance of each food?

One way to answer this question is to plot the mean bout lengths of foods against the frequency with which they were eaten (see fig. 4.16). If frequencies of use were perfect predictors of time budgets, that is, were proportional to them, then the bout lengths of foods would be the same, that is, a constant (the constant of proportionality). Similarly, if mean bout lengths were perfect predictors of time budgets, then bout frequencies would be the same. Finally, if either of these two parameters were a known (univalued) function of the other, then the former parameter, along with the function, could be used to determine the time budget. Clearly, neither of these parameters can even crudely be considered to be either a constant or a function of the other (fig. 4.16). Consequently, although bout length and bout frequency are interesting in their own right and are related to various environmental factors such as food patch size and spatial distribution, as I indicated at the beginning of this chapter, neither the frequency nor the duration alone provides even a crude estimate of the

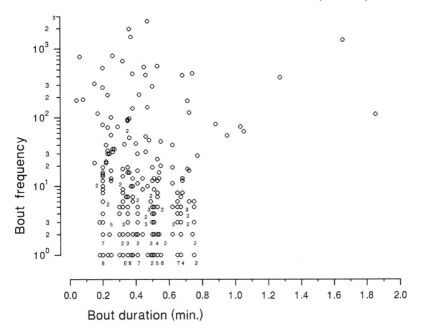

Fig. 4.16. Observed feeding-bout frequencies (all samples pooled) versus mean bout durations of foods. For plotted points representing more than one food, the number of foods is indicated $(2, 3, \ldots, 9, T = \text{ten}, E = \text{eleven})$. Data from table 4.1.

food-specific time budgets, and therefore neither provides an adequate estimate of the relative nutritional values of the various foods that make up an animal's diet.

Figure 4.16 enables us to answer another question about the relation between bout length and bout frequency. Do baboons tend to spend a lot of time per bout on the foods they eat most often, and vice versa? If so, the points in figure 4.16 would cluster along a line starting at the origin and then increasing monotonically. Or conversely, does their diet range from foods eaten quickly and often to those eaten slowly and seldom? Alas, for weanling baboons, no such simple relations exist between bout durations and frequencies (fig. 4.16). Some commonly eaten foods are harvested in long feeding bouts, others in short ones, and similarly for foods that are rarely eaten. Indeed, figure 4.16 shows an entire spectrum of foods having intermediate values and forming no discernible pattern.

Kurland and Gaulin (1987) provide a systematic comparison of some of the methods that have been used to measure primate diets. For five primate species, they regressed components of diet mass, categorized as animal matter, plant reproductive parts, or plant structural parts, each as a percentage of total diet mass, against percentage of feeding time devoted to each of these food categories. Plant structural parts showed very nearly a one-to-one relation between feeding time and food mass estimates, but for plant reproductive parts, relative time underestimated relative mass by about 30% on average and overestimated relative animal matter by 600%. They conclude:

> Although specific research problems may dictate the measurement of time budgets, the functional analysis of the causes and consequences of dietary adaptations requires measurements of the actual quantity of different foods eaten. . . . [Time budget] measures of diet are imprecise and incomparable across species differences in size, producing the most dramatic distortion at the extremes of the diet/body size continuum. This is particularly critical because most statistics are especially sensitive to values at the tails of the distributions. . . . [Time budget] measures must overestimate the actual weight of animal prey eaten unless primates can consume an equivalent weight of animal and plant material in a given amount of time.

Food-specific time budgets alone do not reliably estimate the diet, that is, the consumed quantities, relative or absolute, of various foods. To estimate an animal's mean daily dietary intake from its food-specific time budgets (factors A_j in chapter 3), one must also know the average mass intake rates, which vary from food to food and can in turn be estimated from their mean unit intake rates (B_j) and mean unit masses (C_j). Then, as indicated in chapter 3, the mean number of grams of food j consumed per day can be estimated from $A_j B_j C_j$ (assuming negligible covariances), or in dimensional terms, grams per day = (minutes per day) (units per minute) (grams per unit).

In sum, the time devoted to feeding on a given food cannot be judged even crudely by how often it is eaten or by how long sessions of feeding on it last. Feeding bout frequency and bout duration are essentially independent characteristics of foods, and neither without the other will accurately estimate the total time spent on each food. In turn, food-specific time budgets do not reliably estimate an animal's diet—the amount of each food consumed. That is true whether the time budgets of foods are estimated from these two component factors, as I have done; from point (instantaneous) samples of activities (J. Altmann 1974), a common alternative; or in other ways. Yet food-specific time budgets probably have been the most common way of assessing the contributions made by various foods to the diets of primates (e.g., see most of the chapters in Clutton-Brock and Harvey 1977 or the tabulations in Robinson and Janson 1987). Such assessments should be regarded as only rough approximation of the animals' actual diets. Fortunately, methods are now at hand for estimating the intakes of foods and food components under field conditions.

5

Science is built up with facts as a house is with stones. But a collection of facts is no more a science than a heap of stones is a house.

—Henry Poincaré, 1913

Theory without fact is phantasy, but fact without theory is chaos. Divorced, both are useless; united, they are equally essential and fruitful.

—Charles O. Whitman, 1896

Adequate and Optimal Diets

Strategies and Tactics

This book centers on a general model of optimal diets and its application to yearling baboons in Amboseli. The model will be presented in this chapter, the empirical evidence in those that follow. However, before turning to the model itself, I consider first the general relation of an animal's dietary choices to other aspects of its foraging behavior and then the central problem in foraging research.

Animals' foraging involves selecting among alternative courses of action at several hierarchically arranged levels. The relationship among adjacent levels is that between ends and means, between strategies and tactics. The primary function of foraging is to obtain an array of foods that will provide the material and energy needed to run the machinery of the body. Equivalently we can say, without any implication of conscious planning by animals, that achieving such a diet is the *goal* of foraging.

Obtaining any given diet entails selecting among various modes of foraging. Thus foraging patterns constitute a second level of selection by the animal. A pattern of foraging has many components: time budgeting, spatial deployment, mode of patch utilization, methods of searching and harvesting, spatial deployment relative to potentially competing individuals, and so on, each selected from among the numerous alternative courses of action of which the animal is capable.

An animal's pattern of foraging determines the diet it achieves; it is the means for which the diet obtained is the end product, but it is not the diet itself. The distinction between means and ends is important. For example, an animal might achieve a given diet by any of several foraging patterns. From the standpoint of diets, foraging patterns are tactics.

109

An animal's foraging pattern, in turn, is achieved through its choice of specific foraging behavior: its actual sensory perceptions, information processing, and motor behavior. For this reason, patterns of foraging are commonly referred to as foraging strategies. Foraging strategies have been the focus of much recent research, and a variety of models of optimal foraging have been developed. Pyke, Pulliam, and Charnov (1977) and Stephens and Krebs (1986) provide excellent surveys of developments in this area.

That foraging strategies are not foraging behavior itself may not be immediately obvious, partly because behavioral strategies are often described in terms that suggest behavior and even conscious planning: *select, abandon, allocate, harvest,* and so on. Strategic rules such as *select those foods with the highest net energy yield,* or *abandon a food patch when your intake falls below the mean of all patches,* or the extreme variance rule (see below) do not specify the behavior needed to accomplish these ends—not in a motoric sense—only the ends themselves. Whether, say, a monkey plucks berries from a bush by mouth or by hand, whether it uses the right hand or the left, whether the hand uses a precision grip or a power grip, and so on would ordinarily not be specified in a model of foraging strategies, and for some purposes the choice may be largely irrelevant. Again, a particular end may be achieved by more than one means.

To date, most foraging models have dealt with foraging strategies, some with questions of optimal diets. Foraging behavior per se, which lies at the interface between descriptive ethology and functional anatomy, has received considerably less attention from those involved in formal modeling. Ultimately, foraging theory deals with all three of these levels and their relationships.

The reader may have been puzzled because I described tactics as *preceding* their corresponding strategies, yet they must precede them, since strategies are what tactics achieve, and consequences are always sequelae. In the military, however, the order is the reverse. A general's strategic plan to use a pincer movement to remove an occupying army from an area precedes the decision to land three battalions each at points A and B on specific dates, and that decision in turn comes before the many tactical maneuvers used to drive the occupying army out of each town, each street, each house. The confusion arises because of a failure to distinguish between strategic plans and strategic events, between goals and their achievement. We call them both strategies, so that term carries an implication of consciousness. A similar distinction must be made between tactical plans and tactical maneuvers. The former are cognitive; the latter, physical.

Humans have the ability to plan their future actions and to make a general plan (e.g., drive out the occupying army) without yet having selected specific tactical activities that can execute the plan or even knowing whether any such exist. So, for example, the strategic plan to drive the enemy out of a particular town is executed by numerous subsequent actions: driving them out of one house after another, blowing up bridges, and so on. These tactical maneuvers themselves may be planned before they are executed. If the tactical actions are successful, they will have accomplished a strategic outcome: routing the enemy.

Do animals other than humans have an ability to make strategic or tactical plans? This book is not the place to review the evidence on that question, but let me mention one suggestive fact. In some primate species (rarely in baboons), observers can distinguish food-specific searching patterns even when the food is remote in space and time, and they can thereby record total foraging time, searching plus feeding, devoted to different foods (the *time-foraging method* of Robinson and Janson 1986). Such behavior indicates that the animal's diet selection can be displaced in time and space from its actual ingestion and suggests the existence of an overall plan that governs the foraging behavior. Miller, Galanter, and Pribran (1960) provide an illuminating discussion of how strategic planning can be evidenced by overt behavior.

In the current literature on animal behavior, the words *strategy* and *tactic* are commonly misused: no distinction is made between them, and their implication of conscious planning is ignored. They have been reduced to buzzwords, used for effect rather than meaning.

The Central Problem

Schoener (1971), in his pivotal review of early developments in foraging theory, argued that "the primary task of a theory of feeding strategies is to specify for a given animal that complex of behavior and morphology best suited to gather food energy in a particular environment." Now, a quarter of a century later, we can profitably reconsider this issue.

Food is not merely a source of energy. Many animals may indeed be energy limited, yet even for them, energy maximization per se cannot be the sole objective of foraging: a diet that maximizes the energy intake may leave the animal deficient in other nutrients or exceed its tolerance for toxins or other hazards or its limit on foraging time. For other animals, fitness may be limited by other dietary components, such as protein intake or time expenditure. For many nocturnal, subterranean desert rodents, the problem is not to maximize energy but to minimize foraging time spent aboveground, where a wide variety of avian and mammalian carnivores prey on them (Schoener 1971).

Second, to specify the suite of morphology and behavior that would be best suited for gathering food energy is to specify a fiction. To understand the foraging adaptations of a real animal, we must specify the optimal feeding strategy for the organism as it is, not as it would be if it were ideally designed. The competence, capacities, and behavioral repertoire of the organism, its anatomy, physiology, and so forth, are the givens. They are constraints on a theory of foraging, not its objectives. For example, an interesting fact of natural history and anatomy is that baboons cannot open their jaws wide enough to eat ostrich eggs. On occasions we have watched with amusement as they have tried. Thus, as these animals discover, their problem—and the problem of anyone who wants to understand the adaptive significance of their foraging—is to find the right combination of foods *without* including ostrich eggs, however nutritious the eggs may be. Furthermore, the relevant capacities are not restricted to behavior and anatomy. If an animal does not have enzymes or gut microbes capable of digesting arabinose or galactose, an optimal diet should not include a tree gum that is composed primarily of these sugars and has no other redeeming value.

Of course, I am speaking here of a theory of short-term foraging strategies. On an evolutionary time scale, the capacities of organisms themselves are subject to change. Over generations, jaw gape size may increase, and the capacity to synthesize new enzymes may evolve. Indeed, even within the lifetime of individuals, differences in morphology and physiology, often induced by feeding itself, can result in changes in capacity. Within limits, jaw muscles can increase in strength and gape size can be increased. The production of some enzymes can be induced. Yet much behavior is appreciably more ontogenetically flexible than is the animal's morphology or physiology, so the latter capacities of the organism are usually treated by students of behavior as given.

Finally, an adequate theory of foraging must deal with departures from the optimum, not just the optimum itself. For a wide variety of reasons, a natural population may not be at its fitness maximum, even on average (S. Altmann in prep.; Dawkins 1982; Futuyma 1986; Maynard Smith 1978; Van Valen 1960). The foraging patterns and thus the diets of animals may deviate from the optimum. A theory of feeding strategies that specified only optima, with no indication of the consequences of various possible departures from them, would be a sterile theory.

An adequate foraging theory would indicate the consequences of all actual or potential feeding patterns of which the animal is capable. These consequences include not only short-term effects, such as the intake of nutrients and toxins and the expenditure of time, but also long-term consequences, particularly effects of foraging on life history parameters, adaptation, and biological fitness. From an evolutionary perspective, the primary task of a general theory of foraging is to specify the consequences of each pattern of foraging that any given animal is competent to perform in the habitats in which it lives.[2]

Multiple Constraints

The effect of natural selection on the traits of an organism is commonly treated as though each trait was disembodied and had evolved independently, but of course this is not so. Organisms are integrated systems, and the extent to which any one trait can develop without jeopardizing the fitness of the organism as a whole is constrained by the effects of such development on other traits. This phenomenon of coadaptation (or at the genetic level, epistasis) is well known. "An increase in body size is maladaptive unless compensatory changes in the surface area of gills, lungs, intestines and other organs maintain a constant ratio between the mass of metabolizing tissue and the area over which gases, nutrients, and wastes are exchanged" (Futuyma 1979, 49).

Although multiple constraints on traits are widely recognized, they have seldom been incorporated into formal models of well-adapted organisms, in part because of the technical difficulty in doing so. A striking exception is provided by the application of linear optimization (linear programming) to optimal diet problems. The basic idea is very simple. Costs and benefits are inextricably mixed: a foraging animal cannot obtain its essential nutrients without simultaneously exposing itself to increased risks: from thorns, toxic secondary compounds, predation, dehydration, or whatever. This is the so-called packaging problem, discussed in chapter 2. Let us assume as an approximation that for each nutrient the animal has a minimum, its *requirement* for that nutrient, and for each hazard it has a maximum, its *tolerance*. If the animal's fitness is, say, energy limited, we ask, what amount of each available food should the animal eat so as to maximize its energy intake while simultaneously staying above its minimum for each nutrient and below its maximum for every hazard? The beauty of linear optimization is that it enables us to answer such questions explicitly, and thus to generate verifiable predictions about what well-adapted animals ought to eat.

S. Altmann and Wagner (1978) presented a linear optimization model for finding optimal diets, given certain information about the composition of foods and the nutritional requirements of the animals. Our model is a straightforward extension of Stigler's (1945), although at the time we developed it we were not aware of Stigler's work. We were also then unaware of Dantzig's (1963) method for solving such problems, which is far better than ours. In the years since the work of Stigler and Dantzig, linear optimization has been applied to a variety of optimal diet problems, in both humans and other animals; for example, by Belovsky (1978, 1981, 1984, 1986), Belovsky and Schmitz (1994), Carmel (1976), Chamberlin and Stickney (1973), Dent (1966), Pulliam (1975), Reidhead (1980), Reiss (1986), and Smith (1992). Least-cost optimal diet models are now the most commonly used textbook illustration of linear optimization. Linear optimization is not a single model but a framework for models with certain formal attributes in common, along with methods for finding solutions to them.

In this chapter I summarize our optimal diet model, then use it to calculate optimal diets for Amboseli baboons under several assumptions about what is to be optimized.

A General Model of Optimal Diets

The question I pose is this: Given the foods available to an animal, how much of each should it consume so as to be above its minimum for every nutrient, below its maximum for every toxin, and concurrently, either to minimize some cost function, such as time expended in foraging or exposure to predators while feeding, or else maximize some benefit obtained from feeding, such as protein or energy intake? I now describe how in general such optimal diets can be determined.

I consider food components not only in the conventional sense of chemical compounds (e.g., riboflavin) or classes of compounds (e.g., lipids, proteins), for each of which the amount ingested is some fraction of the mass of each food type consumed, but also in a more general sense to include properties, such as energy content, cost, foraging time, and so forth, that are expressed in units other than mass but that nonetheless are linear functions of the amount of the various foods that are ingested or the amounts of time required to obtain them. (For such components, the term *intake* is metaphoric.) Note also that requirements need not be mutually exclusive. So, for example, an animal can be limited in the amounts of sodium and calcium it can safely ingest and can have another limit, smaller than the sum of the other two, on total cation intake.

For every food component a minimum required intake rate exists, which in some cases may be small enough (e.g., zero) to have no coercive significance. A *nutrient* is a chemical component of food with a non-zero minimum, such that sickness or some other major reduction in fitness occurs if an individual falls below this minimum for a sufficient period of time. Similarly, for every food component, an upper limit exists on the amount per unit time that the animal can consume with impunity. Chemical components for which the upper limit is less than gut capacity are called *toxins* in the narrow sense. Many nutrients are known to be toxins at high intake levels. Beyond that, many toxic secondary compounds occur naturally in wild plants and may protect plants from herbivores and microorganisms (Foley and McArthur 1994; Freeland and Janzen 1974; Fritz and Simms 1992; Huges and Genast 1973; Levin 1976; Rosenthal and Janzen 1979; Wink 1993). For nonchemical components of foods such as foraging time and cost, other factors will establish an upper limit. (As indicated in appendix 1, the animals' nutrient requirements and toxin tolerances are actually probability distributions, not constants; the minima and maxima discussed here can be thought of as the outer limits of those distributions, say, the mean plus three standard deviations.)

For every component except one, intake variations within the range of adequate intakes—that is, strictly above the minimum but below the maximum—are assumed for the purpose of this model to have no significant effects on fitness. I further assume that within these upper and lower bounds, fitness is maximized only if the one exceptional component is maximized (if it is beneficial) or minimized (if it is detrimental). The potential amount of that component in the diet will be referred to as the *objective function*.

The following method, based on S. Altmann and Wagner (1978), treats in a single analytic framework the problems of meeting the requirements of food component minima (for nutrients, etc.) and maxima (for toxins, etc.) and simultaneously satisfying the requirements of a cost or profit function.

Let d_{ij} be the amount per unit mass (e.g., proportion, calories per gram, minutes per gram, etc.) of the ith component in the jth available food and let M_i and T_i be respectively the minimum (typically, a nutrient requirement) and the maximum (typically a tolerance) for the ith component. A diet is an ordered n-tuple $F = \{F_1, \ldots, F_n\}$ of non-negative numbers, where F_j represents the amount of food j that is eaten and n is the number of available foods. Obviously each F_j is the amount of food j obtained in some specified amount of time, say per day, so that literally we are dealing with intake *rates* (mass per unit time), but for many purposes dropping this common time base is convenient and should cause no confusion. (The relationships between the notation used here and that used in chapter 4 are as follows: $d_{ij} \equiv D_{ij}$ and $F_j \equiv A_j B_j C_j$.) Thus in any diet F, the amount of, say, the first component will be $d_{11}F_1 + \ldots + d_{1n}F_n$.

If m components are to be considered and the animal has n foods to choose from, its adequate diets (i.e., those within its upper and lower bounds) are the set of all points satisfying the constraints

$$M_1 \le d_{11}F_1 + \ldots + d_{1n}F_n \le T_1$$

$$M_2 \le d_{21}F_1 + \ldots + d_{2n}F_n \le T_2$$

(5.1)

$$\cdot$$

$$\cdot$$

$$\cdot$$

$$M_m \le d_{m1}F_1 + \ldots + d_{mn}F_n \le T_m.$$

By convention, the minima and maxima are written on the right-hand side of such constraint equations, and all variables are written on the left, so that, for example, the first row of equations (5.1) becomes two equations:

$$d_{11}F_1 + d_{12}F_2 + \ldots + d_{1n}F_n \le T_1$$

$$d_{11}F_1 + d_{12}F_2 + \ldots + d_{1n}F_n \ge M_1,$$

and I shall have occasion to refer to the minima and maxima as the *right-hand-side limits* of equations (5.1).

If the rth component of equations (5.1) is the one to be optimized (maximized or minimized), define a function $G(x)$ equal to the summation in the rth row of equations (5.1):

(5.2) $$G(F) = d_{r1}F_1 + d_{r2}F_2 + \ldots + d_{rn}F_n.$$

Conventionally, the coefficients of the component to be optimized are designated by a distinct letter, so letting $e_j = d_{rj}$ in equation (5.2), that equation becomes:

$$G(F) = e_1F_1 + e_2F_2 + \ldots + e_nF_n.$$

G is called the *objective function* and is a linear function of the F's. The problem of finding an optimal diet is that of finding values $F_1^*, \ldots, F_n^*$ of $F_1, \ldots, F_n$ that maximize (minimize) G subject to the constraints above. If, for example, an animal's fitness is energy limited within the range of adequate diets, then the optimal diet is that food combination $F_1^*, \ldots, F_n^*$ that maximizes energy intake subject to certain limitations, imposed by nutritional requirements, toxin tolerances, gut capacity, time-budget constraints, or whatever. Constraints express not just the competence of the organism to obtain particular diets, limited, for example, by its gut capacity or its ability to harvest certain foods, but also its ability to tolerate the consequences of any realizable diet.

The linear diet model has a simple geometric interpretation. Suppose, for simplicity, that we are dealing with diets made up of just two foods. If each axis of a graph is used to represent the amount (F_1 or F_2) of one food in the diet (fig. 5.1), every possible diet (every combination of F_1 and F_2) can be represented by a point. The set of all diets that contain the minimum acceptable amount of any given nutrient is represented by a solid line, and likewise for the maximum of any given toxin. Consequently, nutritionally adequate diets will form a convex polygon consisting of points that lie above all minima and below all maxima. The objective function—for the food component to be, say, maximized—is represented in the figure by the dotted line. To find the optimal diet graphically, translate this line, that is, move it upward without changing its slope, until it cannot be moved higher without going outside the region of adequate diets. (If the component is to be minimized, the line is translated downward in the same manner.) The optimal diet is the adequate diet point (or points) that the dashed line then touches.

The constraint equations (5.1) determine a closed, bounded, and convex subset of Euclidean n-space, representing all nutritionally adequate diets. The optimum is a corner point of this convex set or an edge or facet defined by several such points. The optimum can be found by brute-force computation, for example, finding all corner points, then choosing the one(s) for which G is

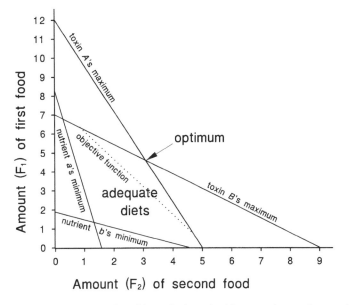

Fig. 5.1. Hypothetical diet, with two foods, each with two nutrients and two toxins, and an objective function.

minimal or maximal, as appropriate (see S. Altmann and Wagner 1978). However, that procedure is too cumbersome for all but the very simplest situations. Under any other conditions, one should use the well-developed algorithms of linear optimization (e.g., Hillier and Lieberman 1974), for which an excellent computer program is available (Schrage 1981).

Scope and Limits of the Model

Although most current models of foraging are based on maximization of the energy intake rate, it has been known for over half a century that a source of energy alone is not adequate to run the machinery of the body. How can one simultaneously capture the ecologist's conviction that many animals are energy limited, the physiologist's demonstration of a variety of essential nutrients, botanists' growing evidence that many wild plants produce toxic secondary compounds, and naturalists' evidence of many other risks faced by foraging animals? That is exactly what we attempt to do in our optimal diet model.

Equations (5.1) provide a model of optimal diet that involves multiple constraints, and these can come not only from nutrients but also from toxins, foraging time, exposure to predation or insolation, or any other consequence of diet that is a function of the amounts of foods eaten or the time devoted to obtaining them.

Some aspects of diet selection are not at present in the model (or in my application of it), for lack of adequate empirical data; others are outside the scope of models of this type. The following characteristics of diet selection could readily be incorporated into the model if only we had the necessary information.

1. The individual's requirement for each nutrient and its tolerance for each toxin are assumed to be independent of the rest of the intake components, but in a number of cases this is known to be false: there are nutrient and toxin interactions (see "Protein Limit" section in appendix 1). Given adequate information about such interactions, we could add to the model a constraint equation indicating the admissible combinations of two or more components, as has already been done implicitly for energy. (In practice, I found it convenient to calculate the energy content of foods ahead of time.)

2. The only toxin for which I had adequate information to write a constraint equation is salt content, but salts are surely not the only significant toxin for Amboseli baboons (see chapters 6 and 9).

3. The analysis of nutrients was carried out primarily at the level of the so-called proximate food components or macronutrients (total proteins, total fats, and so forth), but except for water, nutrient requirements are actually at a finer chemical level. For example, mammals have no dietary requirement for proteins per se but only for the constituent amino acids. Requirements and tolerances for individual cations are discussed later in this chapter (Comment 6, below). Inadequate information about other micronutrients has kept me from writing constraint equations for them; our limited data on them are discussed in chapters 6 and 9 and appendix 1.

4. Feeding time and seasonal availability are the only nonchemical food "components" that have been used in my analyses, but others such as risks of predation and dehydration may be important constraints on diet selection. Each food tends to be available primarily in certain habitat zones (Post 1978), and these often differ in such risks (Cowlishaw 1997). For example, the grass *Cynodon nlemfuensis* grows abundantly around waterholes, where the risk of dehydration is lower but the risk of leopard attack is greater than on the open, shadeless plains where the grass *Sporobolus rangei* is harvested by the baboons.

5. Each food that is eaten contributes a certain amount of any given nutrient to the total intake of that nutrient, and similarly for each toxin, as well as the other costs and benefits. In writing equations (5.1) and (5.2), I assumed that the contributions from each "meal" of each nutrient or hazard were additive. So, for example, I assumed that each milligram of riboflavin, from whatever

source, contributes equally to meeting the total riboflavin requirement. So far as I know, this assumption is correct for nutrients, with two possible types of exceptions. First, in order to be utilized, some nutrients, such as animo acids (appendix 1), must be consumed within certain time limits of each other. Second, nutrients and other food "components" may interact with each other. In some habitats, for example, the probability of being preyed upon may go up exponentially with feeding time spent in that habitat, and in chapter 3, I discussed interactions among nutrients and toxins (Freeland and Janzen 1974; Wise 1980). Methods for optimization with nonlinear constraints make possible the analysis of optimal choice under such conditions; lacking are quantitative data on nonadditive effects.

6. The diet mix that is optimal over a long period may not be optimal on a given day. Indeed, it might not even be available, because of seasonal changes in the habitat. The second problem can be solved by calculating an optimal diet based on the currently available foods, as I did semimonthly for chapter 8. The first cannot. Let me illustrate. Suppose that an animal has had no vitamin A or vitamin C for a fortnight and it now has a choice between the two. Which one should it take? The choice depends not only on the seriousness of the respective deficiencies but also on the lag time to deficiency for each vitamin. Vitamin A is stored in the liver, and the animal may be able to go for many months on those stores; but for vitamin C, no body stores are known. Thus, in the short run, the best choice may depend not only on what is available but on the state of the organism: on nutrient-specific body stores and turnover rates as well as the seriousness of the deficiency. It may also depend on the state of the habitat. If the only appreciable amounts of vitamin A are in a seasonally limited fruit, then the fruit should be consumed when available, even at the risk of temporary deficiencies in other nutrients. Except in some special cases (see example in next sections on risk sensitivity), the optimal diet model could not readily be adapted to such short-term contingent dynamics.

7. In this book I consider five possible objective functions for optimal diets, each of which involves maximizing or minimizing some mean value such as the average daily intake of protein. However, foraging returns always have about them an element of unpredictable variability, so that a forager might benefit from adapting strategies that are adjusted to the variation in foraging yield rather than the expected (mean) value.

A forager that adjusts to the variance in its foraging yield is said to be *risk sensitive,* and several foraging models have been developed to take risk sensitivity into account (e.g., Caraco 1980, 1981; Caraco, Martindale, and Whittam 1980; McNair 1979; Stephens 1981; Stephens and Charnov 1982). For example, to minimize the probability of falling critically short of a required nutrient, say energy, an animal should follow the *extreme variance rule.* If the available foraging patterns have the same mean energy yield, always choose the pattern with the *maximum* variance in energy yield when expected returns are less than the energy requirement (when starving, try anything). Conversely, always choose the one with the *minimum* variance on returns when the expected returns are greater than the requirement, or more generally, select that foraging pattern with the lowest ratio of mean energy shortfall (requirement less mean daily yield) to variance in energy yield (Stephens and Charnov 1982).

In Amboseli, day-to-day variability in diet apparently is considerable (Stacey 1986), and it might indeed be advantageous for baboons to be risk-sensitive. The objective function of a linear optimization diet model could be written in terms of mean/variance ratios, careful attention being paid to the time-to-deficiency of the limiting nutrient. Alternatively, it may be that the foragers' main problem is not that of going bankrupt from lack of a nutrient but of an unfortunate run of one of the costs of foraging, including toxins, predation, and adverse conditions in the physical environment.

Some other aspects of feeding are clearly outside the domain of my model or any straightforward adaptation of it:

1. As indicated above, we are dealing with a model of optimal diet, not optimal feeding or foraging behavior. It specifies *what* the animal should accomplish, not *how*.

2. Similarly, the model does not account for *why* the animal eats what it does, or when. It does not require any assumptions about motivation or perception. It matters not for our present purposes whether a baboon prefers ripe fruit to green because of the color or odor, because it is sensitive to the difference in nutrients, because it eats only what it saw its mother eat, or whatever. Determinants of food choice have been studied in experiments that are designed to incorporate many of the salient features of the natural food choice situation (e.g., Collier 1980, 1982; Marwine and Collier 1979; Galef 1976; Rozin 1976).

3. Many studies of foraging patterns, of determinants of food choice, and of social aspects of foraging turn on the temporal patterns of stimulus and response. The optimal diet model makes neither assumptions nor predictions about the temporal dynamics of diet selection, other than the implicit assumption, discussed above, that for each nutrient, requirements must be met within a certain amount of time. The temporal patterning is also relevant insofar as it presents a practical sampling problem in attempts to estimate the factors (feeding bout duration, unit intake rate, and so forth) that are used to estimate dietary intake, but the method of diet estimation is not part of the model.

However, the temporal dynamics of nutrient intake may be more important than generally realized. For example, if, as has been argued by Holt, Halec, and

Kajdi (1962), there are no body stores for proteins, then utilization of a given amino acid that has been taken up in the diet may require intake of other essential amino acids before the first is catabolized, and the critical time interval may be on the order of three or four hours (Elman 1939; Geiger 1950; Spotter and Harper 1961; Murphy and Pearcy 1993). If so, then for an analysis at the level of amino acids, data pooled over weeks or months, such as we have used in this study, will be inadequate. The periods over which the values are to be obtained are not currently part of the model; perhaps they should be.

Composition of Foods

For each of the 52 *core foods* of the baboons' diet (see chapter 6), table 5.1 indicates the macronutrient composition—that is, the amount of water, ash, lipids, proteins, fibers, and other carbohydrates, each expressed as a percentage (g per 100 g of fresh food)—and also the energetic value of the food (kJ per 100 g of fresh food). Information about the sources of the data in table 5.1 are provided in appendix 5.

Answering Questions about Optimal Diets

In addition to determining what diets would be optimal, linear optimization (linear programming) enables one to answer a surprisingly wide variety of biologically important questions about diets. However, the computer output of a typical optimization program, as well as the corresponding literature, is worded largely in the argot of the marketplace and of economics. Here I provide biological interpretations of the standard outputs of linear optimization programs when applied to optimal diets. These will be helpful in understanding the optimal diet tables that follow. The last part of this section, "A Quick Guide," summarizes the material by interpreting one of these tables. Some readers may find it useful to read the guide first. What may also be useful is table 5.4, which provides a synopsis of linear optimization terminology from economics and the corresponding terminology that I use.

In the next section I present successive approximations to optimal diets that are based respectively on five objectives: to maximize mean daily energy intake, mean energy intake rate, mean daily protein intake or mean protein intake rate, or to minimize mean daily feeding time. That is, I ask:

1. Optimal diet: Given the constraints imposed by the requirements and tolerances for nutrients, toxins, and other food components, what diet—that is, what amount of each food—maximizes energy intake (protein intake)? What diet minimizes feeding time?

As I will explain below, successive approximations to these five optimal diets resulted from attempts to find more realistic constraints than those imposed solely by nutrients and toxins. In this section, my frequent references to toxins should be understood to apply to all of the hazards connected with feeding, or indeed anything entailed by feeding that restricts food intake (e.g., gut capacity, feeding time, dehydration, and so on). For brevity here and in the rest of this section, I sometimes refer to both daily energy intake and energy intake rate as the *amount* or *intake* of energy, or in comparable terms, and similarly for protein and feeding time.

I answer question 1, first for energy and then for protein, by finding the amount F_j^* of each food j that maximizes the objective function (equation 5.2), which represents in turn the amount of energy (first in joules per day, next in joules per minute of feeding) and then the amount of protein (g/day, then g/minute) with the maximization subject in each case to linear constraints of the form shown in equations (5.1). For feeding time as the criterion, the corresponding objective function is minimized. Except for the two objective functions that are rates, whose treatment will be described later, these are standard problems in linear optimization.

Once the composition of an energy-maximizing, protein-maximizing, or time-minimizing diet has been estimated, one would like to know what would prevent the animals from doing better. Which constraints actually limit their objective? That is, which of the constraints are *binding* and how far are the other constraints from being so?

2. Limiting factors, binding constraints: What attributes of the animals or of their environment limit the amount of energy or protein in the diet or set a lower limit on the amount of time needed to harvest it?

3. Slack values, safety margins: How much could each nonlimiting factor change before it became a limiting factor?

By a *limiting factor* in the optimal diet I mean any trait of the animal or any characteristic of its environment for which some change, an increase or decrease as appropriate, would result in an improvement in the objective function value, that is, an increase if that function is to be maximized, a decrease if it is to be minimized. This usage of "limiting factor" stems from the work of Liebig (1840), then Blackman (1905) and Shelford (1911): the rate or extent of a process can be limited by minimal and maximal quantities of different environmental factors; thus a process is said to be limited by an environmental factor if the process increases or decreases as a result of a change in that factor.

Limiting factors for diets are of two types. Some, such as requirements, tolerances, and milk constituents, are characteristics either of the animals or, as with food-specific harvesting time, of an interaction of the animals with aspects of their environment. Other limiting factors, such as the composition of various foods (except milk) and their seasonal availabilities, are not traits of the foragers.

The search for limiting factors that are attributes of the foragers is important for two reasons. First, if the values assigned to these requirements, tolerances, and milk components are correct, they suggest which of these characteristics of the animals limit their foraging success and are therefore subject to natural selection. If, for example, protein intake is limited by the inability of the animals to tolerate a higher salt content in the diet, say because salt and protein co-occur in the diet, then one would expect natural selection to favor better ability to detect and select low-salt variations in foods (or those with a high ratio of protein to salt) and greater renal capacity to excrete excess salt. Beyond that, kidney malfunctions that are characteristic of salt overloads would be expected among the animals that are at or near the optimum protein intake.

Information about limiting factors appears in two places in the computer outputs and in my tables. First, information about each food's energy or protein or harvesting time, as appropriate to the objective, appears in one section of the program output, as will be discussed below (see questions 5 and 6). If, say, some food became much higher in food energy, or if the animals found an enriched local variant, an optimal diet with more energy in it could result.

Information about other limiting factors, including attributes both of the foragers (e.g., nutrient requirements, consumption capacity) and of their environment (e.g., seasonal availabilities of foods), appears in the output section devoted to the constraint equations. Here and under question 4, I consider that information.

Question 2, regarding the identification of limiting factors, was answered for constraints as follows. Consider the two sides of any of the constraint equations (5.1) once the optimal values $F_1^*, \ldots, F_n^*$ of the amounts $F_1, \ldots, F_n$ of the various foods have been entered. If the left and right sides are equal, that is, if their difference (*slack value*) is zero, then the optimal diet calls for the maximum tolerated amount of that toxin or the minimum required amount of that nutrient, and the baboons' requirement for that nutrient or their intolerance for that toxin is a limiting factor for the optimal diet.

Note that although nutrient requirements and toxin tolerances are treated as constants in solving the optimization problem, they are in fact empirically determined random variables, and for some of these values there is, as already indicated, considerable uncertainty. This provides a second reason for the search for limiting factors. The magnitude of the slack values that are used to detect them also provide measures of the *safety margin* of each constraint. If the slack value of a constraint equation is non-zero, that is, if it is strictly positive, we are not dealing with a *limiting (binding) constraint*. All else being unchanged, its slack indicates how much that requirement could be increased or that tolerance could be decreased—or how much the estimated requirement or tolerance could be in error—without degrading the objective, that is, without reducing the energy content (or the protein content) of the diet or increasing its requisite feeding time. Thus, each slack value also answers question 3 for that constraint. It indicates the margin of safety for that nutrient or toxin: the difference between the intake prescribed by the optimal diet and that which would lead, respectively, to a deficiency or to poisoning.

Another use for slack values is that they provide an easy way to tell how much of any nutrient, toxin, or other constraining food component would be obtained from the optimal diet. For example, if the daily protein requirement is four grams and the slack for that constraint is five grams, the optimal diet would provide nine grams.

Of course the energy (protein) content or feeding time requirement of the diet will be affected more by changes in some limiting factors than others, so we would like to be able to measure the *sensitivity* of the diet's objective function to a change in each of the limiting factors.

4. Marginal value of limiting factors: How sensitive is the energy or protein content or feeding time requirement of the optimal diet to the values of the limiting factors?

That is, in an optimal diet, how much would the objective function value of the optimal diet (its energy or protein content, or its required harvesting time) increase or decrease per unit change in the value of any one limiting right-hand-side value in the constraint equations (e.g., consumption capacity), if all other right-hand-side values remained unchanged? This question about the sensitivity of the energy (protein, time) in the optimal diet to changes in the values of those requirements and tolerances that are limiting factors is answered by calculating the *marginal values, y_i^* and z_i^**, of, respectively, the requirement and tolerance for each food component i. The marginal values provide us with sensitivity measures: in an optimal diet, each y_i^* and z_i^* indicates the change in the objective function (energy or protein intake, or time requirement, as appropriate) that would be achieved for each unit change in the corresponding right-hand values of M_i and T_i, respectively, all others remaining unchanged. Of course, in an optimal diet, the energy intake is sensitive only to those right-hand values that are limiting factors, since that is what we mean by a limiting factor; all other constraints have a marginal value of zero.

I shall sometimes speak of a relative marginal value as the *leverage* exerted over the objective function per unit change in a requirement or tolerance. In the economic literature, marginal values are also referred to as "dual prices" or "shadow prices."

Marginal values are of particular significance in evolutionary studies in that they provide *exchange rates* between diet components, such as carbohydrate intake or foraging time, and fitness regulators, such as energy (S. Altmann 1984). To illustrate, let us consider an animal whose fitness is energy limited, so that a diet that is optimal—one that, among available adequate diets, leads to the highest fitness—is a diet that maximizes energy intake, so long as requirements for various nutrients (or other food components) are met and no tolerance level for food-contained toxins (or other hazards) is exceeded. At or near the optimum, each z_i^*, the marginal value for the ith nutrient, is the change in the objective function, the animal's energy intake, per unit change (in the opposite direction) in its requirement for the ith nutrient. Thus, if maximizing energy, in joules, is the objective and if the protein requirement, in grams, has a marginal value of, say, -5.07 (as in diet 8, table 5.11), then each one gram decrease in the animals' protein requirement would enable them to have an optimal diet that was one joule richer.

The marginal value can be regarded as the energy exchange rate for increasing or decreasing requirement M_i by one unit. Figuratively, an animal consuming an optimal diet should be willing to accept a unit increase in the ith nutrient requirement only if compensated by at least the marginal value, z_i^* joules of energy per day; at that rate, the compensation or credit for all M_i units of nutrient i would be $M_i z_i^*$ joules per day. Note that $M_i z_i^*$ is not the energy content of M_i units of nutrient i, which, as in the case of minerals, may have no energy content whatever, nor is it the energy expended by foraging for M_i units of nutrient i (S. Altmann 1984).

Similarly, each y_i^* (the marginal value for the ith toxin or other hazard) is the increase in dietary energy that would result from a unit increase in the ith toxin tolerance T_i. Figuratively, y_i^* is the largest amount of energy that an animal consuming an optimal diet should be willing to pay to raise T_i by one unit (because with the higher tolerance it could obtain that much more energy), and at this energy exchange rate, the total cost of taking in all T_i units of toxin i would be $T_i y_i^*$ joules per day. To make this analogy concrete, suppose the animal's habitat includes an herb that, for each leaf consumed, raises the animal's salt tolerance T_i by one unit, say by one milligram per day, and suppose the herb has no other value, positive or negative. Harvesting this herb requires half of y_i^* joules per leaf. Should the animal expend ("pay") that much energy to feed on this herb? Yes, and indeed, it should be willing to pay up to y_i^* joules per leaf of this herb, because each such leaf will enable the animal to increase its energy intake by y_i^* joules.

Furthermore, one can show (S. Altmann 1984) that the total energy value $(\Sigma_a d_{aj} F_j^* y_a^*)$ of the fees for each food's contribution to toxin tolerances plus the total energy value $(\Sigma_a d_{aj} F_j^* z_a^*)$ for that food's contribution to nutrient requirements must exactly equal the actual energy $(e_i F_j^*)$ that is obtained from that food. Put another way, the energy obtained from a food cannot exceed the total energy fee for that food's toxins and nutrients. As Milton Friedman (1975)

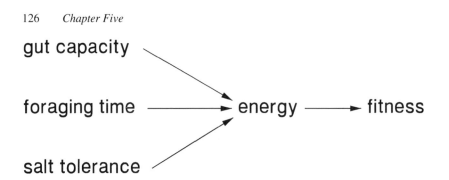

Fig. 5.2. Relations between marginal values.

put it, "There's no such thing as a free lunch." So, in an energy-maximizing diet, the marginal values are *energy exchange rates,* and similarly for other objective functions.

The evolutionary importance of these energy exchange rates is this. For an energy-limited animal, we can in principle find a function that converts energy units into fitness units and thus measure the intensity of selection on the energy yield of the diet. The energy exchange rates, when combined with this function, enable us to do the same for any traits of the organism, such as its toxin tolerances or nutrient requirements, that constrain the energy yield and that thereby affect fitness indirectly (fig. 5.2). Suppose, for example, that the reproductive success of baboons is energy limited, such that each increment of one joule in the diet results in v additional offspring. Suppose also that the energy content of their diet is constrained by their salt intolerance, because the kidneys cannot pump salt out of the blood any faster, and that the marginal value of the salt intolerance is w calories for each milligram per day increase in salt tolerance. If a mutation allowed the kidneys to excrete an additional u milligrams of salt per day and had no other effect, then mutant-bearing individuals could produce uvw additional offspring. If the new mutants differed not only in having more active kidneys but also in having a smaller gut capacity, and if the optimal diet were also constrained by the (old or new) gut capacity, the conjoint effect on energy and thus on fitness could also be calculated.

Beyond that, the marginal values indicate the relative sensitivity of dietary energy, and thereby of fitness, to alternative modifications in the phenotype. If, for example, salt tolerance has a small marginal value per gram but oxalic acid tolerance has a large marginal value, then a small improvement in the organisms' ability to tolerate oxalic acid would have a more marked effect on the energy content of the optimal diet than would a comparable change in salt tolerance.

So the marginal values provide a paradigm for measuring the potential for selection on traits that affect survival and reproductive success only indirectly, via a network of relationships with other traits, and for ascertaining the consequences of a change in any such trait.

The marginal value of a constraint holds only within some interval on either side of its assigned value, and that leads to our next question:

5. Marginal value ranges for requirements and tolerances: How much can each requirement M_i or tolerance T_i be increased or decreased (all others remaining unchanged) without altering the corresponding marginal value?

As I indicated above, after equation (5.1), constraint equations are by convention written with the minima and maxima (e.g., for nutrients and toxins) on their right-hand side. The range in a *right-hand-side value* over which its marginal value holds is bounded by its *allowable increase* and *allowable decrease*. For nonlimiting factors, the range's value on one side will be the slack value, and it will be unlimited on the other (infinity), so no additional information is conveyed.

So far I have discussed ways that feeding time and energy (protein) intake are affected by errors or changes in the right-hand values of the constraint equations, that is, in the requirements and tolerances. Yet surely the optimal mix itself, and thus its yield of protein, energy, or whatever, depends quite directly on the energy (protein, time) densities of the various foods, represented by the coefficients of the objective function. Those values too are subject to experimental error and to natural variability that could form the basis of selective foraging. We therefore ask:

6. Energy, protein, or time leeway: How much could the energy (protein, time) density of a food change or be in error without changing the composition of the optimal diet, that is, the amounts $F_1^*, \ldots, F_n^*$ of the various foods in the optimal diet?

For an optimal diet that maximizes energy, consider the energy density (J/g) of each food, and similarly for optimal diets with other objective functions. In the economic literature, such values are referred to as "cost coefficients" and are labeled as such by linear optimization programs, because minimizing total cost is a common objective in business and the cost coefficients would be the unit costs of, say, the raw materials. (Where the objective is to *maximize* some function, as in an energy- or protein-maximizing diet, they are actually benefit coefficients.)

For each food's energy density, linear optimization programs provide a pair of values commonly labeled "cost coefficient range" but that I shall refer to as its *energy leeway* (and similarly, *protein leeway, time leeway*). Each such leeway is bounded by an *allowable increase* and an *allowable decrease*. The answer to the sixth question above is provided by these leeways. So long as, for each food, the energy density, protein density, or harvesting time requirement (depending on objective) stays within its leeway, the amount of each food that is called for in the optimal diet will not change, nor will the slack values. Of course, with different coefficients, the objective function—the total energy content, protein content, or harvesting time requirement of the diet—will be increased or decreased accordingly. On the other hand, if the energy/protein/time density of any food is changed or is in error by an amount that equals or exceeds its leeway, then in a *nondegenerate* case (see below) the optimal diet mix will be different.

A closely related question is this:

7. Energy or protein shortfall; feeding time excess: How much would the energy (protein) density of a food need to increase or its harvesting time requirement need to decrease before that food would be included in the optimal diet?

Of course for a food already in the optimal diet, the answer is zero. For every other food, the answer in the *nondegenerate case* (see below) is indicated by the so-called *reduced cost*. (In the degenerate case, the reduced cost of a food sets a lower bound on the increase in energy or protein density or decrease in unit harvesting time that would be needed for that food to enter the optimal diet.) Thus all foods that are included in the optimal diet have a reduced cost of zero. For all others, it is the same as the allowable increase in the cost coefficient in maximization problems and the allowable decrease in the cost coefficient requirement in minimization problems.

8. Degeneracy: Is the solution to the optimal diet problem degenerate?

Under unusual conditions, the number of binding constraints at the optimum, those with zero slack, is greater than the number of foods in the optimal diet, or equivalently, the total number of constraint equations exceeds the number of foods in the optimal diet plus the number of positive slack values. If so, the solution is said to be *degenerate*.

For example, if an energy maximizing optimal diet contains just two foods and any further increase in energy is simultaneously limited by three constraints (say, gut capacity and two toxins), then the optimal diet is said to be degenerate. This can be illustrated in figure 5.1, where the diet includes just two foods and correspondingly the constraint equations for the two toxins *A* and *B* pass through the optimum point. So each of the two toxins has zero slack, and on this optimal diet an animal would consume the most that it could tolerate of these two toxins, but the optimal diet is not degenerate. Suppose, however, that the equation for gut capacity passed through that same point. It too would have zero slack, and the same two-food combination would still be optimal; but on that diet the animal would now be at its limit for three toxins: with two foods and three toxins consumed at their limit, the optimal diet is said to be degenerate.

In a degenerate case, if one of the binding constraints is relaxed or removed entirely, say, because the animal becomes immune to one of the toxins, the optimal point does not necessarily remain unchanged, contrary to what one might think by analogy with the number of lines needed to determine the locus of a point on a plane. That this is so can be seen in figure 5.1. If the graph line for the constraint equation of gut capacity lay strictly between the two toxin lines that are shown, and the animal became immune to the effects of toxin *A*, the new optimal point would lie at the intersection of the gut capacity line and the *y*-axis.

Of course the epithet "degenerate," carried over from algebra, is not a comment on the nutritional value of the diet. Degenerate solutions seem to have limited biological importance except perhaps this: the greater the number of binding constraints, the smaller the chance that relaxation or elimination of one selected at random will result in a better diet, that is, one with a higher objective value.

The leeway values, question 6, enable us to answer another important question:

9. Uniqueness: Is the optimal diet unique?

A nondegenerate solution is unique unless the cost coefficient (the energy, protein, or time density) of some food has a zero allowable increase or decrease, in which case some inequality constraint with a zero marginal value will have zero slack or a food not called for in the diet will have a reduced cost of zero. If a unique optimal diet exists, it will lie at a corner point of the multidimensional polyhedron defined by the constraint equations. Even if not, a corner point will always be among the equivalent solutions.

The biological significance of optimal diet uniqueness is this. If the solution is not unique, then various linear combinations of two or more foods actually or potentially in the optimal diet (i.e., those with zero allowable increase or decrease) could be included and the diet would have the same energy yield (protein yield, time cost). On the other hand, if the optimal diet is unique, then any other combination of the available foods would be suboptimal: it would provide less energy or protein or require more harvesting time, would fail to satisfy one or more nutrient requirements, would exceed one or more (toxin) intolerances, or would have some combination of these defects.

A Quick Guide

I now summarize this terminology by looking ahead and providing an interpretation of diet 1, the first approximation to an energy-maximizing diet, shown in table 5.2. In the table, the core foods are ordered by their contribution to the time budget; their four-letter codes and numbers are as in table 6.2. *Intake* refers to the amount of that food, in grams per day, in the optimal diet. So this diet would contain 735 g of milk (PCMX), 201 g of fever tree gum from the tree (FTGX), 66 g of dry fever tree seeds from the ground (FTDG), and nothing else. It would provide the animal with 5.75 MJ of utilizable energy per day, as indicated in the table's note by the *objective function value,* and would also satisfy the animal's requirements for all other macronutrients and all dietary maxima that are specified in table 5.3—rather surprising for such a simple diet.

The *reduced cost* (third column) tells us how much the energy density (concentration) of a food would have to be increased before it could be included in the optimal diet, and indeed, would be included, so long as the solution to our equations is, as in this case, not *degenerate* (see below). For grass corms (food 2) the reduced cost of 0.004 MJ/g, so if corms were richer by at least 0.004 MJ/g (and nothing else were altered), then corms would be part of the optimal diet. Of course, foods 1, 4, and 12 have reduced costs of zero, since they are in the optimal diet. (Other foods with zero reduced cost will be discussed below.)

For each food's energy density, an *energy leeway,* with an *allowable increase* and *allowable decrease* indicates how much (how many MJ/g) the energy density of each food could vary above or below the estimated value I used (table 5.1) without affecting the amount of each food in the optimal diet. So, for example, assuming that you could change just the energy density of milk, its energy content could increase by as much as 0.011 MJ/g or decrease by 0.0005 MJ/g and the optimal diet would still be as shown under *Intake.* Because these energy leeways for milk represent changes of $+329\%$ and -16% in the tabulated values, ordinary laboratory errors in determination of milk's energy content are unlikely to alter the optimal diet. The energy density of the second food, grass corms (GRCX)—which is not in the diet—could be increased by 0.004 MJ/g (by 39%) or decreased without limit, without altering the amount of each food in the optimal diet.

Is the diet *unique?* No: several foods or potential foods have a zero *allowable increase* or *allowable decrease* in their energy density, and if even one such is zero in a nondegenerate case, the solution is not unique. Also, it would not be unique if some constraint had zero values for both its marginal value and its slack.

These zero values, combined with the energy densities of foods (table 5.1), enable us to tell what trade-offs can be made. In this diet, trade-offs, gram for gram, are possible among three sources of fever tree gum (foods 4, 15, 32, and 34) as well as unidentified items picked from logs, probably also gum (food 23). These are foods with a zero *reduced cost,* yet they are not called for in the optimal diet. Gums from all of these sources are thought to have the same energy density (table 5.1) and so from that standpoint are interchangeable, gram for gram. Beyond that, none would aggravate the binding constraints. For the same reason, dry fever tree seeds from the ground and from dry pods (foods 12 and 14) are interchangeable in this diet.

Next, in a separate subtable of table 5.2, I provide information about the nine nutrient and toxin constraints. First, under *Slack,* we see that the number of constraints with no slack (three) is not greater than the number of foods in the optimal diet (three), so the solution to our equations is *nondegenerate.*

The slack values are zero for the baboons' consumption capacity, their second water requirement (for renal clearance), and their total mineral requirement. These zero-slack values indicate that a baboon eating the optimal diet would be eating as much food as it could (1002.4 g/day) and that it would barely be meeting its total mineral requirement (1.73% of food mass) and second water requirement (a function of protein and mineral content). These three constraints, but no others, are what limit the diet to 5.75 MJ/day. They are the binding constraints.

For all other constraints, the slack values in diet 1 are non-zero and indicate a safety margin for that constraint, that is, how far above (or below) the baboons' limit each nutrient (or toxin) would be if the optimal diet were consumed. For example, the slack of 22.60 g/day for the protein requirement indicates that the optimal diet provides 22.60 g more protein than the baboons require. Since their requirement is 3.94 g/day, the optimal diet would provide 26.54 g protein per day.

To the right of each slack value, a *marginal value* is given, then information about the *right-hand side* of each constraint equation: its *current value* (what I used to calculate the optimal diet), along with its *allowable increase* and *allowable decrease.* The marginal values are non-zero only for the limiting food components (the binding constraints, those with zero slack) and indicate the sensitivity of the objective function to changes in that constraint. For example, for consumption capacity, the marginal value is 0.00574 MJ (5.74 kJ) per gram of capacity. This indicates that if the animal could somehow change its consumption capacity, then for each gram per day increase in capacity it could, by a suitable modification of the optimal diet, gain 5.74 kJ energy per day without violating any of its other constraints. The upper limit (*allowable increase*) on this exchange rate is infinite: if unhampered by other constraints, any increase in capacity would increase energy correspondingly, that is, at the rate of 5.74 kJ per gram increase in capacity, unhampered by any other constraint. (If the diet mass increased without limit, with all foods in the same ratios as in diet 1, wouldn't the ash maximum eventually be exceeded? No. According to table 5.3, and as reflected in the corresponding constraint equation, the ash maximum is a constant *percentage* of the diet mass.) Conversely, if consumption capacity decreased, energy would decrease by 5.74 kJ per day for every gram decrease in capacity; but in this case this exchange rate holds only down to 650.7 g capacity below the current value (1002.4 g, table 5.3), as indicated by the *allowable decrease,* that is, down to 351.7 g.

For nonlimiting food components, the *right-hand-side range* adds no new information: it is always bounded on one side by the slack value and is unlimited on the other. For example, in this diet the protein requirement, which is nonlimiting, could increase by 22.60 g/day before it affected the energy content of the optimal diet, but it has no lower limit. Since the optimal diet already provides 22.60 g more protein than required, this is just what one would expect.

The computer program I used (LINDO) follows the convention that if the marginal value of a constraint is positive, an increase in the right-hand side of the constraint equation *improves* the optimal value of the objective function (increases it for a maximization problem, decreases it for a minimization problem) and *depreciates* the optimal value if it is negative. (This requires that upward-limiting constraints, such as toxins, be entered as inequalities of the form "<" or "≤," and the opposite, ">" or "≥," for downward-limiting constraints such as nutrients. This can be accomplished by multiplying constraint equations by -1 where necessary.)

So, for example, the negative marginal value for the mineral requirement ("Ash minimum") in diet 1 indicates that increasing its right-hand side (requiring more minerals in the diet) would *decrease* the energy value of the diet, as one would expect. In particular, for every gram per day increase in the mineral requirement (currently 1.73% of diet mass), the energy content of the optimal diet would decrease by 0.06 MJ per day, and that exchange rate would hold for increases in the mineral requirement up to 1.82 g/day and decreases by as much as 5.05 g/day. Since the optimal diet has a mass of 1002.4 g, this range is equivalent, in percentage, to an allowable increase of 0.18% and an allowable decrease of 0.05%.

If the objective function is a rate, then each marginal value is an instantaneous rate of change between the corresponding constraint and the objective, not an exchange that remains valid within some range of right-hand values. So, for models based on maximizing or minimizing a rate, no allowable right-hand ranges are given. Therefore, for the four models involving maximizing intake rates of energy and protein—models 4, 5, 9, and 10—the corresponding tables (tables 5.7, 5.8, 5.12, and 5.15) do not include right-hand-side ranges.

Optimal Diets in Amboseli

Objective Functions

In chapter 1, I summarized evidence, admittedly inadequate, suggesting that the Amboseli baboon population is primarily regulated by changes in its food supply, but just what components of food alter the survivorship and reproductive success of the baboons? Is their fitness protein limited? Is it energy limited? Toxin limited? Are the baboons limited by time available for other, nonfeeding activities and are they therefore feeding-time minimizers? Unfortunately I do not know, although several studies of wild primates, particularly baboons, have suggested that some vital activities may be time limited (Bronikowski and J. Altmann 1996 and references therein). Indeed, determining whether an animal is, say, an energy maximizer or a foraging time minimizer is no simple task (Hixon 1982). White (1978) suggested that a relative shortage of nitrogenous food for the very young is the single most important limiting factor for most animals. On the other hand, a common assumption in ecology, particularly in the area of optimal foraging theory, is that populations and fitness are energy limited. As early as 1925, Lotka proposed that evolution maximizes energy uptake and metabolism. "The net effect is to maximize . . . energy flux through the system of organic nature." Townsend and Calow (1981) wrote that "the input of energy to organisms must be limited and those organisms which optimize their use of energy should also optimize their fitness." (This is a dubious argument, because [1] it applies as well to all other essential food components, [2] showing that a nutrient is limited—that is, not always available or at least, finite—is not equivalent to showing that it is limiting, and [3] optimizing the use of energy surely includes taking in that level of energy that maximizes fitness, so that the claim is true by definition.) Van Valen (1976) proposed as a general law of evolution that natural selection, at any level or time scale, maximizes energy available for growth and reproduction. Despite these various claims, empirical evidence for the universality of either protein or energy as limiting factors is sparse.

Of course, limiting factors are not unique: a population can be simultaneously limited by more than one factor, which simply means that an increase in any of them lead to increases in the population. One is therefore tempted to try to model diets that maximize all of them (or minimize them, as appropriate). "Find the diet that minimizes feeding time and simultaneously yields energy and protein at the highest rates," and so forth. Unfortunately, no general solution exists to the problem of simultaneously maximizing (or minimizing) several functions. The diet that maximizes energy need not be one that maximizes protein or that minimizes feeding time. I shall take up this problem again later in the chapter (comment 9, below).

What we require for an objective function is some beneficial (or hazardous) food component such that fitness increases (decreases) monotonically with intake of that component, at least within the range of diets that are otherwise adequate. (As before, by a food component I mean not only chemical components but anything that is a function of the mass of each food in the diet or the time expended on it, such as energy, exposure to the sun or to predation, and so forth.) For most nutrients, it has been argued that increases above some threshold value, the *requirement,* are of no value, but evidence suggests this is not true of proteins (see appendix 1), and current ecological wisdom suggests that, within the range of otherwise adequate diets, many animals are energy limited. For the naturally occurring hazards associated with foods, the *dose-response* relationships are at present largely unknown. Clearly, at this early stage in the research on feeding ecology, we cannot say which of the potential objective functions best describes the "objective" of natural selection. Under the circumstances, my approach has been to single out the diet constituents that seemed a priori to be the three most likely candidates: energy, protein, and time, and to use them to write five alternative objective functions for an optimal diet, namely:

1. maximize daily energy intake;
2. maximize daily protein intake;
3. minimize daily feeding time;
4. maximize energy intake per minute of feeding;
5. maximize protein intake per minute of feeding.

(Later in this chapter I shall have more to say about several of these objective functions.) In each case, these are expected values. Explicitly, for each of the above, respectively, an objective function was written in the following way.

Consider objective functions $G_1, \ldots, G_5$:

(5.3) $$G_1(F) = \Sigma\, e_j F_j$$

(5.4) $$G_2(F) = \Sigma\, p_j F_j$$

(5.5) $$G_3(F) = \Sigma\, (F_j/B_j C_j)$$

(5.6)
$$G_4(F) = \frac{\Sigma\, e_j F_j}{\Sigma\, (F_j/B_j C_j)} \quad \text{(pooled mean; time-weighted mean)}$$

$$G_4'(F) = \frac{\Sigma\, e_j B_j C_j F_j}{\Sigma\, F_j} \quad \text{(mass-weighted mean)}$$

(5.7)
$$G_5(F) = \frac{\Sigma\, p_j F_j}{\Sigma\, (F_j/B_j C_j)} \quad \text{(pooled mean; time-weighted mean)}$$

$$G_5'(F) = \frac{\Sigma\, p_j B_j C_j F_j}{\Sigma\, F_j} \quad \text{(mass-weighted mean)}$$

where for each food j, A_j is its time budget (minutes per day spent eating food j); B_j is its unit intake rate (units consumed per minute of feeding on food j);

C_j is its unit mass (g/unit); $F_j = A_j B_j C_j$ is the mass of food j in the diet (g/day); $e_j = D_{ej}$ is its energy concentration (J/g); and $p_j = D_{pj}$ is its protein concentration (grams of protein per gram of food). For G_1, G_2, G_4, and G_5, find values of $F_1, \ldots, F_n$ that maximize it; for G_3, find values of $F_1, \ldots, F_n$ that minimize it.

In each case, the objective function is taken to be a mean (over days, across individuals, or whatever, as needed). As indicated in chapter 3, the product of the means of random variables is equivalent to the population mean of individual products if the random variables are independent; such independence is here assumed.

Objective functions 3–5 above involve food-specific mean values for each B_j (unit intake rates) and each C_j (unit weights). Because $F_j = A_j B_j C_j$, any choice of diet, $F_1, \ldots, F_n$, at specified values of the B_j's and C_j's is simultaneously a choice of the amount of feeding time A_j that will be devoted to each food.

The fourth and fifth objective functions warrant further discussion. That the time-weighted mean G_4 and the mass-weighted mean G_4' are not equivalent is algebraically clear, but the following example may clarify the implications of their differences. Consider an animal that feeds on just two foods, one (say, brie) that is rich in energy and easy to consume, the other (grass) that has a low energy density and, because of its toughness, can be consumed only slowly. Suppose the animal quickly eats a large quantity of brie, and then, perhaps to balance its diet, it slowly consumes a small quantity of grass. An advocate of time-weighted averages would say that the animal's average energy intake rate is low. It spent most of its time feeding on grass, which has a low yield rate, and little of its time feeding on brie, the food with a high energy yield rate, so its mean yield rate is only slightly better than the grass yield rate. Conversely, an advocate of mass-weighted averages would say that the animal did very well. Most of the food it ate, brie, was taken in at a high energy yield rate, so that gram for gram (bite for bite) it did very well: on average, only a little below the mean rate for brie. Can these two methods of evaluating the mean lead to different diets? Yes, for if a third food were available whose energy yield rate and consumption rate both were intermediate in value to brie and grass and whose nutritional value were equivalent to the above brie-grass mixture, a diet consisting of just the intermediate food would be preferable to the mixture if the time-weighted mean were being maximized, but the brie-grass mixture would be preferred if the mass-weighted mean were being maximized. That is, the time-weighted mean but not the mass-weighted mean is very sensitive to even small quantities of *time-consuming foods,* those energy-poor foods that can be eaten only slowly. Conversely, the mass-weighted mean but not the time-weighted mean is very sensitive to *capacity-consuming foods,* those energy-poor foods that are consumed in large quantities, say to satisfy a particular nutrient requirement. Furthermore, maximizing the time-weighted mean G_4 does not always result in a diet of less mass than maximizing the mass-weighted mean G_4', and conversely, the mass-weighted mean does not always result in a more time-consuming diet. So the choice between the two weightings cannot be resolved by deciding whether the animals are time-limited or capacity-limited. I do not know which of these weighting procedures is preferable, that is, which weighting is more important to the animals, nor have I seen this problem addressed in the literature. Somewhat arbitrarily, I have decided to maximize the second, mass-weighted mean rates, in that, if baboons are intake-rate limited, their problem more likely is one of finding the best allocation of their consumption capacity rather than their feeding time.

Maximization of rates, such as is required for the fourth and fifth objective functions, presents a special problem. Among the objective functions defined above, neither G_4 nor G_4' (and similarly, neither G_5 nor G_5') is a linear function of the diet (the F's), so neither can be maximized by standard linear optimization techniques. One might assume—as indeed I did at first—that the maximal value of G_4 can be obtained by maximizing its numerator (which is G_1), then dividing by the time required to consume that diet, and similarly for the numerators of G_4', G_5 and G_5'. Unfortunately, this assumption is false. The problem of maximizing objective functions that are rates was solved by Michael Altmann (appendix 6); on that basis I used the procedure described in appendix 7. For earlier work on this problem, see Charnes and Cooper (1962), Wagner (1969), and Bitran and Novaes (1973).

For simplicity, I used only the core foods in calculating the optimal diets, thus requiring the optimally foraging baboon to meet its requirements from just those. Two of the fifty-two core foods (wood: SPWX and XXWX) were deleted because, so far as I know, they have no nutritional value and were merely gnawed on by the young baboons. The remaining fifty core foods accounted for 94.66% of the baboons' feeding time. Calculations of optimal diets were carried out by means of an optimization program, LINDO (Schrage 1981).

Constraints

Now let's consider characteristics of the baboons that may limit their energy intake, protein intake, or harvesting time.

Let F_j be the consumed amount (g/day) of the jth food, and let w_j (the jth food's water coefficient) be the proportion of water in that food. Then in any diet with total mass $\Phi = \Sigma_a F_a$, the water content is $w = \Sigma_a w_a F_a$ g per day, and similarly for each of the other nutrients or nutrient classes. In what follows, I shall repeatedly use the following designations for these total quantities per diem of various macronutrients in a specified diet: w = water, a = minerals ("ash"), l = lipids ("fat"), p = protein, f = fiber, c = other carbohydrate,

e = energy. Correspondingly, the coefficients (amounts per gram) of these food components for the *j*th food will be designated w_j, a_j, etc. Throughout this section on optimal diets, energy will, except where noted, be scaled to megajoules (MJ)—hence e is in MJ/day—and other macronutrients will be scaled to grams except where noted, so that w, a, l, p, f, and c are in g/day.

As can be seen in equations (5.1) and (5.2), calculating an optimal diet requires values for nutrient requirements, toxin tolerances, and food compositions. Contrary to common belief, the nutrient requirements and toxin tolerances of primates—humans and otherwise—are not well established, so one cannot simply refer to a standard source for accurate quantitative values of these essential parameters. Estimates of nutrient requirement and toxin tolerances, and the basis for the values I adopted, are too much of an aside to intrude on the purpose of this chapter, so they have been relegated to appendix 1.

Letting M be body mass (kg), G be consumption capacity (g/day), and $\Phi = \Sigma F_a$ be diet mass, we can use the values given in tables 5.3 and A1.1 to write the following constraint equations for the macronutrients: (1) $w(1) \geq 120M^{0.84}$, (2) $w(2) \geq 4.79p + 32.58a$, (3) $w(3) \geq 22.6 + 100.2M^{3/4}$, (4) $a \geq 0.0173\Phi$, (5) $l \geq 0.8M^{3/4}$, (6) $p \geq 1.14 + 1.51M^{3/4}$, (7) $f \geq 0.0035(\Phi - w)$, (8) $c \geq 0$, (9) $e \geq 0.419M^{3/4}$, (10) $w \leq G$, (11) $a \leq 0.131\Phi$, (12) $l \leq G$, (13) $p \leq G$, (14) $f \leq G$, (15) $c \leq G$, (16) $e \leq 1.45$).

Some simplification is possible. Constraint 8 is trivially true and can be omitted. Next, note that intake of every macronutrient class except minerals and energy is limited only by total consumption capacity, and that each of these (constraints 10, 12–15) is a special case of $\Phi \leq G$, in that, so long as total food intake Φ is below daily consumption capacity G, so must be the intake of each food component that has mass. Furthermore, the energy maximum (constraint 16) says only that the energy content of the diet is at most what would be obtained if only the most energy-rich food were eaten, and that food, to capacity; that constraint too is a special case of $\Phi \leq G$. Consumption capacity (G) was estimated above to be 1002.4 g/day. Finally, for baboons that are at the mean age in my samples, 47.5 weeks, and that have grown at an average rate of 4.5 g/day, body mass is, on average, 2.273 kg, in which case the right sides of constraints 1 and 3 are 239.2 and 208.1 g/day, respectively—that is, at this body mass, constraint 3 is a special case of constraint 1 and can be omitted.

This leaves nine constraint equations for the macronutrients and total diet mass, the constraints of diets 1 to 5. Several of these include one or two variables on the right side of the equation, whereas the linear optimization program that I used, LINDO, requires that there be only constant terms on the right side. In addition, units of measurement had to be scaled in such a way as to keep within LINDO's range: seven orders of magnitude. The scaling and rearranging of the nine constraint equations resulted in the following constraints on the diet of baboons at the mean sample age and mass of my subjects.

1. First water requirement, for turnover: $w(1) \geq 239.2$, scaled g/day.
2. Second water requirement, for renal clearance:
   $\Sigma(w_j - 4.79p_j - 32.58a_j)F_j \geq 0$, scaled g/day.
3. Mineral minimum: $\Sigma\ (a_j - 0.0173)F_j \geq 0$, scaled g/day.
4. Lipid minimum: $l \geq 1.48$, scaled g/day.
5. Protein minimum: $p \geq 3.94$, scaled g/day.
6. Fiber minimum: $\Sigma[f_j - 0.0035(1 - w_j)]F_j \geq 0$, scaled g/day.
7. Energy minimum: $e \geq 0.871$, scaled MJ/day.
8. Mineral maximum: $\Sigma(a_j - 0.131)F_j \leq 0$, scaled g/day
9. Consumption capacity: $\Phi \leq 1002.4$ g/day, scaled dekagrams/day for calculations, then converted to g/day for presentation herein.

Note that optimal diets were calculated for baboons of mean (not minimal viable) growth rate, 4.5 g/day (appendix 1). Although animals of minimal size require less food, they can hardly have been consuming optimal diets in Amboseli, where much of the variability in growth rate probably is due to differences in diet. Whether baboons in Amboseli on an optimal diet—optimal in terms of fitness in that habitat—would grow at 4.5 g/day is unknown.

The constraints and objective functions of diets 6–10, which are described below, required several additional rearrangements or scaling considerations, as follows.

10. Energy intake rate: scaled kJ/minute.
11. Protein intake rate: scaled g/minute.
12. Feeding time: scaled minutes/day.
13. Seasonality/availability restrictions (fifty constraints): scaled g/day.
14. Cation maxima: scaled as percentage for Na, Mg, Ca, K, and P, as parts per 100,000 for Al, Fe, Sr, Zn, Mn, and Cr, and as parts per million for Ba, B, and Cu.

Optimal Diets with Macronutrient Constraints

Optimal Diet 1: Maximize Energy (Macronutrient Constraints)

The reader is forewarned that diets 1–5 are not very realistic, for reasons that will be given presently. Yet these diets are based on assumptions that seemed at the time to be as realistic as I could make them. In the early stages of research on optimal diets, an important reason for calculating them is to examine the implications of our assumptions, rather than to provide a null hypothesis for an animal's feeding behavior. Initially, the optimal diets are primarily tests of our competence, not the animals'. Some of the shortcomings of diets 1–5 were removed in diets 6–10. Thus, diets 1–5 serve an important heuristic function.

As indicated above, diet 1 (table 5.2) would yield 5.75 MJ/day, more than six times the requirement of 0.871 MJ/day, and would accomplish this by combining 735 g/day of milk, 201 g/day of fever tree gum, and 66 g/day of dry fever tree seeds. Energy intake would be limited by consumption capacity, by the second water constraint (for renal clearance), and by mineral requirements. This optimal diet is nondegenerate but is not unique: several sources of fever tree gum can be substituted for each other, on the assumption that they are nutritionally equivalent (table 5.1), as can fever tree seeds either from pods or from the ground, given the same assumption.

Optimal Diet 2: Maximize Protein (Macronutrient Constraints)

If the baboons are protein limited instead of energy limited, what diet should they consume? The answer, given in diet 2 (table 5.5), is 279.3 g of trumpet-flower leaves (TFLX), 314.6 g dung-beetle larvae (DUKU), and 408.4 g *Cynodon nlemfuensis* stolons (CDSX). The yield would be 61.4 g protein. By comparison, diet 1 would provide 44.5 g protein. Conversely, this diet provides only 2.4 MJ/day, whereas diet 1 yields 5.75 MJ/day. Except under extraordinary conditions, a baboon that maximizes its protein intake will get less dietary energy than one that maximizes energy intake, and one that maximizes energy will get less protein. As indicated earlier, no general solution exists to the problem of simultaneously maximizing more than one function: doing so is possible only in unusual circumstances.

Diet 2 is nondegenerate and unique. Protein intake is limited by consumption capacity, by the water requirement for renal clearance, and by the total mineral requirement. The marginal value for the mineral requirement is considerably larger than for any other constraint, that is, the protein content of the diet is particularly sensitive to this requirement, and any changes in this requirement (or errors in estimating its value) would have the largest impact, per unit change, on the protein content. Thus for every one gram per day change in the mineral requirement, the protein content of the resulting optimal diet would change by 5.86 g/day, and that would be the exchange rate for increases of as much as 3.4 g/day and for decreases of as much as 4.8 g/day.

Optimal Diet 3: Minimize Feeding Time (Macronutrient Constraints)

The third diet (table 5.6) is nondegenerate and unique. It contains just two foods: milk (257.7 g/day) and fever tree gum from the ground (FTHG, 82.0 g/day). Amazingly, this simple diet meets all macronutrient requirements (table 5.3) and does so in just 88.7 minutes of feeding per day. Feeding time cannot be further reduced because such a diet would already be at the baboons' lower limit for water (first requirement, for turnover rate) and for minerals. As in diets 1 and 2, the total mineral requirement has the greatest marginal value: feeding time is particularly sensitive to changes in it.

Optimal Diet 4: Maximize Energy Intake Rate (Macronutrient Constraint)

The objective function of diet 4 has a value of 18.95 kJ for each minute of feeding, equivalent to 315.8 watts. Further increases in the energy intake rate are limited by consumption capacity and the water requirement for renal clearance.

Diet 4 (table 5.7), like diet 3, contains just milk (729.1 g/day) and fever tree gum from the ground (273.3 g/day). The solution is nondegenerate: there are three positive food variables (β and two foods) and three constraints with slack values of zero, including the constraint $\Sigma\ y_a = 1$, which is not shown in the table (see appendix 6). The optimal diet is not unique: β has a zero leeway (for increasing), and since no potential food excluded from the diet has a reduced cost of zero, there is some constraint with zero for both its slack and marginal values, namely, consumption capacity. That is, although further increases in energy intake rate are limited by the baboons' consumption capacity, an increase in that capacity would not result in an increase in their energy intake rate. How such a counterintuitive result can occur is nicely illustrated by Eppen and Gould (1979, fig. 9.11).

In diet 4, only one constraint has a non-zero marginal value, the water requirement for renal clearance. Its marginal value, an instantaneous rate of change, is -0.005; that is, at the optimum, the energy harvest rate would decrease at the rate of 0.00506 kJ/minute per gram increase in this daily water requirement. This may seem like a minute effect, but consider this. As indicated above, the water requirement for renal clearance is $4.79p + 32.58a$, where p and a are the amounts of protein and ash in the diet. Those amounts can be calculated from the requirement + slack of protein and the tolerance slack of ash. Requirements and tolerances are given in the text earlier in this chapter; slack values are in table 5.7. The result, 686.01 g/day, is the same, rounding errors aside, as the amount of water in this diet (as can be determined from the first water requirement + slack)—which is as it should be, since the second water requirement is binding. A one gram increase out of 686.01 g is a change of 0.146%, and it would reduce the optimal energy intake rate by 0.00506 kJ/minute, a reduction of 0.0267%, for a 0.183:1 leverage ratio.

Small changes in the other constraints would have no effect whatever on the baboons' energy intake rate: their marginal values are all zero.

In diet 4, as in diets 5, 9, and 10, the objective function is a rate, and so the marginal values are instantaneous rates, valid only at the optimum, not over some range of constraint values.

Optimal Diet 5: Maximize Protein Intake Rate (Macronutrient Constraints)

Like diet 4, diet 5 (table 5.8) is utterly simple: just two foods, in this case milk (384.8 g/day) plus an extraordinary quantity of ripe *Azima* berries (617.6 g/day). That diet is nondegenerate (three positive food variables including β, and three zero-slack values, including that for $\Sigma \; y_a = 1$) but it is not unique: β has a zero leeway. And since no food has a zero allowable change in its protein density, there must be at least one constraint—in this case, for consumption capacity—that has zero values for both slack and marginal values.

Diet 5 would yield protein at the rate of 0.101 g per minute of feeding. Further increases would be limited by the baboons' consumption capacity (the diet is at the limit, 1002.4 g) and by their second water requirement, for renal clearance.

As in diet 4, this water requirement is the only constraint that has a nonzero marginal value, in this case -3.6×10^{-5}, which indicates that at the optimum, the protein intake rate would decrease at the instantaneous rate of only 3.6×10^{-5} g/minute for each gram increase in this daily water requirement. The objective function may seem remarkably insensitive to the one constraint that has any effect on it at all, but consider the following. Diet 5 contains $p = 36.42$ g protein and $a = 19.06$ g ash (as can be deduced from the corresponding requirements plus slack values for these nutrients), and the renal-clearance requirement is for $4.79p + 32.58a = 795.4$ g water per day. As it should be in this water-bound diet, that is exactly the amount in the diet, as indicated by the first water requirement plus its slack, 239.2 g + 556.17 g. Now, a one gram change in this water requirement is a change of only 0.125%, and such a change would alter the protein intake rate by $3.6 \times 10^{-5}/0.1013 = 0.035\%$, a 0.28 : 1 leverage ratio.

As in diet 4, no small changes in any of the other constraints would alter the yield rate: they all have marginal values of zero.

Comment 1: Are Diets 1–5 Realizable?

All five of these diets would satisfy all macronutrient requirements, and each optimizes some objective function based on time, energy, or protein. But do they describe diets that any real baboon could eat? Consider diet 4. It would require 729 g milk per day, a level not achieved by any infant in this study; the closest was Eno, and only at one age, 30–40 weeks (table 6.6, 634.5 ml = 652 g). Diet 4's 729 g milk per day would require appreciable increases not only in nursing time but also in the mothers' energetic costs of producing that much milk and the increased foraging time be required to do so. In view of the high nutritive value of milk and its ease of consumption, it seems reasonable to suppose that milk intake is limited by supply, not demand. For the purpose of subsequent computer runs (computing diets 6 onward), I established an upper limit on milk intake based on the observed intake distribution, a method that will be discussed in comment 2 below.

Diet 4 also requires 273 g fever tree gum from the ground, and diet 3 calls for 82 g/day. However, for this food (no. 15) the mean plus three standard deviations of the observed intakes in ten-week age blocks was 28.2 g/day, the average amount of gum taken in by the baboons from all five gum sources combined was 6.6 g/day (table 6.1), and the sum of their extreme values (table A8.1, col. 6) was 122.6 g/day. Thus, even if infants devote themselves entirely to feeding solely on milk and fever tree gum, they are very unlikely to find 273 g per day of fever tree gum in deposits, and if for a time they should, the available deposits probably would soon be exhausted. My response to the problem of limited availability in this and similar cases will be given in comment 2 below.

Comment 2: Seasonality and Availability Constraints

Diets 1–5 raise the problem of seasonal foods. Several of the foods called for in these diets are available for only a fraction of the year. Even if baboons ate the large specified quantities of seasonal foods when they were available, such foods would inevitably account for less than called for over a year. Indeed, this might happen even if the baboons ate a seasonal food to their gut's limit on every day that it was available. As an example, consider dung-beetle larvae, which according to diet 2 should be consumed at a mean rate of 314.6 g/day. During 1975–76, these grubs were eaten for only about 3/48 of the year (table A8.2), perhaps because at other times of the year the ground was too hard for digging. Even if the baboons ate nothing but grubs during the time they were available and each day at that time filled themselves to capacity, these grubs could at most amount to 3/48 of the total annual consumption capacity, that is, at most (3/48) (365 days/year) (1002.4 g/day) = 22.9 kg per annum, whereas diet 2 calls for (365) (314.6 g) = 114.8 kg per annum, an impossibly large intake by a factor of five! Similarly, the requirement for a year-round average of 385 g/day of *Azima tetracantha* fruit is unrealistic.

For the purpose of calculating diets 6–10, I dealt with the problems resulting from phenological changes in the following way. I set up fifty constraint equations, one for each core food, each of the form $F_j \leq \hat{s}_j$, where F_j is as before the per diem consumed mass of food j and the $\hat{s}_j$'s are seasonally adjusted, food-specific constraints, obtained as follows. I began by calculating, for each core food, the mean plus three standard deviations of the observed actual intakes of the various infants sampled in the four ten-week age blocks, given in table 5.17, twenty-eight intake values per food, one for each infant × age block. (The pooled means, given in table 6.1, differ somewhat from these means of individual means; standard deviations for the former are unknown.) If the intakes were normally distributed, the mean plus three standard deviations would include 99.87% of the population. However, many foods have asymmetric intake distributions, with long right tails: of the fifty-two foods in table 5.1, thirty-nine had at least one extreme value beyond the mean plus three standard deviations.

Clearly, for many foods the mean plus three standard deviations is too low to represent the highest realizable intake, which is needed for the constraint equations. I therefore established, as the (uncorrected) upper limit s_j on the distribution of individual intakes of each core food j, either the largest actual per diem intake of that food by any infant in any ten-week age block between 30 and 70 weeks of age or the mean plus three standard deviations of these intakes, whichever was the larger.

For an intake distribution, the upper limit, like the mean, is subject to a bias that results from differences in the amount of sampling I did when the food was in season and when it was not. I therefore applied to each s_j a correction factor f_j/t_j, where f_j is the fraction of the year that the jth food was available and t_j is the proportion of total sample time that occurred when food j was available. The rationale for using f_j/t_j is given in appendix 8 for means of the distribution, and the same argument holds mutatis mutandis for the mean plus three standard deviations or any specified percentile of the distribution. For each core food j, the value of f_j/t_j is given in table A8.1, as are the values of $\hat{s}_j = s_j f_j/t_j$ that were used to form the fifty constraint equations $F_j \leq \hat{s}_j$, one for each food j. In addition, total per diem food intake ΣF_a was limited to 1002.4 g/day, as described earlier in this chapter.

This approach to the problems that result from phenological changes in food availability is not perfect. While it eliminates the possibility of computing an "optimal" diet that calls for eating two or more foods each to capacity every day of the year, it does not preclude requiring amounts of two or more seasonal foods that could be met only by such feeding during the overlap in their seasons of availability, so long as, on average over the year, the constraints on seasonality and consumption capacity are not violated. One advantage of this method for establishing seasonality constraints on individual foods is that it also places restrictions—I hope realistic ones—on intakes of those foods, such as fever tree gum and milk, that are essentially nonseasonal but can be obtained only in limited quantities.

Comment 3: What Is a Constraint?

Implicit in this method of establishing constraints on optimal diets is an assumption about the type of models I am trying to construct. Suppose, for a given food, that the mean plus three standard deviations of intake is x grams per day, and that I use this value as an upper limit on the quantity that can be in the optimal diet. In so doing, what I am looking for is the highest realizable intake of a baboon in Amboseli. I am saying that real baboons in Amboseli can find and eat x grams of this food per day on average, but no more.

What would be the consequences of using values other than the highest realizable one? Suppose that in calculating the optimal diet I used a value considerably lower than x, say, the mean value of the daily intake distribution. Since for each food about half the individuals in a population will have intakes above the mean, I would be in the embarrassing position of having a large fraction of individuals doing better than the putative maximum! At a cutoff of three standard deviations above the mean, such embarrassments should be very rare and of small magnitude.

Alternatively, suppose I used a value much larger than x for a food, say, $10x$. If, with a limit of x grams, the food had a non-zero allowable increase, then at $10x$ grams, even more would be called for (assuming that the reason no baboon eats $10x$ grams was not incorporated into any other constraint equation), and the resulting optimal diet could not be achieved by any baboon, because it would place impracticable demands on them. Such a dietary specification would say: If only there were a superbaboon, with capabilities beyond those we can expect to find in any actual baboon, here is what it could achieve. From the standpoint of natural selection, however, a more enlightening question is to ask, Given the capabilities that are within the range of actual baboons (albeit at the extremes of their phenotypes), what is the best diet for them to select, and how far from this optimum are various individuals? Similarly, When has the animal exceeded its capacity? What diets, for example, would include toxic quantities of salts? If $10x$ had been used as the upper limit, rather than x, I would be embarrassed to find that as individuals approached the putative optimal diet but well before reaching it, they would die from toxins or other upper limit constraints. A diet that is lethal can hardly be considered wonderful!

In chapter 2, I considered two approaches that have been taken to modeling optimal diets. Here I want to relate that discussion to the constraint equations. In the top-down approach, an animal's foods are arranged linearly according to some scalable criterion of nutritive value, most commonly energy. Thus, in this approach, the animal's task is to decide whether each potential food would yield sufficient energy to be consumed profitably. The alternative approach, which I follow here, is to recognize that even for an energy maximizer, there is more to diet decisions than energy. In my models, the additional information is contained in the constraint equations. With good reason, much attention has been paid to developing appropriate constraint equations for optimality models. For numerous examples, see the survey by Stephens and Krebs (1986).

The difference between models of energy maximization with and without constraints is not merely the difference between high-energy diets with or without a bellyache. The two approaches lead to quite different expectations and predictions. A hypothetical example may serve to clarify why this is so. Suppose an animal leaves a dense patch of high-energy food to feed on another food that requires more work and provides less energy. By the criterion of energy maximization alone, the animal has made a mistake. But perhaps the high-energy food also contains a toxin, and the animal has consumed about as much as it can tolerate; or perhaps the low-energy food contains a vital nutrient that is missing from the energy-rich food. A baboon can eat only so many acacia seeds without regurgitating from the trypsin inhibitor, but even without the presence of this toxin, it would have to eat other foods to get nutrients that are not supplied in adequate amounts by these seeds.

A pure energy maximization model looks upon various foods as alternatives to each other, each better or worse than others as their energy yields dictate, whereas in my models the constraints identify limits on food constituents that produce complementarity among foods. The latter perspective is more consistent with what animals require in the way of nutrients, what they can tolerate in the way of toxins, and the nonuniform distribution of each among potential foods. Faunivores, for which many of the optimal foraging models were developed, are nearly an exception: the tissues of other animals come close to satisfying their nutritional requirements, so they almost escape the need to search for complementary foods. For herbivores or omnivores, however, finding an appropriate mix of complementary foods is the basic diet selection problem.

One further remark about the nature of constraints: they are not just statements about the capacity of the organism or about the habitat in which it lives but are about the relation between the two. If an optimal diet is constrained by, say, salt, this is at once a statement about the saltiness of the available foods and about the capacity of the animal to tolerate and excrete salt. Constraint equations are predictions, and it is not obvious to me that we can, in principle, predict the behavior of a well-adapted animal in a given environment solely from measurements made on the animal outside that environment and on the environment independent of the animal. For example, an animal's ability to tolerate salty foods varies with its access to water, with diuretics in its foods and thus its rate of salt excretion, with its exposure to elevated temperatures, and so on—all of which involve relations between the animal and its environment. What information about a baboon's anatomy, physiology, perceptual and manipulative abilities, and so forth, and about the distribution and other characteristics of dung-beetle larvae would enable us to establish a realistic seasonality constraint for this food and thus to be able to state whether, for example, a baboon could satisfy the requirement of diet 2 for 315 g of larvae per day if its only other foraging requirements were for trumpet-flower leaves and Bermuda-grass stolons in the quantities specified by that diet? Even if, over a short period, the baboons were capable of digging up this quantity of larvae, they probably would deplete their supply well before the next rainy season, when once again the adults emerge, gather their dung balls, and lay more eggs. Animal-habitat relationships have a history, and that history, too, serves as a constraint on current options (Levins and Lewontin 1985).

Furthermore, for gregarious animals, history is bound up in the current state not just of the individual but of its entire social group. Suppose we find some highly nutritious food on the lava outcrops within the baboons' home range that is not eaten by them but that would be a major constituent of an optimal diet. We then watch a yearling baboon who, unlike the other members of its group, goes onto the outcrops to feed on this food. As a result, it is away from the safety of its group and gets killed by a predator—something that would not have happened if all the members of its group had learned about the food and gone onto the outcrop together.

Thus the history of an individual's group, such as whether the group has become familiar with a particular food source, sets limits on the options open to the members of the group at any given time. A course of action that is optimal at one time may not be so at another. Looking for optimal strategies in any absolute sense, based on independent measurements on the organism and its environment, may prove fruitless. We can, however, make a stronger case for a somewhat weaker type of statement: Considering the limits on competence that can be estimated from the past performance of the animal in this environment, here is what it should do to maximize its fitness.

Comment 4: Water Sources

The water constraints, which affected several of the first five diets because they require that water be obtained from the water content of food, probably can be discounted as limiting dietary factors for Amboseli baboons, since they also get water daily from swamps, rain pools, and waterholes. Water has an indirect effect on diet, in that the spatial distribution of water affects the viable day-journey routes of the baboons, and that in turn affects the array of foods available to them on any particular day. Except in unusual circumstances, however, baboons in the study area can obtain adequate drinking water every day. For that reason, I have calculated all diets after diet 5 with the water constraints removed.

Comment 5: Time-Budget Limitations

If a putative optimal diet is calculated without a time constraint, it could require more foraging time than is available to baboons, or even more than 24 hours per day! As it is, baboons in Amboseli spend about 70% of the daylight hours foraging, a figure based on feeding plus moving time in Post (1981, table II). (I include all moving time as a component of foraging because the search for food and water is the only regular activity of baboons that requires any appreciable amount of locomotion.) If we assume that feeding time cannot be increased without a corresponding increase in time spent on locomotion and that the ratio of locomotion time to feeding time is as in Post's annual mean (1981, table II, last row), namely, 0.514:1, then the baboons could not feed for more than 11 hours/1.514 = 436 minutes per day without exhausting all available foraging time (eleven hours per day).[3] (Post demonstrated that the ratio of feeding time to locomotion time varied seasonally, but I am here considering only the annual mean value.) The feeding time required per gram of the *j*th food is $B_j^{-1}C_j^{-1}$ minutes, where B_j is the average unit intake rate of the *j*th food (units per minute of feeding on food *j*) and C_j is that food's average unit mass (g/unit), so the constraint on total hands-on feeding time is $\Sigma F_a B_a^{-1} C_a^{-1} \leq 436$ minutes/day. This time-budget constraint was added to diets 6 onward. Even this time limit may not be strict enough, since it allows no time during those eleven hours for any other vital activities. However, at this latitude, daylight lasts about twelve hours a day, with but small seasonal variations, so this objection is mitigated.

Bronikowski and J. Altmann (1996) provide evidence that the rest time of Amboseli baboons is more responsive to differences in foraging time than is their social time, as follows. By comparison with wild-feeding groups, those in a food-enhanced group spent about 55% as much time foraging. The vast majority of the time freed from foraging activity accrued to resting, not to socializing. The fitness consequences of such trade-offs between time devoted to feeding versus time devoted to other activities is unknown.

Comment 6: Mineral Constraints

I have three comments about minerals. First, the upper-limit constraint on total ash that was used in diets 1–5 (13.1% of the diet) cannot possibly be binding on a diet composed of the core foods, because no single core food has such a high mineral content (table 5.1), and thus no combination of core foods would: diets 1–5 would have been the same if the total mineral maximum had not been included. The mineral maximum was retained in the calculations in order to obtain its slack value.

Second, the total mineral requirement and tolerance I used for diets 1–5 are based on the assumption of an ideal mineral mix, whereas judging by the three core foods for which analyses of individual minerals are now available (table 6.11), daily intakes of individual minerals are not in just those proportions.

Finally, after the section on mineral tolerances (appendix 1) was written and diets 1–5 were calculated, I located a thorough and extensive survey of mammalian mineral tolerances, based on data from domestic and laboratory animals (NRC 1980). As will be shown in chapter 6, the tolerances given in that publication—if they can be extrapolated to baboons and if I can extrapolate from the three Amboseli foods for which elemental analysis has been carried out—suggest that the mean diet of the weanlings was 127% of the maximum for aluminum, 100% for barium, and 62% for magnesium. Intakes of all other analyzed cations were below those values. Of course, seasonal and individual variations occur in diets, as well as in tolerances, and so the possibility of mineral toxicosis must be considered.

Published estimates of cation requirements (Nicolosi and Hunt 1979) and tolerances (NRC 1980) were used to establish upper and lower limits (the smallest RDA value and the largest MTL value) for each cation; values for these cation limits are indicated in table A1.2 of appendix 1.

At present I have elemental analyses for just three foods, albeit very important ones: green leaves of *Salvadora persica,* green seeds of *Acacia tortilis,* and corms of *Sporobolus cordofanus* (respectively, foods 29, 5, and 16). Because I have elemental analysis for only three foods, only the upper limits on cations could be applied to my data. In calculating diets 6–10, constraints were added for those cations for which a quantitative analysis (table 6.13) and a maximum (table A1.2) are available. Of the seventeen cations in table A1.2, I do not have any analysis of three: iodine, selenium, and fluorine. The remaining fourteen elements added fourteen constraints for calculating diets 6–10.

In summary, the constraints for diets 6–10 differ from those used for diets 1–5 in having, in addition to the macronutrient and total consumption constraints, these fourteen upper-limit mineral constraints, the time-budget constraint (comment 5), and the fifty seasonality/availability constraints (comment 2), and in having the water requirements removed (comment 4). This is a total of seventy-two constraint equations—a seemingly insurmountable computational difficulty but in fact handled quite easily by the LINDO linear optimization program. For brevity, this set of seventy-two constraints will sometimes be referred to below as the *extended constraints.*

For diets 6–10, results of linear optimization are given in tables 5.9–5.12 and 5.15. Salient features of these five optimal diets are summarized in tables 5.13, 5.14, 5.16, and 5.18.

Optimal Diets with Extended Constraints

Optimal Diet 6: Maximize Energy (Extended Constraints)

Diet 6 and the four other diets in this series (diets 6–10) contain far more foods than their corresponding predecessors. Diet 6 (table 5.9) contains twenty-five core foods, whereas its predecessor, diet 1, contained only three. Twenty-two of these foods have zero slack, and these binding seasonality/availability constraints are primarily responsible for the diversity of diet 6. This result is consistent with Westoby's (1978) commentary on seasonal shifts as a primary source of diet diversity in omnivores. Three other constraints have zero slack: the baboons' consumption capacity, their total mineral requirement, and their limited time available for feeding. All marginal values are quite small: the diet would change only very slightly per unit change in any of the various minima or maxima.

Diet 6 would yield 4.69 MJ/day, which is about 82% of the energy yield of diet 1. The difference is the energy cost imposed by the additional binding constraints set by limited (seasonal) availability and feeding time. Diet 6 is nondegenerate (twenty-five foods and twenty-five no-slack constraints) and unique (all cost coefficients are non-zero).

Diet 6, for all its diversity, is dominated by a few foods. In particular, milk (668 g/day) would constitute about 67% of the annual diet. Other major components of the diet would be green grass leaves of unidentified species (108.7 g/day*), fever tree gum from the tree (66.7 g/day*), ripe fruit from the bush *Azima tetracantha* (27 g/day*), fever tree gum from the ground (24 g/day*), and green leaves of the grass *Cynodon nlemfuensis* (23 g/day*).[4] Fever tree gum, from its several sources, would make up about 10% of the diet and would account for 26% of the energy. (Here and in what follows, an asterisk indicates that the prescribed amount of a food has a zero seasonality slack and thus the prescribed amount is at the maximum.)

Optimal Diet 7: Maximize Protein (Extended Constraints)

Diet 7 (table 5.10) would include twenty-one foods. Again, milk would be the largest single component: at 630 g/day it would make up 63% of the diet. Green grass blades would, as in diet 6, rank second (109 g/day*). At less than half this value, the third-ranking food would be fever tree gum from the tree (42.4 g/day*). Umbrella tree blossoms picked from the tree (35 g/day*) would be next, then semiripe fruit of the bush *Salvadora persica* (34 g/day*).

Diet 7 would yield 31.0 g protein per day; that value is half the protein yield in diet 2, which included only macronutrient constraints. The difference is due to the added constraints on availability and feeding time. Diet 7, like diet 2, is also limited by the baboons' consumption capacity and their total mineral requirement. Beyond that, their protein intake would be limited by eighteen foods whose seasonality would preclude greater consumption.

One food, fever tree gum from logs, is not in the diet but has a reduced cost of zero and a zero cost coefficient. Thus the diet is not unique: this food is an alternative to fever tree gum from the tree, which also has a zero cost coefficient and which I presume to have the same protein content. There are twenty-one constraints with zero slack and twenty-one foods in the diet, so the diet is nondegenerate. As in diet 6, all marginal values are small. The largest, for the total mineral requirement, indicates an improvement of just 0.36 g protein per day for every gram decrease in the mineral requirement.

The position of acacia seeds in diet 7 presents an enigma. Maximum available quantities are specified for five sources (8.2 g* of food 12 = FTDG, 3.0 g* of 18 = TODR, 1.36* g of 42 = TODG, 0.064 g* of 50 = ACDD, and 7.3 g* of 10 = XXXG)—just what one would expect from their high protein contents (18.7%, 20.8%, 20.8%, 19.1%, and 19.1%, respectively)—yet acacia seeds from three other sources are not called for in any quantity (14 = FTDR, 19 = FTDU and 5 = TODU). The latter two, seeds from the green pods of fever trees and umbrella trees, have lower protein concentrations (10.7% and 5.3%, respectively), primarily because of their higher water content. However, fever tree seeds from dry pods (14 = FTDR) have the same protein content as do loose fever tree seeds picked up from the ground (12 = FTDG), namely, 18.7%; yet only the latter are called for in diet 7. Why? The difference is in processing times. Seeds from dry pods (excluded) took 7.7 minutes per gram of yield, whereas loose seeds picked up from the ground (included) took only 3.5 minutes per gram, and diet 7 is time-limited.

This example nicely illustrates two important points: first, that the value of a food cannot adequately be judged on a single constituent (protein content, energy content, or whatever), and second, that the value of a food is not an independent trait but depends instead on what it offers relative to what other potential foods in the diet have to offer (S. Altmann 1985).

Optimal Diet 8: Minimize Feeding Time (Extended Constraints)

Without the water-from-food requirement that limited diet 3, a baboon could satisfy its other macronutrient requirements in only 50.9 minutes of feeding per day (table 5.11) by eating a combination of just six foods: milk (119 g/day), fever tree gum from the trees (6 g/day) and from the ground (25 g/day*), ripe fruit of *Azima tetracantha* (20 g/day), fruit of *Commicarpus pedunculosus* (3.0 g/day*), and umbrella tree seeds from green pods (1.4 g/day*). Since the baboons average much more than fifty-one minutes of feeding per day, either they or this diet are very much in error.

This diet is nondegenerate but not unique: fever tree gum from logs could be substituted gram for gram for gum from the trees, assuming as before that the gum from the two sources is nutritionally equivalent. Feeding time would be constrained by the limited availability of the three foods marked above with an asterisk and by the energy, protein, and total mineral content of the diet, which are all minimal. Diet 8 is the only one in this series (diets 6–10) for which the baboons' consumption capacity would not be a binding constraint. The slack value for consumption capacity is 828.3 g and the capacity is 1002.4 g, so diet 8 contains 174.1 g. The diet is quite sensitive to changes in the energy requirement: every MJ/day decrease in the energy requirement would (within the narrow range indicated by the right-hand-side range) result in a 42 minute decrease in feeding time.

Optimal Diet 9: Maximize Energy Intake Rate (Extended Constraints)

Diet 9 (table 5.12) would include just nine foods—considerably fewer than in diet 6, which would maximize total energy instead of the energy rate. Diet 9 is nondegenerate (ten zero-slack values and ten non-zero variables: β and the nine foods). It is also unique. In diet 9, any further increase in the baboons' energy intake rate would be limited both by their protein requirement and by seasonal availability limits on those foods (marked with asterisks in table 5.12) that are already called for at their maxima.

Earlier in this chapter, when discussing objective functions, I pointed out that maximizing the *rate* of energy (or protein) intake can lead to ridiculous results. Diet 9 seems to be a case in point. The energy intake rate would be very impressive: its mass-weighted mean would be 20.65 kJ per minute of feeding, equivalent to 344 watts. Fast food, indeed! By comparison, the mass-weighted mean rate of a baboon on diet 6, for maximizing total energy per day, would be slower: 10.77 kJ/minute (179.5 watts). However, diet 6 would provide 4.7 MJ of energy per day, whereas diet 9 would provide only 1.4 MJ per day—less than a third as much. Note the paradoxical nature of this result. A baboon on diet 6, though taking in energy more slowly than one on diet 9, would nonetheless obtain more energy in total. Of course, doing so would require more feeding time. By putting a premium on a high *rate* of energy consumption, diet 9 sacrifices total energy intake. No general solution exists for the problem of simultaneously maximizing both total energy and energy intake rate. I shall have more to say about this problem shortly (comment 7, below).

Diet 9 would involve a very large waste in the baboons' capacity. Note the slack values for consumption mass and feeding time: 888.5 g and 353.6 minutes of slack per day. Since the baboons' upper limits on consumption and feeding time are 1002.4 g and 436 minutes, diet 9 would take up only 113.9 g and would require only 82.4 minutes to consume. That is, a baboon that consumed diet 9 would make use of only 19% of its available feeding time, and the diet would take up a mere 11% of the food a baboon can process in a day. Surely an animal that then stopped feeding would be making a mistake, unless feeding was a particularly hazardous activity—which is not the case with baboons.

Such an animal has eaten fast, but has it eaten well? I think not. An animal that consumed diet 9 and then stopped would be giving up much energy, protein, and other potentially fitness-limiting nutrients for the sake of speedy consumption. Diet 9 would provide 1.4 MJ per day, just 30% of what the baboons could obtain from diet 6. For protein, slack in diet 9 is zero: the animals on this diet would just barely meet their daily protein requirement, 3.99 g, whereas even an *energy*-maximizing diet (diet 6) would provide them with more than seven times that amount of protein and abundant quantities of other nutrients as well.

What are the foods that make up diet 9's quick-energy bonanza? Primarily, a lot of fever tree gum: from the trees (66.7 g/day*), from the ground (24.2 g/day*), and from logs (8.2 g/day*), all at their limits of availability. The diet would also include a lot of acacia seeds: umbrella tree seeds from dry pods (2.5 g/day), individual umbrella tree seeds picked up from the ground (1.4 g/day*), acacia seeds picked from dung (0.06 g/day*), and individual items, presumably mostly acacia seeds, from the ground (7.3 g/day*). Finally, the diet would be completed by maximum available intakes of insects (0.6 g/day*) and sticky-fruit (3.0 g/day*).

The inclusion of acacia gum and seeds, insects, and sticky-fruit in this diet are not surprising. Energy intake rates for these sources of fever tree gum range from 20.6 to 28.3 kJ/minute, and the latter is the highest energy yield rate of any core food (table 6.10). Many of the acacia seed sources too have high energy yield rates. They range from 3.2 to 11.1 kJ/minute. The rate for sticky-fruit is 12.0 kJ/minute, that for insects is 3.4 kJ/minute. What seems surprising is the exclusion of certain foods that have comparably high energy yield rates. The most extreme case is milk: 12.8 kJ/minute (second approximation, table 6.10), yet not included. Other high-energy omissions include green grass blades (8.3 kJ/minute) and the fruit of *Azima tetracantha,* ripe (9.5 kJ/minute), of unknown ripeness (assumed to be 8.1 kJ/minute), and unripe (4.9 kJ/minute).

All together, five foods in the diet (nos. 10, 18, 37, 42, and 50) yield energy more slowly than at least one excluded food (table 5.13). Why, then, are these five included in diet 9? The answer is, they are included primarily to meet dietary requirements other than for energy. In particular, the diet is protein-limited, yet proteins cannot be added from time-consuming foods without compromising the average energy intake rate. Among core foods, numbers 42 and 37 have the two highest protein yield rates (table 5.14).

That leaves three foods, 10, 18, and 50, whose presence in the diet has not yet been justified. While each has a high protein and energy yield rate, each is exceeded in one or the other by at least one excluded food. The outstanding merit of these three foods is their fiber content, which is among the highest (table 5.3) and is not exceeded by the fiber content of any food that yields either protein or energy faster. By including small amounts of these three foods, the baboons readily satisfy their low fiber requirement (table 5.3). Then too, perhaps none of the alternative foods would yield energy at high rates without violating some constraint that is not binding with the present diet.

As I mentioned, diet 9 would have the baboons at their lower limit for proteins: the protein requirement has zero slack. How much could a change in this requirement alter their energy yield rate? The marginal value indicates that per gram increase in the protein requirement, their energy harvesting rate at the optimum would decrease by 0.737 kJ/minute. The latter would be a shift of 3.57%, the former, 25.4%, so the leverage on the energy harvesting rate exerted by the protein requirement is 0.14:1. (Note in appendix 6 that since the objective function of diet 9 is a rate, its marginal values are instantaneous rates, valid only at the optimum.)

Optimal Diet 10: Maximize Protein Intake Rate (Extended Constraints)

This diet is even simpler than diet 9: it contains just eight foods (table 5.15). From the protein yield-rate ranking presented in table 5.14, the ingredients are what one would expect, including green grass blades, by far the largest component and consumed to their limit (108.7 g/day*); *Azima tetracantha* fruit, both ripe (27.2 g/day*) and of unknown ripeness (10.4 g/day*); insects (0.6 g/day*); and of course acacia seeds from various sources: loose fever tree seeds and small unidentified objects, presumably mostly acacia seeds, picked up from the ground (6.1 g/day and 7.3 g/day*), loose umbrella tree seeds from the ground (1.4 g/day*), and acacia seeds picked out of mammalian dung (0.1 g/day*). This diet is unique and is not degenerate.

The diet would yield protein at the rate of 112.6 mg per minute of harvesting (mass-weighted mean rate). Further increases would be limited by the baboons' energy requirement, which this diet would barely meet, and by limits on the seasonal availability of seven of the eight foods. The marginal value for the energy requirement is 0.0254 g protein per minute of feeding for each unit change in the energy requirement. That is, a 22.6% decrease in the protein harvesting rate would result from an increase of 1 MJ per day (114.8%) in the energy requirement. Thus in diet 9, the energy requirement has a 0.20:1 leverage on the protein yield rate. As with diet 9, the objective function of diet 10 is a ratio, so its marginal value is an instantaneous rate, which applies only at the optimum.

Again, the use of a rate criterion seems to have led to an unrealistic diet. The slack values for consumption mass and feeding time (84.1 g and 321.9 minutes) indicate that diet 10 would require only 26% of the available feeding time and would fill only 16% of the baboons' consumption capacity.

Recurring Foods

For each of the core foods, table 5.16 shows in which of diets 6–10 it would be present and in which diets it is called for in the largest available quantities. A number of foods occur repeatedly in these five diets. By grouping the foods into food classes (leaves, fruits, etc.) we can see how often foods of each type are called for to satisfy the requirements of these five optimal diets, and how frequently foods of various types are called for at the limit of their availability and would therefore limit the objective functions of the optimal diets.

Milk

Milk is an interesting special case. In diets 6–8 (maximize total energy or protein, minimize harvesting time), milk is by far the largest component, constituting more than half of the food in each case, yet never called for in the largest possible quantity—the baboons' consumption capacity, estimated to be 1002.4 g/day at the mean age of my subjects. Clearly, milk is a highly nutritious food by any of several criteria, though not a perfect one; other foods are required to round out these diets. The puzzle is not why baboons are still nursing at a year of age, but why they should ever quit. The answer seems to be that baboon mothers cannot produce enough milk to keep up with the ever-increasing nutritional requirements of their growing offspring (J. Altmann 1980).

Interestingly, no milk at all is called for by the two rate-maximizing diets, diets 9 (energy rate) and 10 (protein rate). This omission may be largely an artifact of how the milk yield rate was calculated in this study: as quantity of milk per minute on the nipple. In baboons as in many other primates, during a considerable (but unknown) amount of the time infants spend on the nipple, they are not sucking and presumably are not taking in any appreciable amount of milk. Whether this prolonged nipple contact is necessary to maintain the mother's level of milk production is unknown. Even with this potential "time dilution" in the estimated rate of milk consumption, the protein and energy intake levels for milk were among the baboons' highest—though not quite high enough for milk to be incorporated into the two "fast food" diets. According to milk's cost coefficient ranges in diets 9 and 10, in order to be included in these diets, milk's energy yield rate would need to be 25% higher than its assumed value of 15.5 kJ/minute (PCMX-1 in table 6.12), its protein yield rate, 29% higher.

Gum

Fever tree gum, in one form or another (core foods 4, 15, 23, 32, and 34), shows up at maximum available quantities in every optimal diet except that for maximizing protein intake rate. (In this section, foods of each class are listed in order of decreasing frequency of consumption [table 4.1, col. a] and are identified in table 6.2. The food numbers themselves are their time-budget ranks in the baboons' observed diet, from table 6.1.) Because of its extraordinarily high carbohydrate concentration, this gum is an outstanding source of high-density, quick-yield energy. Its long-term effect on dental caries is unknown.

Acacia Seeds

The seeds of acacia trees, from both local species and obtained by the baboons from various sources (core foods 5, 10, 12, 14, 18, 19, 42, 47, 50) have energy densities comparable to that of fever tree gum (though they require far more harvesting time per unit of energy), and the seeds have protein concentrations second only to insects. Not surprisingly, then, they are called for from at least one source by every one of the five optimal diets, usually in maximum available quantities.

Leaves of Grass

Green blades of grasses and sedges (core foods 3 and 9, but not 33) are called for in maximum available quantities in three of the five optimal diets, attesting to their high quantities of both energy and protein.

Corms

Among the corms of grasses and sedges (core foods 2, 6, 16, and 36), only those from two species, *Sporobolus cordofanus* and *Cyperus obtusiflorus,* are called for in maximum quantities in any diet, namely, in the energy-maximizing diet (both species) and in the protein-maximizing diet (*S. cordofanus* only). Far more common than the corms of any named species were corms from unidentified grasses and sedges. At the time of my study, relatively little was known about the local flora, and the grasses, which always present taxonomic problems, were seldom identifiable. Based on more recent observations in Amboseli, I believe that a large component of this "corms, species unknown" category consisted of corms from *Cynodon plectostachyus.* Other important corms that were not recognized during my study include those of the palisade grass *Brachiaria pubifolia.* A study with a considerably higher proportion of identified corm species would now be possible in Amboseli.

Fruits

The baboons' core foods include a variety of fruits (core foods 8, 13, 17, 21, 22, 24, 26, 30, 31, 35, 41, 48, 51, and 52). At least one fruit in maximum available quantities is called for by the diets for energy maximization (five fruits), for protein maximization (six fruits), for time minimization and energy rate maximization (one, sticky-fruit), and for protein rate maximization (two fruits). Among the fruits eaten by baboons in Amboseli, none (except *Capparis tomentosa,* whose nutrient content was not analyzed) has a large, sweet, fleshy aril, so, unlike our domestic fruits, none readily supplies large quantities of soluble carbohydrates. Six fruits from the baboons' core diet are not called for in any quantity in any optimal diet.

Flowers

Among the baboons' core foods, the only flowers are those of the two acacia trees and of trumpet-flower plants (foods 11, 20, 43, and 49). Fever tree blossoms from the tree are required at their limit to maximize total energy, and umbrella tree blossoms are an alternative in the protein-maximizing diet. Flowers are not otherwise called for in any quantity in any of the five optimal diets.

Meat

Animal matter appears among the core foods as dung-beetle larvae (food 38) and unidentified insects and other arthropods (37). Only the latter appear in any optimal diet; they are assumed to have the same mean nutrient qualities as grasshoppers, which I believe they often were. Arthropods are called for at the limit in every diet except that for minimizing harvest time. They, along with acacia seeds (from the ground and from dung) and fever tree gum (from the ground), are the only foods that are called for at the maximum in four out of five optimal diets; none are in all five.

At the opposite extreme from foods that would be incorporated at the maximum amounts in several of the optimal diets are core foods that would not be included in any of them. These uncalled-for foods will be discussed in chapter 6.

Nonfood Limiting Constraints

In diets 6–10, what factors other than food availability limit the objective functions: the amounts of energy, protein, and feeding time? The binding constraints in diets 6–10 are shown in table 5.16. All five diets would provide more than adequate quantities of fiber and lipids, and the three foods for which cation analyses are available would yield less than toxic quantities of all cations, so none of these constraints limited any of the objective functions. As might be expected, available feeding time would be exhausted by the diets that maximize total energy and total protein (diets 6 and 7), but not by any of the other diets.

Emphasis on foods that are rich in energy and protein (diets 6 and 7) or that would quickly satisfy the baboons' requirements (diet 8) would make meeting their total mineral requirement, 1.73% of the diet, difficult for them. Put the other way around, the baboons' mineral requirement is barely met by diets 6, 7, and 8, and that requirement would limit further increase in protein and energy intakes. This shortage of minerals is strange in view of the extensive invasion of Amboseli in recent years by halophytic plants. The possibility that the actual diet of the yearlings was deficient in essential minerals is discussed in chapter 6.

In diet 9, further increases in the baboons' energy intake rate would be limited by the need to satisfy their protein requirement, which that diet would barely meet; conversely, in diet 10, further increases in their protein intake rate would be limited by their energy requirement, which that diet would barely meet. Both of these requirements would limit further decreases in feeding time (diet 8).

Comment 7: Energy Rate Maximization

Until now, the prevailing approach in theoretical work on optimal foraging has been "to assume that predators choose diets which maximize the net yield of energy per unit foraging time" (Hughes 1979; and similarly in Charnov 1976; Emlen 1966; Estabrook and Dunham 1976; Lehman 1976; Pearson 1974; Pulliam 1975; Schoener 1971; Sih 1980; Stenseth and Hansson 1979; Stephens and Krebs 1986; Timmen 1973; Werner and Hall 1974; etc.).

As I mentioned above, comparison of diets 6 and 9 indicates that for Amboseli baboons, maximizing energy and maximizing energy rate are incompatible goals. That the first objective (maximize energy intake per day, as in diets 1 and 6) and the fourth (maximize energy intake rate while feeding, as in diets 4 and 9) can lead to different diets may not be immediately apparent (and similarly for total protein versus protein intake rate). After all, if an animal maximizes its energy intake and has a limited amount of time to do so, hasn't it thereby maximized its *rate* of energy intake? The answer is no. These are two different objectives, though under special conditions the results may be equivalent. Which objective is preferable? Consider an animal that first satisfies all of its nutrient requirements for the day by eating the smallest suitable quantities of various foods, regardless of their energy density. It then begins feeding on the most energy-rich food—rich in terms of energy yield per minute, either of feeding on that food or of foraging for it. It thereby increases both its mean energy intake rate and its total energy consumption. (For clarity, I am wording this hypothetical situation as a sequence of choices, but the temporal order is irrelevant to the basic argument.) Suppose that after eating a small quantity of that energy-rich food, the animal can eat no more of it, say because of a toxin in it or because it is available only in very limited quantities. Should the animal now switch to the second-richest food? If its goal is to maximize its total energy intake, then yes. However, if its goal is to maximize its energy intake *rate* (whether time-weighted or mass-weighted), then it should not do so if the second-richest food yields energy at a rate below the average of what the animal has consumed so far, even if, as a consequence, its total energy intake for the day is barely above the minimum and its dietary volume is far below its gut capacity.

In retrospect, this hypothetical situation, while obviously an extreme case, calls into question the goal of maximizing the mean energy intake rate, an objective that has dominated the literature on optimal foraging but that, as our example shows, can lead to absurd putative optima: diets that are quickly consumed but well below the animal's consumption capacity for foods and their components, including energy. Diets 9 and 10 (for energy rate and protein rate, respectively) are examples. Maximizing the mean rate of energy consumption, rather than the total amount, puts a premium on the speed with which the most energy-rich foods are consumed. It does so to the exclusion of foods that yield energy at rates less than the average of the available amount of the richest food plus the minimal diet necessary to meet other nutrient requirements, even if that leads to a meager diet with a paltry energy content. I can think of very few situations in which such an overriding emphasis on "fast food" is advanta-geous. (Doing so is not equivalent to minimizing feeding time, for which see the third objective function, above.) I do not know of any evidence from the nu-trition literature for an advantage to the rapid intake of energy. Energy-limited animals are more likely to be limited by the quantity of energy in their diet than by the rate at which it is obtained.

One could argue that energy intake per day is also a rate, albeit calculated by using a different time scale than energy per minute of feeding or foraging, and by including nonfeeding intervals, as well. Surely, one might say, this is just a difference between short-term and long-term rate maximization, and so just a matter of selecting an appropriate time scale (Stephens and Krebs 1986). Not exactly. The choice of an appropriate time scale for the rate cannot be made arbitrarily. Any interval less than the time required to fill the gut—or at least to fill it to the largest attainable level, considering competing activities—runs the risk of the above-mentioned fast-food fallacy of feeding primarily on high-energy, rapidly consumable foods. Furthermore, the shorter the interval that is selected, the more ludicrous the results can be. Conversely, intervals longer than required to fill the gut to capacity are not subject to this objection. Since the time needed to fill the gut is a random variable that depends on the diet se-lected, and so usually will not be known prior to the analysis, that time interval usually will not be an appropriate choice for the rate scale. Any value longer than that would suffice. Because of the long overnight fast, during which the gut is largely emptied, and because nutrient requirements are scaled per diem, a day is the obvious choice.

For these reasons I have elected to maximize total daily energy intake, not the intake rate, as the basis for comparison with the observed diets of individ-ual baboons (chapter 7).

Comment 8: Foraging-Time Minimization

Essentially the same problem affects diet 8. Some animals may indeed be foraging-time limited. For example, many desert rodents are subterranean dur-ing the day, come out at night primarily during the dark of the moon, very rapidly stuff their cheek pouches with seeds, then return to their burrows to store or eat their gleanings. Their haste while aboveground is justified: numer-ous owls, coyotes, and other predators feed extensively on these small mam-mals. Baboons, however, are much less susceptible to predation, and the difference in predation risk between their feeding and nonfeeding periods prob-ably is relatively small, so that the gain in reduced predation that could be achieved by eating a time-minimizing diet probably would be outweighed by the detrimental effects of a nutritionally minimal diet. The twin goals of si-multaneously minimizing harvesting time and maximizing the intake of energy or some other food component will almost never be attainable, for as I noted before, there is no general solution to the problem of simultaneously maximizing or minimizing more than one function.

Comment 9: Competing Demands

A comparison of diet 10 for maximizing the protein intake rate with diet 9 for maximizing the energy intake rate makes clear that these are incompatible goals. An animal that maximizes protein would get less energy than an energy maximizer, and the latter, in turn, would get less protein than the former.

Nonetheless, an animal's activities or fitness, or the growth of a population, may in some circumstances be simultaneously limited by two or more factors: limiting factors need not be unique. One approach to such situations is to incorporate the requirements of each limiting activity into the constraint set of the others. So, for example, if grooming, sleeping, and other vital activities each have minimal time requirements, their total time requirement establishes an upper limit on the time available for foraging. For an illustration of this approach based on Amboseli baboons, see Bronikowski and J. Altmann (1996). A second method is to evaluate the effect of trade-offs between the incompatible goals so that, ideally, a fitness value could be assigned to any possible combination of, say, protein yield, energy yield, and time consumption. (For linear combinations, that approach is pursued in chapter 7.) In economic terminology, the set of fitness values would form a *utility function,* and in some cases methods for maximizing such functions exist.

A third approach is to seek a diet that is *Pareto optimal* for these limiting factors, that is, an adequate diet (meaning one that satisfies all the relevant constraints) such that no other adequate diet exists at which the objective values for limiting factors are at least as good and at which at least one is better. Put another way, if a unique Pareto optimal diet exists for a set of limiting factors, an individual that shifted from that diet to any other adequate diet would be worse off for at least one of the limiting factors. Suppose, for example, that the fitness of an animal is limited by both its energy intake and its protein intake. If a unique Pareto optimal diet exists, then any other adequate diet would contain less energy or less protein, or less of both.

These approaches to multiple objectives go beyond the scope of the present work, but they will doubtless be important for future development of foraging models and for relating the requirements of foraging to those of other vital activities.

6

The lower animals select with unerring precision, as long as they are in a natural environment, from the materials around them those best fitted to their wants, and they do this by instinctive discernment inherited from a long line of naturally selected ancestors, while they are checked in their consumption by a sense of repletion of coeval origin. We unhesitatingly infer that the articles they choose are, of all the nutrient materials accessible to them, those best adapted to the special needs of their economy, and that their consumption of them is proportioned to their needs for the time being.

—James Crichton-Browne, 1910

'Tis not the Quantity, but the Quality, that makes the Feast.

—Oswald Dykes, *English Proverbs*, 1709

Real versus Ideal Diets

In the preceding chapter, optimal diets for Amboseli baboons were computed both for protein and energy maximizers and for harvesting-time minimizers. Now, what did the baboons actually eat, and what nutrients did each food in their diet contribute? How much time was taken to harvest each food? Was their diet nutritionally adequate? How close was their diet to the optimal diets?

In this chapter I combine food-specific time budgets, unit intake rates, and unit masses to estimate the baboons' mean annual diet, that is, the average amount that they ate of each food. Next, these food masses are multiplied successively by their various nutrient densities to estimate, for each nutrient, the baboons' mean daily intake. (The special method used to estimate milk intake will be presented below.) These food and nutrient intakes are then used to see whether the baboons' diet is nutritionally adequate and whether it approximates the optimal diets. This chapter focuses on the average yearling's diet; the next, on individual differences.

Unit Intake Rates

For any food, mean daily consumption in food units (e.g., blossoms, leaves) per day can be obtained from the product of the mean time, in minutes per day, spent harvesting that food (its time budget), and the mean unit intake rate (food units consumed per minute spent harvesting that food), that is, from $A_j B_j$ of

160

equation (3.1). In turn, the mean unit mass of a food (factor C_j of equation 3.1) times its mean daily consumption rate gives the average amount (mass) consumed per day. (This conversion to mass is necessary because nutrient analyses are reported on a mass basis.) So, for example, the daily intake rate of *Azima* berries is the number of berries eaten per minute times the number of minutes per day spent eating them. In turn, the average number of berries consumed per day, multiplied by the average berry mass, gives the mean number of grams of berries eaten per day. The time budgets of the core foods were presented in chapter 4 as pooled mean values, that is, with data pooled over all yearlings during every season in which each was sampled. Now I present the mean unit intake rate of each food, then its mean unit mass, again using pooled values.

As indicated in chapter 3, I sampled unit intake rates opportunistically, and only when observation conditions were such that I could count every item that entered the baboons' mouths. Most foods they ate were consumed in distinct units, such as leaves, blossoms, and berries. These were counted in timed bouts.

Core Foods

It seemed impractical and unproductive at this stage in the research to analyze the many foods that baboons seldom eat and on which they spend little time. Although some of these foods may make important contributions to the baboons' requirements for vitamins or trace elements, few such chemical analyses of Amboseli foods have been carried out to date. I therefore concentrated my statistical analyses on those foods that the baboons eat often and that make up the largest part of their foraging time, and I assumed that on average the small remainder of their foraging time was equally productive. I began by rank-ordering the foods by their time budgets in my samples. This indicated that just the fifty-four most time-consuming foods accounted for 95% of the baboons' observed feeding time. Of these, I have empirical estimates of the baboons' intake rate—units of food consumed per minute of feeding on that food—for all except three: fever tree gum droplets picked from dead wood, such as logs (FTHW), water from rain pools and waterholes (WAVG), and "unidentified" (XXXX), which of course is not a single food but rather, a "wastebasket" category. I have assumed that fever tree gum was picked from dead wood at the same rate as from the tree (no. 29) but could make no estimate for the other two. The remaining fifty-two foods, which I shall refer to as the "core foods," accounted for 93.37% of the baboons' time budget. For each of these fifty-two core foods, table 6.1 gives the mean unit intake rate (units consumed per minute of feeding on that food in all unit intake samples pooled) and its standard error. These fifty-two core foods will be referred to repeatedly in the rest of this book.

For the convenience of the reader, their four-letter abbreviations are identified in table 6.2 and on the last page of the book.

Some foods were consumed in clusters of units. For example, baboons plucked several stem units ("tillers," in botanical terminology) of green leaves from the grass *Cynodon nlemfuensis* before bringing the handful to the mouth. In 25 such hand-to-mouth clusters, I counted 79 tillers, or an average of 3.165 tillers per handful. A sample of 28 tillers of this grass averaged 4.07 leaves per tiller. This gives 12.88 leaves per handful, and that conversion rate was used for other hand-to-mouth clusters of this food.

How many seeds do baboons get from each pod on an umbrella tree? One July day, I collected a sample of pods from beneath a tree in which the study group had fed three hours earlier. I collected all pods, regardless of condition, from a small area beneath the tree. All but two were green and fresh. Of these, sixty-one, which were used for the following calculations, showed some sign, typically tooth marks, of use by the baboons. An additional eight showed no sign of baboon handling and were discarded; so too were two pods that were so badly chewed up that I could not count seeds or seed sites. I then counted the number of beans remaining in each pod and the total number that were present before the baboons opened the pod, as judged by the conspicuous bulge at the position of each bean. I assume that the difference between these values is the number of seeds consumed by the baboons. Before the baboons fed on the pods, each contained on average 10.33 seeds (SE 1.32, SD 2.41). Of these, on average the baboons ate 8.21 seeds per pod (SE 1.05, SD 3.44). That is, on average they left behind 20.4% of the seeds. Many of these were quite small seeds near the distal tip of the pod and would have required more work for a smaller return. Three of the 129 uneaten seeds were blackish and may have been infested.

In July 1976 a much larger sample was obtained. By collecting three hundred green umbrella tree pods that the baboons had just discarded and counting the number of freshly opened seed spaces, I estimated that they removed an average of 6.72 seeds per pod; that conversion factor was used whenever the baboons fed on that food. (Pods were smaller that year, averaging 8.14 seeds per pod, but the percentage of seeds eaten was hardly different: 79.5% in 1974, 82.6% in 1976.)

As a partial check on the conversion rate for green blades of unidentified grasses (GRLU), I tallied just those GRLU bouts in which the number of tillers were counted. This gave 131.5 tillers in 16.75 minutes, or 7.85 tillers per minute. Using the tiller size of the commonest grass in the baboons' diet, *Cynodon nlemfuensis* (mean 4.07 leaves per tiller) I got 32.0 leaves per minute, which is not far from the estimate in table 6.1 of 28.1 leaves per minute.

Amorphous foods presented special difficulties when it came to estimating intake rates. The three most common amorphous foods consumed by baboons were fever tree gum, water, and milk. Milk intake is discussed in a separate section below. Although I recorded and timed all drinking bouts, I have no way to estimate the amount of water the baboons drank. Beyond that, their foods contain some water (table 5.1; S. Altmann, Post, and Klein 1987), and other water is produced metabolically in the catabolism of carbohydrates. I will assume that all the baboons consumed an adequate but not an excessive amount of water. A special study of water balance in baboons and its effects on their pattern of home range utilization would be worthwhile.

Fever tree gum was consumed either as individual droplets or as larger chunks. I estimate that the droplets, which are what the baboons usually found, averaged 0.05 g, that is, about the mass of a water drop from a burette. When the baboons found larger chunks, I estimated the mass or volume of each in units that were familiar to me (e.g., ounces, teaspoons), then converted to grams. Conversion factors for other foods are shown in table 6.3.

For very few foods are my samples of unit intake rates adequate to detect individual differences, and perforce I have relied on other components of intake (equation 3.1) for that information.

By far the highest unit intake rate was for the very small fruits of *Commicarpus pedunculosus* (table 6.1, no. 41), which we nicknamed "sticky-fruit." The baboons ate over two hundred fruits per minute while feeding on this very seasonal food source. They striped the fruits from the terminal growth by pulling the branches through their teeth. Since each fruit of this plant weighs only 0.0069 g (table 6.4), a very large number must be eaten if they are to make any appreciable nutritional contribution. An annual mean daily intake of over one hundred fruits (table 6.1) gained the baboons only three-quarters of a gram of material per year. Other foods with high unit intake rates include green grass blades of *Cynodon nlemfuensis* (no. 9) and of unidentified grasses (no. 3); the fruit of devil's thorn, *Tribulus terrestris* (no. 13), and of the toothbrush bush, *Salvadora persica* (nos. 21, 22, 24 and 48); the dry seeds of acacia trees picked from the ground (nos. 12 and 42) or from dung (no. 50); leaves and flowers of the trumpet-flower bush, *Lycium "europaeum"* (nos. 28 and 49); the flowers of umbrella trees, *Acacia tortilis,* picked from the tree (no. 11); and various unidentified small items that the baboons picked from the ground, probably acacia seeds for the most part (no. 10). All of these foods were consumed at rates exceeding one unit for every four seconds devoted to that food.

For several foods, we can compare the unit intake rates of yearlings (this study) and adults (Post 1978, table 31). In figure 6.1 I have plotted the mean intake rates of yearlings and of adult females when feeding on the eleven foods for which both Post and I have tabulated values for the same units. I did not include any food for which the unit was a handful, because of potential difference in size. (For none of the foods do adult males and females differ, according to Post 1978.)

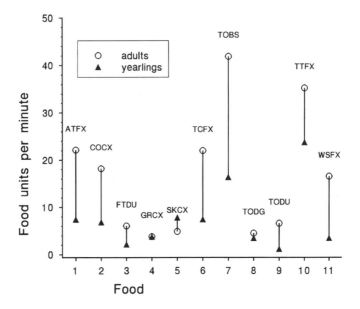

Fig. 6.1. Comparison of unit intake rates for yearlings (this study) versus adult females (Post 1978, table 31). All units are as indicated in table 4.1 except for acacia seeds (FTDU, TODG, TODU), which are in units of pods, converted as shown in table 6.3. Foods: 1 (ATFX) = fruit of *Azima tetracantha;* 2 (COCX) = corms of *Cyperus obtusiflorus;* 3 (FTDU) = (seeds from) green pods of fever trees; 4 (GRCX) = corms of unidentified grasses and sedges; 5 (SKCX) = corms of *Sporobolus rangei;* 6 (TCFX) = fruit of *Trianthema ceratosepala;* 7 (TOBS) = flowers of umbrella trees; 8 (TODG) = (seeds from) umbrella tree pods on the ground; 9 (TODU) = (seeds from) green umbrella tree pods; 10 (TTFX) = fruit of devil's thorn; 11 (WSFX) = fruit of *Withania somnifera.*

In all cases except one, *Sporobolus rangei* corms, the feeding rates of the adults are higher than those of the yearlings, often two or three times higher (figure 6.1). In fact, yearlings fed somewhat faster than adults on *Sporobolus rangei* corms. This is puzzling, in view of the work involved in harvesting corms. Perhaps the yearlings are selecting smaller and more readily harvested plants (see chapter 9), but that does not explain the considerably greater rate at which adults harvest the corms of *Cyperus obtusiflorus,* which grows inter-mingled with *Sporobolus rangei* and is of similar size and shape. On corms of unidentified grasses and sedges, the feeding rates of adults and yearlings were essentially the same. The faster feeding rates of adults on almost all foods is not surprising considering both the greater strength and experience of the adults and their need to satisfy higher nutrient requirements in about the same amount of time (see also the discussion following equation 6.4).

Unit Masses

Our available data on unit masses ("weights") of Amboseli core foods are compiled in table 6.4. Unit mass samples are available for twenty-seven of the fifty-two foods in the core diet (i.e., the fifty-four foods that account for 95% of the foraging time budget minus water and unidentified food, as indicated above). For corms of unidentified grasses (no. 3 in table 6.4), unit mass was taken to be the mean, weighted by time budgets, of the three top-ranking species of grass corms. Weighted means were also used for leaves of unidentified plants (no. 48) and acacia seeds extracted from dung (no. 52), as indicated in the table. For seven foods, intake was estimated in grams, so that the unit mass is automatically one. For the remaining foods, various assumptions and extrapolations were made, as indicated in table 6.4.

No data are available on differences in the unit masses of the foods selected by individual baboons. The unit mass value in table 6.4 are based on food samples that were as representative as we could get of what, on average, the animals fed on. I do not know how to determine the standard errors of these unit masses. Some are based on a single pooled sample of food units that were all massed together; in other cases several subsamples, up to six (ATFR), were counted and weighed separately. In almost all cases the total number of items weighed was in the hundreds or thousands. The largest sample was 9,042 fruits of *Commicarpus pedunculosus*.

Some of the samples took considerable effort for us to obtain and made us appreciate the agility of monkeys. The most laborious of all were the green seeds of *Acacia tortilis* pods, which took the joint efforts of the Altmann and Kummer families and David Klein, first to dislodge the green pods from the trees (fig. 6.2), followed by almost an entire day for all of us to shell the seeds from the pods and then remove their seed coats.

Daily Intake of Foods

In order to estimate the yearlings' average daily dietary intake over the course of the year, I first estimated their average diet during the days when they were under observation (their "observed diet"), then adjusted these values for seasonal fluctuations in my sample time. I begin, then, by presenting their observed diet.

Observed Diet

For each of the fifty-two core foods, table 6.1 gives the mean unit intake rate (food units per minute), time budget (minutes per day), unit mass (grams per

Fig. 6.2. Collecting green pods of umbrella trees. The author (right), Rudy Kummer (top), and Hans Kummer. (Terrestriality has its disadvantages.) Photo by Jeanne Altmann.

unit), and daily intake (units per day and grams per day). Each is a pooled mean based on all relevant samples during days that the baboons' feeding behavior was sampled. The pooled bout rate of each food was computed, as in equation (4.2), by dividing the total number of bouts of that food (n_j) by $t = \Sigma\, n_j L_j = 22{,}299.84$ minutes of sample time. The standard error of these rates was computed, as indicated in chapter 3, from $n^{1/2}t^{-1}$. Because the time budgets (A_j) and the daily intake of each food ($A_j B_j C_j$) are each estimated from the product of random variables, their standard errors (table 6.1) were estimated by equation (3.3).

Seasonally Adjusted Diet

The mean diet presented in table 6.1, which I shall refer to as the average *observed* diet, is the baboons' pooled mean intake over a year during the days on which I sampled them. However, my samples were not uniformly distributed over the year, and the array of available foods changed markedly with the seasons. A better indicator of the yearling's *actual* mean daily intake over the year can be obtained by adjusting the diet values in table 6.1 for seasonal differences in food availability and sampling intensity. The adjustment factors, f/t, are derived in appendix 8 and are listed in table A8.1. The seasonally adjusted, pooled mean diet is presented in table 6.5 (from table A8.1, cols. 16 and 17).

The time required to harvest and consume the specified amount of each food j in the seasonally adjusted diet was calculated from the seasonally adjusted quantity of that food divided by the intake rate for that food, in grams per minute, estimated from B_jC_j (values in table 6.1). That is, I took the mass intake rate of each food to be that of the observed diet, since in that rate estimate, the only time considered is hands-on feeding time.

Milk was a special case, and my way of estimating individual milk intakes will be discussed in the next section. I do not know of any seasonal fluctuations in nursing time or milk intake. Mean daily nursing time in the seasonally adjusted diet was taken as estimated in the observed diet, 65.9 minutes per day (table 6.1). Mean daily intake of milk over the 30–70-week age range was obtained from the four milk intakes (ml per day) given for the four ten-week age blocks in table 6.6 (for which the rationale will be provided in the next section of this chapter). To do so, I converted the four estimated volume intakes to masses, using 1.027 for the specific gravity of baboon milk (Buss 1968, table 1). The four values were then averaged, using as weighting coefficients the total sample times for the four age blocks (column totals, table 3.1). The value I obtained is 316.9 g per day (first approximation, table 6.1). Using a different estimate of growth rates for Amboseli baboons (see below), I obtained 260.9 g milk per day (second approximation, table 6.5). That milk output is about 2.2% of the body mass of a nursing adult female baboon, using an average body mass of 11.9 kg (J. Altmann, Schoeller, Altmann, Muruthi, and Sapolsky 1993) and is roughly comparable to that of humans: an adult female weighing about one hundred pounds with a large infant produces about a pint of milk a day (Davidson, Passmore, and Brock 1972), that is, about 1% of her body mass.

Nursing took up 23.74% of the young baboons' observed feeding time.[5] The other fifty-one core foods accounted for 69.63% of their observed feeding time, for a total of 93.37%. If we assume that the remaining 6.63% of their observed feeding time (all non-milk) was as productive, on the average, as the time spent on the fifty-one non-milk core foods, then the observed intake from all non-milk sources, core and otherwise, can be estimated by multiplying the total intake from the fifty-one non-milk core foods by $(0.6963 + 0.0663)/0.6963 = 1.095$, equivalent to adding 9.5%. This method of estimating the contribution of non-core foods will be used at several places in the rest of this book.

By means of this factor, I estimated the yearlings' average foraging yield, that is, their total daily intake of foods other than milk. Mean daily consumption of the fifty-one core foods other than milk totaled 119.3 g (seasonally adjusted diet, table 6.5). If we assume a comparable average consumption rate (grams per minute of feeding) for the remaining foods, then the average daily consumption of non-milk food was 119.3 × 1.095, that is, 130.6 g/day. If we use a growth rate of 4.5 g/day, then at the age of 47.5 weeks—the mean age of the yearlings when sampled—a baboon weighs 2.27 kg on the average (see appendix 1). It is therefore foraging for about 5.7% of its body mass in food per day. Average daily milk intake at this age (see section on milk consumption) was considerably greater: 260.9 g, bringing total food consumption to 391.5 g, which is on average 17% of the yearling's body mass. On a mass basis, milk made up 66.6% of the yearling's seasonally adjusted diet at this age.

To evaluate the relative contributions of various categories of food to the baboon's total daily food consumption, I have grouped the core foods according to the classification given in table 4.4, except that unknown material from wood has been grouped with fever tree gum droplets, since I believe that is what most such material was. The daily consumption values (g/day) for core foods other than milk were then multiplied by 1.095, as discussed above, to allow for the small contribution from non-core foods. The results are given at the bottom of table 6.5 and are shown graphically in figure 6.3. The intakes of food classes fit fairly well ($R^2_{adj} = 0.91$) to the relationship log $g = 2.82 - 0.37r$, where g is the baboons' daily consumption of foods from a given food class, in grams, and r is the consumption rank order of that class. (The position of water in the ranking shown in figure 6.3 is unknown.)

Further simplification is possible because the corms of grasses and sedges, the growing ends of their stolons and rhizomes (the part on which the baboons concentrate their feeding), and grass/sedge blade bases are all meristem tissue. An additional lumping conflates flowers, fruits, and seeds: they are reproductive organs of plants. Together, they made up nearly a third by mass of the baboons' non-milk diet. Aside from water and milk, the baboons' diet at 30–70 weeks contained, by mass:

31.0% reproductive parts (flowers, seeds, fruits);
33.9% green leaves (primarily grass);
28.5% fever tree gum;
5.4% grass/sedge meristem (corms, stolons, rhizomes, blade bases);
0.22% animal matter; and
0.005% wood, cambium, bark.

The baboons' dietary intake was obtained predominantly from acacias, especially fever trees, and from grasses and sedges. By mass, acacias made up

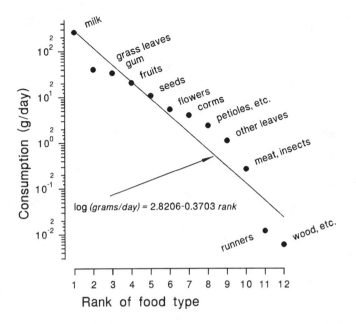

Fig. 6.3. Daily consumption of food types versus their consumption rank. Based on seasonally adjusted mean diet (table 6.5). Regression: $R^2_{adj} = 0.91, p \leq 0.001$.

38.4% (50.2 g/day) of the non-milk part of their diet, or actually somewhat more than that if one takes into account my underestimate of the baboons' intake of umbrella tree blossoms from the ground (chapter 7). Grasses and sedges account for another 39.3% of their diet (51.4 g/day). All together, that comes to 77.7% of the non-milk part of the baboons' diet. The success of baboons in the widespread grass-acacia savannahs of Africa depends on their ability to exploit these plants as major food sources. Similarly, as Norton, Rhine, Wynn, and Wynn (1987) note, based on their observations on yellow baboons in Mikumi National Park, Tanzania, "More than 75% of the stable food species were from the two most common and widely distributed plant families, the grasses and legumes."

Seasonally Adjusted Time Budgets

The amounts of time the yearling baboons of Amboseli devoted to various food classes, each averaged over the year, can be seen in figure 6.4, based on values in table 6.5. At this age milk—or at least holding mother's nipple in the mouth—was the yearlings' most time-consuming feeding activity. Their four

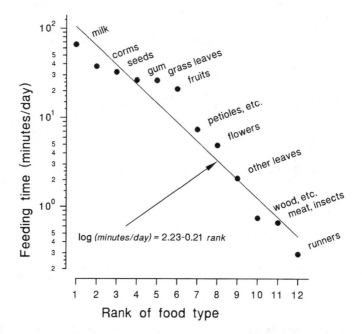

Fig. 6.4. Time spent feeding on foods of each type versus the time ranking of that type. Based on seasonally adjusted mean diet (table 6.5). Regression: $R^2_{adj} = 0.93$, $p \leq 0.001$.

most time-consuming food categories, milk, corms, seeds, and gums, together accounted for 71% of their feeding time; 93% of their feeding time was accounted for with the addition of the next two food categories, leaves of grass and fruits.

Corms ranked second only to milk, yet by contribution to the mass of the diet, they ranked seventh, an indication of their relatively low yield rate. Similarly, wood, cambium, and bark, which were off the bottom of figure 6.3, ranked ahead of meat and insects in consumption time. Other major food classes ranked nearly the same by time as by mass.

The baboons' marked concentration of their feeding time on grasses and acacias is apparent from the graph. The corms, leaves, meristem, and runners of grasses and sedges were respectively the first, fourth, sixth, and eleventh most time-consuming non-milk food types. Seeds (almost entirely of acacias), acacia gum, and flowers (mostly acacia) ranked second, third, and seventh. These grass and acacia categories accounted for 45% and 40%, respectively, of the baboons' non-milk feeding time; thus only 15% of their time was spent feeding on foods from other sources.

The daily food intakes and feeding time budgets that I have summarized above are based on samples taken over a yearlong period, during which numerous phenological changes occurred, and they are therefore not representative of the baboons' intake on any one day. Stacey (1986) provides information on the day-to-day variance. For baboons no day is an average day. However, deficiencies in nutrients on one day can be made up on another, so long as the periods of deficiency do not overextend the body stores of the nutrients and so long as the excess does not overload the capacity of the body to absorb and utilize them. Thus, mean intakes can be used to tell us whether in sum these baboons were fulfilling their nutritional requirements. A more detailed dietary analysis would take into account differences in turnover rates and body stores of the various nutrients.

Similarly, my feeding samples were taken on eleven quite different individuals. Individual differences in diet will be discussed in chapters 7 and 8.

In order to appreciate the nutritional contribution made by the baboons' various foods and to know whether their diet is nutritionally optimal or even adequate, one must estimate the nutrients contributed by each food in their diet, a task I shall return to after a digression on the yearlings' milk intake.

Milk Consumption

> Weaning is a singularly ill-defined term. . . . [It] is frequently represented as though it were a point in time, when it is more usefully viewed as a process. For most mammals, weaning is gradual and involves a progressive reduction in the rate of milk transfer from mother to young, accompanied by an increasing intake of solid food by the offspring and profound behavioural changes in the parent-offspring relationship.
>
> —P. Martin, 1984

For the 30–70-week yearlings in this study, milk was a major though decreasing component of their diet. As mentioned above, I estimate that the baboons in my study, at a mean age of 47.5 weeks, averaged 316.9 g (first approximation) or 260.9 g (second approximation) of milk per day. The method I used to make these estimates is explained below. Overall, milk accounted for about two-thirds of their dietary mass (260.9/391.5, table 6.5). Nursing bouts ranked fourth among foods by frequency in my samples (fig. 4.4) and were on average the longest bouts of any food or food class, except umbrella tree flowers (table 4.5). Nursing occupied far more of the weanlings' time than did any other food: the 30–70-week yearlings spent 66 minutes per daytime (0700 h to 1800 h) on the mother's nipple, equivalent to 27% of their feeding time (65.9/240.5, table 6.5), and they doubtless also sometimes suckled at night.

From direct observation, I do not know for how much of that time the yearlings were actually obtaining milk or how much milk they got; this section is devoted to estimating the latter. Although suckling movements could sometimes be seen in the yearlings' cheeks, such movements could not be scored reliably, and I therefore recorded the amount of time the yearlings spent with the mother's nipple in the mouth. (Baboon suckling movements seem to me to be much subtler than those of rhesus monkeys, *Macaca mulatta.* So too is their "lip smacking," a common cercopithecine facial expression. Perhaps these species differences reflect effects on facial mobility of the baboons' elongated muzzle.)

Estimates of milk intake are of particular importance in this study, since an unweaned baboon may partially compensate for its foraging inadequacy by consuming extra milk. The mother's output, in turn, probably accommodates to the yearling's demand: in humans and presumably also in baboons, the quantity of milk secreted is affected by suckling stimulation. In humans, the lactational response to increased suckling occurs several days later (Macy, Hunscher, Donelson, and Nims 1930; Newton and Egli 1958). Just how well the increased milk output compensates for the shortfall in other components of an infant's diet seems not to have been studied in any mammal. In order to estimate the baboons' milk intake rate, I turned to the literature. What I hoped for but did not find was an allometric relationship between mammalian size and milk intake rate, which I could then scale to baboon size.

In the hope of being able to extrapolate from data on other mammals, I turned to the surprisingly small literature on milk intake by wild mammals. Linzell (1972) and Hanwell and Peaker (1977) regressed mammalian body masses against the peak daily rate of milk secretion, but their values do not take into account variations in the rate of milk secretion at different infant ages, and even the youngest infants that I studied probably were past the period of peak milk intake. Beyond that, one of the methods commonly used in these studies to measure milk yield—determination of the amount of milk expressible from the mammary glands—may not accurately reflect diurnal intake, especially for those mammals, such as primates, that nurse several times a day. This situation is analogous to the problem in ecology of trying to measure turnover rate from the instantaneous standing crop. The problem is particularly acute in primates.

> Direct measurement of the daily milk yield of nonhuman primates is difficult. The mammary glands have a low capacity for storing milk, and the infants . . . nurse frequently—especially Old-World primate infants which are carried by their mothers in a ventral-ventral position. . . . Manual milking rarely gives more than 5 ml, except from apes, where over 20 ml can be expressed. (Buss 1971)

(Note that lactating humans commonly can express more than ten times the latter amount, i.e., over 200 ml. This suggests longer intervals between nursing bouts in early humans than in the great apes.)

MacFarlane, Howard, and Siebert (1969) have developed a method, using tritiated water (^{3}HOH), by which total milk intake can be estimated. Improvements in this method have been suggested by Coward (1980) and by Rath and Thenen (1979).

The method of MacFarlane et al. was used by Buss and Voss (1971) in their study of the milk intake of nursing cynocephalus baboons living in outdoor colonies in San Antonio, Texas. Over the first eighteen weeks of life, milk intake rose from about 260 ml/day in the second week to 450 in the seventeenth week (their fig. 1A). During this same period daily milk intake per kilogram of infant body mass decreased slowly and fairly linearly from about 250 ml in the second week to 210 in the seventeenth, with slightly higher values for females at most ages (Buss and Voss 1971, fig. 1B). Unfortunately, Buss and Voss's study was stopped at eighteen weeks of age, because of limitations on the method used. Furthermore, they give no information on the amount of time spent suckling by their baboons or on any other way of estimating individual differences in milk intake, which are important in my study. Roberts, Cole, and Coward (1985) provide data on milk intake up to ten weeks of age by yellow and anubis baboons; again, no data are provided on nursing time.

The studies by Buss and Voss and by Roberts et al. appear to be unique. At present, there seem to be no other published data on daily milk intake in nursing baboons as a function of age. For that matter, few data on milk intakes of other mammals are available. Oftedal (1984a, 1984b) found just twenty-one species for which reliable data are available.

Although average daily milk intake values are interesting, I particularly needed information on the baboons' individual intakes, which I hoped to get from my individual nursing time values multiplied by the mean milk intake rate of baboons (ml milk per minute of nursing). However, very few studies provide data on milk intake rates in wild mammals, even in captivity, hardly enough to provide an allometric regression that might be interpolated to baboons. I came to this conclusion after reviewing several surveys of milk and lactation (Kon and Cowie 1961; Larson and Smith 1974; Peaker 1977; Pond 1977; Schmidt 1971; Vorherr 1974), a search in *Index Medicus* (under "lactation" and "milk"), and replies to an inquiry sent to a number of people active in research on mammalian lactation and on the behavior of primate infants. This lack of information is astonishing in view of the importance of milk in the life history of mammals (Blaxter 1961, 1971; Gittleman and Oftedal 1987; Oftedal 1980, 1985; Pond 1977) and the directness with which milk intake rates could be determined, say in zoo animals, by weighing the mother or infant before and after timed nursing bouts (Linzell 1972).

When I realized that the comparative mammalian data were inadequate to provide an estimate of milk intake per minute of nursing, I devised a more indirect way of estimating daily milk intake of individual wild baboons. The method I used is described below.

Milk Intake of Individuals

Here I first estimate daily milk consumption (ml per day and ml milk per kg body mass per day) of individual Amboseli baboons from their nursing times at various ages, then compare my data on nursing time with data for baboons from several other sources. In the next section, varied assumptions are used to provide two estimates (m_1, m_2) of daily milk intake (ml per day) at each age and two estimates (d_1, d_2) of total daily food intake at each age. The four possible differences $d_i - m_i$ provide four predictions ($f_1, \ldots, f_4$) of age changes in the amount of food obtained by foraging. (The validity of these predictions will be checked in the next chapter when age changes in diet are discussed.) Finally, a few paragraphs are devoted to "weaning behavior," in particular to whether nursing bouts are terminated by the mother or by the infant.

I estimated age-specific milk intake of the baboons in my study (ages 30–70 weeks) from their diurnal nursing time and, by extrapolation, from Buss and Voss's data on milk intake at younger ages. In brief, my method for doing so was as follows. In each of the four age blocks, I first estimated individual nursing time budgets (step 1). From these I calculated mean daily nursing time, regressed these linearly on age, then extrapolated to zero to get an estimated age of last nursing (step 2). In step 3, assuming that mean daily milk consumption per unit body mass falls off linearly after eighteen weeks, the last age in Buss and Voss's study, I extrapolated linearly from their last body-mass-specific intake value to this age of last nursing. At each age, I assumed in step 4a that a baboon that nursed for the average amount of time for that age would get this average amount of milk per unit body mass. From our Amboseli data, I estimated the mean body mass of an infant at the midpoint of each age block (step 4b). Mean daily milk intake was obtained from the product of these two age-specific values, body mass and milk intake per unit mass. An individual's daily milk intake was assumed to differ from the mean intake for infants of that age by the same percentage as its nursing time during the day differed from mean nursing time for that age, so every individual's milk intake per twenty-four-hour day could be calculated at each age in step 4c. Details of this method are provided in appendix 9.

Steps 2 to 4 above can be illustrated graphically. Figure 6.5 shows Buss and Voss's data on milk consumption in the seventeenth week, the estimated suckling time budget of each yearling in each age block, the linear extrapolation from the seventeenth-week milk intake in Buss and Voss's study to the zero point of the linear regression of nursing times, and the estimated milk intake of

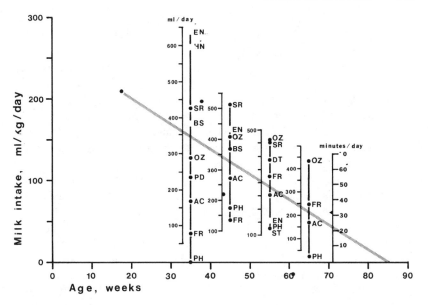

Fig. 6.5. Estimation of individual milk intakes. Mean daily intake per kilogram of body mass (gray line) is assumed to decrease linearly with age from the value shown at 17.5 weeks (from Buss and Voss 1971) to zero at 85 weeks, the average age at last daytime nursing. Mean milk intake (ml/day) for baboons of a given age is estimated from that linear decline rate and the mean size of infants of that age. Individual variations from that mean are assumed to be proportional to deviations of each individual's daylight nursing time from the mean time for that age. Two-letter abbreviations for infant names as in table 3.1.

each yearling at each age. For example, consider the top right data point in figure 6.5, for Ozzie (OZ) at 60–70 weeks of age. At that age, estimated mean daily milk intake was 218.7 ml (table 6.6). That value was obtained by extrapolating from a linear regression of nursing time on age to $y = 0$ (no diurnal suckling time), which occurred at about 85 weeks of age, then assuming that the mass-specific daily intake of milk decreased linearly to zero at that age from the value given by Buss and Voss for the seventeenth week (gray line in figure). These mass-specific values were converted to average daily intakes by multiplying them by mean body mass at that age (equation A9.2 and note 6). However, at 60–70 weeks, Ozzie nursed during daylight hours for 64.8 minutes per day, which is nearly twice the average (32.64 minutes, triangle on graph), so he is assumed to have taken in nearly twice as much milk, that is, 434.4 ml per day.

The estimated daily milk intake (ml/24 hours) for each yearling (step 4 of appendix 9) and the mean for each age block are given in table 6.6. Mean daily intake decreased from 349 ml at 30–40 weeks of age to 219 ml at 60–70 weeks. The mean of the four age-specific intake values, each weighted by the sample time for that age (table 3.1, excluding the small sample time from Pooh at 30–40 weeks, which was not used), was 308.57 ml, which, converted to grams at 1.027 g/ml, comes to 316.94 g per day,[6] or 316.47 g with Pooh's small sample.

The data in table 6.6 on suckling time budgets have been summarized by yearling's sex in table 6.7. In the first two age classes, females nursed more; in the second two, males did. Because I have few yearlings in any category, it is difficult to make too much of these results.

Step 4 requires individual data on suckling time budgets, which I estimated from suckling bout lengths and bout rates. For yearlings 30–70 weeks of age, suckling bouts ranged from 0.01 minute to 39.4 minutes and had a pooled mean of 1.65 minutes. The median duration was 0.40 minutes, and 75% of all suckling bouts lasted 1.18 minutes or less. When data for all yearlings are pooled into ten-week age blocks, the suckling bout length distributions of the four age blocks have similar means with overlapping standard errors (table 6.6) and are not significantly different ($p = 0.33$ and 0.92 in Breslow and Mantel-Cox tests, respectively). Within these four age blocks, there are significant individual differences ($p < 0.05$) in bout length distributions only at ages 30–40 weeks and 60–70 weeks according to both the Breslow test and the Mantel-Cox test. Therefore at the other two ages studied, individual differences in suckling time budgets and thus in milk intake are attributed entirely to differences in the frequency of suckling bouts. At all four ages, the suckling bout rates showed highly significant individual differences ($p \ll 0.001$). Because of these differences in this component of the estimated milk intake and significant differences at two of four ages in the other component, bout lengths, estimates of individual milk intakes were calculated, using in step 4a the mean of mean nursing time budgets, and these intake values were used in subsequent calculations of total dietary intakes of individuals.

At this point, comparisons with the results of several other baboon studies will be useful. Another source of information on the last age of nursing in Amboseli baboons is our monitoring program, in which specific behavioral, demographic, ecological, and life history data are gathered on a routine basis. One of the questions about each baboon under two years of age is whether the yearling was seen suckling since the last physical development sample, which, during the second year of life, is taken monthly. For the eight yearlings for which the available data cover the age of last observed suckling, the average age at the midpoint of the last month in which suckling was observed is 605 days (SD 110, SE 39), equivalent to 86.4 weeks. This estimate is surprisingly close to our previous estimate (step 2) of 85.1 weeks. Nicolson (1982)

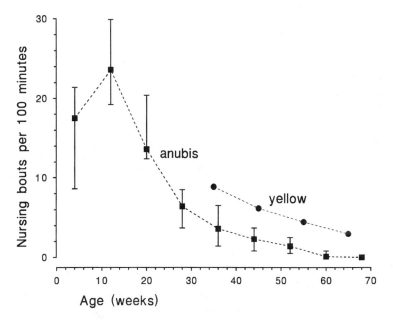

Fig. 6.6. Rates of daytime nursing bouts in anubis and yellow baboons. Yellow baboon values, based on focal animal samples, are pooled means from table 6.6. Anubis baboon values are group medians and interquartile ranges, from Nicolson (1982, fig. 5b). Values plotted at the midpoint of the corresponding age intervals of the two species (eight and ten weeks wide, respectively).

notes that she was able to express milk from wild-caught, cycling olive baboon females with yearlings up to 60 weeks of age.

Several other sources of data are available for baboon suckling time at various ages, but none for milk intake rate, that is, intake of milk per minute of suckling. Of particular interest in comparison with my data are those for olive baboons (*Papio anubis*), gathered during 1977–79 by Nancy Nicolson (1982) at Kekopey Ranch, near Gilgil, Kenya.

First, consider Nicolson's estimates and mine of nursing frequency: of age changes in how often yearlings nurse during the daytime. (Neither of us has data on nocturnal suckling.) In figure 6.6, Nicolson's estimates on suckling bout rates and mine are superimposed; both sets of data are from focal individual samples. Nicolson's estimates rates are consistently lower than mine. The difference may be attributable to a difference in our definitions. Nicolson defined a suckling bout as "a period of continuous attachment, separated from other such periods by at least 30 seconds." I had no such thirty-second criterion and recorded the end of a bout whenever the yearling terminated oral (and

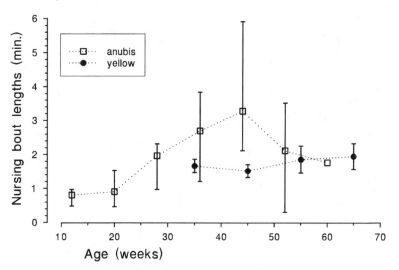

Fig. 6.7. Duration of daytime nursing bouts in anubis and yellow baboons. Anubis baboon values, from Nicolson (1982, fig. 6), are medians and interquartile ranges, from focal-infant samples. Yellow baboon values, from this study (table 6.6), are means and standard errors, based on product-limit estimates of focal-individual samples.

manual) contact with the nipple, except for the very brief gaps, a second or two long, during which it switched from one nipple to the other. However, according to Nicolson (pers. comm., May 20, 1981), recalculation of her data, counting each period of nipple attachment, including those ended by brief interruptions, made little difference in her results, since only about 2% of all nursing bouts ended that way. Thus we apparently have a population (perhaps species) differences: the olive baboons of Gilgil did not nurse as often as the cynocephalus baboons of Amboseli.

Conversely, however, the Gilgil anubis nursed for considerably longer periods once they got on the nipple. Figure 6.7 shows Nicolson's data on mean bout duration and mine superimposed.

I do not know why the baboons of one population nursed less often but for a longer time than those of another. J. Altmann has suggested one possibility. Perhaps Amboseli baboons are more susceptible to predators and therefore less widely dispersed. If so, maintaining proximity to other group members would require more frequent moves by Amboseli females, and that in turn would often interrupt their yearlings' nursing bouts. Alternatively, it may be that Nicolson sampled more extensively than I did during those times of the day when longer nursing bouts prevail, for example, when resting is most common.

Finally, consider the time budget of nursing, that is, the proportion of the

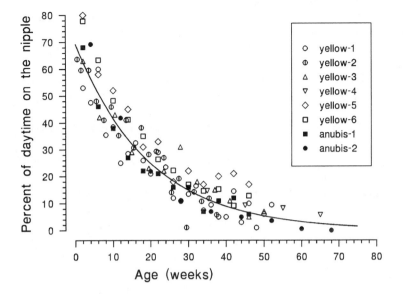

Fig. 6.8. Percentage of daytime spent on the nipple by anubis and yellow baboons. Sources: anubis (1) = Nicolson (1982, fig. 5a), from Gilgil, Kenya; anubis (2) = (Muruthi 1997, fig. 3II.2), from Laikipia, Kenya; yellow (1) = Rhine, Norton, Wynn, and Wynn (1985, fig. 1), from Mikumi National Park, Tanzania. All other data from Amboseli National Park, Kenya, as follows: yellow (2) = Alto's Group, 1974–76, ages 30–70 weeks, this study; yellow (3) = Alto's Group, 1975–76, ages 0–50 weeks, from Jeanne Altmann (pers. comm.); yellow (4) = Hook's Group, monitoring data 1986–92, Muruthi data 1992–93 (Muruthi 1997). Values for yellow (2), this study, are from focal-animal samples; all others from point samples. Values for anubis (1) are group medians; all others are means. Trend line is for linear regression of y = log (percentage of daytime on nipple) on x = age (weeks), N = 106, R^2_{adj} = 0.72, $p \leq 0.00005$, y = 1.841 − 0.0222x.

day that yearling baboons spend on the nipple, and how it changes with age. For wild yellow and olive baboons, several samples of this ontogenetic change are available. In figure 6.8, I have superimposed all systematic samples known to me on age changes in nursing time (literally, time on the nipple) in wild baboons, omitting only data on one Amboseli group (Lodge Group) that feeds extensively on human refuse. The general tendency of nursing time to decline with age but at a decreasing rate of decline is clear from the figure.

The curved trend line in figure 6.8 shows a linear regression of the log of nursing time on age. The square of the regression coefficient indicates that age accounts for about 72% of the variability in nursing time. What about the scatter around the regression line? Two studies of baboons have analyzed other sources of variability in nursing time. Muruthi (1997) investigated potential sources of the remaining variability by comparisons within and between three groups, one of olive anubis, two of yellow baboons. Nursing time (nipple contact) was higher in low-ranking mothers and was highest in the habitat of intermediate food availability. Male and female infants did not differ in time spent on the nipple.

Wasser and Wasser (1995) correlated the slopes and intercepts of individual nursing time age functions during the first three months and of several other developmental functions with several independent variables. Nipple contact time declined significantly faster in one of three study groups; that group was the largest and most competitive. Independence from nipple contact developed faster in infants born later in the birth year (from December, the first month of the rainy season, to November). No significant differences in the rate of decline in nipple contact were detected between sexes or with the amount of play, maternal style as in J. Altmann (1980), mother's dominance rank, frequency with which other baboons handled the infant, or degree of physical maturity at birth. Rate of development of nipple independence was not a predictor of survivorship to one year of age.

Alternative Estimates

In this section I present alternative estimates of mean daily intake of milk and total dietary intake, then look at some of their implications for age changes in diet.

Note that in appendix 9, equation (A9.3) gives expected daily milk intake *per unit body mass* (ml $kg^{-1}day^{-1}$). *Actual* mean daily milk intake, at various ages, in milliliters per day, was obtained (appendix 9, step 4) from the product of mean body mass and mean intake per unit of body mass, each expressed as a function of age, and in appendix 9, I used a growth rate of 6 g/day (equation A9.4). If instead I use in equation (A9.4) our more recent growth rate of 4.5 g/day (see appendix 1, equation A1.1), then multiply equations (A9.3) and (A9.4), mean daily milk intake can be estimated from the following polynomial:

$$(6.1) \qquad m_1 = (264.38 - 3.11w)(0.775 + 0.0315w)$$
$$= 204.89 + 5.918w - 0.0980w^2,$$

where m_1 is daily milk consumption (ml per day) and w is age in weeks. For comparison, equation (6.1) gives an intake of 216 ml/day at 2 weeks of age, whereas estimates by Buss and Voss (1971, fig. 1A) indicate a mean intake of about 270 ml/day in the second week of life. For a baboon of the mean sample age, 47.5 weeks, equation (6.1) gives a milk intake value (with ml milk converted to mass at 1.027 g/ml) that is about 4% higher (271.8 vs. 260.9 g/day) than I obtained earlier in this chapter from the sample time-weighted average

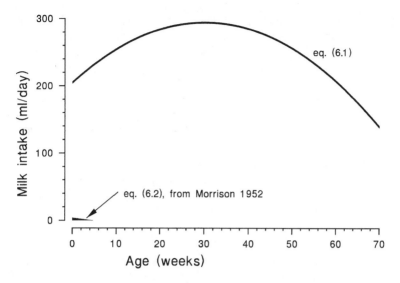

Fig. 6.9. Changes with age in milk intake, as predicted by equation (6.1) (this study) and equation (6.2) (Morrison 1952, based on data in Smith and Merritt 1922).

of the four age-specific values, each of which was, in turn, the unweighted mean of individual milk intake values. (Presumably the difference is due to the weightings.)

To my knowledge, humans are the only wild primates other than baboons for which a quantitative relationship between nursing volume against age has been estimated. Morrison (1952) gives the following equation, based on human data from Smith and Merritt (1922):

$$(6.2) \qquad z = -0.45 + (0.506 \pm 0.099)x - (0.189 \pm 0.05)(4.6)y,$$

where z is the mass in ounces of milk produced, x is the mass of the child in pounds, and y is the age of the child in weeks. Clearly equations (6.1) and (6.2) are not the same, even if in equation (6.2) we try to eliminate body mass based on its being a linear function of age, because equation (6.1) is a polynomial function of age whereas equation (6.2) is linear. But do they approximate each other over the age span we are dealing with? No. Equation (6.1) is shown graphically in figure 6.9. On this same graph, equation (6.2) is nearly invisible, descending linearly and precipitously from a mere 12 ml of milk at birth to less than nothing by 5 weeks of age. Clearly this equation is not applicable to baboons. Examination of equation (6.2) indicates why. The equation has a large coefficient for body size, and by human standards all baboon yearlings are extremely small for their age.

Returning to m_1, my first estimate of daily milk intake (equation 6.1), I ask,

Why should milk consumption be a polynomial function of age? Why should it increase, then decrease? By examining the components of equation (6.1) we see that the initial rise (fig. 6.9) is due entirely to the predominant effects of early growth, that is, to the positive term $264.38(0.0315w)$, which dominates the value of m at small values of w. The subsequent decline, after 35 weeks of age, results from the term $(-3.11w)(0.0315w)$, that is, from an interaction between the $0.0315w$ kg of mass that the infant has added by age w and the change in the mass-specific milk intake, which declines by $3.11w$ ml between birth and age w.

Because infant mammals are sustained for long periods on little more than milk, we can use milk intake to suggest an adequate dietary intake for infants. If the entire dietary requirement were met by milk alone, which it nearly is in neonates, then daily milk intake per kilogram of body mass would be 264.38 ml milk per kg (based on the y-intercepts of equations A9.3 and 6.1). During the first year of life, the infant baboon's body mass is increasing linearly with age, as discussed in chapter 5. If we first assume that food (nutrient) requirements of immature baboons are in proportion to body size, then food requirements too must be increasing linearly, so that for a yearling at age w and thus weighing $0.775 + 0.315w$ kg (second approximation, appendix 1), the expected total daily dietary intake d_1, expressed in milliliters of milk or its nutritional equivalent in other foods, is

$$(6.3) \qquad d_1 = 264.38 \, (0.775 + 0.0315w)$$
$$= 204.89 + 8.33w,$$

where, as before, w is age in weeks. The difference between these values for total diet and those of equation (6.1) for milk intake represents the contribution to its own diet made by the yearling's foraging: total intake = milk intake + intake from foraging. Subtracting equation (6.1) from equation (6.3), we get the following equation for the yearling's daily foraging intake, expressed in equivalents of milliliters of milk:

$$(6.4) \qquad f_1 = 2.41w + 0.098w^2.$$

This suggests that the yearling's increased foraging success has two additive components: a linear component, equivalent to an increase of 2.4 ml milk per week, and an exponential component, equivalent to 0.1 ml milk per week per week (i.e., an additional 0.1 ml the first week, twice that the second, etc.). The latter, accelerating component may represent the effects of learning to learn, or "deutero-learning" (Harlow 1951). An alternative explanation for the acceleration component of equation (6.4) is this. Limited strength restricts young yearlings to the smallest forms of some foods, for example, the corms of certain grasses. As the yearling increases in strength, it can feed on material of ever larger size. However, the size distributions of many foods, like most such

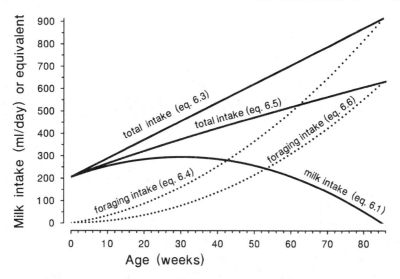

Fig. 6.10. Foraging intake (equations 6.4 and 6.6) predicted as the difference between milk intake, predicted by equation (6.1), and total intake, as predicted by both equations (6.3) and (6.5).

empirical distributions, probably have a left tail that is approximately exponential, so that a linear increase in strength leads initially to an exponential increase in available food.

The relationships among equations (6.1), (6.3), and (6.4) are shown graphically in figure 6.10. From equation (6.4) we can predict the total average dietary intake of the infants at the ages of those included in our study. At the ten-week intervals 35, ..., 65 weeks, equation (6.4) predicts mean foraging intakes equivalent to 204.4, 306.9, 429.0, and 570.7 ml of milk per day, if baboons obtain, via foraging, an amount of food that is nutritionally equivalent at each age to their declining milk intake.

I assumed in equation (6.3) that nutrient requirements are proportional to body mass. Suppose instead that nutrient requirements are proportional to basal metabolic rate (as suggested by Kleiber, 1947, 1961; Miller and Payne 1963; and Munro 1976; see also appendix 1), and therefore proportional to the three-quarter power of body mass (Kleiber 1947, 1961; but see Hayssen and Lacy 1985).[7] (Actually, ontogenetic scaling of metabolic rates is poorly understood.) Then the total daily food requirement (milliliters of milk or equivalent) per basal kilocalorie, based on the y-intercept of equation (6.1) and on 70 basal kilocalories per three-quarter power of body mass expressed in kilograms (Kleiber 1947; Bartholomew 1972), is $204.89/70(0.775)^{3/4} = 3.54$. The number of basal kilocalories for a yearling at age w and thus on average at mass $0.775 + 0.0315w$ kg is $70(0.775 + 0.0315w)^{3/4}$. The product of these two gives a second estimate of the yearling's total dietary requirement, again expressed in equivalents of milliliters of milk:

$$(6.5) \qquad d_2 = 248.05(0.775 + 0.0315w)^{3/4}.$$

The difference between equations (6.5) and (6.1) gives us a second estimate of the average yearling's daily foraging intake, expressed in equivalents of milliliters of milk:

$$(6.6) \quad f_2 = -204.89 + 248.05(0.775 + 0.0315w)^{3/4} - 5.918w + 0.098w^2.$$

The relationships among equations (6.1), (6.5), and (6.6) are shown in figure 6.10. At the ten-week age intervals, 35, . . . , 65 weeks, equation (6.6) predicts, respectively, mean foraging intakes equivalent to 105.9, 174.2, 260.3, and 364.6 ml milk per day.

In view of the assumption of step 4a, step 3 is nearly equivalent to assuming a constant, age-invariant mean mass-dependent milk intake per minute of suckling (ml kg^{-1} min^{-1}) and therefore, because the yearling is growing, an increase in the actual milk flow per minute of nursing: older yearlings drink faster. (It would be equivalent if the best fit regression line of step 2 went through the means at all ages [$r^2 = 1$], which with my data it did not. The equivalence results from the fact that the regressions of mean suckling time and of ml milk/kg day are both straight lines with a common x-intercept [step 3] and therefore have a constant ratio for all values of x.) I suspect that this assumption of an ever increasing rate of milk flow is not correct, but rather that milk flow per minute of suckling declines near the end of lactation (referring both to bouts and to suckling age). However, in the absence of data on this topic from any wild mammal there is no way to take into account any such change.

J. Altmann (pers. comm.) has questioned the assumption that milk intake per minute of suckling is higher for older and larger yearlings, except perhaps, during the first few weeks of life. She has suggested that milk intake per minute of suckling is essentially invariant with age for infants from at least the seventeenth week, which is the last age class in Buss and Voss's data. That assumption leads to a second estimate of milk intake, based on milk intake and nursing time. Buss and Voss (1971, fig. 1B) indicate daily milk intake during the seventeenth week as 460 ml, but they do not indicate nursing time. However, nursing time at that age can be estimated from J. Altmann's point samples of nursing in Amboseli baboons. Her data were tabulated by months. The seventeenth week is at the boundary between the fourth and fifth months of life. In the fourth month, suckling took up 28.3% of the daytime (ten hours, 0800–1800 h), based on the mean of four infants for which sample sizes were judged adequate. In the fifth month the mean was 23.7%, based on seven infants with adequate samples. I will take the midpoint of these two values, 26.0%, as an estimate for the seventeenth week. (Similarly, in Nicolson's 1982 study of anubis baboons, suckling occupied 28% of the daylight hours during weeks 17–24.) Using the Amboseli data for nursing time and Buss and Voss's data for milk intake, I get a daily milk intake rate of $460/0.26(600) = 2.95$ ml milk per twenty-four hours for each minute of suckling during the ten hours 0800–1800 h. On this basis and using the nursing time-budget data in table 6.6 (means of individual means, reduced by a factor of 11/10, since I sampled during eleven hours per day, 0700–1800 h), I calculated mean daily milk intakes for each age block in my study:

(6.7) $m_2 = (240.3, 167.8, 159.5, 87.5)$ ml milk/day

in the same four ten-week age blocks, respectively. (These values are all well below the estimates in table 6.6, which was based on step 4, above.)

I then subtracted m_2 successively from d_1 and d_2 (each evaluated at ages 35, . . . , 65) to get two additional estimates of f, respectively:

(6.8) $f_3 = (256.1, 411.9, 503.5, 658.8)$ ml/day

and

(6.9) $f_4 = (157.6, 279.1, 334.8, 452.7)$ ml/day,

at ages 35, . . . , 65, expressed, as before, in terms of nutritional equivalents of milk. In chapter 7, the four sets of predictions $f_1, . . . , f_4$ will be tested.

Pooled Means versus Means of Means

Before leaving the topic of age changes in milk intake, let me mention a statistical problem I may not have resolved satisfactorily. A rule of thumb in statistics is that when calculating the mean of a population, data from subpopulations are pooled only if the subpopulations are known to be statistically homogeneous or can reasonably be assumed so; otherwise the mean of the subpopulation means is used. In practice, however, this rule is not always unambiguous, as I shall show, and in a large study like this one takes on the extra computational burden and loss of sample size that are imposed by the mean of means only if convinced that doing so is decisively preferable.

In calculating the milk intake of each individual at each age (table 6.6), I made use of values for nursing bout lengths and bout rates that were calculated separately for each individual in each of the four age blocks in which it was observed. Subsequent statistical analysis (chapter 7) showed that nursing bout rates indeed differed very significantly across age blocks (table 7.6D), and within each age block the rates differed very significantly between individuals. Nursing bout lengths differed between individuals in the first and fourth of the four age blocks but not the other two (table 7.6). However, in view of the other demonstrated differences, it seemed reasonable to suppose that in reality bout lengths differ among individuals at all four ages and that a larger sample would show this to be so. Yet surprisingly, when bout lengths within each age block were pooled and their distribution compared with bout length distributions in the other three age blocks, no significant age difference was revealed: there are within-age differences, but not between-age differences.

Given these characteristics of nursing bout lengths and rates, which values of each should one use in calculating the average nursing time budget at each age? (Recall from equation 3.2 that the time budget is obtained from the product: bout length × bout rate × potential foraging time.) One can justify several choices of values. For example, when calculating the values in table 7.7, I used age-specific values for any food that showed significant age differences in bout length (table 7.2); otherwise pooled values (for 30–70 weeks) were used (table 4.1). Similarly for bout rates: if they differed significantly by age, age-specific values were used (table 7.6A–D); otherwise pooled values for 30–70 weeks (table 7.6E) were used. In each case the age-specific values used were the results of pooling all yearlings sampled at that age; but if individual differences within an age block have been demonstrated, shouldn't one use the mean of individual means for that age block? And suppose, as in the case for bout lengths, that individuals differ within an age but age differences are not demonstrated. What then? What about other possible ways of partitioning a population that might yield more homogeneous subpopulations, such as by sex or by maternal rank? Is one obliged to try many potential partition criteria? Age-class differences are confounded by changes in the cast of characters that are available to sample at each age. Should age-class comparisons be based only on those individuals that were sampled in all four (or, say, at least three) age classes? If so, what age-specific value should one use if the pooled age values of these well-sampled yearlings differ but the age values pooled over all yearlings do not?

Similar problems occur when characterizing each individual at each age, and when nutrient intakes were determined (chapter 7) I used the milk intake values shown in table 6.6. In table 6.1, which supposedly depicts the average of the baboons in my samples, bout length and rate values other than those for milk were obtained from data pooled over the entire 30–70-week age interval. Would the means of individual means have been preferable? Or the means of age-specific means, each based on the mean of individual means?

Perhaps these questions are of interest only to statisticians. Empirically, do the results differ appreciably? To answer this question in a particular case, I have calculated the average nursing time budget at each age, in four different ways: (1) using the mean bout length and bout rate of each individual at each age, calculating each individual × age time budget, then averaging the individual time budgets within each age; (2) for each age, using the mean of these individual $\times$ age bout rates and lengths to calculate the time budget; (3) for each age, using pooled mean values for bout rates and lengths; and (4) for each age, using these pooled mean rates and lengths only if they differ significantly, otherwise using values pooled over the entire 30–70 weeks. The results are shown in table 6.8. In the best case (30–40 weeks) all values are within 4.2% of the median. However, in the worst case (age 50–60 weeks) the spread is 17.2% on either side of the median. Clearly, the choice of methods matters. I have tried throughout this book to indicate the source of values that were used in each calculation.

A Monte Carlo simulation by Machlis, Dodd, and Fentress (1985) of the effects of taking repeated measures on each of several individuals and pooling the results versus taking a single observation per individual suggests that pooling does not have any untoward effect on estimates of the mean of the population of individuals. However, in pooled samples, the variance error of the mean (and thus its standard error) is reduced unless the variability between individuals is high relative to total variance (within- plus between-individual), and correspondingly, the probability of a type I error—of rejecting a true null hypothesis—is almost always substantially greater than the stated alpha level. The alternative, making just one observation per individual, is often impracticable. (Oh that I could sample the behavior of a random 18,460 baboons!) Moses, Gale, and J. Altmann (1992) provide an illustration of an appropriate statistical analysis of repeated measures, growth of body mass, taken on the same individuals.

Who Terminates Nursing Bouts?

I mentioned above that there were no significant differences with age in the lengths of nursing bouts. Who decides when to terminate a nursing bout, the infant or the mother? I have data to answer this question. Opportunistically during my 1975–76 focal samples, I recorded whether a yearling terminated its suckling "voluntarily," that is, as a result of its own action, or "involuntarily," as a result of some action by the mother, for example, pushing or hitting the yearling or moving so as to pull the nipple from its mouth. The results of 594 such observations (45% of the 1975–76 nursing bouts) are given in table 6.9. From 30 to 70 weeks of age, 38% of all nursing bouts were terminated involuntarily. The overall proportion of involuntary terminations, all infants pooled, showed no detectable age changes or differences from 30 to 70 weeks of age (homogeneity test, $P(\chi_3^2 = 3.42) > 0.05$). By contrast, in a study of anubis baboons (*Papio anubis*) covering the first 64 weeks of age, Nicolson (1982) found changes in the percentage of mother-terminated nursing bouts, which peaked at 40–48 weeks.

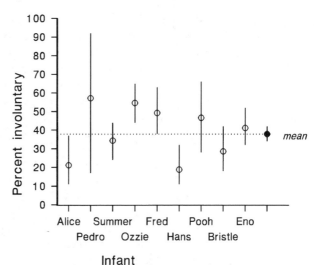

Fig. 6.11. Percentage of nursing bouts terminated by the mother (i.e., involuntarily by the infant), 1975–76 data. Vertical lines are 95% binomial confidence limits. Overall mean is from pooled data. Infants ordered based on relative dominance ranks of their mothers, highest to the left.

In my sample, significant individual differences occurred in the proportions of involuntary terminations, both overall, for which $P(\chi_8^2 = 55.26) \ll 0.005$, and in each ten-week age block from 30 to 70 weeks of age (table 6.9). These individual differences indicate that the mean of individual means should be used to characterize the population, rather than the pooled result above. However, when one does use it, the value is only one percentage point higher: 39% involuntary terminations.

Individual proportions of mother-terminated nursing bouts and their 95% binomial confidence limits are shown in figure 6.11. These differences have an interesting relation to whether the mothers of these infants were "restrictive" or "laissez-faire" in the infants' first few months, using the criteria of J. Altmann (1980). The mothers of Bristle, Hans, and Summer were restrictive (J. Altmann 1980), and these three are all below the mean in figure 6.11. Conversely, Pedro's mother was laissez-faire, and she terminated a higher percentage of nursing bouts than any other mother, but because of Pedro's early death my sample on him is very small. (The maternal style of the other mothers is unknown.) This finding of fewer mother-terminated nursing bouts for infants of restrictive mothers is consistent with J. Altmann's findings that such infants spend more time in contact with their mothers than other infants of the same age (1980,

145) and in general develop more slowly (1980, 187–88). In figure 6.11 the infants are ordered by their mothers' relative dominance ranks, which appear unrelated to the proportion of an infant's nursing bouts that are terminated by the mother.

Because of the possibility of sex-determined differences in the relationships of yearlings with their mothers (e.g., Simpson 1983), I retabulated the nursing termination data by sex. Only a small difference was found. For males, 60.1% of nursing bouts were terminated voluntarily ($n = 311$); for females, 64.3% ($n = 280$). Nicolson (1982), in her study of anubis baboons, reported that "no differences were found in any measure of suckling activity, contact, proximity, maternal rejection or aggression."

Nutrient Intakes

Estimation

The mean daily intake of the yearlings was estimated for each of the six standard nutrient classes of macronutrient analysis: water, minerals (ash), lipids, proteins, fibers, and other carbohydrates. In addition, mean daily energy intake was estimated from the intakes of lipids, proteins, fibers, and other carbohydrates, as described in appendix 1.

To estimate the mean daily intake of any specified nutrient or other food component for a given individual or population, I multiplied together the mean values (for that individual or population) of the following six factors for each food j (see equations 3.1 and 3.2): (1) *bout length,* L_j (minutes of feeding on food j per bout on that food); (2) *bout rate,* R_j (number of bouts on food j per potential foraging minute); (3) *potential foraging time, S* (potential foraging minutes per day; the product $A_j = L_j R_j S$ is the number of minutes per day spent feeding on the jth food); (4) *unit intake rate,* B_j (units of food j eaten per minute of feeding on that food); (5) *unit mass,* C_j (grams of food j per unit of that food); (6) *composition,* D_{ij} (grams of component i per gram of food j). To adjust for seasonal shifts in sampling intensity and food availability, an additional factor, f_j/t_j, was included, as discussed in appendix 8.

Ideally, these products, $A_j B_j C_j D_{ij} f_j/t_j$, each of which gives the daily intake of the specified nutrient from the jth food, would then be summed over all of the baboons' foods to get the total daily intake of the nutrient in question. However, for practical reasons discussed earlier, the values of $A_j B_j C_j D_{ij} f_j/t_j$ were calculated only for the fifty-two core foods. Then allowance was made, as described above, for the contribution of the non-core foods: for each nutrient, I multiplied the total yield from non-milk core foods by 1.095, then added the contribution made by milk. For optimal diets or others in which the daily consumption of each food is specified or known, daily intake of the ith nutrient or other component was computed from $\Sigma_j D_{ij}F_j$, where for each food j, F_j is the specified amount to be consumed per day.

For each of the fifty-two core foods, values for A_j, B_j, and C_j and for their product $A_jB_jC_j$ are given in table 6.1. These are pooled mean values, based on samples taken on the eleven yearlings and using all samples taken between their thirtieth and seventieth weeks of age. For these same fifty-two foods, composition values D_{ij} are given in table 5.1 for the six macronutrients plus energy. For milk, the revised mean uptake, 260.9 g/day, was used to calculate nutrient values in the remainder of this chapter, except where noted. The contribution $A_jB_jC_jD_{ij}$ of each core food j to the baboons' pooled mean daily intake of proximate nutrients is given in table 6.5 (seasonally adjusted diet, data pooled from all eleven infants 30–70 weeks old). In table 6.5 these nutrient contributions are grouped by food class (seeds, corms, etc.), then an allowance is made for the contribution of non-core foods in order to estimate the total daily intake of each nutrient.

Nutrient Intakes: Seasonally Adjusted Diet

The baboons' seasonally adjusted, pooled mean diet (table 6.5) had a mass of 391.5 g per day, including 260.9 g milk, and required 240.5 minutes per day to consume (36.4% of daytime). It contained 293.4 g water (74.9% of diet mass), 6.3 g minerals (1.6%), 15.1 g lipids (3.9%), 11.1 g proteins (2.8%), 10.1 g fiber (2.6%), and 55.2 g other carbohydrates (14.1%). It had an average energy density of 4.57 kJ per gram and yielded 1.79 MJ of energy per day, on average. If requirements for these nutrients scale, like basal metabolism, to body mass to the three-quarter power (appendix 1, but see also note 7), then at a mean body mass of 2.27 kg, these daily intakes can be expressed as a function of body size, as follows: $158\,M^{3/4}$ g water, $3.4\,M^{3/4}$ g minerals, $8.2\,M^{3/4}$ g lipids, $6.0\,M^{3/4}$ g proteins, $5.5\,M^{3/4}$ g fiber, $29.8\,M^{3/4}$ g other carbohydrates, and $967\,M^{3/4}$ kJ energy, where M is body mass in kg.

Nutrient yield rates of individual foods are indicated in table 6.10. The jth food's yield rate of any nutrient i (table 6.10) was obtained from the product of its unit intake rate, its unit mass, and its concentration of that nutrient—that is, from $B_jC_jD_{ij}$ of equation (3.1). (These rates are not known to vary seasonally and so received no seasonal adjustment.) Overall intake rates for energy and protein were as follows. The pooled mean energy intake rate (pooled over all foods) was 7.45 kJ per minute of feeding; the mass-weighted mean rate was 11.1 kJ/minute (0.67 watt). For protein, the pooled mean intake rate was 0.046 g of protein per minute of feeding; the mass-weighted mean was 0.059 g/minute.

Major Sources of Nutrients

Of the foods that are available to yearling Amboseli baboons, in which are nutrients most concentrated? Which foods contributed the most to their actual intake of each nutrient? In this section I answer these two questions for macronutrient classes (proteins, carbohydrates, etc.), combining information in table 5.1 for the compositions of the baboons' core foods, in table 6.5 for the baboons' nutrient intakes (seasonally adjusted, pooled mean annual diet), and table 6.10 for their nutrient intake rates. I also summarize more limited data on rich sources of four vitamins.

Figure 6.12 shows the frequency distribution of the core foods for their concentrations (percentage of mass in fresh matter) of six macronutrient classes of nutrients. In that figure, the richest foods in each class are identified by their four-letter abbreviations (table 6.2). The same abbreviations are used in tables 6.5, 6.10, and in the following text.

Proteins: Foods with Highest Concentrations

Of the baboons' core foods, insects (INKX) have by far the highest protein content (27%). Other animal material that baboons eat but that is not among the core set, such as birds, bird eggs, and various mammals that baboons capture, have comparably high values. Of the plant material they eat, seeds of the two acacia trees (TODR, TODG, ACDD, FTDG, FTDR, and by inference most of XXXG and XXXD) are the only foods providing more than 16% protein. Just behind them, with 13% protein, is the fruit of the sticky-fruit plant (*Commicarpus pedunculosus,* SFFX), which is available only after the rains. After that come the moist green seeds of fever trees and the larvae of dung beetles (both 11% protein), then the corms of various grasses (SKCX, SMCX, and GRCX, 10% protein). All other foods contain less than 9% protein.

Proteins: Sources of Highest Intakes

If one looks at the weanling baboons' actual intake of protein (table 6.5), the overwhelming contribution made by milk to their seasonally adjusted diet is apparent: 5.1 g/day. Even at this age (mean age when sampled, 47.5 weeks), milk contributed nearly twice as much protein as the next highest food, and much more than any beyond that. Somewhat surprising is the identity of the second-largest source of protein: green leaves of unidentified grasses and sedges (GRLU). On average, baboons ate these leaves for about 20 minutes per day, and except for milk they were the bulkiest food in the diet (35.5 g/day), with a yield of 2.3 g of protein per day. My impression was that the largest source of this unidentified green grass was *Cynodon nlemfuensis,* which is

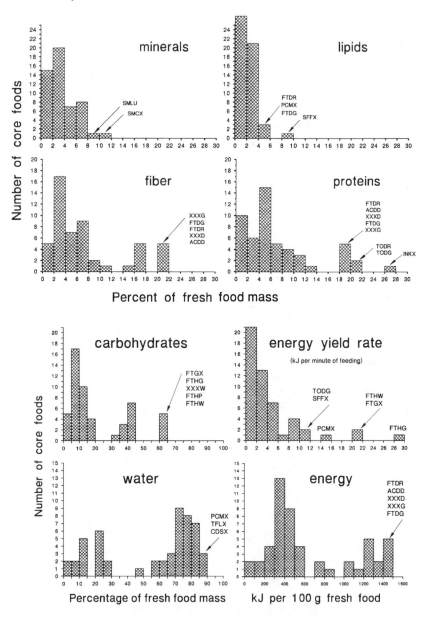

Fig. 6.12. Frequency distribution among core foods of macronutrients and energy concentrations, and of energy yield rates. Four-letter food codes as in table 6.2.

available year-round on the edges of swamps and waterholes but was usually recorded as unidentified grass when the telltale seedheads were not present on the plants being eaten. In addition, various grasses that are green only during the rainy seasons and that could not always be identified on sight contributed to this category. No food other than milk and green grass blades contributed more than a gram of protein per day, and the protein obtained from just these two sources accounted for 59% of the young baboons' total.

Although milk and green grass were the sources of by far the largest amounts of protein in the yearlings' diet, the protein yield rates of these two foods (63 and 115 mg/minute, respectively) were not the fastest. In this, both were exceeded by sticky-fruit (SFFX in table 6.10, 200 mg/minute), umbrella tree seeds from the ground (TODG, 168 mg/minute), insects (INKX, 128 mg/minute), and ripe fruit of the shrub *Azima tetracantha* (ATFR, 116 mg/minute). Milk's protein yield rate was exceeded by several additional foods, namely, acacia seeds (ACDD and by inference most of the small unidentified items picked up from the ground, XXXG, 87 and 112 mg/minute, respectively), *Azima tetracantha* fruit both green and of unidentified ripeness (ATFU and ATFX, 77 and 104 mg/minute), and green fruit of the shrub *Salvadora persica* (SPFU, 90 mg/minute).

For young baboons, the great advantage of milk and green grass is their accessibility: all foods that have greater protein concentrations or that yield protein more rapidly either are very seasonal, like sticky-fruit, or are sparse or otherwise difficult to obtain, like animal foods.

Energy: Foods with Highest Concentrations

The energy value of foods is a linear function of the amounts of lipids, carbohydrates, and proteins in them, and since the baboons' foods are low in lipids, the caloric value depends primarily on proteins and carbohydrates. The highest energy values among the baboons' core foods are the seeds of acacia trees (ACDD, FTDG, FTDR, FTDU, and by inference XXXD and XXXG), with 14–15 kJ/g, and the gum of these trees (FTGX, FTHG, FTHP, FTHW, and by inference XXXW), with 12 kJ/g. These are the only core foods with at least 12 kJ of utilizable energy per gram of fresh matter. Next came the corms of their favorite grasses and sedges (SMCX, SKCX, COCX, and GRCX), with values exceeding 8 kJ/g. All other core foods contain less than 8 kJ per gram of fresh matter.

Energy: Sources of Highest Intakes

Milk was the young baboons' major source of energy, 841 kJ per day. The food making the second-highest energy contribution was gum picked from fever trees (FTGX, 293 kJ/day). However, to appreciate the total energy contribution made by fever tree gum, which the baboons pick from various places on and under the trees and at various degrees of aging, we need pooled results (below). Only two other foods contributed more than 100 kJ/day: green leaves of un-identified grasses (GRLU, 165 kJ/day) and fever tree gum picked up from the ground, which is typically more fermented than gum picked from the tree (FTHG, 103 kJ/day).

As was true of protein, so too for energy: core foods with higher yields were not necessarily ones with higher yield rates. Although milk contributed far more energy to the baboons' diet, fever tree gum yielded energy more rapidly: fever tree gum from the ground (FTHG, 28.3 kJ/minute of feeding, equivalent to 471 watts), from trees and from logs (FTGX and FTHW, both 20.6 kJ/minute, 344 watts), whereas milk yielded energy at 12.8 kJ per minute on the nipple (table 6.10), equivalent to 213 watts. (The recorded time on the nipple includes both suckling and non-suckling time; the extent to which the latter is required to maintain milk flow is unknown.) Next in order of the year-lings' energy intake rate are sticky-fruit (SFFX, 12.0 kJ/minute) and umbrella tree seeds from the ground (TODG, 11.1 kJ/minute). No other core food yielded energy at more than 10 kJ per minute of feeding.

For energy as for proteins, the discrepancies between the richest foods and the greatest contributors to the yearlings' intake appear to be the result of dif-ferences in accessibility. Milk is available and accessible daily, year-round; acacia seeds are not. Although green acacia seeds are a seasonal favorite of the baboons, milk ends up contributing more energy on a yearly basis.

Carbohydrates: Foods with Highest Concentrations

The food with by far the highest concentration of non-fiber carbohydrates is the gum of the fever tree (FTGX, FTHG, FTHP, FTHW, and by inference most of XXXW), with 63% sugar. No other food of Amboseli baboons comes close to this carbohydrate level. In five analyses of this gum from sources in Am-boseli, carbohydrate has ranged from 43% to 82% of fresh mass (S. Altmann, Post, and Klein 1987; Hausfater and Bearce 1976). The primary sugars are D-galactose and D-arabinose (Hausfater and Bearce 1976). The fresh exudate is usually transparent and colorless, or nearly so. Once it is exposed outside the tree, the gum's composition and color slowly change; as it ages it becomes pro-gressively browner and more opaque. At all stages, fever tree gum is the ba-boons' candy, and as I mentioned in chapter 4, they will go considerably out of their way to get at rich sources of it. On the other hand, gum from the local um-brella trees contains less than 1% carbohydrate (Hausfater and Bearce 1976), has higher levels of phenols and tannins (Wrangham and Waterman 1981), and is very sparse; not surprisingly, baboons rarely eat it.

Seven other core foods have carbohydrate concentrations in excess of 40%:

dry fever tree seeds and dry umbrella tree seeds from various sources (FTDG, FTDR, ACDD, TODR, TODG, and by inference XXXG and XXXD), ranging from 43% down to 40% carbohydrate. Also rich in carbohydrates are grass and sedge corms (SKCX, SMCX, GRCX, and COCX), with 39% to 33% carbohydrate.

Carbohydrates: Sources of Highest Intakes

Milk was the yearling baboons' greatest source of non-fiber carbohydrates (18.5 g/day). Again, accessibility has a major impact on intake. Gum picked from fever trees (FTGX) was a close second (15.3 g/day), while fever tree gum picked up from the ground (FTHG, 5.4 g/day) and green blades of un-identified grasses (GRLU, 3.6 g/day) ranked next. After that came the carbo-hydrate yields of unidentified objects (presumably mostly acacia seeds) from the ground (XXXG, 1.6 g/day), corms of unidentified grasses and sedges (GRCX, 1.0 g/day), and the seeds of umbrella and fever trees (TODU at 0.75 g/day, FTDG at 0.66 g/day). All other foods provided the baboons with less carbohydrate.

How fast did the young baboons harvest carbohydrates? The three foods from which they obtained the most carbohydrates—milk, fever tree gum, and green grass—were also among the foods that yielded carbohydrates most rapidly. In particular, fever tree gum picked from trees (FTGX) and from where it had dripped and collected on wood (FTHW) and on the ground (FTHG) all yielded sugars at rates over a gram per minute of feeding: 1.5, 1.1, and 1.1 g/minute, respectively. (By contrast, the isolated drops of gum that the ba-boons sometimes got from the leaves of understory plants [FTHP] yielded sug-ars at an order of magnitude more slowly, 0.1 g/minute.) Well below the best of the gum sources came acacia seeds picked from the ground and from dung (TODG, ACDD, and FTDG, ranging from 325 down to 120 mg/minute), milk at 281 mg/minute, ripe fruit of *Azima tetracantha* (259 mg), acacia blossoms picked from the trees (fever trees, FTBS, at 187 mg, umbrella trees, TOBS, at 106 mg), unidentified green grass leaves (GRLU, 182 mg), "sticky-fruit" (SFFX, 166 mg), and *Salvador persica* fruit (SPF*, 146 to 143 mg/minute). All other core foods yielded carbohydrates at less than 100 mg per minute of feeding.

Lipids: Foods with Highest Concentrations

When it comes to lipids, "sticky-fruit" (SFFX) leads the list of core foods, at 9.6%. Next comes milk (PCMX), at 4.9%. The only other food with an appre-ciable amount of lipid is dry fever tree seeds, at 4% (FTDG, FTDR, ACDD, and also, by inference, XXXG and XXXD). Close behind these are the green fruits of a shrub, *Azima tetracantha* (ATFU), and insects (INKX), with 3.8%

lipid, dry umbrella tree seeds (TODR, TODG), with 3.6%, and the fruit of another shrub, *Withania somnifera* (WSFX), with 3.5%. All other core foods contained less than 3.5% lipids.

Lipids: Sources of Highest Intakes

For fats (lipids), the baboons had no major source other than milk (12.8 g/day). Green grass leaves (GRLU) were a distant second (781 mg/day) and gum picked from fever trees (FTGX), with less than half of that, was third (267 mg/day). All other core foods provided less than 144 mg of lipids per day.

Milk was the yearlings' predominant source of lipids both because of its relatively high lipid concentration and because of the volume they consumed. The other high-lipid foods listed above, such as sticky-fruit, are too seasonal to have averaged much over the year, though each was seasonally important. Indeed, sticky-fruit ranked twelfth, contributing an average of 50 mg/day to the annual diet.

That sticky-fruit's relatively low contribution to the baboons' overall lipid intake was an artifact of its seasonality is evident when one looks at the rates per minute of feeding at which various foods yielded energy. Sticky-fruit (SFFX) yielded lipids at 147 mg per minute of feeding on it, a rate second only to milk (PCMX, 194 mg/minute). After these two came the fruit of the common shrub *Azima tetracantha* (ATFR, ATFX, ATFU, 64 to 41 mg/minute). Ranking below these came green grass leaves (GRLU, 40 mg/minute), acacia seeds picked up from the ground (TODG and by inference much of XXXG, 29 and 23 mg/minute), fever tree gum picked up from the ground (FTHG, 26 mg/minute), and the flowers of umbrella trees (TOBS and TOBR, 24 and 20 mg/minute). No other core food yielded lipids at more than 19 mg per minute of feeding on it.

Minerals: Foods with Highest Concentrations

As for minerals ("ash"), the annual grass *Sporobolus cordofanus,* which was eaten so extensively by Dotty and Striper during 1974, has the highest mineral content: 10% and 8.7% in the corms and green leaves, respectively (SMCX, SMLU). These values may be inflated by adhering dust. The only other core foods with at least 8% mineral content are the leaves of *Salvadora persica* (SPLU, SPLX), corms of unidentified grasses (GRCX, but based in part on SMCX), and fever tree gum (FTGX, FTHG, FTHP, FTHW, and by inference most of XXXW). For Amboseli fever tree gum, analyzed mineral content has varied widely, from 1.6% (Hausfater and Bearce 1976) to 15.1% (S. Altmann, Post, and Klein 1987), perhaps as a result of included dust and dirt.

Of particular interest is the mineral content of *Suaeda monoica,* for although it is not one of the baboons' core foods, its terminal growth, including the cryptic inflorescence, is occasionally eaten, yet this plant is an obligatory halophyte

(Waisel 1972). In Amboseli, one sample of the terminal growth (flowers, stem tips, leaves) of this plant and two samples of green leaves contained 0.9%, 3.9%, and 5.8% ash, respectively. These values are unexpectedly low in view of Ovadia's (1967) demonstration (quoted in Waisel 1972) that this plant tends to move sodium into young leaves and buds. Additional samples are desirable.

Minerals: Sources of Highest Intakes

Among foods, the primary source of minerals in the baboons' diet were fever tree gum from the trees (FTGX, 1.5 g/day) and the ground (FTHG, 0.52 g/day), green grass blades (1.3 g/day), and milk (0.78 g/day). Below them came unidentified grass/sedge corms (GRCX, 0.2 g/day), a variety of fruits (ATFR, SPFU, TTFX, ATFX, SPFS, SMCX, TCFX), which yielded from 0.15 to 0.07 g minerals per day, and various sources of acacia seeds (TODU, FTDG, and by inference much of the unidentified objects from the ground, XXXG), which yielded 0.14 down to 0.6 g minerals per day. All other foods provided 0.04 g ash per day or less.

As for rates of mineral intake, the young baboons harvested minerals most rapidly from fever tree gum (FTHG, FTHW, FTGX at 143 to 104 mg/minute), *Salvador persica* fruit (SPFU, SPFX, SPFS at 87 to 53 mg/minute), *Azima tetracantha* fruit (ATFR, ATFX, ATFU at 69 to 43 mg/minute), and "sticky-fruit" (SFFX, 66 mg/minute). All other foods yielded minerals at rates below 43 mg per minute of feeding.

Fiber: Foods with Highest Concentrations

Fibers make up 18–22% of the dry seeds of fever trees and umbrella trees (FTDG, FTDR, TODG, TODR). These fiber values are considerably higher than those of legume cultivars, for which fiber values almost all fall in the range of 4–12% (e.g., Leung 1968). As indicated in appendix 5, the digestibility of fiber by baboons is unknown but probably is appreciable. The only other core foods with high fiber content (over 16%) include various grass corms (GRCX, SKCX, SMCX). On the other hand, the corms of the sedge *Cyperus obtusiflorus* (COCX, COCU), which are eaten by baboons and grow intermixed with the perennial grass *Sporobolus rangei* (SKCX) but at much lower densities, have a lower fiber content (6.4–12.5%, S. Altmann, Post, and Klein 1987).

Milk is apparently the only fiber-free food in the baboons' diet (table 5.1). (Other zero fiber values in tables 6.5 and 6.10 are artifacts of rounding values less than 50 mg or from wood, which is gnawed but seldom swallowed by the infants and is scored as having no nutritive value.)

Fiber: Sources of Highest Intakes

The two primary sources of fiber in the infant baboons' diet were green grass blades (GRLX, 2.8 g/day) and fever tree gum from the tree (FTGX, 1.8 g/day), with all other foods contributing well under a gram per day. For example, the next highest sources were fever tree gum from the ground (FTHG, which provided 621 mg of fiber per day), grass and sedge corms (GRCX, 436 mg), and the green leaves of Bermuda grass (CDLU, 384 mg).

Yield rates of fiber were highest from fever tree gum from the ground, from wood, and from the tree (FTHG, FTHW, FTGX at 171 down to 125 mg/minute), but again, lower in the droplets picked from the leaves of understory plants (FTHP, 14 mg/minute). Other foods from which the yearlings rapidly harvested fiber include acacia seeds (TODG, ACDD, FTDG, and probably much of XXXG, at 143 down to 60 mg/minute), green leaves of Bermuda grass (CDLU, 64 mg), unidentified green grass (GRLU, 142 mg/minute), numerous fruits (ATFR, SFFX, ATFX, SPFU, SPFX, SPFS, SPFR, at 83 down to 53 mg/minute), the corms of the grass *Sporobolus cordofanus* (SMCX, 63 mg/minute) and the grass seedheads of grasses and sedges (GRDU, 53 mg/minute). All other foods yielded fiber at less than 50 mg per minute of feeding.

Water: Foods with Highest Concentrations

Even more succulent than milk are the stolons of Bermuda grass (CDSX): 87% water. Milk was second with 86%. Then in order came the leaves of trumpet-flower plants (TFLX, 85%), ripe *Salvador persica* fruit (SPFR, 84%), devil's thorn fruit (TTFX, 82%), green leaves of the shrub *Suaeda monoica* (SULU, 82%), dung-beetle larvae (DUKU, 81%), then the flowers of trumpet-flower plants and umbrella trees (TFBS, TOBS and TOBR, all 80%). No other food contained more than 79% water.

The water content of the baboons' core foods varies relatively little compared with other nutrients; the most succulent contain 87% water, but half contain more than 70% water. Yet these variations are important because water is a diluent of other nutrients: the water content of the core foods is negatively correlated with the concentrations of all other macronutrient classes.

Water: Sources of Highest Intakes

At the ages in my samples, 30 to 70 weeks, the baboons were already drinking water from the same sources as the adults: swamps, waterholes, and rain pools, but I could not estimate the quantity or the rate at which the youngsters imbibed from these sources. Among their core foods, milk contributed by far the greatest amount of water: 224 g/day—nearly an order of magnitude more than the solid food from which they got the most water, green grass blades (GRLX, 25 g/day). Grass blades in turn contributed about an order of magnitude more water than the third-ranking food, fever tree gum from the trees (FTGX), from which the baboons got 5.2 g/day. Gum dries out as it ages: the fever tree gum that the baboons got from the ground yielded only 1.8 g of water per day. Other foods that ranked high included green leaves of Bermuda grass (3.4 g) and the fruit of several plants (ATFR, TTFX, ATFX, SPFS, SPFU, TCFX, and SPFR), ranging from 3.8 down to 1.2 g/day. Green umbrella tree seeds provided 2.6 g/day. The only other foods contributing at least a gram of water per day were the blossoms of the two acacias (TOBS at 2.7 g and FTBS at 1.4 g).

Milk provided water at the highest rate (again, setting aside drinking water), 3.4 g/minute on the nipple, and this despite long periods without any apparent sucking movements. Various fruits were also very high (ATFR, ATFX, SPFR, SPFS, SPFX, SPFU, SFFX, TTFX, and ATFU), contributing water at 1,786 down to 777 mg/minute of feeding. Other foods that provided water quickly include green leaves of some grasses (GRLU, CDLU at 1,254 and 930 mg/minute, respectively), the green leaves of trumpet-flower plants (TFLX, 882 mg/minute), the blossoms of acacias (TOBS, TOBR, FTBS, at 954 to 789 mg/minute), green grass seeds (GRDU, 581 mg/minute) and fever tree gum from the ground (FTHG, 501 mg/minute). All other foods yielded water more slowly.

Ascorbic Acid

Data on vitamins in Amboseli primate foods are not as extensive as for macronutrients and are inadequate to document selective foraging. Here I describe the richest known sources of the four vitamins that have been analyzed to date in the core foods of yearling baboons in Amboseli. Data are from S. Altmann, Post, and Klein (1987).

Of thirty analyses for ascorbic acid (vitamin C) in foods that are commonly eaten by baboons, the highest value is for umbrella tree flowers: 178 mg (All vitamin concentrations are given here in mg or μg per 100 g fresh food mass.) By comparison, domestic oranges average 50 mg (Watt and Merrill 1963). The flowers of fever trees too are high in ascorbic acid (111 mg), as are green seeds of umbrella trees (49 mg), the fruit of *Azima tetracantha* (42–49 mg), green leaves of a lily, *Chlorophytum* sp. nr. *bakeri* (52 mg), green tillers of Bermuda grass, *Cynodon nlemfuensis* (88 mg) and of the grass *Setaria verticillata* (57 mg), ripe fruit of trumpet-flower plants, *Lycium "europaeum"* (63 mg), and of the bush *Withania somnifera* (62 mg), the soft fruits of devil's thorn, *Tribulus terrestris* (58 mg), and the young leaves of the bush *Salvadora persica* (128 mg). Scurvy seems unlikely in Amboseli baboons.

Riboflavin

So far, thirty analyses of riboflavin have been carried out on the foods of Amboseli baboons (S. Altmann, Post, and Klein 1987). The highest concentrations are in umbrella tree seed coats (1.7 mg/100 g), which the young baboons discard! By contrast, the naked seeds, which yearling baboons eat, have much less

(0.2 mg). Other high concentrations of riboflavin are in the flowers (1.3 mg) and dry seeds (0.90 mg) of fever trees, the flowers of *Lycium "europaeum"* (1.1 mg), and the fruit of *Trianthema ceratosepala* (0.87 mg). In fact, fully a third of all baboon foods analyzed so far from Amboseli have riboflavin concentrations of at least 0.5 mg/100 g, which, by comparison with human foods (e.g., Davidson, Passmore, and Brock 1972, table 12.5), puts them within the range of what are considered to be "good sources" of riboflavin. This is remarkable in view of the instability of riboflavin in sunlight. In view of these results, a riboflavin deficiency seems unlikely in Amboseli baboons.

Folic Acid

We obtained thirty analyses of folic acid (S. Altmann, Post, and Klein 1987). (Subsequently, I have received conflicting comments from chemists on the stability of this vitamin in our preservative medium, oxalic acid at a pH of about 2, so the values I report should be considered lower bounds.) By far the richest sources of folic acid are the flowers of the trumpet-flower plant, *Lycium "europaeum"* (85 μg/100 g fresh mass). Other high concentrations are in the dry seeds and flowers of fever trees (52 and 48 μg), the green tillers of *Setaria verticillata* (50 μg), the soft fruit of devil's thorn (44 μg), the young, green leaves of *Lycium "europaeum"* (37 μg), the flowers of umbrella trees (35 μg), and the ripe fruit of sticky-fruit (33 μg), all per 100 g fresh food mass. Among human foods, those with at least 20 μg per 100 g are considered to be good sources. By this criterion, about a third of baboon foods are good sources of folic acid.

Vitamin A

To date, only one analysis of provitamin A has been carried out on Amboseli primate foods: on the fruit of *Lycium "europaeum,"* which was selected for analysis because of its deep orange color and because, when the fruit ripens, it is consumed in considerable quantities by the baboons, who move long distances to local concentrations of this plant. The ripe fruit of *Lycium "europaeum"* from Amboseli contains 1.04 mg beta-carotene per 100 g, equivalent to 173 μg retinol per 100 g. By comparison, raw domestic carrots range from 1.3 to 28 mg beta-carotene per 100 g (Watt and Merrill 1963, 175).

Since the average mass of *Lycium* fruits is 37.4 mg (S. Altmann, Post, and Klein 1987), a baboon weighing 3 kg, about 55 weeks old, can satisfy a daily requirement of 12 μg retinol by consuming 126 *Lycium* fruits. This amount of retinol is the minimum for normal growth and freedom from clinical symptoms but allows for little or no storage; three times that amount (i.e., 12 μg/kg body mass) is required for significant, optimal dark adaptation and reproduction (Guilbert, Howell, and Hart 1940). Since the young baboons took in about

15 fruits per minute of feeding on this food and gorged on them until their feces turned orange, they could easily satisfy either of these requirement levels during the season when *Lycium* fruit was available. Beyond that, because retinol can be stored in the liver, baboons in Amboseli probably stored sufficient excess retinol after these *Lycium* fruit binges to last for several months. Although vitamin A is toxic at about fifteen times the RDA, provitamin A carotenoids are not (NRC 1980), so the baboons' enthusiasm for this fruit does not present a health hazard.

Another commonly eaten fruit in Amboseli, that of *Withania somnifera,* also is bright orange and may therefore be high in carotenoid compounds.

Food Classes

Let us turn from the nutritive value of individual core foods to that of food classes (seeds, leaves, etc.), again using the categories introduced in table 4.4. In examining various aspects of the baboons' selective foraging, I have repeatedly commented on the importance of acacias and grasses in their diet. In what follows, I show the nutritional significance of this selectivity: in several respects, grasses and acacias provide the baboons with some of their most nutritious foods.

Nutrient values (daily intakes, intake rates, and time budgets) for food classes are given for the seasonally adjusted mean diet in table 6.5. The time devoted to each food class, rank-ordered by time, is shown graphically in figure 6.4. The rank discrepancies of some foods in this figure compared with their rank by consumed amount in figure 6.3 are revealing. For example, the baboons devoted more time to grass and sedge corms than to any other class of non-milk foods, figure 6.4, yet the amount of food they obtained from corms ranked seventh among these same food classes (fig. 6.3). Not surprisingly, then, corms' yield rate of food mass and of various nutrients are among the lowest, as discussed in the next section.

We now get a clearer picture of the major contribution made by fever tree gum to the baboons' energy intake. If we combine the energy contribution made by fever tree gum picked from the trees (FTGX) and from places where it had dripped onto logs (FTHW), onto other plants (FTHP), and onto the ground (FTHG)—all usually older and more fermented gum—and include unidentified small objects picked from logs and thought to be primarily fever tree droplets (XXXW), we see that the energy contribution of fever tree gum to the young baboons' diet (448.7 kJ/day) is greater than that of any other food or food class except milk (fig. 6.13). Indeed, fever tree gum contributed 47.2% of the energy in the non-milk portion of the diet.[8]

Below milk and gum, the other major food classes, ordered by their mean daily energy contribution, ranked as follows from the top: green grass and

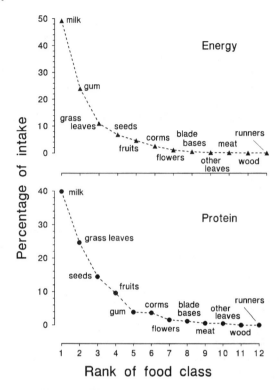

Fig. 6.13. Relative contributions of food classes to seasonally adjusted mean intake of energy and protein. Based on values in table 6.5B.

sedge blades, seeds, fruits, corms, flowers, and grass blade bases, with all other classes below that (fig. 6.13). When ordered by their protein contribution, these eight foods ranked the same way with just one exception: fever tree gum ranked fifth, not second (fig. 6.13). This is consistent with its composition: primarily sugars. Indeed, fever tree gum ranked first among food classes as a source of non-fiber carbohydrate, outranking even milk.

Fever tree gum was the largest contributor of minerals to the baboons' diet (2.3 g/day); green grass blades ranked second (1.6 g/day), and fruits were third (708 mg/day). Gum was also a major source of fiber, exceeded only by green grass blades. This result is surprising, since one does not think of a clear, transparent, viscous liquid as containing any "fiber" (primarily celluloses, hemicelluloses, and lignins). Some fiber compounds, such as methyl cellulose, are water soluble, but the standard methods of analysis do not include such materials in the fiber fraction. An alternative is that the fiber is material such as leaves, sawdust from insect tunnels, and insect exoskeletons that was trapped at the exudate site. In either case, it would be consumed by the baboons when they ate the gum.

The importance of gums in the diets of primates has been emphasized in an excellent review by Nash (1986). My study appears to be the first to document its nutritional contribution to the total diet of a primate. For some primates, gum is the major diet component. A few primate species (*Phaner, Euoticus, Cebuella, Callithrix*) have dental adaptations for gumivory that allow them to gouge or scrape trees to stimulate the flow of gum, and nails modified into claws that help them cling to large trunks while gum feeding. Few nonprimate vertebrates are gumivorous. Marsupials appear to be the only other mammalian group that includes several gum specialists (Smith 1984, 1985).

Among food classes as with individual foods, milk was the baboons' only appreciable source of fats: 12.8 g/day (84.3% of the total). Milk was also the major source of food-bound water 224 g/day (76.5% of the total in food). Green grass leaves and fruits were other major sources of food-bound water (31 and 17 g/day, respectively).

Yield Rates and Time Budgets

If an animal's vital functions, including growth and reproduction, are energy limited, and if energy intake is in turn limited by available feeding or foraging time, then the energy yield rate of a food—that is, energy consumed per unit of time devoted to that food—can be considered a benefit-cost ratio. For example, in an experiment designed to test an optimal foraging model, Emlen and Emlen (1975) defined benefit in terms of the energy content of food items and cost in terms of the time required to consume each item. (In their study and mine, including the costs of foraging, not just of consumption, might have been preferable. I have no data on travel time, but even if I did, it is not obvious how to apportion such time among the foods that are harvested in a day's journey.) If an animal's sole task in foraging is to maximize its rate of energy intake, it should not consume the various foods in proportion to their benefit-cost ratio (energy yield rate). To do so would be an example of the gambler's fallacy of distributing bets in proportion to odds (S. Altmann 1979b). Rather, it should feed exclusively on whichever available food has the highest net energy yield rate, moving on to others only when the richest is exhausted, and similarly for any other nutrient if maximization of its intake rate is the sole objective of foraging. Even in the more realistic situation of an animal coping simultaneously with a variety of nutrients and toxins, some foods can be considered more nutritious than others if they have a higher yield rate for one or more nutrients, a lower yield rate for one or more toxins, or both of these, without in either case sacrificing any nutrients or increasing any toxins.

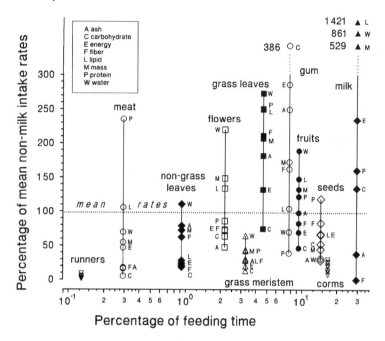

Fig. 6.14. Relative intake rates and feeding time budgets of nutrients and food mass. Based on the seasonally corrected pooled mean diet, with feeding time budgets calculated from values in table 6.5 and relative intake rates calculated from values in table 6.10. Food classes as in table 4.4.

Do baboons concentrate their feeding time on those foods that provide the largest amount of nutrient per minute spent feeding on them? To answer this question I plotted, for core foods in the seasonally adjusted annual diet, the percentage of feeding time that the baboons devoted to each food class (corms, seeds, etc.) against the relative yield rates of the seven macronutrient categories (proteins, lipids, etc.), that is, with each rate as a fraction of the mean rate for all non-milk core foods together (table 6.10). Thus, for any food class on the graph (fig. 6.14), any nutrient below the 100% line is provided by that food class at a less than average rate. Pooled mean nutrient yield rates (g/minute, kJ/minute) for the core food component of any given food class (table 6.10) were obtained from the total daily nutrient yields for all foods of that class (g/day or kJ/day in the seasonally adjusted diet, table 6.5) divided by the corresponding time budget (mean daily minutes of feeding on the core members of that food class, table 6.5). As indicated in chapter 5, the pooled mean rates used here are equivalent to time-weighted (not mass-weighted) mean rates.

The answer to the question at the beginning of the preceding paragraph is no, not invariably. While the baboons had a tendency to concentrate their feeding time on food classes that are high in several nutrients and to devote relatively little time to the less nutritious food classes, some marked exceptions warrant comment. Most striking are grass/sedge corms. Corms take up more time than any other food class save milk. Yet they yield every class of nutrients at rates that are less than a third of the average (fig. 6.14), and this despite being rich in carbohydrates, fiber, and energy (table 5.1). The low nutrient yield rates of corms result from the amount of time required to dig them up and remove them from their roots and sheaths, rather than from low nutrient concentrations. This confirms Post's (1982) speculation that corms "yield less energy per unit of feeding time" and is consistent with my characterization of grass corms in chapter 4 (written before I had these results) as a major fallback food. They sustain the baboons at those times of the year when high yield rate foods are not available and are eaten out of necessity, not out of choice.

What foods have the highest nutrient yield rates? Consider proteins and energy. Seven core foods yielded proteins at rates exceeding 100 mg per minute of feeding: the fruits of a common shrub, *Azima tetracantha* (ATFX, ATFR) and of a prostrate herb, *Commicarpus pedunculosus* (SFFX), acacia seeds picked from the ground (TODU, XXXG), and finally insects (INKX) and the green leaves of grasses (GRLU). The highest energy yield rates among core foods were obtained from milk (PCMX), from fever tree gum (FTHG, FTGX, FTHW), from the fruit of *Commicarpus pedunculosus* (SFFX), and from acacia seeds picked up from the ground (TODG). No other core foods yielded food energy at rates exceeding 10 kJ per minute of feeding.

Now, what about milk? The milk intake rates in the observed diet, shown in table 6.10 and figure 6.14, are based on an assumption—which may be wrong—that infants in the age range of my samples, 30–70 weeks, obtain milk only during the daylight hours. I have no way of knowing whether this is true. (Earlier in this chapter, when I estimated each infant's milk intake per twenty-four hours, no such assumption was made, only that each infant's milk intake per twenty-four hours was *proportional to* its time on the nipple during the daylight hours.) Except for humans, almost no information seems to be available for mammals on the time of day milk transfer occurs (as opposed to when infants are on the nipple); yet in view of the frequency of non-nutritive sucking in primate infants (Benjamin 1961, 1967), one cannot automatically assume that the temporal distribution of milk intake is the same as the distribution of time on the nipple. Human neonates initially have wake/sleep cycles that are apparently endogenous, but they gradually become entrained to their mothers' activity cycles. In the absence of contrary evidence, let me assume that the same is true of baboons by 30–70 weeks of age, and that by that age, milk, like other foods, is consumed only during the daytime. If in fact baboon infants are taking in some milk during the night, then my estimates of milk intake rates are overestimates. Conversely, milk intake per minute of *suckling* is underestimated by my values, which are calculated per minute on the nipple, whether suckling or not.

The exceptional nutritional value of milk is clear from figure 6.14. Except for salts and fiber, milk yielded all classes of nutrients at rates well above the average of other classes of foods that are available to the baboons. To appreciate the nutritional richness of milk, consider its yield rates of energy and of macronutrients per minute of feeding, compared with yield rates of the rest of the diet (core plus non-core foods). Milk provided 841 kJ of energy in 65.9 minutes of nipple time per day, that is, at a rate of 12.8 kJ per minute (table 6.10), whereas the remaining foods yielded 10.2 kJ per minute of feeding (mass-weighted mean). That is, nursing yielded energy 1.3 times faster than the average for the baboons' other foods. Corresponding calculations for the other nutrient classes give the following yield rate ratios for milk compared with the mean of the rest of the diet: energy 1.3 to 1, water 4.6 to 1, minerals 0.19 to 1, fats 7.7 to 1, proteins 1.0 to 1, fibers 0 to 1, carbohydrate 0.67 to 1, and mass 2.8 to 1.

Seeds—which in this case means almost entirely seeds of acacia trees—yield proteins at somewhat higher than average rates, but they yield other nutrients more slowly than average. Again, the deleterious effect of long harvesting time on nutrient intake rates are apparent, because acacia seeds are second only to insects in their concentrations of protein and second only to fever tree gum in their carbohydrate density. During the dry season, green umbrella tree seeds provide a preferable and more nutritious alternative to corms. They are harvested in the shade and safety of an umbrella tree canopy, where time may not be as critical, in terms of temperature regulation and predators, as it is in the open.

The high nutritional value of the green leaves of grass is clear from figure 6.14. Doubtless the baboons' consumption of this food would be greater if new grass shoots were available during a larger portion of the year.

The fruits available to Amboseli baboons are all quite small. As a group, they provide minerals, proteins, fats, and water at above average rates. As indicated earlier, the core foods that yielded proteins at rates exceeding 100 mg per minute of feeding include some fruits, namely, the fruits of a common shrub, *Azima tetracantha* (ATFX, ATFR), and of a prostrate herb, *Commicarpus pedunculosus* (SFFX). Surprisingly, fruits are not exceptional as sources of carbohydrates, in terms of either concentration (table 5.1) or yield rate (figure 6.14).

Among food types, fever tree gum provides the baboons with the highest yield rates of fiber, ash, carbohydrates, and energy (FTGX, FTHG, FTHW).

The only other foods that yielded energy at rates exceeding 10 kJ per minute of feeding are milk (PCMX), the fruit of *Commicarpus pedunculosus* (SFFX), and acacia seeds picked up from the ground (TODG). Again, if fever tree gum were available more, it probably would be consumed more. As it is, individual baboons often go out of their way to get it, and the entire group often makes long treks to the edge of Lake Amboseli, whose primary goal appears to be exploitation of the abundant gum exudation sites on the young, regenerating fever trees along the edge of the lake.

The baboons devoted less time to each of the remaining classes of foods, and none of these plant classes have any exceptional nutritional yield rates, compared with those already mentioned, except for the lipid yield rate of flowers, which is more than 30% above average. This does not mean that these other foods are "poor" foods, only that they yield nutrients at slower rates than the richest foods. They are worth harvesting when abundant and near at hand, especially when alternatives are scarce. Some may be good sources of micronutrients, for example, provitamin A in the fruit of *Lycium "europaeum"* (see chapter 9).

Animal tissue is an outstanding source of protein, with a yield rate over twice the average of the plant sources, but exceeded by green grass blades (250% of the average protein yield rate). I believe the young monkeys are limited in their intake of animal tissue—at this age, primarily insects—only by their inability to catch more of them.

These rates of nutrient yield reveal some familiar patterns: grass corms that yield neither energy nor nutrients quickly but on which the baboons invest much time when better foods were not available, quick-energy foods like fever tree gum that are not correspondingly rich in proteins, and proteinaceous foods like *Azima* berries that have a relatively low energy content.

Trade-offs: Energy Density versus Harvesting Rates

The mean energy yield rate of a food (kJ/minute) is the product of two factors, the mean energy density of the food (kJ/g) and the mean rate at which a baboon can harvest it (g/minute), assuming that these factors are independent. The first factor is a property of the food; the second, of the abilities of the baboon, and it is interesting to look at the relative contributions of these two factors to variations in the energy intake rates of foods. Suppose two foods differ in that, when eating the first, the baboons harvest and consume twice as much food per minute, but that this food has only half the energy density of the second food. Then these two foods would have the same energy yield rate. If baboons were pure energy rate maximizers and ignored other attributes of food, they would be indifferent between the two, as indeed they would be for any two foods with the same product of energy density and harvest rate.

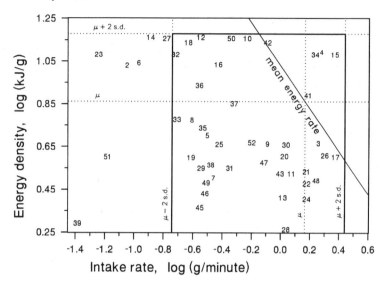

Fig. 6.15. Relationships between factors of mean energy yield rates of foods: their energy densities and their mass yield rates. Core food numbers as in table 6.1, that is, ranked by time budgets. To avoid overlapping numerals, 4, 7, 38, and 48 have been shifted to the right.

Therefore a particularly revealing way to look at these two components of energy intake rates is to plot them against each other, each scaled logarithmically. On such a plot, food with the same energy yield rate would fall on a straight line of negative slope, the slope itself depending on the scaling of energy, time, and mass. Such a plot is shown for the non-milk core foods in figure 6.15. The harvesting rates were obtained from $(B_j C_j)^{-1}$, where B_j is the unit intake rate of the jth food and C_j is its unit mass, both in table 6.1. Energy densities are given in table 5.1. The solid line of negative slope in figure 6.15 indicates the mean energy intake rate, 10.7 kJ/minute (mass-weighted mean, non-milk core foods). The deviation of each food whose energy yield rate differed from this mean has two vector components, one due to the energy density of the food, the other due to its harvesting rate. The number by each point of figure 6.15 is that food's rank order for time spent harvesting that food in the observed diet, with lower numbers designating the more time-consuming food.

Figure 6.15 reveals no tendency for the baboons' primary foods to cluster along an isopleth of high energy yield rates. Indeed, the energy yield rates of the core foods span three orders of magnitude (table 6.10). Nonetheless, did the baboons tend to devote more time to the quick-energy foods? Clearly not: in figure 6.15 the lowest food numbers (largest time commitments) show no tendency to concentrate near the top right part of the distribution (highest energy yield rates). Judging by the coefficients of variation for these two components,

43.8% for harvest rates and 54.9% for energy densities, differences in the energy densities of the foods themselves contributed slightly more to the food-to-food differences in energy yield rates than did the baboons' harvesting abilities. Put another way, foods of a given energy yield rate differed more, on average, in their energy densities than in their harvest rates.

Similarly, the baboons showed no tendency to adjust their harvesting rates to the energy densities of their foods. As one would expect from figure 6.15, the linear regression of the baboons' harvesting rates of their core foods (g/minute of feeding) on the energy densities (kJ/g) of these foods has an adjusted R^2 of zero ($p = 0.36$).

The most deviant foods, in terms of energy yield rates, are those that lie beyond the mean plus or minus two standard deviations of harvest rates and energy densities. The solid-lined box in figure 6.15 indicates the mean plus and minus two standard deviations of harvest rate and of energy density. No non-milk core food lies above either of the two upper cutoff points. (Milk does, but it has not been included because its yield per minute on the nipple, even including periods of non-nutritive suckling, is extraordinarily high and distorts the distribution.) No food can fall below $\mu(kJ/g) - 2\sigma$ because my estimate of that value is less than zero. The seven foods that are below $\mu(g/minute) - 2\sigma$ of intake rates in figure 6.15 are as follows: 2 = corms of unidentified grass/sedge, 6 = corms of *Sporobolus rangei*, 14 = fever tree seeds from dry pods; 23 = unidentified items (arthropods? fever tree gum?) picked from wood, 27 = unidentified items (mostly acacia seeds?) picked from dung, 39 = stolons of Bermuda grass, and 51 = fruit of *Capparis tomentosa*. Corms are, as mentioned before, a fallback food, a resource baboons resort to primarily during those times of year when more nutritious or less labor-intensive foods are not available. My information on "unidentified items" is too scant to warrant further comment in this context. The actual energy density of *Capparis* fruit is not known; my assumed value (table 5.1) may be a gross underestimate. In recent years I have seen baboons of another group, Hook's Group, gorge on this slightly sweet fruit in an area—outside the range of Alto's Group—where these clamberers are very abundant. The stolons of Bermuda grass (*Cynodon nlemfuensis*) are the most deviant food of all: they have the lowest harvest rate of all core foods, the second-lowest energy density, and the lowest energy yield rate. One wonders why the baboons bother to eat them. Perhaps because of other nutrients they provide, described above.

Is the Baboons' Diet Adequate?

How do the yearlings' mean intakes of macronutrients compare with the nutrient requirements and tolerances of baboons at the mean age and body mass of those that I sampled (table 5.3)? For comparisons with requirements and tolerances, the seasonally adjusted mean diet must be used. The nutrient yields of the seasonally adjusted annual diet are given in table 6.5. The total non-milk mass, total non-milk yield of each nutrient, and total non-suckle feeding time were each multiplied by 1.095 to adjust for the contributions of non-core foods. The same was done for each of the food classes (leaves, corms, etc.), though they are more susceptible than the total diet to small-sample fluctuations.

Energy

Let's begin with energy. The mean sampling age of my subjects was 47.5 weeks. (Bear in mind throughout this section that the intake data being used were obtained not just at this mean age but from 30 to 70 weeks of age.) At this age average body mass is $0.775 + 0.0315(47.5) = 2.273$ kg, assuming a growth rate of 4.5 g/day. Daily energy requirement for a baboon of this size (table 5.3) is $419.3(2.273)^{3/4} + 94.5 = 870.6$ kJ/day. On average, the baboons took in 1791 kJ/day (table 6.5), that is, more than twice their minimum requirement. However, about half of this energy was supplied by milk, without which they would, on the same diet, be taking in only 950 kJ—still an adequate amount, but with a much smaller margin of safety (9%) or room for variability.

A mean daily intake of 1,791 kJ at the mean body mass of my subjects, 2.27 kg, results in a mass-specific intake of 788 kJ kg^{-1} day^{-1}. (If the assumed growth rate is off by one gram per day one way or the other, the mass-specific intake could be as low as 687 or as high as 923 kJ kg^{-1} day^{-1}.) By comparison, Stacey (1986) estimated the intake of adults in the same group during a season of high food availability (1979–80) to be 387 kJ kg^{-1} day^{-1}. Errors aside, the difference probably reflects the greater mass-specific basal metabolism in the yearlings combined with their greater activity.

Proteins

The baboons' protein intake averaged 11.1 g/day (table 6.5). The requirement (table 5.3) is for $1.51 M^{3/4} + 1.14$ g/day, which at the mean age and body mass comes to 3.94 g/day. That is, on average the baboons were getting considerably more than the minimal amount of protein, indeed, nearly three times the minimum. Even without milk, mean protein intake (6.9 g/day) would be more than adequate.

So these weanling baboons were self-sufficient for both energy and protein by an average age of 47.5 weeks—about the age of the youngest baboons that have survived their mothers' death (J. Altmann, pers. comm.).

Minerals

The baboons' mineral needs could be met by $0.0173t$ g per day, where t is the mean daily grams of food in the diet (table 5.3), that is, by food that on average yielded 1.73% minerals. This would require the right balance among the cation elements in the diet, as described in chapter 5. The mean mass of the baboons' diet, core plus non-core foods, averaged 391.5 g (table 6.5), which leads to a daily mineral requirement of 6.77 g. However, actual mineral intake was 6.3 g/day (1.6% of the diet), an amount that would be inadequate even if the right balance of minerals was present. The discrepancy is not large: total mineral intake was 93% of the recommended daily allowance (table 5.3, based on Nicolosi and Hunt 1979). Perhaps the non-core foods are, on average, richer in minerals than the core foods, even though in allowing for their contribution I considered them to be equal. Nonetheless, we must consider the possibility that the young baboons are subject to one or more mineral deficiencies. Milk is much less salty than the rest of the baboons' diet. Without milk, the mineral content would be 4.2% salts, more than adequate. So if there is any mineral deficiency, it probably disappears as the baboons stop drinking milk. (Conversely, since baboon milk contains only 0.3% ash, infants on a straight milk diet would be mineral deficient according to the requirement above.)

Noteworthy is the result that among the optimal diets (diets 6–10 in chapter 5), three out of the five could be achieved only if the baboons were at their minimum for total salts (table 5.3), and that minimum would apply literally only if the minerals in the diet were in the ideal ratios (tables 6.11 and A1.2).

The total salt intake of the baboons is surprisingly low, considering the extensive encroachment of halophytes in Amboseli during recent years (Western and Van Praet 1973), and testifies to the baboons' capacity for selective foraging. Although to date I have elemental analyses for just three baboon foods, they are important components of the diet (all are core foods), and if we consider their average mineral composition to be representative—doubtless not literally correct but only a first approximation—we can estimate total intake for each of the fourteen elements that have been analyzed. For the young, green leaves of the toothbrush bush, *Salvadora persica* (SPLU, SPLX), the green seeds, in their coats, of umbrella trees, *Acacia tortilis* (TODU), and the corms of *Sporobolus cordofanus* (SMCX), table 6.11 (after S. Altmann, Post, and Klein 1987) gives the results of quantitative elemental analysis. The average quantities of these three foods in the seasonally adjusted daily diet of the yearlings were 0.342 g, 3.807 g, and 0.734 g, respectively (table 6.5), and on that basis the elemental contributions of these foods were calculated. The total consumed mass of these three foods, 4.88 g, is 1.247% of the mass of the diet. I therefore divided the total amount of each cation in these three foods by 0.01247 to estimate the amount of that mineral in the total diet. The results are shown at the bottom of table 6.11, along with recommended daily allowances (RDA's) from table 5.4. The latter are given by Nicolosi and Hunt (1979) as proportions of the diet, and I have multiplied each by the mean mass of the diet, 391.5 g/day. (Wherever a range of RDA values are given in table 5.4 for an element, I have used the smallest.) If these three foods are typical in their salt content, then the average diet of the yearlings is below the RDA for phosphorus (49% of the RDA), calcium (68% of the RDA), sodium (66% of the RDA), and zinc (37% of the RDA) but is providing adequate amounts of potassium, magnesium, iron, manganese, copper, and chromium.

The magnitude of some of these putative deficiencies would be important if primate RDA's were well established, but a reading of Nicolosi and Hunt (1979) and the literature they cite indicate that this is not so and that the RDA values they give include a large margin of safety. Furthermore, the figures above are extrapolated from just three foods. I am therefore reluctant to conclude, on present evidence alone, that the baboons were actually deficient in any of these minerals, though the possibility cannot be dismissed.

The consequences of major deficiencies in these four cations would be severe. A review of phosphorus deficiencies in animals by Schryver and Hirtz (1978) mentions

> depressed feeding intake coupled with depraved appetite, poor growth, weight
> loss, lameness, reproductive inefficiency and early death of the newborn, . . . ,
> hair coats dull, ragged and rough, . . . , often an alteration of hair color or loss of
> hair, . . . , stiff, stilted gait, . . . , joints are enlarged and painful, . . . , rickets in
> young animals.

Of particular interest are their comments on the binding of phosphorus by phytate:

> Phosphorus deficiency may also result when the dietary phosphorus is composed
> largely of phytate . . . [which] is not well utilized as a phosphorus source . . . and
> may reduce the availability of calcium by forming insoluble calcium phytate.
> Two-thirds or more of the phosphorus content of most cereal grains [grass seeds]
> and their products is phytate phosphorus.

To date, no analyses are available of phytate in the grass seeds or other parts of grass plants eaten by baboons.

Many of the overt symptoms of calcium deficiency in animals (Roland 1978) are like those of phosphorus. They include reduced appetite, retarded growth, muscular stiffness or paralysis, joint enlargement, rickets, and osteoporosis.

For zinc, the primary deficiency symptoms are decreased food intake and cessation of growth (Apgar 1978); these would tend to aggravate a phosphorus and calcium deficiency and increase the chance of other nutritional deficiencies. So too would a sodium deficiency, for which overt symptoms include decreased appetite, decreased weight gain, and growth depression (Pike 1978). However, the resulting salt appetite (Richter 1942) would make a deficiency of sodium unlikely for animals in Amboseli.

Of the baboons in this study, female Pooh and male Pedro (see chapter 7) seemed to have several of the mineral deficiency symptoms indicated above, but these symptoms are too nonspecific for a definite diagnosis to be made from our field observations.

What about the toxicity of these minerals? Data on the toxicity of many cation elements, based on studies of domestic and laboratory animals, are compiled in NRC (1980). In table A1.2, I have summarized values, obtained from studies of cattle, sheep, swine, horse, and rabbit, for the maximum tolerable levels of each element for which I have chemical analyses for any Amboseli foods. In table 6.11 these tolerances are compared with the baboons' mean consumption of these elements, calculated as described above. (Wherever a range of values for the tolerances is given in NRC 1980 I have used the largest for each element.) For none of these fourteen metallic elements did the mean diet of the baboons exceed the toxic limit. The only two cations for which mean consumption was at least half the maximum are aluminum (65% of tolerance) and barium (57%). Next came magnesium (37%), then iron (19%).

The high aluminum intake may be particularly important because "aluminum toxicosis is expressed largely as a secondary phosphorus deficiency, presumably because it binds phosphorus in an unabsorbable complex in the intestine" (NRC 1980, 10). Note (table 6.11) that almost all of the aluminum comes from the corms of *Sporobolus cordofanus,* one of the baboons' primary "fallback" foods during those times in the dry season when preferred foods are not available. During those times in the dry season when the baboons rely heavily on corms, phosphorus deficiency, from both excess aluminum and inadequate phosphorus, may be a health problem for Amboseli baboons.

Carbohydrates

For non-fiber carbohydrate, no dietary requirement is known, and the baboons seem never to fill their bellies with their richest carbohydrate source, fever tree gum: it just isn't that abundant. Of course, the significance of carbohydrates is as a source of energy.

Water

As for water, the baboons do not rely solely on food-bound water or metabolic water: they also drink from ground sources. Yet it is interesting to see what portion of their water requirements is met by the water in their food. Three water requirements are given in table 5.3. The first, based on ad lib water intake of mammals, is an upper bound of $120 \, M^{0.84}$ g/day, which at the mean age and body mass in my sample comes to 239 g/day. The baboons' actual intake of food-bound water was 293 g/day (table 6.1); that is, food-bound water (not including metabolic water produced from the catabolism of food) exceeded the average amount of drinking water taken ad lib by mammals of this size (but bear in mind that those mammals also get food-bound water). By far the largest source of water for the infants was milk. The non-milk portion of their diet provided 69.0 g water per day, which is 29% of the above "requirement."

The third water requirement, based on renal clearances, is $4.8p + 32.6a$ g/day, where p and a are the grams of dietary protein and ash respectively, values for which are $p = 11.6$ g and $a = 6.4$ g (table 6.5), giving a daily water requirement of 264 g. This is an amount that the baboons meet by means of food-bound water, but only because of the large amount of water in milk. The non-milk part of their diet leaves them with a deficiency of 195 g water per day, to be made up from metabolic and drinking water.

The second water requirement calls for $100.2 M^{3/4} + 22.6 = 208.1$ g/day. The baboons' intake of food-bound water would be adequate to meet this requirement, but again, only because of the large contribution made by milk. The non-milk part of their diet was deficient by 139 g water per day. The difference apparently was provided by metabolic water, by milk, and by drinking water.

Young infant baboons often sit, quietly waiting, or continue to cling to the mother when she and other members of her group drink from rain pools, swamps, or waterholes. In the transition to nutritional independence, young baboons must become self-sufficient in water intake, and that involves learning where to drink. Waterholes along the edge of the dry bed of Lake Amboseli are fed by very localized seepages from underground. As the water flows from a seepage across the ground, it dissolves the alkaline and saline deposits that abound in these areas. When the adults are all crowded at the source, inexperienced infants sometimes try to drink from the uncrowded but saline end of these pools. One taste of the brackish water teaches them the merits of drinking where the adults do, even if they must wait.

Fiber

The baboons' fiber requirement (table 5.3) is estimated to be 3.5×10^{-3} times the dry-matter mass of the diet. The total mass of the diet averaged 391.5 g (table 6.5). The average water content of the diet (table 6.5) was 293.4 g, leaving a dry-matter mass of 98.1 g, on the basis of which the fiber requirement is 0.34 g per day—a requirement that the baboons' diet, with 10.1 g fiber per day, easily meets. None of this fiber is from milk.

Lipids

The baboons require $0.8 M^{3/4}$ g of lipids per day in order to meet their requirement for essential fatty acids (table 5.3). At a mean sample body mass of

2.27 kg, this requirement comes to 1.5 g fat/day. Actual intake (table 6.5) was 15.2 g/day, far above their requirement. Most of this fat (12.8 g) was from milk, but even without milk, the average diet would contain 2.38 g fat, more than enough to satisfy requirements.

In sum, for energy, water, lipids, protein, and fiber, the pooled mean dietary intake of the eleven yearlings that I sampled at 30–70 weeks of age was adequate to satisfy requirements at their mean sample age, 47.5 weeks, but was not excessive. No requirement for non-fiber carbohydrate is known. Total mineral content was 93% of the requirement, and judging by extrapolation from three foods on which cation analyses have been made, intake levels of phosphorus, calcium, sodium, and zinc were below the currently available recommended daily allowances.

Is the Baboons' Diet Optimal?

> While discussion of the optimal in nutritional intake may seem currently unrealistic, this is no reason why we should not seek to obtain more knowledge on the subject.
>
> —A. R. P. Walker, B. F. Walker, and B. D. Richardson, 1976

Our story so far: optimal diets for Amboseli baboons were calculated in chapter 5 (diets 6–10), and the mean diet of my eleven yearlings was presented above. I now ask, How well do they match? That is really two questions, and both will be pursued in the remainder of this chapter: Are the diets of real baboons optimal? And are the calculated optimal diets realizable? This pair of questions calls for some clarification.

Suppose that discrepancies are discovered between the animals' real diet and the ideal. What should one conclude? That depends on whether one is convinced that the nominally optimal diet actually is so, for if it is, then such discrepancies indicate that the animal's actual diets are not ideal and that a search for the causes and consequences of their shortcomings is in order. However, as indicated at various places in chapter 5, diets 6–10 are susceptible to several potential sources of error, including seasonal changes in each food's availability, interactions among nutrients and toxins, errors in the nutrient and toxin values of the foods, errors in the animal's assumed nutrient requirements and toxin tolerances, and unrealistic assumptions about the animals' foraging capacity, time requirements, and other organism-based constraints. If some of the errors are ours, then we should alter the optimal diet accordingly. If, however, the errors are entirely the animals', then we should not: our task here is not to model the animals' actual diets, which may or may not be optimal, but to model those realizable diets that would maximize fitness. Discrepancies between such diets and the animals' actual diets indicate the potential for natural selection to act on foraging behavior.

In principle, the causes of such discrepancies could be determined experimentally: we would somehow induce the animals to abandon their current diet and forage for our nominally optimal diet or something closer to it than they now get, see whether they could accomplish this, and if so, see whether their fitness increased. (Experimentally provisioning the animals with the optimal diet would not suffice because that would bypass the time and effort required to obtain the diet on their own and the risks they would run in doing so.) We would relate the nutritional value of the current and "improved" diets to differences in fitness, in extreme cases by noting the consequences of specific forms of malnutrition. Unfortunately, this strategy is not available to me. I do not know how to bring about the requisite alterations in the animals' behavior. Furthermore, I have detected few externally visible signs of malnutrition (but see the cases of Pooh and Pedro, in chapter 7), and I have no physiological measurements on the animals that could be used to detect more subtle forms.

Another strategy is available in an observational study such as this, and indeed it was used in chapter 5 when I rejected the constraint set of diets 1–5 in favor of that used to calculate diets 6–10. Based on what is known about the competence of the animals and the nature of their habitat, we can consider possible errors of omission in the constraint set; that is, we can ask whether any potentially binding constraints have been overlooked. For example, a constraint on total feeding time was added to the constraint set before calculating diets 6–10 to avoid an optimal diet that required more time than is available to the animals (see comment 5 in chapter 5). One can also consider possible errors of commission, both in the constraint set and in the objective functions themselves. Does it make sense for an animal to maximize its energy intake rate if doing so means eating a diet that is only a fraction of its consumption capacity and provides a bare minimum amount of protein? Is the optimal diet realizable? For example, could yearling baboons in Amboseli, on a sustained basis, each find 273 g of fever tree gum per day (with comparably larger amounts for older individuals), as diet 4 would require, and not exhaust the local supply? In some cases errors of omission or commission in the constraints or objectives can be detected by examining their dietary consequences, taking advantage of locally occurring differences. Do those baboons whose diets have the highest energy content have the highest fitness? Can animals survive and remain apparently healthy on diets that are below the putative minima for various nutrients? The methods I am describing here are the classical approach to testing models in science, by examining both their assumptions and their implications (Kaplan 1964). In these ways, progressively better models of optimal diets can be developed—better in the sense that they improve our ability to predict fitness differences from differences between the optimal and the observed diets.

The reader may wonder whether there is a circular argument in this approach. If proximities of animals' actual diets to a putative optimal diet are used to predict individual differences in the dietary component of fitness, how can the fitness of individuals, or indirect indicators thereof, such as nutritional deficiencies, be used to decide on an optimal diet? The answer is that the process is iterative, not circular. The failure of one set of assumptions (constraints and objectives) to produce realizable optimal diets or to find a fitness-maximizing diet in one case can be used to establish more realistic assumptions and make better predictions in the next. Particularly in the early stages in the development of a field of science, a primary function of a model is to stimulate the development of a more adequate one.

Real versus Ideal Diets

In figure 6.16 the food array of the yearlings' seasonally adjusted mean annual diet is compared with those of the five diets that maximize energy intake (diet 6), energy intake rate (diet 9), protein intake (diet 7), and protein intake rate (diet 10) or that minimize feeding time (diet 8) and satisfy the extended constraint set of diets 6–10. In figure 6.16 the vertical axis has been scaled logarithmically, so that vertical deviations of equal length represent equal percentage deviations between any two intake values. Values below 0.10 g per day

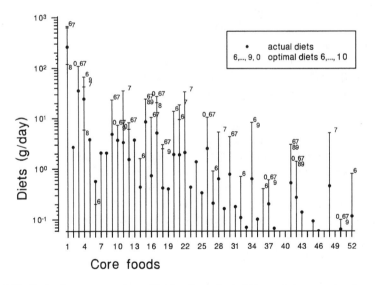

Fig. 6.16. Optimal versus average actual intake of core foods. Values not shown are less than 0.06 g/day, mean annual intake. Foods are identified in table 6.2. Values from table 6.5 (actual diet) and tables 5.9–5.12, 5.15 (optimal diets).

are not shown; in the optimal diets, intakes of many core foods would be zero (tables 5.9–5.12 and 5.15), though of course none of the actual intakes of core foods are.

Clearly, at the level of individual foods, the baboons' mean diet was far from that specified by any of these optimal diets. However, that alone does not imply that their intake was *nutritionally* far from optimum: identical or nearly identical nutritional yields can, in certain circumstances, be obtained from different diets. In chapter 5, the possibility that several diets could be made up of different combinations of foods but yield identical amounts of protein or energy was investigated when I asked whether each optimal diet was unique. (That was always the case except where two foods of nominally identical nutritional values could be interchanged.) Similarly, any two diets that have zero slack for the same nutrient constraint would yield identical, limiting quantities of that nutrient, no matter how different their array of foods. However, diets that are *nearly* identical nutritionally cannot be detected in this way. I have therefore calculated the proximate nutrient yields and time requirement of each of the five optimal diets (diets 6–10) and compared them with the nutrient yields of the yearling's actual, seasonally adjusted diet. In table 6.12 each such value is compared with the requirement or tolerance (table 5.3) of a baboon at the mean age of those in my samples. The comparisons take the form of relative slack values, that is, as the percentage deviation of the diet from the baboons' minimum (requirement) or maximum (tolerance) for that nutrient. Perusal of table 6.12 reveals that at the level of proximate nutrients, two of the diets are very similar, namely diet 6 (maximize energy) and diet 7 (maximize protein). The total amount of food taken in would be the same, as would the quantity of minerals; both diets would take 436 minutes per day. Each of these diet components has zero slack; that is, each is a binding constraint on the magnitude of its respective objective function. Beyond that, the quantity of water in these two diets would be nearly the same (744 vs. 763 g), as would lipids (39.5 vs. 38.2 g), proteins (29.1 vs. 31.0 g), fiber (27.1 vs. 26.7 g), carbohydrates (145.4 vs. 126.0 g) and energy (4,695 vs. 4,347 kJ).

We already know that all of the optimal diets are also nutritionally adequate diets: each satisfies all analyzed nutrient requirements, toxin tolerances, and all other constraints of the extended constraint set (diets 6–10, chapter 5). Consequently, for all optimal diets, the deviations will be non-negative for all nutrients that were in the constraint set for these diets. For diets 6–10, the only negative values in table 6.12 are for water content, which was not a constraint for diets 6–10. (The negative values for water's relative slack in diets 8–10 indicate that these diets would require drinking or metabolic water supplements.) We also know, from the last section above, that on average the baboons' actual diet was nutritionally adequate, with the possible exception of being below requirements for some minerals. So far as is known, variations in nutrient intakes within the range of adequacy are irrelevant to fitness (appendix 1), with the presumed exception of the objective function components: protein, energy, and time. The following discussion will therefore center on how close the baboons' mean diet came to the objective functions: to that component of each diet that would be maximized or minimized by that diet and that supposedly is a limiting fitness factor for the baboons.

From a nutritional standpoint, the yearlings' mean diet clearly fell short of what they presumably could have achieved in Amboseli (table 6.12). Were they attempting to maximize their energy consumption? Their total energy intake averaged 42% of what they could have gotten had they eaten diet 6. Protein? Their protein intake was 36% of what diet 7 would have provided. They took 241 minutes of feeding per day to satisfy their requirements, whereas only 51 minutes are needed to do so (diet 8). That is, the baboons took 4.7 times the amount of time needed to do the job. Similarly, their average energy and protein intake rates were 59% and 56% of the rates (all mass-weighted means) that they could have achieved on diets 9 and 10, respectively.

So, whether judged by the array of foods in the baboons' diet or the nutrient yields of that diet, the yearling baboons' mean intake was well below optimum by any of the five criteria I used. Let us look more closely at the sources of these shortcomings.

Energy Maximization

The diet of the yearlings totaled 1.79 MJ per day (table 6.5), and though that amount is twice their requirement of 0.87 MJ per day, it falls far short of the amount of energy, 4.69 MJ/day, that diet 6 would provide. Where did they err? In this case the largest single discrepancy between the real and the ideal, in terms of both mass and energy, is in milk, for although the baboons averaged 260.9 g per day, diet 6 calls for 667.7 g/day. The difference, 406.8 g/day, at 3.22 kJ/g (table 5.1), results in a shortfall of 1.31 MJ/day. That accounts for 45% of the discrepancy. I do not know why the baboons did not, on average, consume more milk than they did—perhaps because of limitations on the mother's time or lactational capacities (J. Altmann 1980) or other factors that would limit her "willingness" to invest further in her present offspring at the cost of her future reproductive success. However, as can be seen in table 6.6, one individual, Eno, consumed an estimated 634.5 ml/day (652 g/day) during one ten-week age block, judging by the amount of time spent on the nipple, and the mean plus three standard deviations of milk intakes (all yearlings, all four ages) was 819.8 g/day.

The baboons probably would have eaten more fresh green leaves of grasses and sedges if these foods had been available more widely and for more of the year. As it is, their green blade intake—4.86 g/day from *Cynodon nlemfuensis* (CDLU), 0.070 g/day from *Sporobolus cordofanus* (SMLU), and 35.5 g/day from various unidentified grasses and sedges (GRLU), plus white meristem tissue from the bases of green leaves of *S. consimilis* (SCTX, 2.1 g/day) and of other grasses or sedges (GRTU, 0.33 g/day)—was only 32% of the amount (total 132.0 g/day) that diet 6 specifies. The differences between the actual and the ideal amounts of grass/sedge leaves (respectively, 18.4 g at 4.60 kJ/g, 0.07 g at −6.00 kJ/g, 73.2 g at 4.63 kJ/g, 2.1 g at −3.12 kJ/g, and 0.33 g at −4.58 kJ/g) represent a net shortfall of 0.42 MJ/day. (In this section, negative values indicate foods for which the baboons' non-optimal intakes actually tended to alleviate their shortfall in the objective function, more than would the amount specified in the optimal diet, but always at the cost of getting them closer than they would have been to one or more of the binding constraints. Here and in the remainder of this discussion of the baboons' dietary shortcomings, a portion of the calculated shortfalls results from comparing the core foods that make up 93.37% of the baboons' actual diet with 100% of the optimal diet, albeit composed entirely of core foods. The difference is negligible, however. For energy, the difference between core foods and all foods, 82.4 kJ/day [from table 6.5], is only 2.8% of the difference, 2,986.2 kJ/day, between the core diet and diet 6.) Three of these grasses (foods SCTX, GRTU, and SMLU) were eaten by the baboons even though these grasses would not be included in diet 6, hence their energy content (indicated as negative, above) actually betters the baboons' energy intake, albeit at a cost: all three foods contribute unnecessarily to the binding constraints affecting this diet. (The four-letter abbreviations for foods are decoded in table 6.2.)

Fever tree gum from its various sources is another food that made an appreciable contribution to the baboons' energy shortfall. The baboons' average daily gum intake totaled 34.0 g (foods FTGX, FTHG, XXXW, FTHP, and FTHW), whereas their energy-maximizing intake of gum from these five sources would be 99.9 g/day. The difference, at 12.07 kJ/g, represents a deficit of 0.80 MJ/day.

The fruits of the spiny bush *Azima tetracantha* (ATFR, ATFX, and ATFU, ripe, unknown ripeness, and unripe), were eaten by the baboons in quantities of 5.12, 2.52, and 0.78 g/day, respectively. According to diet 6, however, the corresponding amounts should have been 27.2, 10.4, and 4.27 g/day. The differences, at their corresponding energy densities (4.00, 4.07, and 4.56 kJ/g), represent a deficit of 0.14 MJ/day.

The wrong quantities of these foods (milk, grass blades, fever tree gum, and *Azima* berries) account for 2.67 MJ/day, which is 98% of the 2.72 MJ shortfall—the difference between the yearling baboons' actual diet over the year and the diet that would maximize their energy intake. That is, the baboons failed to maximize their energy intake primarily because they did not drink enough milk and did not eat enough fever tree gum or *Azima* fruit, or the right quantities of green grass leaves of various sorts.

However, judging by the zero slack values for diet 6 (table 6.12), harvesting diet 6 or anything approaching it would bring its own problems. Baboons on diet 6 would have a barely adequate mineral intake, foraging for diet 6 would take all of their available time, from sunrise to sunset, and the baboons would be consuming food at their estimated maximum gut capacity. (All other relative slack values for diet 6, water aside, are well above zero.) For diet 6 as for any other optimal diet, small errors in selection might make an individual subject to potentially large deleterious effects of constraining factors in the diet.

Protein Maximization

The yearling baboons' seasonally adjusted diet yielded 11.1 g protein per day, on average, which is comfortably (181%) above their requirement of 3.94 g per day but is well below the yield of the optimal selection, diet 7, which would have yielded an impressive 31.0 g protein per day. What accounts for the difference? Diets 6 and 7 call for nearly the same quantity of milk, in this case 630.4 g/day, but the baboons' actual intake, 260.9 g/day, was only 41% of this amount. However, compared with many other baboon foods, their milk is not very high in protein: 1.6%. Thus the baboons' lower-than-optimum milk intake accounted for only 5.91 g (29.7%) of the 19.9 g difference between the protein-maximizing diet and the actual diet—their protein shortfall.

Diet 7, like diet 6, calls for consumption of green grass blades from *Cynodon nlemfuensis* (CDLU, 23.3 g/day) and from other, unidentified grasses and sedges (GRLU, 108.7 g) at their limits of availability, but for no green blades from the grass *Sporobolus cordofanus* (SMLU) and for no grass blade bases (SCTU, GRTU). Differences between these optimal intakes and the baboons' actual intakes of these four foods, at protein concentrations of 6.3%, 6.4%, −8.5%, −4.6%, and −8.3%, respectively, result in a net protein shortfall of 5.7 g/day.

According to diet 7, the blossoms of umbrella trees (TOBS) should have been consumed, when available, in a quantity large enough to average 35.3 g/day over the year; the baboons' actual consumption averaged only 3.33 g/day. The difference, at a protein density of 4.1%, represents a shortfall of 1.3 g protein per day. On the other hand, once the blossoms have matured and have fallen from the trees (TOBR), they are contraindicated by diet 7; yet the baboons consumed 0.14 g/day, and they also ate the contraindicated fever tree blossoms (FTBS, 1.91 g/day); but these sources of extra protein reduced the protein shortfall by a negligible 0.026 g/day.

In diet 7, the berries of *Azima tetracantha* (ATFR, ATFX, and ATFU) are all called for at the limits of their availability to the baboons, as they were in diet 6; actual consumption, however, was less than optimal in each case, and at protein concentrations of 4.9%, 5.2%, and 7.1%, respectively, that amounts to a shortfall of 1.7 g protein per day from *Azima* berries.

Salvadora persica fruit, unripe (SPFU), semiripe (SPFS), and of unknown ripeness (SPFX), are all called for in maximum quantities (18.5, 33.5, and 5.1 g/day, with, respectively, 6.0, 3.4, and 3.6% protein) in diet 7, though ripe fruit (SPFR) was not called for at all: it has a much lower protein concentration (−0.7%). The baboons did not eat any of these three types of *Salvadora* fruit in optimal quantities (1.5, 2.1, 0.45, and 1.37 g/day). In total, these *Salvadora* fruit discrepancies resulted in a shortfall of 2.2 g protein per day.

Acacia seeds are very high in protein, so not surprisingly diet 7 calls for these seeds from several sources in maximum available amounts: foods FTDG (8.18 g with 18.7% protein), TODR (3.02 g, 20.8% protein), TODG (1.36 g, 20.8% protein), ACDD (0.064 g, 19.1% protein), and XXXG (7.27 g, −19.1% protein). However, actual intakes from these sources were, respectively, 1.54, 0.42, 0.27, 0.064, and 3.69 g/day; all of these except seeds from dung (no. 50) are below the optimal levels, resulting in a total shortfall of 2.7 g protein per day. Four other sources of acacia seeds would not be included in diet 7, though the baboons ate them: foods TODU (5.3% protein; 3.8 g/day consumed), FTDR (14.5% protein; 0.44 g), FTDU (10.7% protein; 0.40 g), and XXXD (19.1% protein; 0.20 g), and these four reduced the protein shortfall from acacia seeds to 2.4 g protein per day.

All together, these six food classes—milk, green grass blades, umbrella tree blossoms, *Azima* berries, *Salvadora* fruit, and acacia seeds—accounted for 19.7 g/day (99%) of the total 19.9 g/day protein shortfall between the baboons' actual diet and the protein-maximizing optimal diet.

The zero slack values of diet 7 (table 6.12) reveal the potential risks of foraging for this diet: consumption at gut capacity and requiring all available time, resulting in a diet barely adequate for minerals. For all other constraining proximate nutrients, intake would be well away from requirements and tolerances. Again, this statement does not include water, since the baboons regularly supplement their food-bound water with drinking water. Yet, diets 6 and 7, like the baboons' actual diet but unlike the other three optimal diets, would satisfy the baboons' water requirements without such supplements.

Feeding Time Minimization

Diet 8 is a diet that would enable the baboons to satisfy all their proximate nutrient requirements in the least amount of time, 50.9 minutes per day, whereas the actual diet took 240.5 minutes per day, on the average. Clearly, if merely satisfying proximate nutrient requirements (total protein, minerals, fiber, lipids, and energy) in the least amount of time is the baboons' goal, they are wasting a lot of time!

Milk would be by far the largest component of a time-minimizing diet: it would be 119.2 g of the 174 g of food in this diet. The baboons' actual milk intake, 260.9 g/day, was far in excess of this. As a consequence, by the criterion of this diet's objective, they were spending too much time nursing. With a time density of 0.208 minute on the nipple per gram of milk, the baboons apparently were spending 29.5 minutes per day longer nursing than was necessary.

They also wasted much time eating green leaves of grasses and sedges— foods GRLU (33.5 g at 0.556 minute per gram), CDLU (4.86 g, 1.24 minutes/g), SMLU (0.070 g, 5.136 minutes/g), SCTX (2.07 g, 3.12 minutes/g), and GRTU (0.33 g, 2.64 minutes/g), total 33.4 minutes—none of which would be required in a time-minimizing diet.

According to diet 8, acacia seeds were another major time waster for the baboons, including foods TODU (12.14 minutes per day actual), XXXG (6.3 minutes/day), FTDR (3.36 minutes/day), TODR (1.80 minute/day), FTDU (1.64 minute/day), XXXD (1.26 minute/day), and ACDD (0.14 minute/day), all of which are completely contraindicated in diet 8 and in total represent 26.64 minutes wasted per day. One source of acacia seeds, umbrella tree seeds from the ground (food TODG), took 0.34 minute per day, versus 1.70 minutes recommended, for a saving of 1.36 minutes per day. So the overall time wasted on acacia seeds comes to 25.28 minutes per day.

The baboons' consumption of grasses and sedge corms—foods GRCX (30.15 minutes/day), SKCX (5.26 minutes/day), SMCX (1.97 minutes/day), and COCX (0.18 minute/day), all contraindicated in diet 8—represent in total a waste of 37.6 minutes per day.

Excess time that the baboons spent feeding on these foods—milk, green grass blades, acacia seeds, and corms—amounts to 125.8 minutes per day, 66% of their wasted time. Because of the very limited array of foods in diet 8, just six foods, most of the baboons' actual foods contribute to the discrepancy between the real and ideal in this case, so the larger number of foods listed above account for a smaller percentage of the shortfall than was the case with diets 6 and 7.

Like diets 6 and 7, diet 8 would require taking in the bare minimum of minerals. It would also entail the minimum intake of protein that is compatible with normal health and growth.

Energy and Protein Rate Maximizations

In chapter 5 I suggested that maximizing intake rates, because of the emphasis on "fast foods," could lead to unreasonable diets, whether the rates are computed per minute of feeding on each food, as in this study, or per minute of foraging for it. Diets 9 and 10 appear to be cases in point. If the baboons ate either of these diets, they would sacrifice energy, protein, and other nutrients for the sake of speed, leaving them with much time on their hands and nearly empty bellies. Can it be that this is just what baboons do? Not at all. Their actual diet has little resemblance to diets 9 or 10 (cf. the seasonally adjusted diet in table 6.5 vs. diets 9 and 10 in tables 5.12 and 5.15, respectively).

An animal on diet 9 would take in energy at a remarkable 20.7 kJ/minute (mass-weighted mean). The yearlings were more leisurely: their mass-weighted mean energy intake rate was 12.1 kJ/minute. Similarly, an animal on diet 10 would take in protein at an impressive 113 mg per minute (mass-weighted), whereas in reality the baboons' mass-weighted mean intake rate was 63.5 mg per minute. Clearly, the baboons are not "living in the fast lane."

Like diet 8, diets 9 and 10 would include few foods, so most of the foods in the baboons' diverse diet contribute to the discrepancies between their actual mean yield rates of energy and proteins, respectively, and the rates that could be achieved via these optimal diets. Since neither diet 9 nor diet 10 calls for any milk, that food, which still played a major role in the actual diets of the yearlings I sampled, is a major contributor to the baboons' failure to live up to the expectations of those two diets. On the other hand, the emphasis of diet 9 on fever tree gum and of diet 10 on grass leaves, acacia seeds, and *Azima* fruit means that these foods contribute relatively little to the baboons' shortfalls in these two diets. In view of my remarks in chapter 5, comment 7, about the unrealistic nature of the objectives of diets 9 and 10, little is to be gained by detailing the contributions of individual foods to the baboons' pseudo-deficiencies in living up to the requirements of these two diets.

To maximize the rate of energy consumption, a baboon on diet 9 would risk taking in its minimum of protein. Conversely, on diet 10 it would maximize its rate of protein intake but be required to function on a bare minimum of energy.

Are the Optimal Diets Realizable?

Excluded Core Foods

That the baboons' actual diet was not optimal became clear once the optimal diets were calculated (chapter 5) without any quantitative comparisons' being made, for after all, the optimal diets, diets 6–10,[9] each excluded several of the baboons' core foods, and some foods were excluded from all optimal diets (table 5.16). Those uncalled-for foods seem like the best place to begin looking for unrealistic specifications in the optimal diets. Their nutrient contents are given in table 5.1. Of the fifty-two core foods, two (SPWX and XXWX) are primarily small pieces of wood the yearlings gnawed on, thereby qualifying them to be recorded as "food," though they seldom swallowed any of this material and it should not be a part of any optimal diet for Amboseli baboons.

These two foods were not included in the calculations of diets 6–10. Of the remaining fifty core foods, twenty-one were not called for in any optimal diet. Yet each is a core food: one of the foods to which the baboons devoted the most time. In so doing, were they making a mistake? This is not just a question of whether they ate the right quantities of foods in the optimal mix, but of whether they ate many of the wrong foods. Their score, recall, is twenty-nine "right" foods and twenty-one "wrong" ones—wrong by all five of the objective function criteria I tried!

At this stage in the development of optimal diet models, it is more likely that the models are defective than that the animals are. We must ask not only whether the diets of real animals are ideal, but also whether our versions of ideal diets are realizable. I am reminded of Thomas Paine's question, "Is it more probable that nature should go awry, or that a man should tell a lie?" Just as I do not expect that many real baboons will have an optimal diet, for reasons that I spelled out in chapter 5, so I do not believe that baboons that ate the twenty-one uncalled-for foods would be better off forsaking all of them. Indeed, as I shall now show, most of these contraindications probably are "false negatives."

Acacia Seeds

What are these core foods that are not called for in any optimal diet? The diets all exclude the *green* seeds of acacia trees, TODU and FTDU, respectively, which rank 5 and 19 by feeding time in the baboons' observed mean diet, yet the *dry* seeds are called for in maximum amounts by several diets. Why? The answer is that nutrients are far less concentrated in the green seeds than in the dry ones: 2.0 versus 3.9% minerals, 2.6 versus 4.0% lipids, 10.7 versus 18.7% protein, 6.7 versus 42.6% carbohydrate, 2.7 versus 21.3% fiber, and 4.0 versus 14.5 kJ per gram in green versus dry fever tree seeds. Beyond that, a toxin, antitrypsin, is less concentrated in the dry seeds (see chapter 9), and these seeds are abundant in dry pods, in the dung of several large mammals, and lying loose on the ground. Are the baboons making a mistake, then, by gorging on green acacia seeds when they could substitute the more nutritious dry ones? I think not, for several reasons. First, the dry, lentil-like seeds of acacias are extremely hard, and considerable masticatory force is required. From my own experience chewing them, I would not be surprised if the baboons sometimes chip their molar crowns on them. Second, the green seeds are seasonally available in enormous quantities, and a baboon can obtain large numbers from just a few sites in the shady canopy of a single acacia tree. Neither search time nor processing effort was measured in my study, but green acacia seeds obviously require less of each. Third, the greater nutrient concentrations in the dry seeds are almost entirely the result of evaporation, which provides no advantage, since excess water is hardly a problem for baboons in this arid region! (Does this mean that nutrient concentrations should be calculated per gram of dry matter? I am not sure.)

Gum Drops

Small items picked from wood (XXXW) are excluded from all the optimal diets because of their large impact on the time budget. Their nominal composition is the same as that of fever tree gum, which most of them probably were, yet they take thirty times longer per gram of yield. These gum drops are typically picked up one by one en passant by the baboons, with seldom any appreciable deflection in their day journey. Surely these items should be eaten so long as the time required to harvest them is negligible and the energy they provide exceeds the energy expended in harvesting them. Near the end of this chapter, I discuss the problem of deciding how much of each day journey's time and energy to allocate to each food harvested along the way. The exclusion of these gum drops from all optimal diets is a consequence of not resolving that problem.

Corms

The role of various grass and sedge corms (ranks GRCX, SKCX, SMCX, and COCX) in the optimal diets is enigmatic. *Sporobolus cordofanus* corms (SMCX) are called for in two optimal diets, those of *Cyperus obtusiflorus* (COCX) in one, in each case in maximum available quantities, whereas corms of *Sporobolus rangei* would be an alternative in one optimal diet (in that, at less than maximum quantity) and are otherwise uncalled for. Yet this is counterintuitive, since *Sporobolus rangei* corms are one of the baboons' favorites, fed on for more time than the other two species combined, and are higher than the other two in concentration of protein (a binding constraint in diets 8 and 9), and energy (binding in diets 8 and 10). *Sporobolus rangei* is more abundant than either of the other two, and where it occurs, its tussocks represent a major local concentration of food. The problem with *Sporobolus rangei* corms seems to be that they are labor intensive and time consuming. Each gram required 9.3 minutes of effort, whereas the other two species required only 2.7 and 3.6 minutes per gram, respectively. On the other hand, *Sporobolus rangei*, with its large, dense tussocks, is particularly suited for sustained periods of feeding, in that a baboon, having finished eating the most readily excavated corms at one *S. rangei* site, need move only a short distance away, usually just a few paces, before it is at the next. Such a sustained yield would appear in my records as a series of brief interbout intervals, a factor I did not take into account in my analyses.

The corms of all other grasses and sedges combined (GRCX) provide a similar problem: they are nutritious but time consuming (the highest among corms, 11.2 minutes per gram), yet second only to milk in the amount of time the baboons actually devoted to them: half an hour per day, averaged over the year. Optimal diets 6 and 7 would be time limited, and minimizing feeding time is the objective of diet 8; unidentified corms apparently did not have enough of any nutrient in short supply to overcome this disadvantage. Similarly, because diets 9 and 10 center on rate maximization, the high time requirement of these corms would work against them. So are the baboons making a mistake by feeding extensively on these corms? Based on the available evidence they are, but that evidence is markedly deficient in at least one respect: the actual nutritive value of these corms is unknown, and I have assumed that on average their nutritive value is approximately that of the mass-weighted mean of the other two (table 5.1). If the actual nutritive value of these corms is better than this, or if they have other unmeasured merits, such as micronutrients that are otherwise in short supply, then they may indeed warrant a role in the baboons' diet. Clearly a more detailed analysis of corm feeding and of the nutritive value of corms would be welcome.

Meristem

Meristem tissue other than corms predominates in the blade bases of grasses and in the growing ends of rhizomes and stolons, all of which the baboons ate (core foods SCTX, GRTU, and CDSX). Yet these foods are not called for in any optimal diet. *Cynodon nlemfuensis* stolons are handicapped by a high time requirement, 24.3 minutes per gram (the highest among the fifty core foods). However, for the first two of these foods the problem is not harvest time (respectively 3.1 and 2.6 minutes/g) but low concentrations of nutrients. For example, the food in energy-maximizing diet 6 would average 14.1 kJ/g (mass-weighted mean), but the lateral meristem of the tall, tough grass *Sporobolus consimilis* (SCTX) contains just 3.1 kJ/g, and green grass blade bases (GRTU) contain just 2.6 kJ/g. The exclusion of these two foods from protein-maximizing diet 7 is more difficult to explain, since both have protein densities (4.6% and 8.3%) above the average for the food in that diet (3.1%, mass-weighted). However, diet 7 is markedly affected by non-protein considerations. For example, milk, its major constituent, is only 1.6% protein. Diet 7 is limited by consumption capacity, foraging time, and mineral requirements. The first two factors in particular work against inclusion of these two foods: in diet 7, only three foods have higher "time densities" (minutes of feeding required to obtain one gram) than do grass blade bases, only two are higher than *S. consimilis* meristem, and each of these foods has compensating advantages that these two meristem sources do not. I suspect that new green grass blade bases are among the most profitable of diet components during their limited season: abundant, readily harvested and consumed, and moderately nutritious.

Unfortunately, the seasonality constraints that were used in calculating diets 6–10 limit only the total quatity of each food, not the array of alternative foods each competes against for a place in the diet. Thus seasonal foods were pitted against other seasonal foods that in reality are not available at the same time of year, and the result of this "unfair competition" is the exclusion from the optimal diets of some foods that are in reality among the best foods available *in their season.* For example, diet 7 would include *Acacia tortilis* blossoms from the tree and semiripe fruit of *Salvadora persica* in preference to grass blade bases; yet during most of the time when grass blade bases are being eaten by the baboons, these other foods are not available.

That argument does not apply to the lateral meristem of *Sporobolus consimilis,* however. This material was eaten during most semimonths and probably was available during all. During this study it was fed on much more often than during 1963–64 (S. Altmann and J. Altmann 1970), when the baboon population of Amboseli was much higher (J. Altmann, Hausfater, and S. Altmann 1985) and the food situation presumably more favorable. Nonetheless, other foods in combination, as in diet 7, appear to be a better way for Amboseli baboons to maximize their protein intake, and by this criterion, eating the tough, bulky, time-consuming lateral meristem of *Sporobolus consimilis* appears to be a mistake.

Fruits and So Forth

The baboons (and I) prefer the ripe fruit of *Salvadora persica* (SPFR) to the unripe fruit (SPFU) or semiripe fruit (SPFS). Yet only the latter two are called for in any of the optimal diets. If the chemical analyses of these fruits are correct, the unripe fruit of *Salvadora* contains more minerals, much more protein, and more energy. Although the carbohydrate contents of the ripe fruit and of unripe fruit are about equal, the ripe fruit is more flavorful and tastes sweeter to me. The fruit's ripening process appears to be deceiving the baboons into preferring it at a stage when it is not most nutritious but when the tiny seeds are mature and ready for dispersal. Verification of the nutrient composition of these fruits is desirable. Since either unripe or semiripe *Salvadora* fruit usually is available whenever the ripe fruit is, eating the ripe fruit appears, on present evidence, to be a poor choice.

Similarly, the unripe fruit of *Azima tetracantha* (ATFU) appears to be more nutritious than the ripe fruit (ATFR): the former contains more minerals, fat, protein, fiber, carbohydrate, and energy. However, none of these differences is as large as in the case of *Salvadora* fruit and so, perhaps not surprisingly, either ripe or unripe *Azima* fruit (or both) is called for in maximum quantities in four of the five optimal diets.

Several others of the baboons' favorite fruits are not called for in any of the optimal diets: the fruits of two prostrate plants, devil's thorn (TTFX) and *Tri-*

anthema ceratosepala (TCFX), and fruits of the scandent *Capparis tomentosa* (CTFX) and of the bush *Withania somnifera* (WSFX). In addition, the leaves of *Salvadora persica* (SPLU and SPLX) are out, as are those of *Suaeda monoica* (SULU) and of various unidentified plants (XXLX). Finally, neither the unopened seedheads of unidentified grasses or sedges (GRDU) nor the green *Sporobolus cordofanus* grass leaves (SMLU) made the lists, nor did umbrella tree blossoms from the ground (TOBR), trumpet flowers (TFBS), or dung-beetle larvae (DUKU). Most of these foods are highly seasonal, and either their concentrations of nutrients are too low or their time requirements are too high, considering their contribution to the binding constraints, for them to compete against those foods that are called for in the five optimal diets, even though the latter may not be available at the same time of year—the computational artifact of unfair competition, which I described above. The same artifact also affected many foods that were excluded from somewhat fewer than all five optimal diets.

A few foods mentioned in the previous paragraph are not particularly seasonal, and they warrant further comment. First, feeding on leaves of the shrubs *Salvadora persica* and *Suaeda monoica* and of various unidentified plants went on much of the year, and some such leaves probably were available throughout the year. The nutritional value assigned to unidentified leaves is of course a surmise, and no firm conclusions can be made about such food. The composition of *Salvadora* leaves is very similar to that of its fruit, but the yield rate is much slower: 3.5 minutes are required to obtain one gram of leaf. This low yield rate reflects the fact that yearling baboons sometimes held the leaves of this plant in their lips while playing on this springy bush, and by my criteria all of that counts as feeding time. I cannot consider such playful mouthing of food a mistake. (*Salvadora persica* leaves were eaten more often during 1974 than during 1975–76. Perhaps the leaves vary in composition from year to year.)

Baboons sometimes nibbled on the terminal growth of *Suaeda monoica*—leaves and perhaps cryptic flowers—but this material appears to have rather low nutritional value, except for modest concentration of protein (3.5%). Other foods that are available at the same time of year appear to provide more nutritious alternatives.

Finally, the fruits of *Trianthema ceratosepala* were eaten during every semimonth of the year. Baboons often spent extended periods plucking and orally hulling these tiny young fruits, which ranked eighth in the observed mean diet based on time expenditure. Yet not one optimal diet would include them. They appear to be an average source of energy (3.1 kJ/g) compared with those foods that warrant inclusion in diet 6, and they are a fairly good source of protein—indeed, above the average of the food that would be included in protein-maximizing diet 7 (4.6 vs. 3.1%, mass-weighted mean). Unfortunately, their yield rate is slow: 4.1 minutes of feeding per gram of yield, and this works against them not only in the diets that are bulk limited and time limited (diets

6 and 7) but in all the others as well. None of the other nutrient densities of these fruits are remarkably high or low. With this combination of merits and demerits, I cannot say whether either their inclusion in real baboon diets or their exclusion from optimal ones is a mistake.

In summary, a few foods in the baboons' array of core foods appear to be mistaken choices in that, based on currently available evidence, other foods that are available at the same time of year provide more nutrients at less risk. These foods are the corms of grasses and sedges other than *Sporobolus cordofanus, S. rangei,* and *Cyperus obtusiflorus,* the lateral meristem of the grass *Sporobolus consimilis,* and the terminal growth of *Suaeda monoica.*

In addition, this survey of uncalled-for core foods revealed several significant factors that were not incorporated into my optimal diet constraints and that would affect which foods are included in the optimal diets. These include time and energy spent searching for and moving to food sites; food site clustering; time and energy spent getting from one site to the next; harvesting and processing effort; evaporative concentration of nutrients; and probably most important, the array of potential foods available at any given time of year, from which the baboons choose their diet.

The last of these deficiencies was circumvented when I calculated diets for individual yearlings: for each yearling, I calculated a separate optimal diet for every semimonth in which the youngster was sampled (chapter 8).

The first three factors above all hinge on the fact that my calculations are based on gross energy yield (assimilable energy obtained from foods), not net (energy obtained minus energy expended), which I do not know, and on time spent harvesting each food but not the time spent searching for and moving to it. Energy is expended at three stages of obtaining food: in the search and travel phase (up to the onset of the feeding bout), in manipulation/handling/ extraction (during the feeding bout), and in subsequent internal processing: mastication, digestion, excretion, and so forth. Some foods require relatively large expenditures of processing energy per unit yield, others take relatively little. Dry acacia seeds are very hard, whereas the green seeds, though not as concentrated in nutrients, are easy to chew. As another example, the soft blossoms of acacia trees require little energy to pick and masticate, and for this reason they make good "weaning foods" (J. Altmann 1980). When the trees are in flower, a baboon can obtain large numbers of blossoms within arm's reach; then a short move within the crown of the tree brings it to the next concentration.

I do not know how much travel effort (time and energy) baboons expend on each food per unit consumed. While travel costs may be appreciable for baboons, with their long day journeys (Alberts, J. Altmann, and Wilson 1996; S. Altmann and J. Altmann 1970; S. Altmann 1987; J. Altmann and Samuels 1992; Bronikowski and J. Altmann 1996), it is not clear just how much of this total cost to attribute to each food, since on any one day many foods are consumed. Consider fever tree gum drops on logs under the trees (FTHW), one of the uncalled-for foods. These droplets are picked up as the baboons come upon them in the course of their daily trek, and the extra travel cost that results from the small deflection in their path is negligible. So the question is not whether other gum sources yield nutrients at a faster rate than these droplets, but whether the baboons would be better off ignoring these droplets when they come upon them. Other "en passant foods"—those that are eaten as the baboons come upon them, with little deflection or extension of their route—include fever tree droplets picked from understory plants (FTHP), gastropods (GAKX), insects and various small, usually unidentified objects picked from the ground or from plants (INKX, XXXG, XXXP, XXDD), and small items (seeds? insects?) taken from within, on, or under pieces of wood and large boluses of dry dung (XXXW, XXXD, and XXDD), both of which the baboons flip over in passing.

At the other extreme, the baboons of Alto's Group sometimes made long journeys from their sleeping trees to the base of Nairabala Hill to feed on *Lycium "europaeum"* fruit and to the edge of Lake Amboseli, primarily to harvest the abundant gum from a band of regenerating fever trees in that area. In the latter case the energy acquired from the gum must have been greater than the extra travel energy required to obtain it if these long trips resulted in a net energy gain, but it is not obvious which part of total travel energy to attribute to these long extensions: other foods were eaten en route. The primary benefit of the *Lycium* fruit probably is its provitamin A, but those trips too extract an energy cost, of unknown magnitude. In this case, costs and benefits are measured in different currencies, as indeed they will be for any non-energy-producing nutrient.

Beyond this problem of allocating energy debits among foods, there is a more difficult problem. If baboons ate one of the specified optimal diets and nothing else, their day-journey routes would be very different but in unknown ways, and consequently so would be the net cost of obtaining each food. Ultimately, theories of optimal diet must be subsumed under more general theories of home range use: of the spatiotemporal distribution of activities ("scheduling," Hockett 1964). To illustrate, diet 3 called for 82 g per day of fever tree gum from the ground and a large quantity of milk. While baboons in Amboseli might have difficultly obtaining that much gum per day on the treks they now take, the question is, Could they do so if they had no foraging requirement other than gum? Beyond that, Would we change our answer to this question if our hypothetical baboon also had to remain with the other members of its group, including the adults, who are not nursing and must therefore have a different array of foods in their optimal diets?

Two potential sources of error in my input values might also alter which foods are included in the optimal diets, and in what quantities. First, the food composition values are subject to various errors, of unknown magnitude. Furthermore, for several core foods I do not have macronutrient analyses, and in those cases I have used presumptive values (table 5.1). Presumptive values were also used for several food classes that result from incomplete observations, such as leaves from unidentified plants. For many uncalled-for foods, the "reduced cost" (tables 5.9–5.12 and 5.15) is small; that is, with just a small change in the nutritive value of these foods, they would enter the optimal diets. Second, the effects of errors in the composition table are confounded by errors and uncertainties in my estimates of nutrient requirements and toxin tolerances (appendix 1), some of which may be even larger.

Included Foods

I have dealt above with possible errors of omission in the optimal diets. What about errors of commission? Are any of the amounts of food that are specified by the optimal diets unrealizable? The availability constraints imposed on diets 6–10 supposedly took care of this problem from the standpoint of the animals' capacity to procure (see chapter 5, comments 1 and 2). What about the ability of the habitat to provide? Recall that a limit on the intake of each food was established based on the largest amount of that food the animals were likely to be able to harvest. Almost all foods in every optimal diet are called for in these extreme amounts. Suppose that the diets of the baboons converged on one of these diets. An implicit assumption is that the habitat could sustain such a high level of harvesting. Whether it could is quite unknown. The information needed would require a detailed, quantitative study of resource turnover rates and sustainable yield levels. Perhaps milk is the only food of young baboons for which an attempt has been made to specify limits on availability (J. Altmann 1980, 1983).

7

All animals are equal, but some animals are more equal than others.

—George Orwell, *Animal Farm*, 1954

Every man is in certain respects like all other men, like some other men, like no other man.

—Clyde Kluckholn and Henry A. Murray, 1953

Individual Differences and Age Changes

In chapter 6, the pooled mean diet of the yearlings I sampled at ages 30–70 weeks was compared with adequate and optimal diets for a hypothetical baboon whose age was the average of those in my samples, 47.5 weeks. Such characterizations of the average individual provide a considerable economy of description. However, no baboon is an average baboon. Even individuals of the same age and sex, living in the same group, differ in diet. Means, modes, and other statistical measures of the "central tendencies" of traits are subject to the same abuse as descriptions of "typical" traits that these statistical descriptions of central tendency are replacing in biology. Both lead too easily to typological thinking that is antithetical to the view of evolution as fundamentally a population process (Mayr 1963).

In the preceding chapter I showed that the average diet of the baboons was nutritionally adequate, except perhaps in minerals, but was not optimal by several criteria. A crucial question remains. How much did individuals deviate from this average, and how close did some of them get to an optimal diet? Conversely, were any individuals consuming a nutritionally inadequate diet? In addition to these questions about variability, we must cope with the fact that the diet of an individual, particularly a very young one, changes with age. In this chapter I first look into some of the sources of age changes and individual differences in diet, particularly, feeding bout rates and durations. Then I ask, Did the yearlings' diets differ significantly? And if so, what were their population distributions?

These demonstrations of individual differences in diets are an essential prelude to the next chapter, in which I compare the nutrient intakes of individuals with the composition of their respective adequate and energy-maximizing optimal diets, then correlate their dietary intakes with several measures of fitness.

233

Calculation of Observed Intakes

For each of the eleven yearlings in each ten-week age block in which it was sampled (twenty-eight blocks all together, table 3.1), I calculated the yearling's mean intake (grams per day) of each of the core foods during the days it was sampled. From these observed diets, I then calculated each food's yield of energy and of each of the six proximate nutrient classes: water, ash, fat, protein, fiber, and other carbohydrate. As in chapter 6, the nutrient intakes that result from an observed diet were obtained by calculating, via equation (4.1), the quantity of a specified nutrient that was obtained from each of the fifty-one core foods other than milk, summing these values, multiplying the total by 1.095 to allow for the contribution of non-core foods, then adding milk's contribution of that nutrient. Summary values (totals for mass, energy, and each nutrient) are given in table 7.8.

In this chapter, individual-specific values were used for bout rates and lengths, as described below; for the other components of equation (4.1), individual differences have not been documented. When calculating the nutrient intake of an individual, I took bout rate and length values for each core food from the most specific (most narrowly partitioned) level—by individual × age, just by age, or all data pooled—for which significant differences were demonstrated for that food. So, for example, in calculating the contribution that abscised *Acacia tortilis* blossoms picked from the ground (TOBR) made to the diet of Alice at 30–40 weeks of age, I noted that the yearlings' bout lengths for these blossoms differed significantly by age (table 7.2), but that within the 30–40-week age class, individual bout lengths did not differ significantly (table 7.3). Therefore, at 30–40 weeks of age, Alice (and each other yearling) was assigned the pooled mean value of these blossoms for that age class, namely 0.08 minute, as given in table 7.2. By contrast, for unripe fruit of *Salvadora persica* (SPFU), differences between individuals were demonstrated at 30–40 weeks (table 7.3), so for that food I used Alice's own mean value at that age (0.81 minute, table 7.3). For any food for which neither individual nor age differences in bout length occurred, I assigned to Alice at 30–40 weeks (and at each of the three other ages) the overall pooled value (table 4.1). In calculating the intakes of various nutrients of a given individual at a specified age, the criterion above led to procedural rules that are spelled out in appendix 10.

Sources of Differences in Diet

Individuals can change or differ in their nutritional intake as a result of any of the components of equation (3.1): the amount of time devoted to each food (feeding bout length times bout rate), the rates at which the food items are consumed, the sizes of the food items, or the nutritional compositions of the selected material. However, I have adequate age-specific and individual data only for the two time-budget components: feeding bout durations and bout rates. For the rest, I assume that bite rate and size did not differ appreciably between individuals and that I was successful in demarcating "homogenous" food classes, that is, classes with consistent nutrient distributions, so that the mean composition of a food can be used for all individuals. (Some data on individual bite rates will be presented below in discussing Pedro's feeding.)

Because I have adequate data for age changes and individual differences only for bout rates and durations, my conclusions about individual differences or age changes in diet or in nutrient intake rest on these two time-budget parameters. A given percentage difference in time spent each day in feeding on a particular food indicates (ceteris paribus) a corresponding difference in the amount of that food ingested (number of grams or food units) and in that food's contribution of each nutrient and other component. For example, if one individual spends 25% more time than another feeding on a particular food, it will, if nothing else is different, be taking in 25% more of that food, thereby getting 25% more protein from it than does the other individual.

Because bout duration and rate varied essentially independently across foods (fig. 4.16), their product cannot be determined from either alone, as explained in chapter 4, and so neither can a food-specific time budget, which is proportional to this product (equation 3.2). I shall therefore consider individual differences and age changes both in bout lengths and in bout rates.

Another reason for examining the two components of time budgets is this. Methods for statistical comparison of rates and durations are available, whereas for their product, time budgets, and similarly for the nutrient intake levels that were estimated from them via sums of products of means of random variables in equation (3.1), no method of statistical comparison is known to me. If, however, one of two individuals eats some food significantly more often and significantly longer, on average, it also seems reasonable to suppose that their product, the time budget of that individual for that food, is significantly longer and thus that the first individual is eating more, assuming no countervailing differences in bite size or rate. I would conclude similarly if just one of the two factors differed significantly (especially if markedly so) and the other is in the same direction. If, on the other hand, the rate is higher and the duration shorter (or vice versa) or if only one of the two differs significantly but the other is in the opposite direction, it is not obvious what to conclude about differences in the time budgets.

Even these guidelines are of no use for comparisons of the total yield of nutrients and other food components. For these I have relied on overlap in standard error intervals. As indicated in chapter 4, these do not always correspond with the desired uncertainty intervals for which nonoverlap would indicate significant differences in means at a specified significance level.

Bout Lengths

How do the amounts of time that the baboons devoted to various foods differ between individuals? How do these time budgets change with age? Are the differences due to differences in how often foods are eaten, how long each time, or both? I begin by examining bout lengths.

In chapter 4 mean bout lengths were calculated based on pooled data from all eleven sampled yearlings at all sampled ages (30–70 weeks). Here I break the largest of these samples down by age and individual. For every primary food class (as defined in chapter 4) that occurred forty times or more in the total sample (table 4.1), the product-limit estimate of its mean bout length was calculated separately within each of the four ten-week age blocks between 30 and 70 weeks, with the data from all yearlings pooled. The cutoff at forty guarantees an average of at least ten occurrences per age block. The sample size cutoff excluded eight of the fifty-four core foods, namely COCX, XXWX, and the six that rank lowest in the time budget among the fifty-four (table 6.1), and only one of the excluded core foods (SPFX) was in an aggregate primary category (G) for which the sample size was above the cutoff. The cutoff at forty included two foods (grasshoppers GHKX and unidentified items picked from *Rhamphicarpa montana* plants RMXX) that are not among these top fifty-four. Above this cutoff were sixty-seven primary food classes, including forty one-food classes (table 7.1) plus twenty-seven aggregate food classes (table 4.1).

For each such primary food class (individual or aggregate), age differences in bout duration between the ten-week age blocks were checked statistically by the method of Mantel and Cox, at a significance level of 0.05. Next, for each such food, if any ten-week age block had a sample size of at least ten times the number of yearlings that fed on that food during that age block, individual product-limit estimates of mean bout duration were calculated, and the distributions were then checked statistically in the same way.

This food-by-food comparison of bout lengths among individuals was also carried out for data from all ages (30–70 weeks) pooled, but to reduce the confounding age effects that result from the unevenness of my sample times (table 3.1), the comparisons were limited to the six "well-sampled" yearlings (Alice, Eno, Fred, Ozzie, Pooh, and Summer), that is, those that were sampled in at least three of the four age blocks (table 3.1).

The results of these analyses of bout length differences are discussed below.

Age Changes in Bout Lengths

Out of sixty-seven primary food classes represented at least forty times in my samples, statistically significant age differences in feeding bout lengths were found in twenty-six, about 39% (table 7.2), too often to be entirely dismissed as pseudosignificance from repeated statistical testing. If food classes represented less often had been analyzed, the results almost surely would be even smaller proportions of significant deviations. Indeed, when the twenty-one foods and food classes that occurred twenty to thirty-nine times were analyzed for age differences in bout length, using just two age categories for them (30–50 and 50–70 weeks) because of their small sample sizes, only three additional food classes (unknown material picked from wood = XXWX, miscellaneous plant parts = PT, and *Azima tetracantha* leaves = ATL) were significantly different by the same criterion.

The data summarized in table 7.2 reveal few secular changes with age: even for most foods that showed significant age differences in bout length, feeding bouts do not follow any simple rule of getting longer or shorter with age. However, feeding bouts on umbrella tree pods (nos. 16 and 18 in table 7.1) became appreciably shorter with age, and since these bouts usually consisted of feeding on just one pod per bout, this change represents a considerable increase in harvesting efficiency. Conversely, the green blade bases of grasses (no. 126) were fed on for progressively longer periods as the animals matured. This suggests an increased ability to cope with tough material, except that green grass blades (no. 121) and the corms of *Sporobolus rangei* (B) and of unidentified grasses (117), both of which are tougher, show no such trend with age. Green grass blades are bulky: perhaps older yearlings, with larger stomachs, can eat more of them at once. For unknown reasons, bouts of feeding on insects (no. 252) and *Salvadora persica* fruit (no. 162 and G) became progressively shorter with age. The marked increase in the length of feeding bouts on dung-beetle larvae (no. 271) almost surely represents maturation of digging ability as the yearlings became stronger. The lengthening of water drinking bouts (no. 276) with age may reflect decreasing milk intake. The increased amount of time spent by older weanlings on *Salvadora persica* wood (aggregate primary class X), which typically is chewed but not swallowed, might represent a teething response. In wild anubis baboons from the Gilgil area of Kenya, the deciduous premolars emerge through the gingiva during the age span covered by my samples (Kahumbu and Eley 1991), then slowly move into alignment with the rest of the teeth.

Individual Differences in Bout Lengths

Foods and food classes for which bout length within an age block differed significantly among individuals are given in table 7.3. In the four ten-week age blocks, individual differences in feeding bout lengths were detected in respectively twelve out of twenty-six, thirteen out of twenty-nine, fifteen out of twenty-eight, and seven out of twenty-four tested foods or food classes, for an average of 44%.

What is the magnitude of those individual differences that were significant? In table 7.3, which lists only significant cases, the coefficients of variation (sample standard deviation expressed as a percentage of the sample mean) of the individual mean bout lengths are given in the last column. The average deviation is 50.3% of the mean. In scanning the coefficients of variation, I have not been able to discern any obvious correlate of those foods on which bout lengths showed large differences between individuals.

Of course, my partitioning data into ten-week age blocks means relatively few foods have adequate sample sizes for analysis. I therefore reanalyzed the data, using the entire 30–70-week age period but restricting the analysis to the six well-sampled yearlings (defined above) to reduce age bias. For comparison with subsequent results, the fifty-two core foods plus water were tested. The sample size criterion was again set at $n \geq 40$ for all six yearlings pooled. Product-limit estimates of the bout length distributions were computed and were tested for individual differences by the Mantel/Cox test at a significance level of 0.05. The analysis (table 7.4) revealed significant individual differences in seventeen of the forty-six core foods (37%) that met the sample size criterion, but not in drinking. (The slight reduction in the proportion of foods for which the yearlings were significantly different, despite the larger sample size, probably occurred because the aberrant Pedro and Pooh were not part of this sample.) Thus, as before, individuals differed in feeding bout length on more than a third of their foods but fewer than half.

The relative magnitudes of these individual differences in feeding bout lengths are revealed by their coefficients of variation. The coefficients of individual variation (table 7.4) ranged up to 96% for the six well-sampled yearlings, and their average was 46%. Coefficients of individual variation within the ten-week age blocks (table 7.3) are of similar magnitude. As in that analysis, there is, so far as I can tell, no distinguishing characteristic of those foods that have high coefficients of variation.

Many of the individual differences in bout length (table 7.3) could be the result of seasonal effects. For example, at 30–40 weeks of age, Alice fed on green fruit of *Salvadora persica* (no. 164) for about twice as long per bout as Bristle and much longer than Eno. But Alice was born on or about January 1, 1975, Bristle was born August 17, 1975, and Eno was born April 22, 1975. Consequently they would be 30–40 weeks of age at very different times of the year—Alice in August–September 1975, Bristle in March–May 1976, Eno during November 1975–January 1976 (fig. 3.2). I do not know whether Alice reached 30 weeks at a time when patches of *Salvadora persica* green fruit were larger than they were when Eno and Bristle did, but this seems very likely, since the peak in feeding on this food (for both frequency and time) occurred in August. Similarly, the baboons fed on this food much more frequently in August–September than at any other time of year. Thus, whatever individual differences may exist in feeding ability for this food are being swamped by birth season effects.

Of course time of birth, like feeding behavior, is a trait that may have a heritable component and be subject to natural selection, but it is a trait of the parents and is otherwise quite a different sort of thing from feeding behavior per se. If differences in birth season are heritable but decisive differences in feeding bout length merely reflect seasonal differences in the size of available food patches, then an important principle has been illustrated: an environmentally induced trait, such as meal length, may in essence evolve if its induction depends on heritable differences in the environment to which the organism is exposed, such as available food patch sizes that are entailed by time of birth. Habitat selection and its consequences provide a more familiar example.

In some cases, birth date effects can be ruled out. Three yearlings in my samples were born within a few days of each other: Alice, Fred, and Ozzie (table 3.2). Similarly, Dotty and Striper were born less than two weeks apart. Finally, two others, Summer and Bristle, were born fairly close together: within forty-one days. (Pedro was born just two days after Summer, but he died shortly after my sampling on him began.) Thus, comparisons among Alice, Fred, and Ozzie, between Dotty and Striper, and to a lesser extent between Summer and Bristle should reveal individual differences that are unlikely to be the result of birth date. I therefore tested for differences in bout length within each of these three groups of yearlings. For this analysis, foods were separated from their aggregate classes. For any of the foods (but not aggregate food classes) that occurred at least forty times in my entire sample (tables 4.1 and 7.1), a Mantel/Cox test of individual differences was run on Alice/Fred/Ozzie in each of the ten-week age blocks in which at least two of the three ate the food and these three yearlings had a combined sample size of thirty, and similarly for Dotty versus Striper (and for Bristle versus Summer) when their combined sample size in any age block was at least twenty. Few Dotty/Striper comparisons can be made: they were sampled only at 50–60 weeks of age. Similarly, because Bristle did not reach fifty weeks before my study terminated, only a few Bristle/Summer comparisons could be made.

The results are as follows. For the Alice/Fred/Ozzie comparison, which is the best one in terms of both sample size and age proximity, sample sizes were adequate, by the criterion above, in fifty-one cases (nonaggregate food $\times$ age classes) involving twenty-four foods. Of these fifty-one, significant individual differences in feeding bout lengths ($p \leq 0.05$) were detected in just nine cases (table 7.5). At this significance level, about two or three cases of pseudosignificance can be expected by chance out of fifty-one tests. Even those foods that showed significant individual differences in bout length at one age failed to show them at others, and the rank order of the yearlings by bout lengths (table 7.5) did not show any consistent pattern.

If my samples on all eleven infants show individual differences for a food on which the like-aged infants differed, the former difference cannot be attributed to differences in season of birth. I therefore compared feeding bout lengths of all yearlings that were sampled in each ten-week age block. Statistical tests were made for every individual food (designated here as elsewhere by a four-letter abbreviation) and every food aggregate that had an adequate sample size, as described above. In each age block, those foods and food aggregates on which the yearlings' bout lengths differed significantly are listed in table 7.3. The presence of twenty-six four-letter food abbreviations in this table indicates that the yearlings' bout lengths differed significantly for twenty-six individual food × age classes, including individual foods that for reasons of sample size were run in their aggregate class.

Of these twenty-six individual foods, eighteen had samples large enough to use in the Alice/Fred/Ozzie comparison shown in table 7.5. In table 7.3 these eighteen are marked with a diamond (♦) next to their mean bout length value in the Alice/Fred/Ozzie columns (if at least two have non-zero values), that is, either in that food's own row if its sample size warranted an independent statistical test or else in the row of its aggregate class. For seven of these eighteen foods, Alice's, Fred's, and Ozzie's bout lengths differed significantly, namely for SPFU-30, XXXD-30, SKCX-60, DUKU-40, PCMX-60, SCTX-60, and GRCX-60—where GRCX-60 indicates food GRCX (that is, grass/sedge corms, table 6.2) eaten during age 60–70 weeks, and so forth.

For the remaining ten food × age classes these three yearlings did not differ significantly in bout length. Among the nine (26 − 17) food × age classes whose sample sizes were below criterion, statistical analyses (Mantel/Cox) of differences in bout length were nonetheless carried out on Alice/Fred/Ozzie for the three food × age classes (CDLU-50, FTDU-40 and XXXD-50) that had at least two individuals with non-zero sample sizes. In the first two of them the bout lengths of the yearlings were significantly different, indeed at a high level ($p \leq 0.01$). Thus, when the bout lengths of foods were tested for individual differences within ten-week age blocks, about 45% (= 9/20) of the comparisons in which all yearlings differed were also ones on which these like-aged yearlings differed and cannot be attributed to birth season effects. (In addition, the bout lengths of these yearlings differed in two other food × age-block classes (FTDG-30 and FTDG-40) on which the eleven yearlings did not differ when analyzed en bloc.) Conversely, for about half (11/20) of the foods on which these eleven weanlings differed in mean feeding bout length, the differences disappeared when the comparisons were limited to those yearlings that were born at the same time of year. So for many foods, an infant's season of birth, and consequently the ages at which it encounters those foods, apparently affects the duration of its feeding bouts.

Bristle and Summer can be compared only in the first two age blocks, during which both were sampled (table 3.1). All together in the Bristle/Summer comparison, sample sizes were adequate to run the statistical test in twenty-two cases, from sixteen foods (table 7.5, group 2), and of these only one, green grass leaves at 40–50 weeks, showed significant individual differences. These twenty-two cases include six individual food × age classes for which bout lengths differed when all available yearlings were compared (foods with four-letter abbreviations marked ♦ in Bristle's column in table 7.3), on none of which Bristle and Summer showed individual differences in bout length.

The Dotty/Striper comparison of bout lengths was much more restricted, because sample sizes were smaller and samples were taken only during age 50–60 weeks. Six foods that they ate at that age had adequate sample sizes. Dotty and Striper differed significantly on two, GRCX-50 and TODU-50 (table 7.5, group 3). These are the two foods with the largest sample sizes. In both cases, Dotty's bouts were longer. Dotty and Striper did not differ significantly in the durations of their nursing bouts, for which their sample sizes (nine each) were just below my cutoff. At 50–60 weeks, sample size was adequate for a Dotty/Striper comparison on just one of the foods on which individual differences were detected when all available yearlings were compared (table 7.3)—corms of unidentified grasses/sedges (GRCX), which was statistically significant. The five other foods on which Dotty and Striper were compared (table 7.5, group 3), including one on which they differed, are foods whose bout lengths were not significant in the comparison of all available yearlings.

The results in tables 7.3 to 7.5 can be summarized as follows. I tested foods for bout length differences between individual baboons in each of the age blocks (table 7.3). Out of 107 food × age classes for which sample size was judged adequate to carry out the statistical test, forty-seven of them (44%) showed significant individual differences (table 7.3). Because of marked phenophase changes in Amboseli, I asked whether date of birth could explain all of these individual differences. When twenty-seven of the significant food × age classes were also tested for individual differences within groups of like-aged yearlings, only ten food × age classes (37%) differed significantly in at least one such liked-aged group. So in 37% of the cases in which a food showed significant individual differences in bout length within a ten-week age block, differences in season of birth can be ruled out as an explanation for these bout length differences. The remaining 63% of the cases are consistent with the hypothesis that an infant's feeding bout duration on a food depends in part on the season of its birth. The like-aged yearlings also differed in four other food × age classes; for these, addition of data from all available yearlings obscured individual differences. In total, some like-aged yearlings differed in bout length in twelve food × age classes, out of eighty-two that were tested, that is, in 15% of the cases.

When data from all four age blocks were combined, and comparisons were limited to the six yearlings that were sampled in at least three of the four age blocks (table 7.4), again with a sample size requirement of at least forty, then seventeen out of forty-six foods (forty-four core foods) showed significant individual differences. Of these, eleven were used at least once in the comparisons of like-aged yearlings (table 7.5), and four of them showed significant individual differences in at least one such comparison.

I am surprised at how few significant individual or age differences in bout length appear in tables 7.2 to 7.5, despite two-, three-, and even four-digit sample sizes and nonoverlapping standard error intervals for the means of many foods and food classes. While some additional significant differences probably would show up in each of these analyses if additional foods were run, the proportion of significant cases would almost surely be smaller in the remaining foods, since they have smaller sample sizes. Perhaps the Mantel/Cox test is unusually conservative. (Of the seventy-three results in tables 7.2 and 7.3, all of which are significant at the 0.05 level on the Mantel/Cox test, sixty were also significant at that level according to the Breslow test [1970], which also indicated ten others as being significant at that level.) If we accept these results, it appears that for most of the yearlings' foods, bout duration does not change appreciably with age from 30 to 70 weeks (table 7.2) or differ appreciably between individuals (tables 7.3 and 7.4), including those individuals born at the same time of year (table 7.5). As I shall discuss further below, I suspect that the major source of differences in feeding bout lengths of yearlings are characteristics not of the baboons but of the food sources themselves, of "patch size" differences in the broad sense of that term.

Bout Rates

Individual Differences in Bout Rates

Do yearling baboons differ in how often they eat? Does the frequency change with age? Because sampling time was not the same for all yearlings, these are questions about individual feeding bout rates (bouts per unit time), not just relative frequencies of bouts, whereas in chapter 4, in which data from all yearlings were pooled, relative frequencies were equivalent to relative feeding rates, and the common time base could be ignored.

For every yearling during each ten-week age block, the bout rate for each core food is given in table 7.6A–D (one panel for each ten-week age block), and table 7.6E (pooled data, 30–70 weeks). As can be seen in these tables, the majority of bout rate comparisons between individuals were statistically significant at the $p \leq 0.05$ level, and indeed most at the 0.01 level. Of the fifty-two core foods, the number for which individual differences (all eleven yearlings) were detected in each of the four age blocks are thirty-six, forty-one, thirty-four, and twenty-five (table 7.6A–D), which overall comes to 65% of all bout rate tests within food × age classes. Not surprisingly, the sample sizes (numbers of bouts) in the four age blocks go up and down correspondingly (4,152, 5,027, 4,382, and 2,630, respectively); that is, the larger the sample, the larger the number of significant differences that can be detected.

The magnitudes of these individual differences are indicated by their coefficients of variation (CV's), which are given on the right side of table 7.6 for every row (every food × age class) that showed significant individual differences at the 0.05 level. Average CV's in the four ten-week age classes were 169%, 143%, 160%, and 103%, respectively, all very high—a number of the coefficients are over 200%, and four of them have a value of 300%—and all considerably higher than for feeding bout durations. Because several of these individual differences are appreciable and include many foods that are major components of the baboons' diet, in the presence of even moderate levels of heritability they will allow for quite strong selection on dietary selectivity.

What is the nature of those foods that are eaten much more often by some individuals than by others? Let me list the ten foods with CV's of 200% or greater for at least two of the age classes:

• Umbrella tree flowers on the tree (TOBS) and the ground (TOBR);
• Green seeds and flowers from fever trees (FTDU, FTBS);
• The fruit of devil's thorn (TTFX), of *Solanum dubium* (SBFX), and of the sticky-fruit plant (SFFX);
• Closed, green seedheads and green stolons or rhizomes of unidentified grasses/sedges (GRDU, GRTU);
• Corms of the sedge *Cyperus obtusiflorus* (COCX).

All of these except the corms of *Cyperus obtusiflorus* are highly seasonal; that is, they are available for only a brief period each year. For such foods, birth dates alone account for some of the individual differences. For example, Hans is the only one of the eleven yearlings that ate sticky-fruit (*Commicarpus pedunculosus*) at 30–40 weeks of age because he was the only one of the eleven yearlings that was 30–40 weeks old when sticky-fruit was available. Although this is an extreme case, similar birth date effects may contribute to other individual differences in feeding frequencies.

The last panel in the set (table 7.6E) covers the entire age range 30–70 weeks and was obtained by pooling the data for each yearling (not by averaging the results in table 7.6A–D). When the bout rates of all eleven yearlings were compared, they were significantly different for all except four foods with small sample sizes (column "All" in table 7.6E). To highlight major sources of individual differences in feeding bout rates, the two highest rates for each food (each row) of table 7.6E are shown in boldface. (Individual differences at the other extreme, very low rates, are much more difficult to detect because of the sample size requirement.)

The two weanlings who ate most frequently (all foods combined, last row) are Striper and Dotty, who were sampled during the 1974 pilot study. They are the only two who fed more than once per minute, on average, during the daytime hours. The rank order below Striper and Dotty is Eno, Ozzie, Summer, Hans, Bristle, Pooh, Alice, Fred, and Pedro. The three yearlings whose mothers are known to have been laissez-faire in their maternal care (J. Altmann 1980), Summer, Hans, and Bristle (table 3.2), are clustered in the center of this order; the one yearling whose mother is known to have been "restrictive," Pedro, is at the bottom.

Striper and Dotty, who had the two highest overall feeding rates, also had the two lowest nursing rates. The high feeding rates of Striper and Dotty are almost entirely attributable to grass corms, those of *Sporobolus cordofanus* (SMCX) and other corms of unidentified species (GRCX), on both of which Striper and Dotty, in that order, had the two highest feeding rates. These corms accounted for 66% of Striper's feeding bouts and 46% of Dotty's. Seeds from the green pods of umbrella trees (TOPU) accounted for an additional 13% of Striper's feeding bouts and 20% of Dotty's, and again they had the two highest rates. The leaves of *Salvadora persica* (SPLX) were eaten much more often than by them than by any of the other weanlings (9% of Dotty's bouts but just 1% of Striper's). These apparently idiosyncratic diets of Dotty and Striper probably largely reflect the fact that they were sampled only during a very circumscribed part of the year, during which the diversity of available foods was low. They may also reflect differences between the severe dry season of 1974 and the twelve months in 1975–76 during which the other samples were obtained. I do not know of any time during the latter period when *Sporobolus cordofanus* was eaten so extensively, and in 1975–76 *Salvadora persica* foliage was out of favor both as a food and as a yearling play site.

Unlike the 1974 subjects, with their strong and consistent emphasis on a few foods, no such simple pattern could be discerned to account for the overall differences in feeding rates among the 1975–76 subjects. Eno, who had the highest overall rate, had only one food, green grass blades (GRLU at 16%), that accounted for more than 6% of her diet, and that was not a food that Ozzie, who ranked just below her in feeding frequency, ate very often. His top-ranking food, grass corms (GRCX), accounted for just 13% of his feeding bouts and ranked second among the non-milk foods in Eno's diet. Ozzie ranked first or second among the yearlings on more foods (sixteen) than did anyone else including Dotty and Striper, and Eno was second (thirteen foods), but Fred, next to the bottom by feeding frequency, was third (eleven foods).

Can it be that the large number of foods on which the individuals differed in feeding frequency (bout rate) is solely an artifact of their having been sampled at somewhat different ages (table 3.1)? Decidedly not! I tested for bout rate differences among just the six well-sampled yearlings (boldface column heads in table 7.6E), for whom my samples were more uniformly distributed by age, and I found that these yearlings differed in feeding frequency on all but twelve core foods, at $p \leq 0.05$ and on most at $p \leq 0.01$ (right side of table 7.6E under "WSI"). Bout rates for green leaves of *Salvadora persica,* which were, as I mentioned above, a specialty of the 1974 yearlings Striper and Dotty, were not significantly different among the six well-sampled yearlings, a group that does not include Dotty or Striper. As expected, some seasonal foods dropped out: the six well-sampled yearlings did not differ significantly in bout rate when feeding on seeds from the green pods of umbrella trees (TOPU) or the green fruit of either *Salvadora persica* (SPFU) or *Azima tetracantha* (ATFU), but the bout rates of these six yearlings differed on many other seasonal foods, including the even more seasonal ripe and semiripe fruits of these two species of bushes (ATFR, ATFX, SPFS, SPFR). Their feeding frequencies differed significantly when they were feeding on green grass seedheads (GRDU), even though all eleven yearlings when tested together did not—perhaps a small-sample artifact.

Bout Rates and Nursing

Although the nursing data indicated individual differences in nursing frequencies at all four ages (PCMX in table 7.6A–D) and in nursing bout lengths at two of four ages (table 7.6A, D), there is no consistent ranking among the individuals across ages: an individual that nursed considerably more often or longer than another at one age did not necessarily do so at another. Some of these differences might be due to seasonal effects: at times of the year when food is more abundant, females may lactate more. Because the birthdays of these animals are widely scattered, such effects would show up at a different age in each. As before, such seasonal effects can be ruled out for Alice, Ozzie, and Fred, who were born within a few days of each other. Yet they too showed many rank-order reversals in estimated nursing frequency, nursing bout lengths, and nursing time budgets (figs. 7.1–7.3) and so, by implication, in their daily milk intakes, whose estimates were based on nursing time (appendix 9). Furthermore, they show no consistent trend with age in these parameters. I do not know whether these irregularities result from peculiarities of the monkeys or of my samples.

The frequency with which weanlings feed should also be related to the proportion of mother-terminated nursing bouts, although one could argue either

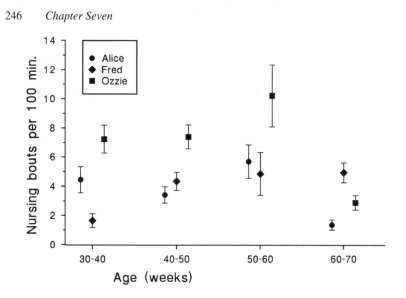

Fig. 7.1. Nursing bout rates of Alice, Fred, and Ozzie at ten-week intervals. Vertical lines are standard errors. Data from table 6.6.

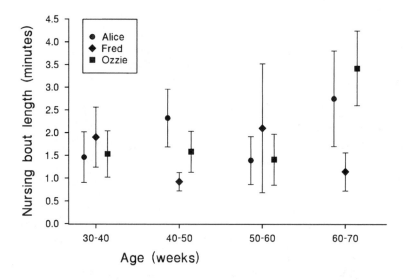

Fig. 7.2. Nursing bout lengths of Alice, Fred, and Ozzie at ten-week intervals. Values are product-limit estimates of means. Vertical lines indicate standard errors. Data from table 6.6.

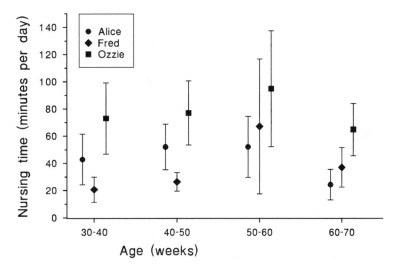

Fig. 7.3. Nursing time per day of Alice, Fred, and Ozzie at ten-week intervals. Vertical lines are standard errors. Data from table 6.6.

way: that infants being rebuffed in their attempts to suckle should more often seek other food, or that infants that get other foods often enough should nurse less often and so be rebuffed less. A graph (fig. 7.4) of percentage of mother-terminated nursing bouts (table 6.9) against overall feeding bout rate (non-milk core foods, table 7.6E) shows only a slight trend for infants to feed on solid foods more often if they are more often rebuffed at the nipple. However, this trend is not apparent in the three yearlings (Fred, Ozzie, Alice) with the closest birthdays. A test of a hypothetical complementarity (inverse correlation) between how often yearlings nurse (whether or not they are rebuffed) and how often they eat other foods is shown in figure 7.5 based on data in table 7.6E. No obvious relationship can be discerned.

Age Changes in Bout Rates

For each core food in each ten-week age block, the pooled bout rate, that is, with data for all available yearlings pooled, is given in the rightmost number in each panel, A–D, of table 7.6. The four values for each food were compared statistically using equation (3.4) to see whether the infants changed significantly with age in how often they ate each food. Those foods for which the infants showed significant changes with age are indicated on the far right of table 7.6E under "By Age." Of the fifty-two core foods listed in table 7.6, all but nine showed significant age changes in bout rate at the 0.05 level, most at

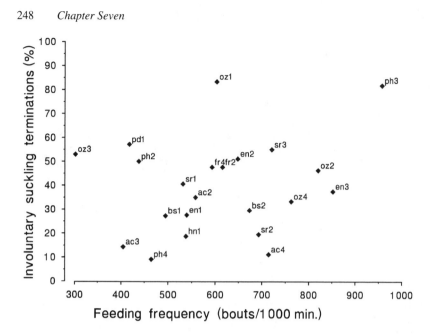

Fig. 7.4. Mothers' weaning behavior versus yearlings' frequency of feeding on non-milk foods. Data on involuntary suckling terminations (i.e., those initiated by the mother) from table 6.9. Data on pooled mean feeding rates (bouts per thousand minutes of daytime) from table 7.6. Data points labeled by infant (abbreviations as in table 3.1) and age block (1, . . . , 4 represent respectively 30–40, . . . , 60–70 weeks). Data for ac1 and fr3 omitted because of small samples of nursing bouts. Regression is not significant.

the 0.01 level, and the nine insignificant cases were concentrated near the bottom of the tables, where sample sizes are smallest. As with bout duration, one might argue here that these apparent age changes are confounded by the sampling schedule, since the age distribution of samples (table 3.1) was not the same for all yearlings.

In similar fashion, I analyzed the four pooled rates for each yearling, that is, pooled over all core foods (table 7.6A–D, column totals). Of the seven yearlings that were sampled in at least two ten-week age blocks and could therefore be tested, all showed highly significant ($p \leq 0.01$) age changes in how often they ate. (These tests, as well as those of the pooled results in table 7.6E, are not independent of the tests of rows in table 7.6A–D, but for these rate comparisons I do not know how to partition the total chi-square into independent components.)

The individual trajectories in age changes (fig. 7.6) show no single pattern. The three same-aged yearlings, Ozzie, Fred, and Alice, tended to change rates similarly from one age to the next. So too did the close-aged yearlings Bristle

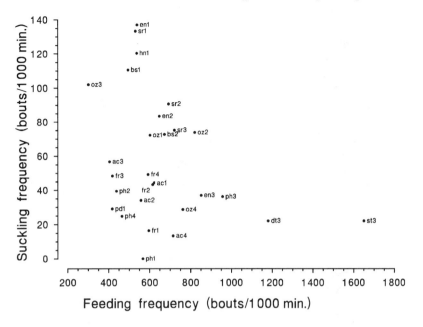

Fig. 7.5. Frequency with which yearlings suckled versus their frequency of feeding on non-milk foods. Data sources and abbreviations as in figure 7.4. Regression is not significant.

and Summer. Thus, seasonal effects may be large. The third close-aged group, Dotty and Striper, had higher feeding bout rates than any other yearlings, but they were sampled at just one age and at a different time. Their high rates (table 7.6C) may reflect different conditions during the year (1974) when they alone were sampled. Feeble Pedro, who was sampled on three days before he died, had the lowest feeding bout rate, and umbrella tree blossoms picked from the ground made up much of what little he ate, other than milk, during that time (see below for further details). Aside from Pedro, Pooh had the lowest feeding rates, except when the rains came and the abundance of fresh green grass seemed to revive her: at that age, she ate green leaves of unidentified grasses (GRLU) at the highest rate (208 times per thousand minutes) of any yearling at any age.

Bout Rates versus Bout Lengths

Bout rates and bout lengths contrast impressively in the extent to which individual baboons differed significantly. The weanling baboons differed much more, and on more foods, in how often they ate various core foods than in the

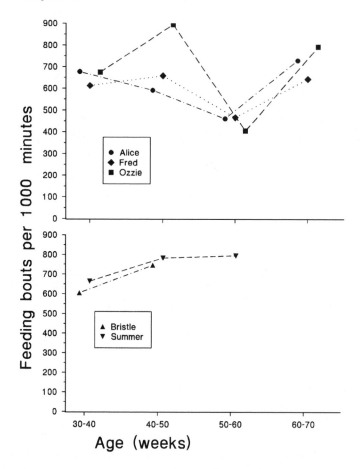

Fig. 7.6. Frequencies of feeding bouts (all core foods) of like-aged infants Alice, Fred, and Ozzie (top) and of like-aged infants Bristle and Summer (bottom) at ten-week intervals. Data from table 7.6.

durations of their feeding bouts. For example, when data for the entire 30–70-week age interval were analyzed for individual differences using a significance level of $p \leq 0.05$, bout length differences between individuals showed up in 37% (seventeen out of forty-six with adequate sample sizes) of the core foods and those that differed had an average coefficient of variation (CV) of 42% (table 7.4), whereas for bout rates, individual differences were detected in all except one of the same forty-six core foods (= 98%) and these forty-five foods had an average CV of 114% (table 7.6E). All together, differences in feeding bout rates among the eleven yearlings were detected in forty-eight of the fifty-two core foods (table 7.6E). This means that bout rates account for much more of the dietary variability between individuals than do bout durations.

How can one explain this greater variability in bout rates than in bout lengths? One possibility I cannot dismiss is that it is a statistical artifact. Although the same significance level was used in both cases, the χ^2 test used for bout rates is a comparison of means, whereas the Mantel/Cox test used for bout lengths is a test of the entire distribution of bout lengths. Perhaps the χ^2 test for rates is more powerful than the Mantel/Cox test, in the sense of having a higher probability of detecting a significant result.

Conversely, if these differences are real, as I believe they are for many foods, then I suggest that baboons have more control over the rate at which they encounter food patches and the decision about whether to feed on any particular patch than they do over characteristics of the food patches they get, and that bout length is determined primarily by characteristics of the food patches, such as the number of items per patch, processing requirements, interpatch distances, and so forth, rather than by characteristics of the baboons that feed on them. Exceptions to this last occur when, for example, an animal is displaced from a food patch before it has finished (Post, Hausfater, and McCuskey 1980; S. Altmann and Shopland in prep.; Shopland 1987) or when, as with a feeble infant like Pedro, bout length probably is an idiosyncratic consequence of slow feeding, fatigue, and other factors.

The foods on which the baboons differed significantly in bout length tend to be ones on which they differed in bout rate as well. That is not surprising, both because the chance of detecting significant differences goes up with sample size in both cases and because most foods differed in bout rate. Yet do foods on which they differ in both duration and rate tend to have differences of comparable magnitude? That is, are the coefficients of variation correlated? In figure 7.7 I have plotted the coefficients of variation for all core foods on which the weanlings differed significantly ($p \leq 0.05$) in both bout duration and rate, using the results of pooled data, 30–70 weeks, as given in tables 7.4 and 7.6E, and results from within each age block, as given in tables 7.3 and 7.6A–D. No pattern is apparent: the magnitudes of individual differences in bout length and in bout rate appear to vary nearly independently. This figure identifies seventeen foods that make an extraordinary contribution to individual differences in diet: foods that are eaten at significantly different frequencies, for significantly different amounts of time, and that make a major contribution to the baboons' diet as judged by the fact that they account for a large part of their feeding time and dietary intake.

Why is it that some foods have large coefficients of variation (CV's) for both feeding bout length and bout rate? The extreme CV value in figure 7.7 is for umbrella tree blossoms picked up from the ground. If we look at the individual rate values for this food (table 7.6E), it becomes clear that these fallen flowers

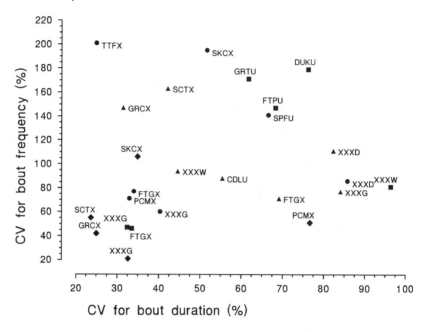

Fig. 7.7. Coefficients of variation (CV) among individuals for core foods on which yearlings differed significantly ($p < 0.05$) both in feeding bout length and rate during each ten-week age block. Symbols (circle, square, triangle, diamond) indicate the ten-week age blocks (beginning, respectively, at 30, . . . , 60 weeks of age). CV data from tables 7.3 and 7.6. Regressions are not significant in any age block or in the aggregate of all points.

were eaten seldom or never by any yearling except Pedro, and for him it was the food he ate by far most often. Pedro was a weak infant who died not long after I began sampling him. He was too weak to climb a tree, so when the other yearlings ascended an umbrella tree to pick blossoms, he remained with his mother on the ground beneath it, very slowly picking up flowers that had dropped from the tree. Dung-beetle larvae (DUKU) show a very high CV for bout rate. This is primarily a reflection of sharp seasonality of this food, combined with the birth seasons of the yearlings (chapter 9), so that only half of the yearlings in my sample fed on them at 30–70 weeks of age (table 7.6E). Other foods on which the yearlings' feeding bout rates and lengths differed considerably include the corms (GRCX), green blades (GRLU) and blade bases (GRTU) of grasses/sedges and the gum (FTGX, FTHP), seeds (FTDU, FTPR), and blossoms (FTBS) of fever trees. This again emphasizes the importance of grasses/sedges and acacias in the diet of the baboons. But there is no obvious reason why they should be major sources of individual differences in diet. Although some of these foods, such as green fever tree seeds (FTDU) and blossoms (FTBS), are very seasonal, others such as fever tree gum show very little seasonality in consumption (appendix 8), so while seasons of birth combined with nonuniform sampling might account for some of the individual differences in these foods, it is unlikely to account for all of them, particularly since the six well-sampled yearlings differed on many of these same foods.

In adult females at Amboseli, bout length is often related to dominance status, but bout length differences are at least partially offset by bout rate differences (J. Altmann, pers. comm.). On the other hand, low-ranking adult female olive baboons did not compensate with additional feeding time for their lower feeding rates (nutrient intakes per minute of feeding; Barton and Whiten 1993).

Time Budgets

Age Changes in Time Budgets

Using data from all eleven yearlings pooled over the year, I computed the time budget (minutes per day) that the weanlings spent on each of the fifty-two core foods during each ten-week age block between 30 and 70 weeks. In these computations, for every core food for which significant age changes in bout length were detected (table 7.2), I used the age-specific bout lengths for each age block; for all others, I used the overall pooled value (that is, based on all 30–70-week data for that food; table 4.1) for every age block. Similarly for bout rates: for those that showed significant age changes, I used age-specific values (table 7.6A–D, right side); for all others, I used their overall pooled values (table 7.6E, right side) for every age block. In each of the four age blocks, the product of these two numbers—minutes of feeding per bout on each core food and bouts per sampled potential foraging minute—was multiplied by 660 (potential foraging minutes per day) to give the time budget, in minutes of feeding per day, for that food at that age. The standard errors of the time budgets were computed from equation (3.3). The results are given in table 7.7.

Time spent nursing (all yearlings pooled) declined steadily from 96 minutes per day at age 30–40 weeks to 32 minutes per day at 60–70 weeks. For each of the fifty-one other core foods, the time per day spent feeding at each age is given in the table and summed at the bottom, along with my standard allowance for non-core foods. On this basis, I estimate that in the four age blocks, 30–40, 40–50, 50–60, and 60–70 weeks, the baboons spent, respectively, 177.0, 195.7, 294.5, and 198.5 minutes per day feeding on all foods other than milk (pooled means). With nursing time added, I get the following time budgets for the four age blocks: 273.4, 262.8, 343.0, and 230.7 minutes per day. These correspond with 41.1%, 39.8%, 52.0%, and 35.0% of the 660 minutes of potential foraging time from 0700 h to 1800 h. The average of these values, weighted by the sample times for each age block (table 3.1) in order to make the results comparable, gives an overall value of 42.0%—virtually identical with the estimate of 42.1% given in chapter 4. The latter estimate was not based on age-specific bout lengths and rates.

The feeding time-budget value in table 7.7 for the 50–60-week yearlings, 269 minutes a day not counting nursing, is considerably higher than for the yearlings at other ages, for which feeding time appears to be approaching an asymptote. The very high value at 50–60 weeks is due primarily to the two yearlings, Dotty and Striper, who were sampled at this age only, during a different year (1974). I have already commented on their high frequency of feeding, particularly on grass corms, green umbrella tree seeds, and leaves of the toothbrush bush. I therefore recalculated feeding time budgets at this age for all yearlings (other than Dotty and Striper) combined, except that for practical reasons bout lengths were not recalculated, only bout rates. The results (table 7.7) show a total of 193 minutes per day spent on non-milk core foods, which comes to 212 minutes per day for all foods except milk. With nursing time added, I get 267 minutes per day (40.4% of potential foraging time) at 50–60 weeks of age. These figures are still somewhat higher than the values for the other age blocks, but not wildly so. The new average, that is, weighted as before by the sample times for each age block but excluding the 1974 data (Dotty and Striper), gives a mean feeding time budget of 40.0% in 1975–76 (mean sample age 47.2 weeks).

The feeding time budget values for the 1975–76 weanlings are below those obtained in any other study on Amboseli baboons,[10] but all these other studies were on adults. It is not surprising that nursing yearlings spend less time foraging than do adults. Milk is, after all, a concentrated food (Pond 1977; Oftedal 1984a), and so long as it is available in adequate quantities, yearlings do not need to provide all of their own sustenance. Of course, as the infant continues to grow, its mother's milk production becomes progressively less adequate (J. Altmann 1980). I noted a tendency, albeit an imperfect one, for time spent on non-milk foods to increase as nursing time decreased with age (fig. 7.8). One might argue the causal relation either way: either yearlings begin switching to solid foods as their mothers' increasing weaning behavior makes it more difficult to get nourishment by nursing than by foraging (Nicolson 1982), or else mothers make up by milk whatever is missing from the yearlings' foraging forays, and the latter improve with age and experience.

The feeding time values I obtained for infant yellow baboons (nursing excluded) are below values for olive baboons (*Papio anubis*) near Gilgil, Kenya, at the same ages (Nicolson 1982). Figure 7.9 shows superimposed results from the two studies. Are these true population differences in feeding times, or are they procedural artifacts? They may result from inherent differences between mean values (my study) and medians (Nicolson), but judging by Nicolson's interquartile ranges and my standard error intervals (fig. 7.9), this seems un-

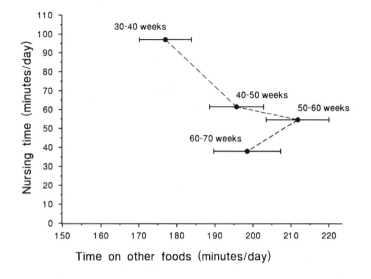

Fig. 7.8. Changes with age in time spent nursing versus time spent feeding on other foods. Data from tables 6.6 and 7.7, respectively.

likely. Nicolson used point samples and I used focal samples, but this alone would not account for the difference, since either can be used to estimate time budgets (J. Altmann 1974). Nicolson's sample period ran from 0700 h to 1900 h, whereas mine ran only to 1800 h. However, the differences in our results (fig. 7.9) are too large to be accounted for in this way; for example, even if yellow baboons fed nonstop from 1800 h to 1900 h, non-milk feeding would only take 36% of daytime at one year of age, still well below the anubis value.

One contributing difference in methods is that none of the others who have estimated time budgets in baboons have corrected for out-of-sight periods. As I indicated in chapter 3, feeding was less susceptible to such "censoring" than other activities: a higher percentage of nonfeeding bouts were interrupted by the subject's going out of sight (24% vs. 10.5%), and the censoring rate during nonfeeding periods was higher (38 vs. 24 disappearances per hundred minutes). Consequently, focal and point samples are systematically biased against nonfeeding periods, and without correction this bias will lead to an overestimate of the time budget for feeding. However, as I shall now explain, the magnitude of this bias can be estimated, and it is inadequate to explain all of the difference between Nicolson's results and mine. From my focal samples, using product-limit estimates of bout lengths to correct for out-of-sight effects, I estimated mean daily feeding time (30–70 weeks pooled) for all foods minus milk (table 4.8) as $277.74 - 65.93 = 211.81$ minutes, which is 32.1% of the 660 minutes from 0700 h to 1800 h. If, however, I had used point samples with

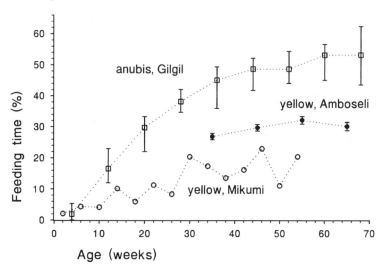

Fig. 7.9. Changes with age in non-nursing feeding time in anubis and yellow baboons. Values for anubis at Gilgil, Kenya, from Nicolson (1982, fig. 9) are medians and interquartile ranges, from focal-infant samples. Yellow baboon values from Amboseli National Park, Kenya (this study), are means and standard errors, based on bout frequencies and product-limit estimates of bout durations in focal-infant samples (table 7.7, 1975–76 values). Values for yellow baboon at Mikumi National Park, Tanzania, from Rhine, Norton, Wynn, and Wynn (1985, fig. 3) are based on point samples. All values plotted at the midpoints of the age intervals for each population (eight, ten, and two weeks wide, respectively).

no correction, my estimate of the non-milk feeding time budget, as a percentage, would have been essentially equal to the ratio of total observed non-milk feeding time in my study to total in-sight sample time. Observed non-milk feeding time was obtained by subtracting from total in-sight time (19,978.74 minutes) both the observed nursing time (1,869.64 minutes within the scheduled sample times) and the in-sight-not-feeding time (11,357.18 minutes). This gives an estimate of 33.8% of daytime for feeding, which is just 5.3% larger than my out-of-sight-corrected estimate of 32.1%, above and is inadequate to account for most of the between-study difference shown in figure 7.9. Differences in feeding time between Amboseli's yellow baboons and Gilgil's anubis may be even greater than these figures suggest. The mean time-budget value given above, 277.4 minutes per day at 30–70 weeks of age, is based on the observed samples corrected only for censoring of bouts. However, when the food-specific time budgets are also adjusted for seasonal differences in sampling intensity (table 6.5), mean feeding time comes to 240.5 minutes per day (core plus non-core foods) or 174.6 minutes for all non-milk foods, and that amounts to just 26.5% of the 660 potential minutes of feeding per day.

So young anubis baboons near Gilgil, Kenya, apparently spent more time feeding than did young baboons in Amboseli: about 60% more by one year of age. This result is very surprising in view of the higher rainfall near Gilgil, with, I assume, correspondingly higher productivity. Perhaps higher rainfall leads to higher floristic diversity and thus to more time spent searching for and selecting among potential foods. Conversely, Amboseli baboons spent more time on the nipple than did Gilgil baboons (chapter 6), suggesting that in the harsher habitat of Amboseli, yearling baboons take longer to achieve the same degree of nutritional independence. That seems to me to be the most likely explanation. The site (and species) differences in the yearlings' feeding time is not reflected in the adults' time budgets. At Gilgil, Kenya, anubis adults of both sexes fed for about 43% of the daytime (Forthman Quick 1986, fig. 2, PHG troop, 1981 data). In Amboseli, yellow baboon adult females of Alto's Group averaged 44.7% of daytime spent feeding (Bronikowski and Altmann 1996, table 2, mean of nine annual means, 1983–90 data).

On figure 7.9, I have also graphed values from a study of infant yellow baboons at Mikumi National Park, Tanzania, by Rhine, Norton, Wynn, and Wynn (1985). In this case the results are the opposite: these infants spent *less* time feeding than did Amboseli baboons of the same age. I can provide no explanation. Based on rainfall differences, I would have expected the Mikumi baboons to deviate from those at Amboseli in the same direction as those at Gilgil.

Examination of table 7.7, 1975–76 data, revealed surprisingly few foods for which there was consistent age change in time per day spent feeding on them. For example, I expected a progressive increase with age in the amount of time spent feeding on grass corms, on the grounds that corm feeding by the youngest yearlings in my samples (30 weeks) is limited by inadequate strength. Yet for the four age blocks in order, the values (1975–76 data) are 33, 40, 36, and 39 minutes per day (sum for GRCX, SKCX, SMCX, and COCX). These are, respectively, 19, 21, 17, and 20% of the baboons' non-milk feeding times. By comparison, Post (1982, fig. 3) found that adult baboons in Amboseli (1974–75) spent about 31% of their feeding time on grass corms, and Muruthi (1988, table 10) found that adult females spent 30% of their feeding time on them, so corm feeding apparently does increase before adulthood. Similarly, one might expect grass blades to be more difficult for the youngest animals to chew, but the four age-block values (sum for GRLU, CDLU, and SMLU) are 25, 31, 35, and 31 minutes per day, respectively, which do not represent a consistent change in percentage of non-milk feeding times: 14, 16, 17, and 16%. These values are close to those obtained by Post (1982) for adults: about 15% (his "leaf" and "blade" categories combined). Perhaps by 30–70 weeks feeding patterns are changing too slowly with age to be readily detected, particularly in the face of other sources of variability.

Individual Differences in Time Budgets

The time-budget differences between ages for any one food (table 7.7) may result variously from age changes, from individual differences (since the cast of sampled individuals shifted with age; table 3.1), from birth season effects, from ecological differences between 1974 and 1975–76, and even from changes in my ability to recognize certain plants. Consider an extreme example. According to my records, no yearling ate fever tree blossoms (FTBS) at 60–70 weeks, although they did at younger ages. However, this apparent difference results not from an aversion of yearlings at this age to fever tree blossoms but from the fact that of the four yearlings I sampled in this age block (table 3.1), none had fever tree blossoms available to them at this age (fig. 1.5 vs. fig. 3.2). Similarly, as I mentioned before, only Hans ate sticky-fruit at 30–40 weeks of age because he is the only one of the yearlings who was that age during the short period when sticky-fruit was available. At a more subtle level, as I indicated earlier in this chapter, even within one age, individual differences in time per bout of feeding on green *Salvadora persica* fruit probably reflected seasonal changes in patch size. As an example of the effects of my abilities, consider the low value at 50–60 weeks of age for the corms of the very important grass *Sporobolus rangei*. It resulted partly from my inability to recognize this species in 1974, when I carried out the pilot study on Dotty and Striper at age 50–60 weeks. Any feeding they did on this food would have been recorded as corms of unidentified grasses/sedges (GRCX). (My impression was, however, that this perennial grass was not often used by the baboons during June–July 1974, perhaps because the annual grass *Sporobolus cordofanus* was more abundant than in 1975–76 and is preferred.)

Some of these factors might be teased apart, as before, by examining individual differences between animals born at the same time of year. This will be done in the next section, based on the amount of each food consumed rather than the time spent on each. Because in my records these differ between individuals only by a scale factor $B_j C_j$ for any food j (see equation 3.1), the comparisons are essentially equivalent, although the comparisons based on mass are somewhat more conservative because of the additional contribution of the standard error of $B_j C_j$ to the standard error of the intake mass.

Differences in Diet and Nutrition

Diet Mass: Age Changes

Let's begin with the overall pattern of age changes in total intake mass (table 7.8 and fig. 7.10). Contrary to expectations, the average amount of food consumed per day was essentially unchanged over the 30–70-week age span, declining slightly but insignificantly (as judged by overlapping standard error

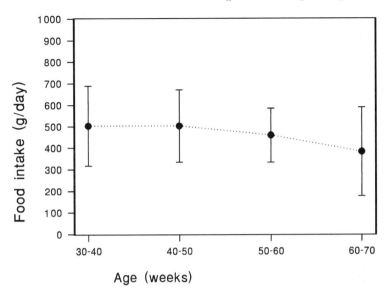

Fig. 7.10. Food consumption (g/day) as a function of age. Values (means and standard deviations) from table 7.9.

intervals) in successive ten-week age blocks. I can offer only one explanation, other than sampling error, for this apparent anomaly: seasonal changes in food availability, combined with differences in birth dates. The overall pattern of slight decreases in food consumption with age (table 7.8) does not show up in the diets of those yearlings that were sampled at more than one age. For Bristle, Fred, Ozzie, and Pooh, the lowest food intakes occurred during the dry season, June–October, and for Alice and Summer their next-to-lowest intakes occurred at that time. (No such comparisons can be made for the other five yearlings.) The highest intakes of Alice, Eno, and Pooh occurred during the short rains (November–December), those of Bristle and Summer during the long rains (March–May). On the other hand, Fred and Ozzie ate the most during the inter-rain period, January–February. (Alice, Fred, and Ozzie were born within a few days of each other; table 3.2.) In short, the yearlings tended to eat most during the rainy seasons and least during the long dry season. These results, based on diet mass, seem inconsistent with Post's (1982) analysis of seasonal changes in feeding time among adults: highest during the long rains and dry season, lowest during the short rains and inter-rains.

In chapter 6, four predictions, there labeled $f_1, \ldots, f_4$, were made about age changes in the baboons' food intake, based on predicted age changes in milk intake (two predictions, m_1 and m_2) and in total dietary requirements (two predictions, d_1 and d_2). First, consider milk intake. Depending on assumptions, it was predicted either to be

$$m_1 = 204.89 + 5.918w - 0.098w^2 \text{ ml milk per day } (w = \text{age in weeks}),$$

which for the four ten-week age classes in this study (centered on 35, . . . , 65 weeks, respectively) comes to

$$m_1 = (292.0, 272.8, 233.9, 175.5) \text{ ml milk per day}$$
$$= (299.9, 280.2, 240.2, 180.2) \text{ g milk per day},$$

or alternatively to be

$$m_2 = (240.3, 167.8, 159.5, 87.5) \text{ ml milk per day}$$
$$= (246.8, 172.3, 163.8, 89.9) \text{ g milk per day}$$

for each of the four age blocks. Conversions from milliliters to grams are based on the specific gravity of baboon milk, 1.027 g/ml (Buss 1968). As indicated in endnote 6, actual milk intake values (means of individual means for these same ten-week age blocks from table 6.6, using the second approximation of mean growth rate, 4.5 g/day) are, respectively, 292.0, 273.1, 234.5, and 176.1 ml/day, equivalent to 299.8, 280.5, 240.8, and 180.9 g/day, which are close to the predicted values for m_1—not surprising considering how m_1 and the actual values were obtained (see chapter 6)—but not at all close to m_2, so predictions f_3 and f_4, which depend on m_2, will not be pursued.

Chapter 6 included predictions d_1 and d_2 for total dietary intakes. If from each of these I subtract the milk intakes m_1, I get two sets of predictions f_1 and f_2 for the animals' foraging requirements, in equivalents of milliliters of milk, at age w weeks:

$$f_1 = 2.41w + 0.098w^2 \text{ ml milk equivalent per day}$$

or

$$f_2 = -204.89 + 248.05(0.775 + 0.0315w)^{3/4} - 5.918w$$
$$+ 0.098w^2 \text{ ml milk equivalent per day.}$$

At the ten-week midpoints (35, . . . , 65 weeks) of the four age blocks in this study, values of f_1 and f_2 are, respectively, as follows:

$$f_1 = (204.4, 306.9, 429.0, 570.7) \text{ ml milk equivalent per day}$$
$$= (209.9, 315.2, 440.6, 586.1) \text{ g milk equivalent per day},$$

or

$$f_2 = (105.9, 174.2, 260.3, 364.6) \text{ ml milk equivalent per day}$$
$$= (108.8, 178.9, 267.3, 374.4) \text{ g milk equivalent per day.}$$

Of course no food has the same composition as milk, so milk equivalents must be considered separately for various nutrients. Here, two are considered; proteins and energy. Baboon milk contains 1.6% protein (table 5.1), so the specified protein values for the four age classes are as follows:

$$f_1 = (3.4, 5.0, 7.0, 9.4) \text{ g protein per day}$$

or

$$f_2 = (1.7, 2.9, 4.3, 6.0) \text{ g protein per day.}$$

The actual protein intakes, averaged across all sampled yearlings in each age class except Pooh at 30–40 weeks (table 7.9) were 13.2, 13.9, 13.6, and 10.8 g/day in the same four age classes, and so it is obvious that the actual protein intake greatly exceeds the predictions of both f_1 and f_2.

With hindsight, I see another reason the predictions of f_1 and f_2 are unreasonable and should have been considered so even without looking at the actual intake values. The protein requirement for baboons is given in appendix 1 as $1.51 \, M^{3/4} + 0.254 \, \Delta\mu$, where M is body mass in kilograms and $\Delta\mu$ is growth rate in g/day. With a birth weight of 0.775 g and a growth rate of 4.5 g/day, baboons at ages 35, . . . , 65 weeks would have corresponding protein requirements of 3.6, 3.9, 4.2, and 4.4 g protein per day. By comparison, f_1 and f_2 increase much too rapidly with age.

As for energy, baboon milk contains 3.22 kJ/g (table 5.1), so predicted energy intakes at the four ages are

$$f_1 = (0.41, 0.66, 0.99, 1.4) \text{ MJ/day}$$

and

$$f_2 = (0.35, 0.58, 0.86, 1.21) \text{ MJ/day.}$$

Actual intakes were 2.1, 2.3, 2.1, and 1.9 MJ/day (table 7.9), and requirements are 0.65, 0.85, 0.93, and 1.0 MJ/day (chapter 5), so again, f_1 and f_2 seem to be unrealistic as predictors of baboons' dietary intakes. What is not clear is which of the assumptions that went into these predictions are incorrect.

Pedro's Intake

As I mentioned earlier, for each yearling's unit intake rates of the core foods, I used mean values based on data from all yearlings pooled. These pooled mean rates were used in calculating the nutrient intakes of individual baboons (table 6.1). However, the bite rate of one yearling, Pedro, was sufficiently different from that of the others that such pooled values are misleading, and they were therefore recalculated separately. Pedro was a very weak, sickly yearling.

His feeding was desultory, and when he was feeding his hand movements were wavy and poorly coordinated. He died at age 246 days. That he survived that long is primarily a tribute to his mother's care; her milk provided his major source of nutrients.

Pedro's major non-milk sources of nutrition were green grass blades, umbrella tree flowers from the tree and the ground, and small unidentified objects (presumed to be mostly acacia seeds) picked up from the ground. In most cases I gave these last items a nominal one second per hand-to-mouth movement, so no correction can be made for Pedro's slowness with them. For the other three foods, however, I have calculated Pedro's unit intake rates separately and have also calculated the rates for all other yearlings pooled. When he was eating flowers on umbrella trees, Pedro's intake rates were only 8% below those of other yearlings. For the other two foods, however, the differences were much greater (table 7.10). He fed on green grass at only 16% of the rate of other yearlings, and on umbrella tree seeds on the ground at only 20% of their rate. Using Pedro's slower bite rates for these three foods, I recalculated his nutrient intakes (table 7.11), and of course his already poor record now looks even worse.

For one food, abscised umbrella tree flowers picked up from the ground (TOBR), Pedro's records turned out to be a large portion of my bite rate samples on all yearlings. After removing his values, the intake rate of this food for all remaining yearlings (table 7.10) was considerably higher: 60.0 flowers per minute instead of 13.61, as in table 6.1. This means that for yearlings other than Pedro, I have underestimated the contribution of this food to their diet, but the magnitude of this bias is negligible (see below). For the other two foods, the discrepancy was 5% or less.

Population Distribution of Intakes

The statistical distribution of dietary intakes is of considerable interest. If dietary intake is a major fitness-limiting factor or correlates well with other traits that are (for example, if genetically sickly yearlings feed poorly), then the distribution of food intakes may approximate the distribution of fitness in the population. In view of the importance of fitness distributions (Wright 1968) but the meager attention they have received in the empirical literature, I hope I will be forgiven for using my small samples of dietary intakes to estimate the size of the zero fitness class.

Wright (1968) suggested that fitness may have a bimodal distribution, with one mode at zero. In human nutritional surveys, intake distributions are used to estimate the number of people below an intake level that defines malnutrition and who therefore are at or near zero fitness. Human energy intakes may have a beta distribution (FAO 1975). My samples are too small to distinguish among many possible statistical distributions. When the twenty-eight intake values (diet mass) in table 7.8 are plotted on a log-probit graph, the fit is poor, so a log-normal distribution can be ruled out.

A linear-probit plot and a rankit plot (Sokal and Rohlf 1981) of diet masses (fig. A1.2) each gave a reasonable fit to the data, so a normal distribution provides a reasonable approximation to the distribution of diet mass. (I am here treating these twenty-eight values as statistically independent, but of course they are not: only eleven yearlings are involved, most sampled at several ages.) The Gaussian distribution also provided reasonable approximations to the population distributions of macronutrient intakes, as revealed by rankit plots of intakes, one for each macronutrient class at each age (ten-week age block). The few cases in which the fit was poor showed no systematic deviations from normal and may well be artifacts of small sample sizes. (These normal or near normal distributions for nutrient intakes are in contrast to the intake distributions of foods, many of which, as noted in chapter 5, have markedly asymmetric distributions.) From these rankit plots, the expected proportion of nutrient-deficient individuals was estimated by finding, for each, the rankit value at which the best fit line, drawn "by eye," reached the minimum for that nutrient at that age (the midpoint of the age class), then finding the corresponding probability in a table of standard normal deviates.

The technique is illustrated in figure 7.11, which shows the rankit plot for energy intakes at 30–40 weeks of age. At 35 weeks, the energy requirement is 672 kJ per day. The regression line reaches that level at -1.95 rankits. The mean minus 1.95 standard deviations of a normal distribution includes 2.6% of the population. I therefore estimate that at 30–40 weeks, 2.6% of baboons living in Alto's Group or under comparable conditions elsewhere will be energy deficient. Of course this value is only a rough approximation, not only because of the usual errors in estimating parameter values but for other reasons as well. Requirements for energy and other food components are variable, not constant, whereas the approach used here assumes that all and only those individuals with intakes below a critical level (672 kJ/day in this example) would be deficient. Also, I have estimated the energy requirement for an individual at the age block midpoint, whereas individuals in that age class have a variety of body mass values and thus different mass-specific requirements. Indeed, those with long-standing low dietary intakes will grow more slowly (appendix 1) and so will require less food. With these caveats in mind, we can use the method described above to obtain population estimates of intake distributions and of the expected proportion of nutritionally deficient individuals in the population. The resulting deficiency estimates are given in table 7.12 for energy, protein,

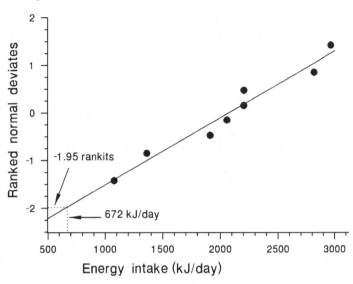

Fig. 7.11. Distribution of individual energy intakes (kJ/day) at age 30–40 weeks, plotted as ranked normal deviates (rankits). The energy requirement of 672 kJ/day at this age corresponds with −1.95 rankits and thus with 2.4% of a standard normal distribution. Values from table 7.8. $R^2_{adj} = 0.95, p \leq 0.001$.

fats, minerals, and fiber and for toxic levels of minerals. Values were not calculated for total dietary mass or for feeding time maxima because the putative limits on these were estimated from observed values in Alto's Group. Tables 7.8 and 7.9 give the calculated means and standard errors of nutrient intakes.

For most nutrients at most ages, the projected fraction of the population below the minimum or above the maximum is less than 1%. That this should be so for yearling baboons in an area as harsh as Amboseli is an indication of the remarkable skills these selective omnivores develop at an early age.

In chapter 6, I noted that the mean mineral intake of the baboons was below their requirement. At each of the four ages at which I sampled, a sizable proportion of mineral-deficient individuals is predicted based on a normal distribution of intakes, and at each age the actual intakes of several yearlings are apparently below their requirements (tables 7.8, 7.12). The accuracy of this estimate is unknown. Clinical symptoms of mineral deficiencies were not noted in those individuals supposedly below their requirement. The published mineral requirements doubtless include a considerable safety margin. On the other hand, the value I used for the total mineral requirement is just the sum of the requirements for individual minerals (cations) and therefore would tend to underestimate the critical value for foods in which the ratio of the various cations differed from the ideal.

One potential source of error in estimated mineral intakes can be eliminated. In discussing Pedro's low intake rates above, I noted that for one common food, umbrella tree flowers from the ground, removing Pedro's abnormally low values resulted in an appreciable difference in the pooled mean value for the remaining yearlings. I have therefore recalculated the intake of abscised umbrella tree flowers for each yearling at each age at which it was sampled and on that basis recalculated each yearling's mineral intakes. Other than Pedro, only three yearlings, each at one age, ate the abscised flowers during my samples: Eno at 40–50 weeks, Fred and Ozzie each at 60–70 weeks. In none of these cases did the revised intake rate for abscised blossoms change the mineral intake value of a baboon by more than 0.4%, and in no case did the proportion of minerals in the diet—which is the way the requirement is expressed—change by as much as 0.01%. So correcting the unit intake rate for abscised acacia blossoms does not eliminate the calculated mineral deficiency that shows up in the diets of about half the baboons.

Deleterious effects of mineral deficiencies are more likely to be manifest if an individual is deficient for a prolonged period. Table 7.13 indicates which yearlings were deficient at each age and shows the percentage of mineral requirement their diets achieved. Alice's diet was mineral deficient at only one of the four ages at which she was sampled, so she probably was not affected. Bristle was deficient at both ages at which he was sampled, Eno at two ages out of three, Fred at two out of four, Dotty at one out of one, and Hans at one out of one. Ozzie was the most deficient yearling at three out of the four ages at which he was sampled and barely above minimum in the fourth. Pedro was mineral deficient at one age out of one, Pooh at one out of three, Summer at three out of three, and Striper at none out of one. The most extensive mineral deficiencies were those of Pedro, Eno, and Hans at 30–40 weeks and of Ozzie at 50–60 weeks, all with mineral contents of less than 1% of diet mass, that is, less than 58% of requirement. No gross effects of mineral deficiency were noted in any of these yearlings; more subtle symptoms would require impracticable clinical tests. None of the diets exceeded the toxic limit, 13.1% minerals.

Except for mineral requirements, for only one food component that I analyzed, energy, was any individual deficient at any age: Pooh at 60–70 weeks of age. Pooh's diet was above energy requirement at each of the two other ages at which she was adequately sampled, and so this one low value may indicate either a short-term deficiency or a small-sample error. (Only 397 minutes of sampling were done on Pooh at age 60–70 weeks.) I suspect that the deficiency was real and was more persistent than my samples indicate. Pooh was a weak, undersized, and apparently malnourished yearling. She rarely played with the other yearlings, even when they tried to enlist her participation in play interactions. The exception was when *Salvadora persica* fruit—sweet and easy to harvest—were ripe.

Here is a final comment on estimating the expected number or proportion in a population that is nutrient deficient. The approach used here was to estimate the proportion of the population falling below a putative threshold or requirement. That is, it considered variability in dietary intake but not variability in requirements. However, the proportion of individuals that are deficient for a nutrient is the sum, over all intake levels, of the proportion of individuals with each intake times the probability of a deficiency at that intake, and similarly for toxic excess intakes. When information on variability in requirements and toxic limits is available, that method is preferable (FNB 1986).

Nutrient Content of Individual Diets

As I indicated in chapter 3, nonoverlap in standard error intervals of the baboons' intakes is used in what follows to indicate significant individual differences, and a difference of at least two standard errors between an observed and an expected (i.e., optimal) intake will be taken to indicate a significant deviation from optimum. To avoid any confusion with the outcomes of formal statistical tests of significance, which in this case I do not know how to do, observed intakes with nonoverlapping standard error intervals will be referred to as *distinguishable* rather than as *significantly different*. I will specifically consider intakes of energy and protein because of their potential significance as limiting factors for fitness. For each yearling, means and standard errors of intakes of the seven classes of macronutrients are tabulated in table 7.8 and are shown graphically in figures 7.12 through 7.15 for the four ages at which the yearlings were sampled. In the figures, intakes are scaled logarithmically; consequently, a vertical difference of any given size represents the same percentage difference in nutrient intake anywhere on the scale.

Energy

At 30–40 weeks, the energy intakes of the four yearlings whose energy intakes rank lowest (Alice, Fred, Pedro, and Pooh, in descending order), were distinguishable from each other except for Fred versus Pooh. The energy intakes of these four are also distinguishable from those of all five yearlings with higher energy intakes except for Alice, whose intake is distinguishably lower than those of just Hans and Eno. Among the top five (Eno, Hans, Summer, Ozzie,

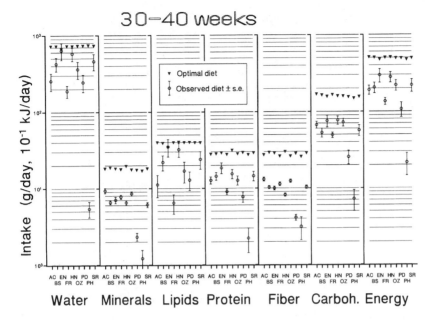

Fig. 7.12. Macronutrient intakes of individuals, 30–40 weeks old. Bar plots show means and standard errors. Triangles indicate the composition of an energy-maximizing optimal diet with extended constraints (diet 6 of chapter 5) with individual-specific values (see text). Two-letter codes for individuals (AC = Alice, etc.) as in table 3.1. Lipid intake of Pooh (not shown) was below baseline of graph. Energy intakes plotted at one-tenth actual values; all other nutrients in grams/day.

and Bristle, with energy intakes in that order), Bristle's energy intake is distinguishably lower than that of Hans, but no other intakes are distinguishable.

Similarly at 40–50 weeks, the lowest energy intakes are the most readily distinguished. The three lowest (those of Bristle, Fred, and Pooh, in that order) are mutually distinguishable. The intakes of Fred and Pooh are also distinguishable from those of the four yearlings with higher intakes, whereas Bristle's intake cannot be distinguished from any of them. Energy intakes of the top four yearlings (Summer, Eno, Ozzie, and Alice, in that order) are mutually indistinguishable based on overlapping standard error intervals.

Results of the samples taken when the baboons were 50–60 weeks of age are discouraging in that only two intakes can be distinguished: Alice's was distinguishably lower than Eno's. Perhaps the small number of distinguishable pairs at this age reflects the fact that at this age, mean sample time per individual (528 minutes, table 3.1) was lower than at any other age (658 to 920 minutes each).

40–50 weeks

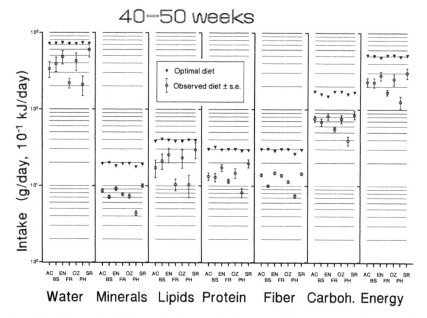

Fig. 7.13. Macronutrient intakes of individuals, 40–50 weeks old. Symbols and scaling as in figure 7.12.

For the four baboons sampled at 60–70 weeks, energy intakes are indistinguishable only among two, Fred and Alice. All other pairs are distinguishable, and the energy-intake ranking is Ozzie, Fred, Alice, then Pooh.

To get an overall evaluation of individual differences in energy consumption, I averaged each yearling's energy intake over all ages at which it was sampled. The standard error of each such mean of means was obtained from $(s_1^2 + \ldots + s_n^2)^{1/2}/n$, where s_1 is the standard error of a yearling's mean intake at age 1 and n is the number of ages at which it was sampled (table 7.8). Distinguishable pairs of individuals are summarized in figure 7.16. Based on overlapping standard error intervals, the baboons' mean energy intakes could seldom be distinguished from the intakes of those whose energy consumption ranked next to theirs, or even two or three ranks away (respectively, just one case out of ten, three out of nine, and four out of eight). However, their energy intakes usually could be distinguished from those of yearlings whose energy intakes ranked four, five, or more ranks above or below them (respectively four cases out of seven, four cases out of six, and always except once for those six or more ranks apart). On average, for those yearlings whose energy intakes were four ranks away, the larger intake was 50% above the smaller.

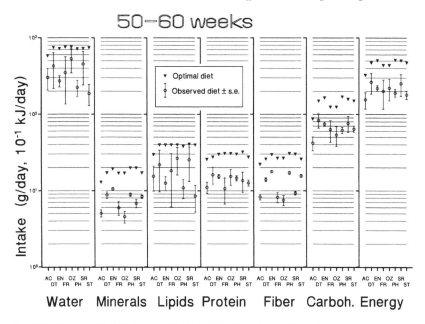

Fig. 7.14. Macronutrient intakes of individuals, 50–60 weeks old. Symbols and scaling as in figure 7.12.

This increase in resolution with magnitude suggests that other yearlings also differed but that the differences could not be resolved with the available samples. Let's look into this further. If *n* baboons are sampled at a given age, then $n(n - 1)/2$ paired comparisons can be made among them. The numbers of baboons sampled at the four ages that I selected are nine, seven, eight, and four, respectively, or all together ninety-one possible paired comparisons. In fifty of these cases (55%) the energy intake levels were indistinguishable by the methods I used. For the overall comparison of 30–70 week means, 24 (44%) were indistinguishable. Yet these samples were my primary research activity for a year—very discouraging! (The possibility that any two yearlings actually had identical energy intakes seems remote.) The obvious solutions, such as sampling fewer yearlings, using longer age intervals, or having more researchers, are either impracticable or undesirable. Within the range of sample sizes in my study (table 3.1), total sample time per pair at a given age had only a small effect on the chance of being able to distinguish their energy intakes. Intakes were distinguishable, 36% of the time in pairs of individuals with a combined sample time, during any age at which both were sampled, of 400–800 minutes, 50% of the time when the combined sample time

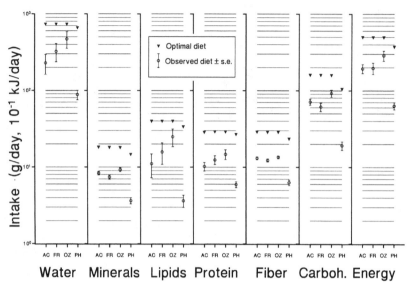

Fig. 7.15. Macronutrient intakes of individuals, 60–70 weeks old. Symbols and scaling as in figure 7.12.

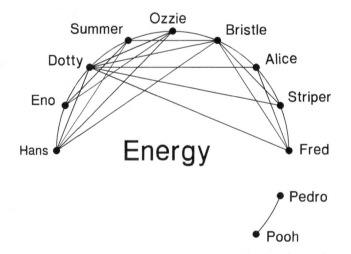

Fig. 7.16. Individual differences in energy intake. Solid lines join pairs of individuals whose mean daily energy intakes are indistinguishable in the sense of having overlapping standard errors. All other pairs are distinguishable. The yearlings are arranged in order of mean daily energy intake, starting on the left with Hans, who had the highest intake. From values in table 7.8.

was 800–1,200 minutes, and 51% of the time for longer samples (1,200–2,048 minutes). Perhaps more individual differences in energy and nutrient intakes would have been detected if I had obtained individual-specific estimates of unit intake rates, at least for the most commonly consumed foods, instead of using a pooled mean for each food. That the problem lies in my choice of statistical method—overlapping standard errors—not in my sample sizes, is suggested by remarkably high correlations between individual energy values, especially deviations from optimum, and several life history measures related to fitness, as demonstrated in the next chapter.

Proteins

As for protein intakes, at 30–40 weeks they are distinguishable among the four yearlings whose daily protein consumption ranked lowest (Ozzie, Fred, Pedro, and Pooh, in order), except for Fred versus Pedro. Protein intakes of the last three of these yearlings are also distinguishable from those of the five high-ranking yearlings (i.e., those with the five highest daily protein intakes: Eno, Hans, Summer, Bristle, and Alice, in that order), whereas the protein intake of Ozzie, as well as that of Alice, can be distinguished only from that of Eno among these top-ranked yearlings. At this age, results for protein intakes are very similar to those given above for energy intakes.

At 40–50 weeks, standard error intervals for the protein intake of the bottom-ranked yearling, Pooh, do not overlap those of any other yearling sampled at that age. Protein intakes of all other yearlings are indistinguishable only from that of the yearlings immediately above them in protein ranking; Bristle and Fred were also indistinguishable from those yearlings two ranks above them. All other pairs are distinguishable.

Results for protein intakes at 50–60 weeks are like those for energy at this age in that few intakes were distinguishable, in this case only Eno versus Striper and Alice, and Pooh versus Alice.

Finally, at 60–70 weeks, protein intakes ranked Ozzie, Fred, Alice, Pooh. Ozzie's intake is indistinguishable from Fred's, and Fred's from Alice's. The other four pairs have nonoverlapping standard error intervals for their protein intakes.

Of the ninety-one possible paired comparisons of protein intakes in the four age classes, forty-six (51%) were indistinguishable in that their standard error intervals overlapped.

When protein intake values for each yearling are averaged over all ten-week periods in which it was sampled, the results (fig. 7.17) are similar to those for energy, in that the two yearlings with the lowest intakes, Pooh and Pedro, had protein intakes distinguishably below those of everyone except each other. All others had protein intakes that are distinguishable from (and only from) those

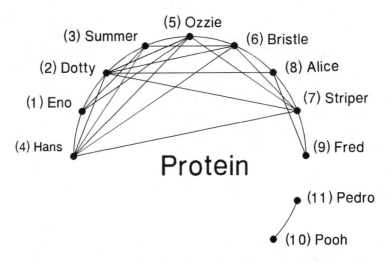

Fig. 7.17. Individual differences in protein intake. Solid lines as in figure 7.16. The parenthetical number next to each yearling indicates its rank order for mean daily protein intake, with the highest intake (Eno) indicated by (1). Yearlings graphed in same positions as in figure 7.16 to facilitate comparisons. From values in table 7.8.

whose protein intakes were several ranks away from theirs: at least two or three ranks sufficed for everyone except Dotty. Of the fifty-five pairs, twenty-three (42%) were indistinguishable.

What are the long-term consequences of these individual differences in nutrient intakes? That is the subject of the next chapter.

8

Baboons are eclectic feeders that appear to be optimizing their diet by selective feeding from among a wide array of available foods in an ever-changing floristic environment.

—G. W. Norton, R. J. Rhine, G. W. Wynn, and R. D. Wynn, 1987

From Food to Fitness

As demonstrated in chapter 7, yearling baboons differ in their diets and thus in their nutrient intakes. How close did the yearlings come to their optimal diets? To what extent are their resulting differences in proximity to optimal diets reflected in what became of these yearlings in their later years? In particular, did those individuals that came closer to optimal diets as yearlings live longer and produce more offspring? The answer to that last question is yes. This chapter provides the evidence.

In chapter 5, I introduced five optimal diet objectives. Here I focus on one of them. On the assumption that energy is the primary fitness-limiting component of the baboons' diets, I took optimal diets to be those that maximize daily energy intake while simultaneously remaining above the minima for nutrients and below the maxima for various constraints such as time limits, gut processing capacity, and so forth.

Computations

In chapter 6 I used seasonality factors to adjust the baboons' observed mean intake over a year to seasonal fluctuations in my sampling intensity, but I acknowledged certain shortcomings in these adjustment factors, stemming particularly from their inability to take into consideration the array of foods that are simultaneously present in a given season. In this chapter those problems are obviated by calculating optimal diets for each semimonth and then comparing these with the yearling's observed intakes at the same times of year. To provide adequate samples, I made comparisons between observed and expected values on a ten-week basis, using the same age blocks as before. To do so for a given yearling in any one age block, I averaged the corresponding semimonthly optimal diets, each weighted by my sampling time on that yearling during those semimonths.

Calculation of Optimal Intakes

For each of the eleven yearlings, an optimal diet that maximizes mean daily energy intake (objective G_1 in chapter 5) was calculated for each ten-week age block during which the yearling was sampled. That was done by calculating an age-specific and semimonth-specific optimal diet for each semimonth during which the yearling was of that age and then, for every age block of each infant, calculating a weighted mean across semimonths for each core food in those diets, where the weighting coefficients were the relative amounts of in-sight sample time on the yearling at that age during each semimonth. So, for example, Alice, when in age block 1 (30–40 weeks of age), was sampled during three semimonths, nos. 16, 17, and 18 (respectively, Aug-2, Sep-1, and Sep-2), with corresponding sample times, here rounded for simplicity, of 141 minutes, 309 minutes, and 54 minutes (total 504 minutes). (The suffixes -1 and -2 indicate the first and second semimonths, that is, the first 15 days of the month and the remaining days.) I therefore calculated three optimal diets that would be suitable for a block 1 baboon, one based on the core foods available in Amboseli during each of those three semimonths. Arranging each of these diets as a fifty-element column vector (one element for the mass of each core food potentially in the diet) and letting $d_{i,j}$ designate the optimal diet vector for a baboon in age block i during semimonth j, the optimal intake for Alice during her first age block (age 30–40 weeks), adjusted for shifts in amount of sampling on her, is $(141d_{1,16} + 309d_{1,17} + 54d_{1,18})/504$, and similarly for other age blocks and other yearlings (twenty-eight combinations all together). Sample times for each yearling at every "age" (ten-week interval) during each semimonth are given in table 8.1; in my study, no yearling was sampled in one semimonth at two ages, so no further partitioning was needed.

For all of these semimonthly optimal diets, the constraints were the same as those used in diets 6–10 (the *extended constraint set*), except as follows: (1) The requirements for energy, protein, and lipid requirements are mass-dependent and so were calculated for animals of average body mass at the midpoint of each ten-week age block (i.e., at 35, 45, 55, and 65 weeks, as needed), using the mass-specific formulas in table 5.3 and the mean growth formula (second approximation) given in appendix 1, equation (A1.1). These are the only known age-specific (size-specific) constraints. (2) I allowed only foods available during the semimonth in question, that is, those "present" or "half present," as explained in appendix 8 and identified in table A8.2. For each semimonth in which any given food j was present, maximum daily intake w_j was calculated by means of equation (A8.8). For this purpose, I used p and h values presented in table A8.2. For each semimonth in which a food was half present, I limited daily consumption to $w_j/2$.

The twenty-eight age- and season-specific individual diets that maximize total daily energy intake are given in table 8.2. These sample-time weighted optimal diets and their nutrient content are compared below with the baboons' observed dietary intakes. The nutritional composition of each of these twenty-eight optimal diets was calculated by multiplying the mass of each food in the diet by the corresponding concentration of each macronutrient class and of energy (table 5.1).

Deviations from Optimum

Observed versus Optimal Diets

Table 7.8 shows the nutritional content of the observed diet and of the energy-maximizing optimal diet for each yearling at each age at which it was sampled. The percentage deviation of a nutrient's observed value from its optimal value, shown in the table, is minus the complement of the percentage of optimum that was achieved. For example, if a baboon took in 40% of the optimum amount of water, then its water intake deviated from optimum by -60%.

If we compare the diet mass of individual baboons at various ages with the mass of the energy-maximizing diet for an Amboseli baboon of that age and at the same time of year (table 7.8), one striking result is immediately apparent: none of these yearlings, at any age, ate as much as they should have, nor are any of their diet masses within two standard errors of the optimum. The closest any of them came to eating the optimal amount of food was Eno, whose diet at 30–40 weeks of age was within 16% of the mass of the optimum. The source of this food deficit is unknown. My suspicion is that the sparse distribution of foods in Amboseli (not their quality, abundance, or seasonality) is the primary factor limiting the baboons' food intake, and that food sparseness, combined with handling-time requirements, translates into requirements for effort, time, and energy needed to get at widely distributed food sources and the attendant problems of fatigue, dehydration, thermal imbalance, predation, and other nondietary risks of foraging. Whatever the causes, the nutritional consequences of these smaller than optimal diets can be anticipated: for virtually all macronutrients, every baboon at each age at which it was sampled took in suboptimal quantities relative to quantities in corresponding energy-maximizing diets. (Indeed, there is only one exception: Eno's lipid intake when she was 30–40 weeks old.) I shall therefore often refer to the deviations from optimum (table 7.8) as "shortfalls."

Some intake shortfall is not surprising, in that twenty-two of the twenty-eight energy-maximizing diets call for consumption at gut capacity every day. However, the baboons' actual deviations in diet mass were of considerable magnitude: on average 52.2% of the mass of their energy-maximizing diets. (In this section, this and all other shortfalls are means of individual $\times$ age-block values, table 7.8, and were calculated without the value from Pooh's very small sample when 30–40 weeks of age.) That is, on average the baboons ate only about half the amount of food they should have. They simply did not eat nearly enough to maximize their energy intake, even though they all could have done considerably better without violating any of the known constraints on their foraging.

Can it be, though, that the baboons ate a balanced diet, in the sense that they ate foods in the ratios of an optimal diet, albeit in too small quantities? If that were so, then at any given age a yearling's deviations from optimum would be essentially the same for all nutrients as for the total mass of the diet. That this is not the case can be seen from a perusal of the deviations in each row of table 7.8. Not only didn't the baboons eat enough food, they didn't, by an energy-maximizing criterion, eat a balanced diet.

However, I find it interesting that the baboons' mean energy shortfall, 55.7%, is very close to their shortfall in total food, 52.2%. That is, on average the baboons took in food of nearly the optimal energy *concentration,* albeit not the optimal *amount* of energy, and they did so despite major phenological changes that resulted in an ever changing food selection. Whatever imbalances are present in the proportion of other food components, their food choices have not, on average, seriously compromised the energy richness of their diet. What about these other food components? Their mean shortfalls from the corresponding twenty-eight energy-maximizing diets are as follows: water 50.2%, minerals 59.8%, lipids 52.7%, proteins 54.7%, fiber 59.1%, and other carbohydrates 58.3%, compared with the shortfall in total mass, 52.2%. That is, the baboons' diets were, on average, slightly moister and less fibrous than the energy-maximizing optimum and contained slightly smaller concentrations of minerals, lipids, proteins, and non-fibrous carbohydrates than they should have. The above shortfalls in lipids, proteins, and carbohydrates conjointly account for the baboons' suboptimal energy intakes. None of the nutrient shortfalls is very different from the shortfall in total mass. The largest difference is less than eight percentage points. That is, many more of the baboons' nutrient shortfalls are attributable to inadequate food consumption than to poor choices of foods—to the quantity of their diet, rather than its quality.

Note that these are all mean values, but as can be seen from their standard deviations or from the individual values (table 7.8), individual differences in shortfalls are very large. Consider energy. Even if I set aside the results of Pooh's very small sample when 30–40 weeks of age, the baboons' energy intakes deviated from optimum—always downward—by anywhere from 37.3% (Eno at 30–40 weeks) to a wretched 82.9% (Pooh, 60–70 weeks). In other words, at best a yearling's energy intake was 62.7% of what it should have been, and at worst a mere 17.1%. None of the energy intakes were within two standard errors of the energy available from each yearling's energy-maximizing diets. Similarly, the ranges in shortfalls were wide for each of the other nutrient classes (table 7.8). Of the twenty-eight diets, the moisture content of seven were within two standard errors of that of the energy-maximizing diet and the same seven were within two standard errors of optimum for lipids. For all other proximate nutrient classes, all other yearlings at all sample ages were below optimum, by this criterion.

By contrast with these observed diets, the energy-maximizing optimal diets show surprisingly little change in energy content through the year, despite radical changes in available baboon foods in Amboseli over the course of a year. The energy content of these optimal diets, which cover every semimonth in which I sampled the infants, would average 4,793 kJ/day, for which the coefficient of variation is 8.8%, whereas the energy content of the baboons' observed diets averaged 2,094 kJ/day with a coefficient of variation of 28.1%. Of course, since most individuals were sampled at several ages, these samples are not independent and so probably do not accurately reflect the actual population distributions of real and ideal diets, but they illustrate that seasonal fluctuations in the habitat are not the sole source of variability in the nutrient content of the baboons' diets. Indeed, seasonal fluctuations in available foods may be of relatively minor nutritional significance, thanks to the baboons' selective omnivory.

The diet of only one yearling at one age, Eno at 30–40 weeks, came within 10% of the energy-maximizing quantity of any macronutrient class (water and lipids, in her case). Beyond that, Eno at that age also had the best intake, relative to her energy-maximizing diet, of both energy and water. Yet Eno was not consistently so skillful a forager. Her energy shortfalls spanned 19 percentage points: her energy intakes ranged from 43.3% to 62.6% of that in the optimal diets for her. How consistent were the other baboons that I sampled in at least three of their four ten-week age blocks? Alice's energy shortfalls spanned 10 percentage points, Fred's spanned 33, Ozzie's 13, Pooh's 20, and Summer's 5.5. That is, judging by these small samples, Summer was the most consistent in her energy intake—and she always did better than average—whereas Fred was the least consistent and did better than average only half the time.

Primary Shortfalls

Do the imbalances of the various baboons tend to be of the same type, say, usually too low in protein? Here for each baboon is the nutrient class for which it had the greatest shortfall, that is, the nutrient for which its observed intake was the lowest percentage of the intake in its energy-maximizing optimal diet: lipids for Alice, fiber for Bristle, minerals for Dotty, fiber for Eno, lipids for Fred, fiber for Hans, fiber for Ozzie, minerals for Pedro, lipids for Pooh, fiber for Summer, and lipids for Striper. That is, for nine out of the eleven yearlings the greatest nutrient shortfall, relative to the content of an energy-maximizing diet, was in either lipids or fiber. One should not make too much of this, how-ever, because for most of these yearlings one or more other nutrients had short-fall values nearly as large, so that appreciable sampling error may well be involved.

Improvement with Age

If the yearling baboons' dietary deviations are due in part to their inexperience, then with age they should on average come closer to their respective optimal intakes. Average energy deviations, calculated from data in table 7.8, are as fol-lows: 57.9% at 30–40 weeks, 55.0% at 40–50 weeks, 53.9% at 50–60 weeks, and 56.4% at 60–70 weeks. These four values are too close to be distinguish-able and show no consistent trend. By 60–70 weeks of age, the baboons' diets were providing about the same proportion of their optimal intake as did their diets at 30–40 weeks of age. This does not mean that the baboons did not im-prove during the interval, only that the results of their increased skill did not exceed their increased requirements.

Consistency

The deviations from optimum shown in table 7.8 suggest major individual dif-ferences in performance. Could any of these differences be small-sample arti-facts? That is, are they statistically significant? Because of the way these values were obtained, each a composite of several other values, I do not know how to answer that question directly, as I explained in chapter 3. However, if over time one individual in a given pair was consistently closer to optimum than the other, our confidence that their differences are real would be increased. Indeed, that was the case: deviations from optimum for yearlings sampled at more than one age usually but not invariably had the same rank order.

This rank-order consistency over time can be evaluated as follows. In fig-ure 8.1, I have plotted deviations in protein content of the baboons' diets at each age from that of their energy-maximizing optimal diets. For clarity, only values for Fred, Ozzie, and Alice are shown. Lines connect adjacent values for each yearling. Each intersection of two such lines represents a pairwise reversal in rank order, that is, an inconsistency. For any two adjacent ages, the largest pos-sible number of intersecting lines is $\binom{r}{2}$, where r is the number of connecting lines between them. If i is the number of intersections (inconsistent pairs) be-tween two adjacent ages, then $100[1 - i/\binom{r}{2}]$ is the percentage of consistent pairs across these two ages. When more than two ages are involved, the values of i and of $\binom{r}{2}$ can each be summed to obtain the overall percentage of consis-tent pairs, which provides a simple index of consistency. One modification

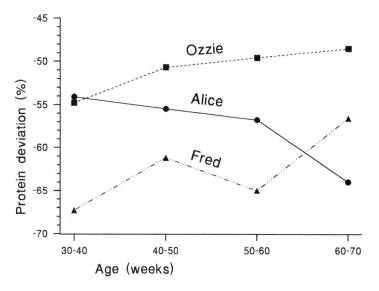

Fig. 8.1. Deviations of protein intakes for like-aged subjects at four ages. Protein deviations measured as percentage deviation of actual mean daily protein intake from the protein intake of a yearling's energy-maximizing optimal diet. Crossed lines indicate inconsistencies across age in the infants' protein rank orders. Data from table 7.8.

of this index was made for near ties. Look at figure 8.1. For Fred, Ozzie, and Alice, $r = 3$ for each of the three age changes, for a total of nine possible inconsistencies, of which two occurred. However, of these two, one (Alice vs. Ozzie, age 30–40 weeks) involves a "near tie"; that is, it hinges on a difference of 1% or less, surely too small to be resolved with my data. (In this case none of the consistent pairs involved a near tie.) If I set aside this unresolvable pair, one inconsistency occurred out of eight possible ones, that is, these three baboons were 88% consistent. In calculating consistency indices (table 8.3), I set aside all pairwise ranks, consistent or inconsistent, that were affected by near ties if reversing the near-tied values would alter the results.

Since some inconsistencies may be due to birth-season effects, I first considered consistency among yearlings whose birthdays were close together. Three such clusters of like-aged yearlings have already been identified: Alice/Fred/Ozzie, Dotty/Striper, and Bristle/Summer. Dotty and Striper were sampled at just one age, so no test of their consistency can be made. Next, I considered all yearlings together. Consistency indices for both sets of yearlings are given in table 8.3 for each class of macronutrients. In both cases the yearlings were 72–73% consistent overall. That is, if at a given age one of two specified yearlings ate a diet that was closer to optimum, that yearling usually still did so ten weeks later, irrespective of the birth dates of the two. This consistency strengthens my conviction that the observed individual differences in proximity to optimum usually reflect real differences, not just sampling artifacts, even for those yearlings for which I do not have samples at several ages.

Diets and Fitness

Does a better diet during infancy result in higher fitness? While my nonexperimental study cannot answer such questions about causality, it can address a more general question: Is lifetime fitness predictable from diet during infancy? To answer this, I here use, as indicators of fitness, survivorship to age six years and several components of female fecundity. As predictors, I use several characteristics of diet that are based primarily on its energy and protein content, as well as several relational variables, all described below. In a previous publication (S. Altmann 1991), I provide a synopsis of this section.

Fitness variables and predictor variables were related by using product-moment correlation as an exploratory device and multiple regression to generate and test predictive models. I used model I regression (Sokal and Rohlf 1981); although both the fitness variables and their predictors are random variables, subject to error, multivariate versions of model II regression apparently are not yet available. Except where indicated otherwise, all correlation coefficients (r) and all regression coefficients of determination (R^2_{adj}) that are discussed below are significant at the $p \leq 0.05$ level. All multivariate regression coefficients have been adjusted (as indicated by R^2_{adj}) for the number of variables involved. Calculations were carried out via NWA/Statpak and SAS/Stat, including the latter's program, MAXR, for stepwise multiple regression. MAXR does not necessarily find the best linear combination of variables, but in this study it certainly found very good ones. Because stepwise regression calculates more regression equations than are used, the regression coefficients of determination from those analyses are biased upward to an unknown degree (Kerlinger and Pedhazur 1978). For subjects that were sampled in more than one ten-week age block, their unweighted mean value for each predictor variable (table 8.4) was used. Descriptive statistics for the variables are given in table 8.5.

Predictor Variables

The primary characteristics of the baboons' diets that I selected as predictor variables are aspects of protein and energy consumption, specifically, the baboons' mean daily intake of these two food components and the relative deviations of their intakes both from their requirements and from the levels of energy and protein in the baboons' energy-maximizing optimal diets. In addition, the same energy and protein intake characteristics also were calculated for milk and for just the non-milk portion of the diet, to give some measure of the subjects' degree of foraging independence (table 8.4, notes). Protein and energy were selected for the same reasons they were selected as objectives in calculating optimal diets (chapter 5): they are likely to be fitness-limiting components of animals' diets.

Other diet components that were used are the number of core foods for which each subject's feeding bout rate is the highest among all subjects and the number for which its rate is second highest, which can be considered as two measures of relative feeding efficiency, and the percentage of nursing bouts that were terminated by the mother, which can be considered the complement of the offspring's independence.

I also included three components of social play, described below, and the dominance rank, among adult females, of the subject's mother at the time the subject was born, on the ground that the mother's dominance relationships potentially influence not only her offspring's diet but many other aspects of its life as well.

Three components of social play—play bout length, play bouts per day, and play time per day—were selected as intervening variables, that is, hopefully predictive of fitness and predictable from the diet.

These selected play components warrant further comment. In long-lived animals, measuring lifetime fitness is often impracticable. However, a reasonable assumption is that healthier, more vigorous individuals have greater life expectancies and higher reproductive potential—or simply put, that physical fitness promotes biological fitness. I therefore looked for a simple indicator of physical fitness. My subjects were not captured, so detailed medical and physical examinations were out of the question. What I sought was something comparable to a step test (Master and Oppenheimer 1929), in which change of heart rate in response to standardized exercise is used as an overall measure of physical condition. I noticed that young baboons often drop out of vigorous play sessions in apparent exhaustion. In the extreme, they lie on the ground panting and for a while are unresponsive to play invitations from other youngsters. I therefore decided to test the mean duration of social play bouts as a physical fitness surrogate for biological fitness. During all of my feeding samples except those in the 1974 pilot study, I recorded the times of onset and termination of all of the focal individual's social play bouts. A product-limit estimate (Kaplan and Meier 1958) of mean play bout length was then calculated for each subject. In addition to play bout length, my field records on play readily yielded two other potential fitness surrogates: the frequency of play (scaled here as play bouts per day) and the amount of time devoted to play (estimated from the product of mean bout length and frequency, scaled as minutes of play per day).

Fitness Indicators

The only advantage to the inordinate amount of time taken to analyze and write up the data from my feeding samples is that, in the meantime, our ongoing research on the baboons of Amboseli has provided long-term life history data on the individuals I sampled. I selected eight life history variables to indicate potential individual differences in biological fitness. These fitness indicators are here related to dietary attributes of my subjects. Each of the first seven fitness indicators is a component of reproductive success; the eighth, of survivorship.

Because the predictor and fitness variables are referred to repeatedly in the following text, tables, and figures—often in an abbreviated form—their labels and their column location in table 8.4 are listed below for convenience. Details of their definitions, computations, and units of measure are given in the footnotes of table 8.4.

COLUMN	ABBREVIATION	VARIABLE
Fitness Variables		
c	REPD	Reproductive life span (days)
d	PUBAGE	Age at menarche (days)
e	CONAGE	Age at first conception (days)
f	INF	Number of liveborn infants
g	INFRATE	Infants per thousand days of reproductive life span
h	YRLG	Number of yearlings produced
i	YRLGRATE	Yearlings per thousand days of reproductive life span
j	SURVIVE6	Survival to age six years
Relational Variables		
k	BRANK	Mother's rank at birth of this individual
l	PLAYDUR	Duration of play bouts in minutes
m	PLAYFREQ	Play bouts per day
n	PLAYTIME	Minutes of play time per day
o	REBUFF	Percentage of nursing bouts terminated by the mother
p	FOOD1	Number of core foods for which this yearling's bout rate was highest among the subjects
q	FOOD2	Number of core foods for which this yearling's bout rate was second highest among the subjects

Energy Variables

r	J	Daily energy intake (kJ/day)
s	JDO	Energy shortfall (as proportion of optimum)
t	JDR	Energy surplus (as proportion of minimum)
u	MILKJ	Daily milk energy intake (kJ/day)
v	NMILKJ	Daily non-milk energy intake (kJ/day)
w	NMILKJDO	Non-milk energy shortfall (as proportion of optimum)
x	NMILKJDR	Non-milk energy surplus (as proportion of minimum)

Protein Variables

y	P	Daily protein intake (g/day)
z	PDO	Protein shortfall (as proportion of level in energy-maximizing diet)
a′	PDR	Protein surplus (as proportion of minimum)
b′	NMILKP	Daily non-milk protein intake (g/day)
c′	NMILKPDO	Non-milk protein shortfall (as proportion of level in energy-maximizing diet)
d′	NMILKPDR	Non-milk protein surplus (as proportion of minimum)

Yellow baboons have a polygynandrous mating system, and almost all males leave their natal group when they become adults. Consequently, the reproductive success of the males in my study is unknown. By contrast, females remain in their natal group. In the case of the females in our main study groups, we know when they reach menarche, when each of their offspring is born, and whether these offspring survive infancy (one year). For each of the females in my study, table 8.4 provides data on seven reproduction-based indicators of fitness: length of their reproductive life span, age at menarche, age at first conception, number of live births, number of offspring that survived one year, and numbers of live births and yearlings[11] produced per thousand days of reproductive life span.

The number of offspring each female produces in her lifetime and especially the number that survive the first year of life probably are the best available estimators of realized fitness. Beyond that, the second measure has the merit of spanning one full generation, yearling to yearling. However, in a study such as this with a small sample size, if the age of females at death is largely independent of their diet as yearlings, then each female's *rate* of offspring production might provide a better estimate of her relative fitness in the dietary population she represents. Age at menarche and at first conception are included on the grounds that age at first reproduction has a powerful influence on the intrinsic rate of natural increase (Cole 1954). Beyond that, age at menarche in humans and probably in other mammals as well is affected by diet (Frisch 1980). Various studies of food-enhanced nonhuman primate groups indicate lower age at first reproduction for females in groups with high food availability, for example, in Japanese macaques, *Macaca fuscata* (Fukuda 1988; Loy 1988; other references in Lyles and Dobson 1988) and in the refuse-feeding Lodge Group baboons in our study population.

Survivorship is the eighth of my selected fitness indicators. Only much larger samples than are now available could adequately estimate survivorship distributions (the probability of surviving to age *x* for all *x*) as a function of diet attributes. Instead, I have settled on a more modest goal: predicting survivorship to age six years from diet attributes. That cutoff age covers the prepubertal period of primary parental influence, though for females subsequent coalitions with their mothers and sisters may have some effect on relative fitness. Six years is also an age up to which we know survivorship for virtually all our subjects, both male and female: emigration of males from their natal groups after that age means that we do not always know their age at death. For both males and females, survivorship through December 31, 1988, and data on the date of death or disappearance are presented in the footnotes of table 8.4. Any subject that disappeared from its natal group and was never seen again is assumed to have died at the end of the last day it was seen.

For four of the six females in my samples, my tabulated values for the lengths of their reproductive life spans are problematic. First, two females, Dotty and Eno, were still alive as of December 31, 1988 (and Dotty as of January 1998), so their tabulated values are only lower bounds on the lengths of their reproductive life spans and on the number of infants and yearlings they will produce in their lifetimes. On the other hand, for *rates* of offspring production, their samples are among the longest. Second, two females in my samples, Alice and Striper, are two of three females of Alto's Group that had persistent pathologies before adulthood (partial hind limb paralysis in Alice's case, a deformed pelvis in Striper's) and whose delayed ages at menarche were outliers in the distribution, thus resulting in appreciably shortened reproductive life spans. Alice and Striper should not be excluded from my analysis on this basis—my predictions are not intended just for the near average baboon—but their presence does mean that my sample of reproductive parameters may be imperfectly representative of the population.

I could have selected fewer fitness indicators than the eight I used. However, a study such as this, with no precedent, is perforce exploratory: we do not yet know which life history variables are most sensitive to diet. For that reason I have used a variety of life history variables, each of which, if it varies with diet, can have a dietary impact on biological fitness.

Relationships among Predictor Variables

Among the potential predictor variables used here, I distinguish two classes: nutrient variables and relational variables, as indicated above. The former (columns r through d' in table 8.4) are those based on either energy intake or protein intake. The others—play time, play frequency and duration, mother's rank, frequency of nursing rebuffs, number of foods on which the yearling's feeding bout frequency ranked first or second among its peers (columns k through q in table 8.4)—all involve either social or rank-order relationships. Within each of these two categories, some of the variables are interrelated. For example, play time was estimated from the product of play frequency and play bout length, so from any two of these variables the third can be determined. Nonetheless, because the relationship is multiplicative rather than additive, each of the three may contribute to multilinear regressions.

Within any ten-week age block—and approximately so for averages across age blocks, as in table 8.4—all the above energy variables of each individual (and likewise, all protein variables) are functionally interrelated in the sense that each can be written as a function of all the others. The energy variables (represented here by the one-letter column headings in table 8.4 and also shown above) are related as follows:

$$(s + 1)(t + 1) = \frac{r^2(w + 1)(x + 1)}{v(r - u)},$$

and similarly for the protein variables.

Some fitness variables too are functionally related. In particular, rates of production of infants and yearlings are the ratios of the numbers of offspring in these two classes divided by the length of the female's reproductive life span. Also, in the absence of menopause, the length of a female's lifetime (not tabulated) is the sum of her age at menarche and the duration of her reproductive life span.

To detect closely correlated variables, as well as to find good one-variable predictors of fitness, I calculated the correlation coefficients among all variables, both fitness variables and predictor variables (table 8.6). In figure 8.2, all significantly and well-correlated pairs of fitness variables, that is, those for which $p \leq 0.05$ and $r \geq 0.80$, are connected by lines. The fitness variables form two clusters: age at menarche and at first conception, which are closely correlated; and four other fitness variables, which are intercorrelated as shown. The

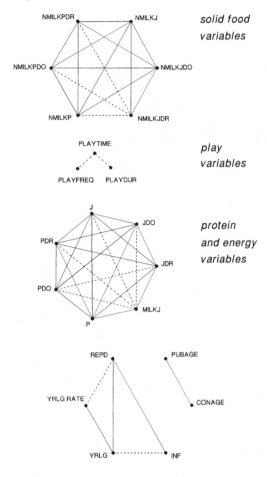

*solid food
variables*

*play
variables*

*protein
and energy
variables*

fitness variables

Fig. 8.2. Correlated pairs of predictor variables (left) and fitness variables (right). Shown by connecting lines on the left for predictor variables (and similarly on the right for fitness variables) are all pairs of variables that were significantly ($p \leq 0.05$) and well correlated ($r \geq 0.80$). Pairs for which $r \geq 0.90$ are shown by solid lines. Data from table 8.6.

remainder of figure 8.2 shows the three clusters of predictor variables that are closely correlated. These three clusters involve, respectively, play; protein and energy in foods other than milk; and all other protein and energy variables, that is, those based on all other core foods.

In perusing the table of correlations, I noticed a common pattern: many of the correlated pairs of variables share a third variable with which each of them was correlated. I therefore undertook a search for "well-connected trios," that is, sets of three variables each of which is well correlated with the other two ($r \geq 0.80$, $p \leq 0.05$). An algorithm was developed to accomplish this search.[12] The resulting well-connected trios of predictor variables are shown in figure 8.3. Only those among fitness variables or among predictor variables

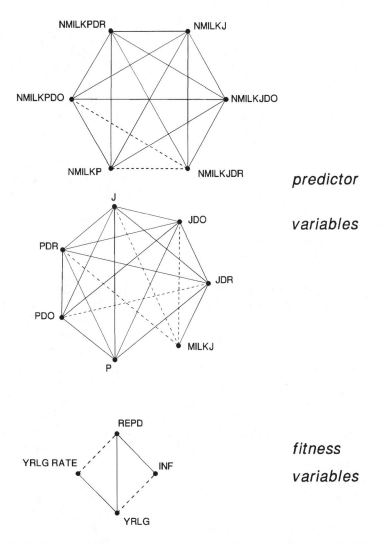

Fig. 8.3. Well-connected trios of variables. Any three variables connected by a triangle of lines form a well-connected trio. See text for details. Solid and dashed lines as in figure 8.2.

(but not between fitness and predictor variables) are displayed in the figure. Several results stand out. First, as with correlated pairs, the predictor variables form two unconnected clusters, the one involving protein and energy in the entire diet, the other only in the non-milk components of the diet—in "solid foods." Furthermore, within each of the three clusters, most of the correlation coefficients exceed 0.90, as indicated by heavy lines in figure 8.3. We can therefore anticipate, for pairs of variables within each cluster, that regression coefficients will be high, that they will be relatively little improved by adding more variables from the same cluster, and that the best two-variable predictions usually will involve one variable from each cluster of predictor variables. Also noteworthy is that the relational variables (play and so forth) are not part of any well-connected trio.

Relationships among Fitness Variables

As shown in figure 8.3, four of the fitness variables form a cluster of overlapping, well-connected trios, as defined above. The four are number of infants per female, number of yearlings, yearlings per thousand days of reproductive period, and length of the reproductive period. The strong correlations among these four fitness indicators suggest that if any of them is predictable from diet, all will be.

The correlations among the fitness variables (table 8.6) reveal several interesting relationships. First, both the number of liveborn infants a female produces in her lifetime and the number that survive the first year are closely correlated with the duration of her reproductive period, indicating that those that live longer produce more offspring. These correlations are consistent with (1) an age-invariant fecundity for adult females, and thereby with the lack of correlation between the rates at which infants are produced and the lengths of reproductive periods (table 8.6). They are also consistent with (2) our earlier demonstration that for females in Alto's Group the numbers of surviving juvenile offspring (to age six) is a linear function of the lengths of the females' reproductive periods (fig. 25.6 in J. Altmann, Hausfater, and S. Altmann 1988) and (3) with results obtained for Japanese macaques (*Macaca fuscata*) by Fedigan, Fedigan, Gouzoules, and Gouzoules (1986) in which the number of reproductive years of females accounted for two-thirds of their variance in reproductive success.

The regressions of number of liveborn infants and of surviving yearlings on the duration of reproductive period in my study are shown in figures 8.4 and 8.5, respectively. The latter also shows the best-fit regression line without the data for Pooh, the one female in my samples that died before reaching sexual maturity. For those females that reached adulthood, YRLG $= -2.29 + 0.00161$ REPD, for which $R^2_{adj} = 0.98$ ($p = 0.001, n = 5$). That is, the number

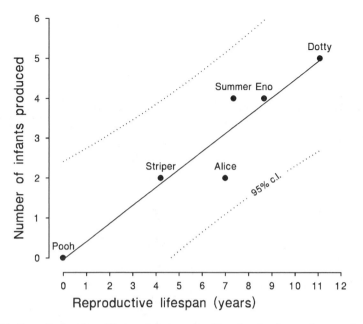

Fig. 8.4. Fecundity (number of liveborn infants produced by a female, y) versus her reproductive life span in years (x). Regression: $N = 6$, $R^2_{adj} = 0.87$, $p \leq 0.01$, $y = -0.0412 + 0.451x$. Data from table 8.4. The slope of the regression estimates mean annual fecundity: infants per female-year.

of reproductive years of these animals accounted for virtually all the variance in their lifetime reproductive success. The regression model indicates that on average each additional day of reproductive competence results in an additional 0.00161 yearlings, or put the other way around, each additional surviving yearling represents the yield of an additional 621 days of adulthood. The implications of age-invariant fecundity will be discussed below, in the section on economical forecasting.

Second, early bloomers conceive earlier: age at menarche is strongly correlated with age at first conception, both in my small sample ($n = 5$, $r = 0.93$; $p = 0.02$; Pearson correlations, here and below) and in all females of Alto's Group for which we have data ($n = 28$, $r = 0.78$, $p \leq 0.001$; J. Altmann, pers. comm.). However, neither early sexual maturation nor early first conception confers any detectable long-term influence on reproductive success: neither age at menarche nor age at first conception is significantly correlated with number or rate of production of infants or of juveniles in my sample (table 8.6), and similarly, life span—the primary determinant of the length of female's reproductive span—is independent of age at menarche ($n = 33$, $r = 0.29$, $p = 0.10$, taking censored intervals at face value). Apparently neither age at sexual

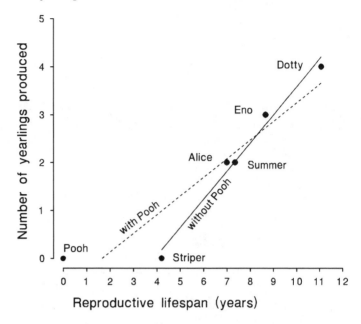

Fig. 8.5. Reproductive success (number of surviving yearlings produced by a female, *y*) versus her reproductive life span in years (*x*). Regression: $N = 6$, $R^2_{adj} = 0.84$, $p \le 0.01$, $y = -0.649 + 0.3896x$; without Pooh, $R^2_{adj} = 0.98$, $p \le 0.001$, $y = -2.292 + 0.588x$ (with *x* in years). Data from table 8.4.

maturation nor age at first conception is a good fitness surrogate, at least so long as one considers fitness to be reflected in reproductive success per se and ignores the influence of age at first conception (and thereby of generation length) on fitness.

Third, no parental investment effect (Trivers 1972) was detected: the mean rate of infant production (infants/day), which is approximately the inverse of mean interbirth interval (days/infant), is not correlated with the number of infants or yearlings produced. Yet it would be if females could swap quality for quantity, that is, if females that invested for longer in their current infant increased survivorship, albeit at a cost of a delay in producing their next offspring.

Correlates of Female Fitness

I now consider how well the lifetime fitness of females correlates with and can be predicted from the energy and protein in what they ate as yearlings, combined when necessary with information from the relational variables described above. Each of the fitness measures is considered in turn, first by examining its univariate correlates, then by using multivariate linear regression to combine information from several predictor variables. Values and scaling for the variables of the multivariate fitness models are given in table 8.4. Correlations among all pairs of variables are given in table 8.6. The regression models are presented in tables 8.7 through 8.12. Each row of this last set of tables represents a multivariate linear model that predicts the indicated fitness measure. For example, the second row of table 8.7 (model 2) indicates that the duration of females' reproductive period (REPD) can be predicted from their energy intake (J) and energy shortfall (JDO) by means of the following equation: REPD = 42171 − 6.857 J + 45770 JDO. Definitions and individual values for REPD, J, JDO, and the other variables are given above and in table 8.4. In the usual manner, significance tests for linear regressions were for the null hypothesis of zero slope, and the tests were two-tailed, at a significance level of 0.05. Some significant models are not tabulated here because preferable models were found. For example, the number of infants a female produces in her lifetime is accurately predicted from a linear combination of variables JDO and FOOD2 (defined above), for which $R^2_{adj} = 0.98$ and $p = 0.002$. However, in making up the tables, I gave preference to models based solely on nutrient (energy and protein) variables.

The correlation coefficients (table 8.6) indicate that, even from one variable, most measures of lifetime fitness in these baboons are surprisingly predictable at the time of infancy, primarily from what the baboons ate at 30–70 weeks of age. If the correlation coefficient between a fitness variable and a predictor variable is at least 0.75 and the correlation coefficient is significantly different from zero ($p \leq 0.05$), I shall refer to the fitness variable as "well correlated." By that criterion, the duration of the reproductive period in females was well correlated with each of six predictor variables. Similarly, age at menarche was well correlated with one variable; age at first conception, with none; number of infants, with seven; rates of infant and yearling production, with none; number of yearlings produced, with four. In many of these cases, the fitness variable was "highly correlated," that is, with correlation coefficients exceeding 0.90. Correlates of each of the fitness measures are discussed further in what follows.

Predicting Females' Reproductive Life Spans

The duration of a female baboon's reproductive period (REPD), which is the primary component of her life span and a major determinant of the number of offspring she produces, is well correlated ($r \geq 0.75$) with several aspects of her diet as a yearling (table 8.6). As indicated above, the length of her reproductive period was well correlated with each of six predictor variables. Indeed, three of these—daily energy intake, energy shortfall, and protein shortfall—were highly correlated with reproductive life span ($r \geq 0.90$), and the others were close to that.

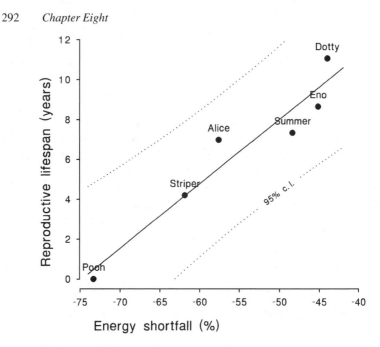

Fig. 8.6. Reproductive life span of a female in years (y) versus her energy shortfall (percentage deviation from optimum, x). Data from table 8.8, model 15. Regression: $N = 6$, $R^2_{adj} = 0.90$, $p \le 0.005$, $y = 24.21 + 0.32x$. Data from table 8.4.

The highest of these is for energy shortfall. The scatterplot of reproductive period, in years, against energy shortfall as a percentage of optimum, is given in figure 8.6. The coefficient of determination (R^2_{adj}) (table 8.6) indicates that childhood energy shortfall alone accounts for 89% of the female-to-female variance in reproductive period. The regression model for predicting reproductive life span (in days) from energy shortfall, model 1, is given in table 8.7. The model's coefficient for energy shortfall is positive, and the energy shortfalls are negative, that is, female yearlings whose dietary energy was *closer* to optimum grew up to have longer reproductive periods. Energy shortfall was scaled as a *proportion* of optimum in this regression. Thus the slope of the regression, 11,707, indicates that, on average, each percentage point decrease in energy shortfall translates into an additional 117 days of reproductive period. For a fictitious female baboon whose energy intake was at the optimum (JDO = 0), reproductive period, according to the y-intercept of the model, would be 8,764 days—more than twice the highest observed value.

As shown in table 8.7 (model 2), adding one more nutrient variable, daily energy intake, to energy shortfall resulted in a better model for reproductive periods ($R^2_{adj} = 0.98$). Adding to energy shortfall two other variables, daily protein intake and the protein shortfall of the non-milk foods, improved the predictions still further, leaving virtually no variance unaccounted for (model 3, $R^2_{adj} > 0.99$).

Predicting Females' Ages at Menarche

Based on a very small sample (three females for whom both age at menarche and play data are available), age at menarche (PUBAGE) was very closely correlated with how often females played as yearlings ($r = 0.997$; model 4, $R^2_{adj} = 0.99$). However, the scatterplot for age at menarche against play bouts per day (fig. 8.7) shows that the regression is completely dominated by one female, Summer, who reached maturity earlier and played more often than the other two. This regression is discussed below, in the section on relations between play and fitness.

Surprisingly, age at menarche was not well correlated at the $r \geq 0.75$ level with any single nutrient variable (five females) and, indeed, showed no significant correlation with any one predictor variable other than play frequency. I say "surprisingly" because of evidence that age at menarche in humans is affected by diet through effects on body fat (Frisch 1980). I did not analyze body

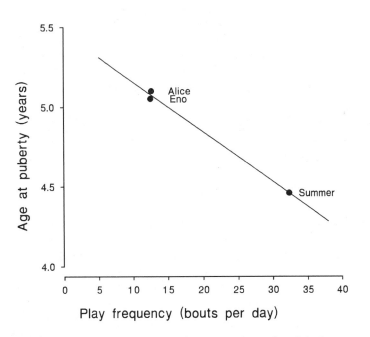

Fig. 8.7. Age at puberty in years (y) versus play frequency (mean number of play bouts per day, x) for females. Regression: $N = 3$, $R^2_{adj} = 0.99$, $p \leq 0.05$, $y = 5.47 - 0.0312x$. Data from table 8.4.

fat in my subjects, but our recent analysis of body fat in adult females of the same group (Alto's Group) indicates extraordinarily low levels (J. Altmann, Schoeller, S. Altmann, Muruthi, and Sapolsky 1993).

Can age at menarche be predicted better from linear combinations of variables? Not from any set of nutrient variables I checked, but a model based on two nutrient variables (energy shortfall and protein surplus), combined with mother's rank, accounted for 99.4% of the variance in age at menarche (model 5, table 8.7). The reasons for this apparent interaction between yearlings' diets and mothers' ranks in delimiting age at menarche are not apparent.

Predicting Females' Ages at First Conception

Like age at menarche, age at first conception (CONAGE) was highly correlated with play frequency ($r = 0.96$), but in this case the correlation was not significantly different from zero. Again, the available sample is very small (three females) and the relationship is dominated by Striper, so not much can be made of it.

Age at females' first conception, like age at menarche, was not well correlated, by my criterion, with any other single predictor variable I used. However, it was well correlated with the female's energy shortfall combined with her mother's rank and her own protein surplus. The resulting three-variable model (model 6, table 8.7) accounts for 99.4% of the variance in age at first conception.

Predicting Females' Fecundity

As I mentioned above, the number of liveborn infants a female produces in her lifetime (INF), her fecundity, is well correlated with any of seven dietary variables, including four energy-based variables and three protein-based variables (table 8.6). Indeed, with six of these seven variables, her fecundity is highly correlated, and in three cases (energy intake, energy shortfall, and protein shortfall) the correlation coefficient is 0.98. For the highest of these, energy shortfall, the scatterplot is shown in figure 8.8; the corresponding regression (model 7) indicates that on average each percentage point improvement in a female's energy shortfall results in (or at least predicts) an additional 0.158 infant; that is, a female, in her lifetime, produces an additional infant for each 6.33 percentage points closer to the optimum her childhood energy intake was (regression $R^2_{adj} = 0.96$). According to this model's y-intercept, females that took in the optimum amount of energy (JDO = 0) would produce an average of 11.5 infants in their lifetime.

Clearly, such extraordinary one-variable predictability of fecundity from diet leaves very little room for significant improvement from multivariate

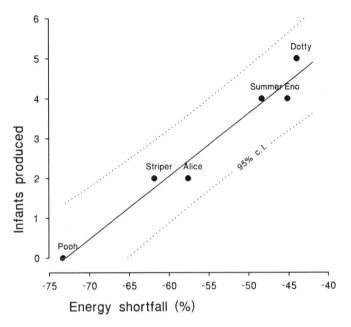

Fig. 8.8. Fecundity (number of liveborn infants produced by a female in her lifetime, y) versus her energy shortfall as a yearling (percentage deviation from her optimum, x). Regression: $N = 6$, $R^2_{adj} = 0.96$, $p \leq 0.0005$, $y = 11.55 + 0.158x$. Data from table 8.4.

models. Just two energy-based variables, daily energy intake and energy surplus (intake above requirement), are all that are needed to bring the regression coefficient above 0.99 (model 8, table 8.7).

Predicting Females' Birth Rates

As indicated above, the number of infants a female produces in her lifetime is well correlated with her diet as a yearling, and so too is the duration of her reproductive period. Yet, curiously, the ratio of these two—the rate at which she produces infants (INFRATE), her "fecundity rate"—is not well correlated at my criterion level ($r \geq 0.75$, $p \leq 0.05$) with any of the single predictor variables I used (table 8.6). This enigma is discussed below in the section on yearling production rates.

The two best multivariate linear models I found for fecundity rate (models 9 and 10, table 8.7) had R^2_{adj} values of 0.989 and 0.990, respectively, but their respective p values (0.068, 0.064) were above my selected significance level ($p \leq 0.05$). Furthermore, because many other linear combinations of variables were tried, even these two values may be statistical artifacts.

Predicting Females' Reproductive Success

For animals that exhibit parental care, fitness depends not only on producing offspring—on fecundity—but also on facilitating offspring survival. I therefore looked at the number of surviving yearlings (YRLG) produced by my female subjects. It was well correlated ($r \geq 0.75$, $p \leq 0.05$) with four different variables; three of these involved energy intake and one involved protein intake (table 8.6). Each of these five was also well correlated with the number of infants the females produced. In every case, the correlation was somewhat poorer for yearling production than for infant production, and three correlates of infant numbers were not significant correlates of numbers of yearlings produced.

Judging by correlation coefficients, the best single predictor of yearling production (YRLG) is energy shortfall (JDO). The scatterplot of YRLG against JDO is shown in figure 8.9. The regression model (model 11 in table 8.7, $R^2_{adj} = 0.76$) indicates that each percentage point difference in energy shortfall during a female's childhood predicts a difference of 0.1256 yearlings during adulthood, that is, that each 7.95 percentage point improvement in energy

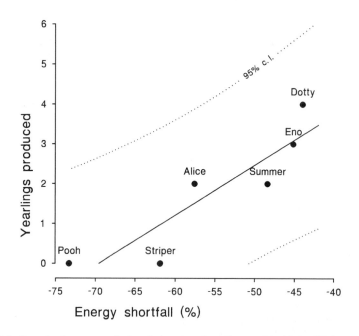

Fig. 8.9. Reproductive success of a female (number of yearlings produced in her lifetime, y) versus her energy shortfall as a yearling (percentage deviation from her optimum, x). Regression: $N = 6$, $R^2_{adj} = 0.76$, $p \leq 0.05$, $y = 8.80 + 0.127x$. Data from table 8.4.

shortfall during childhood translates into an additional surviving yearling over the lifetime of the female. The *y*-intercept indicates that at the optimum energy intake (JDO = 0), females would produce an average of 8.8 yearlings in their lifetimes, more than twice the highest value for any of my subjects. Adding a second variable, protein surplus, to energy shortfall accounts for even more of the variance (model 12 in table 8.7, $R^2_{adj} = 0.94$). Adding protein surplus and energy intake to energy shortfall results in a model that accounts for 99.7% of the variance in lifetime production of surviving offspring (model 13, table 8.7).

Predicting Females' Reproductive Success Rates

Rate of yearling production (YRLGRATE), like rate of infant production, was not significantly predictable from any single variable. Indeed, no predictor variable showed a significant correlation at any *r*-level, even though the *numbers* of yearlings and the durations of reproductive periods, whose ratio is this rate, are each highly predictable. The explanation for this enigma appears to be that the numbers of yearlings and the durations of reproductive periods are highly correlated (table 8.6). In the extreme, if they were *perfectly* correlated (each starting at zero), then their ratio would be constant and thus totally independent of the predictor variables, however dependent each was separately.

Using linear combinations of variables, I found that energy shortfall combined with two other nutrient variables, protein shortfall and non-milk protein shortfall, led to a model that accounts for 99.9% of the variance in the rate at which females produce offspring that survive the first year of life (model 14, table 8.7).

Predicting Survivorship of Males and Females to Age Six

To determine whether early diet affects survivorship to adulthood, I carried out a discriminant analysis. I used data both from the six female subjects and from the five males combined. The yearlings' diets at age 30–70 weeks were used to predict whether each would survive to six years of age (SURVIVE6).

Because of the demonstrated efficacy of energy shortfall and protein surplus at predicting other aspects of fitness, I selected these two diet variables as predictors in the discriminant analyses, and as before I used for each individual its mean value for all ten-week age blocks in which it was sampled. As I indicated earlier in this chapter, nutrient intakes are approximately univariate normal; on the assumption that they are also multivariate normal, a parametric discriminant analysis was carried out (SAS's DISCRIM procedure).

Based on the baboons' energy shortfall and protein surplus, the discriminant function perfectly predicted which individuals would survive to age six. Equality of class means (survive vs. not) was rejected at the 0.005 level by all four tests generated by the SAS program. These results must be regarded with caution, however, because in a sample of eleven bivariate normal random variables, a given group of two points can be completely separated from the remainder 15–16% of the time, according to a simulation run by Peter McCullagh (Department of Statistics, University of Chicago). A logistic regression, which avoids many of the strong assumptions of discriminant analysis such as that the data are multivariate normal, also perfectly predicts which individuals would survive to age six and which would not (McCullagh, pers. comm.). The *y*-intercept constant and the coefficients for the discriminant function and the logistic regression equation are given in table 8.8 (models 22 and 23, respectively).

Although the probability of separating a given two out of eleven points is fairly high, the actual degree of separation is in this case remarkably high, a fact not taken into account in calculating the above-mentioned probability. Indeed, complete separation can be obtained based on either predictor variable (energy shortfall or protein surplus) alone, as can be seen from figure 8.10. That is, the two individuals that did not survive to age six were the two that were farthest from their optimal energy intakes and also were the two that had the smallest protein surplus. For any one predictor, the probability of complete separation

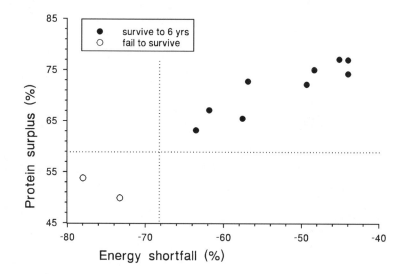

Fig. 8.10. Survival to age six years predicted from energy shortfall and protein surplus. The two dotted lines show the complete separability of survivors (solid circles) from non-survivors (open circles) based on either of these diet characteristics alone. Data from table 8.4.

by chance (i.e., the probability that the two points are the two largest or the two smallest) is 4/110 (McCullagh, pers. comm.).

Causes versus Correlates

In summarizing these results, I must remind the reader of something I myself find difficult to believe. With the possible exception of rates of infant production, each of my seven measures of females' lifetime fitness—of the reproductive success of females *as adults*—is very highly predictable from what these individuals ate when they were yearlings, just 30 to 70 weeks of age. Two of the seven fitness surrogates, age at menarche and age at first conception, required additional information about play frequency and mother's rank, respectively. In all cases, regression coefficients of 0.99 or better were achieved with three or fewer predictors, and except for fecundity rate models ($p_1 = 0.064$, $p_2 = 0.068$), all models were statistically significant ($p \leq 0.05$), often highly so ($p \leq 0.01$). Furthermore, for males and females combined, energy shortfall and protein surplus each predicted without error which individuals would survive to age six. In fact, most of the other seven fitness measures I used were significantly correlated with at least one energy or protein variable, and for four of them—duration of reproductive life span, age at menarche, number of infants produced, and number of surviving yearlings—significant one-variable correlation coefficients exceeded 0.75.

Nutritional differences, particularly energy intake relative to optimal intake, predict not only which individuals survive to adulthood but also the length of their subsequent reproductive life span, that is, their survivorship after becoming adults. In turn, these differences in survival determine individual differences in the number of infants and surviving yearlings that a female produces in her lifetime: those that live longer produce more offspring.

Although one expects some consistency in life histories, this degree of predictability of lifetime fitness based on childhood diet is remarkable. My results clearly call for independent confirmation in other studies, on other populations, and with larger samples. Nonetheless, the results at hand suggest that by 70 weeks of age, a female baboon's lifetime fitness is largely established, as is the likelihood, for both males and females, of surviving to adulthood. Yet at that age they still nurse, still sleep in their mothers' arms at night, and are still several years away from menarche and from the time when they will conceive their own first offspring and thereby demonstrate their own biological fitness.

Do some animals have higher fitness because they eat better, or do they eat better because they are more fit? In many cases, I suspect that both are true. In turn, dietary selection and fitness each are related to many other characteristics of animals. The causal links among traits cannot be measured in a study like this: correlation is not causation. Regardless of whether fitness is affected by diet, either directly through effects on reproductive success or indirectly through effects on other traits, the results at hand suggest that in these wild primates, fitness itself is highly predictable from diet at a remarkably early age.

In this study I related fitness not to motor patterns of feeding behavior per se, but to their dietary and nutritional consequences—to performance. The task of measuring selection on traits, whether morphological, physiological, or behavioral, is simplified by breaking the task into two parts: measurement of the effects of trait variation on performance and measurement of the effects of performance on fitness (Arnold 1983). This study is a contribution to the latter task. Differences in the traits by which these primates achieve their individual diets and the heritability of those traits are virtually unstudied.

Economical Forecasting of Lifetime Fitness

Predicting fitness as I did, by testing large numbers of hard-to-obtain variables, is expensive—in time, money, energy, and commitment. However, the results suggest a natural economy, for although I entered all thirteen of the protein and energy variables into the multiple regressions that generated the nutrient-based models, only two or three variables were needed to get outstanding predictor models of each of the fitness measures.

The variables that appear in the models differed from one fitness measure to the next, yet three of the nutrient-based variables do not appear in any of the fitness models. At the other extreme, two of the variables appear far more often than others: the baboons' so-called protein surplus (PDR) and, especially, their so-called energy shortfall (JDO), which is their relative deviation from their respective energy-maximizing optimal daily intakes. Furthermore, the former, protein surplus, is a component of more models than just dietary protein per se, and similarly the latter, energy shortfall, appears in more models than just dietary energy. This result raises an interesting possibility: Could one achieve a computational economy by using just these two variables, energy shortfall and protein surplus? (The computational economy would result, not so much from computing multilinear models of smaller size than many that I presented here, which would be a trivial saving, but from the computational labor involved in the many steps needed to estimate some of the variables.) For three of the fitness indicators, the answer is yes. Regression of the seven indicators for female fitness on energy shortfall (JDO), on protein surplus (PDR), and on both variables together resulted in seven statistically significant models for three of the fitness measures; the models are given in table 8.8 (models 15–21). For each of these three fitness measures, a linear model accounted for at least 86% of the variance between individuals. The four significant one-

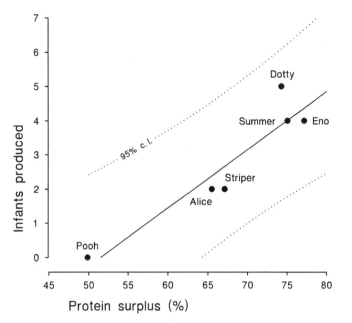

Fig. 8.11. Fecundity of a female (number of liveborn infants produced during her lifetime, y) versus her protein surplus as a yearling (percentage above her requirement, x). Regression: $N = 6$, $R^2_{adj} = 0.85$, $p \leq 0.01$, $y = -8.80 + 0.171x$. Data from table 8.4.

variable regressions in table 8.8 (models 15, 17, 18, and 20) are shown graphically in figures 8.6, 8.8, 8.11, and 8.9, respectively. Figures 8.12 and 8.13 show the predictability of, respectively, fecundity (number of liveborn infants produced) and reproductive success (number of surviving juveniles produced), using linear combinations of energy shortfall and protein surplus; in each case the individual deviations from the plane of regression are barely visible.

A further economy can be achieved by using energy shortfall alone, because each of the three reproductive variables that is predicted by a linear combination of energy shortfall and protein surplus is also significantly predicted by a regression model based solely on energy shortfall—despite the fact that the energy intake of each individual in every ten-week age block in which it was sampled was well below that of its energy-maximizing diet (note negative coefficients in table 7.8). Mean energy shortfall of each female provides excellent predictions of the length of her reproductive period, the number of infants she produced in her lifetime, and the number of her offspring that survived the first year of life (table 8.8). The last two measures probably are the best available estimators of realized fitness, and the last measure has the merit of spanning one full generation, yearling to yearling.

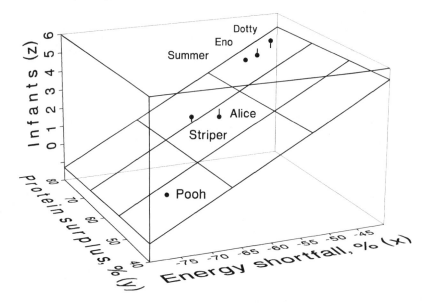

Fig. 8.12. Fecundity of females (number of liveborn infants they produced) predicted from a linear combination of their energy shortfall (percentage deviation from optimum) and their protein surplus (percentage deviation from requirement). Data from table 8.4.

If, as I suggested earlier, birth rate is nearly age-invariant in these primates, then differences in all three of these components of fitness depend primarily on survivorship, and as indicated above, energy shortfall also perfectly predicts which subjects, male and female, survived to age six. On the other hand, these energy shortfalls by themselves do not provide significant predictions of age at menarche or first conception, or of the rates at which infants and yearlings were produced (offspring per thousand days of reproductive period). Individual differences in the first group of fitness variables (those predicted) depend primarily on differences in survival; those in the second (not predicted), on differences in fecundity.

Why not use daily energy intakes per se or energy surplus—that is, energy intake in excess of requirement—as a predictor, instead of deviations from the optimal energy intake? Doing so would avoid the labor of calculating optimal diets. Furthermore, in this study daily energy intake was correlated with the same fitness measures as was energy shortfall, and to about the same degree (table 8.6). (Remember that neither of these values differs by a constant amount from energy-maximizing optima, because optimal diets differed between individuals, even at the same age: at any given season, the subjects were of various

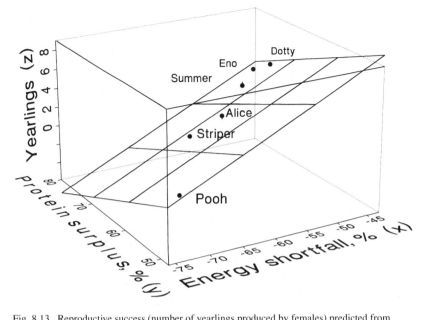

Fig. 8.13. Reproductive success (number of yearlings produced by females) predicted from
a linear combination of their energy shortfall and protein surplus (as in fig. 8.12). Data from
table 8.4.

ages, and at a given age they were feeding at various seasons. So the distinctions here are not just a matter of scaling.)

For several reasons, I do not recommend using daily energy intake or energy surplus by themselves instead of energy shortfall (relative to that of an optimal diet), except as a temporary expedient. First, if animals are sampled at different ages, their requirements are different. That problem could be dealt with by using energy surplus. However, both measures have other deficiencies that result from bypassing optimal diets as benchmarks. An animal's diet may be high in energy and even close to the optimum but be harmful or even lethal if one or more of the constraints is violated, that is, if the diet is outside the region of adequate diets either because of a deficiency of one or more nutrients or because of an excess of one or more toxins.

Second, even if individuals are sampled at exactly the same ages, different arrays of food will be available to them at any given age if their birth dates differ. At some seasons nutritious foods are relatively abundant; at other seasons they are sparse. That means different optimal diets, even for individuals of the same age; I calculated diets for each semimonth. That probably also means differences in the difficulty of achieving any given degree of nutrient intake, reflecting differences in the fitness of those that do so. Finally, note that in this study energy shortfall entered into far more of the multivariate models than did energy intake or any other potential predictor (table 8.7), probably because it incorporates information on food availability that only indirectly influences the values of other potential predictor variables.

Another type of economy would involve a reduction in the number of error-susceptible measurements and weakly justifiable assumptions. So, for example, my estimates of food energy are based on measurements not only of the protein content of foods but also of their lipid and carbohydrate contents. These estimates also require various assumptions about the digestibility of fiber (appendix 1). On those grounds, protein-based variables would be preferable to energy-based variables. I therefore regressed the seven fitness measures against linear combinations of the six protein variables. The protein variables alone (table 8.9, models 24–27) were adequate to provide significant predictions for three of the fitness measures. For these models, the adjusted regression correlations range from 0.600 to essentially 1.000 (rounded to thousandths). Four models are shown in table 8.9; for numbers of infants produced, two additional significant models, of intermediate size to those tabulated, were also generated.

Similarly, the non-milk nutrient variables avoid the many assumptions that I had to make (chapter 5) in order to estimate the baboons' milk intakes. I therefore regressed the fitness variables against the non-milk energy and protein variables (variables v, w, x, b', c', d' in table 8.4). Table 8.10 shows all the significant models (models 28–31) that were found by stepwise multilinear regression. These include models for three of the seven fitness measures: reproductive period, the number of infants produced, and the number of surviving juveniles. All four models accounted for at least 99% of the variance in the fitness indicators.

The three types of economical forecasting discussed above show a common pattern. In all three cases, significant models were found for (and only for) the duration of the reproductive period and for the two fecundity measures. Among the fitness estimators I used, probably the closest approximations to a female's actual fitness are provided by those two fecundity measures—the number of liveborn infants she produces in her lifetime and the number of surviving yearlings she produces.

A last type of economical forecasting is this. In the field, data on time budgets for any activity are far easier to obtain than data on bout lengths and durations, in that time-budget estimates can be obtained from point (instantaneous) samples (J. Altmann 1974), which can be incorporated into ongoing sampling of other activities. Furthermore, if foods of each type are harvested at a (food-specific) constant rate during feeding bouts—which appears usually to be the case, judging by a sample of the most common baboon foods in Amboseli

(S. Altmann and Shopland in prep.)—then mean feeding time on any food type will differ from mean intake of that food (or mean intake of any nutrient in that food) only by some food-specific constant (grams of intake per minute of feeding). On this basis, time devoted to each food, if effective as a predictor, would provide very economical predictors.

A further refinement might involve eliminating those foods that reveal little about the variability of individuals. A food that is consumed in nearly the same amounts by all subjects tells us little about individual differences in the intake of any given nutrient or in its consequences. The same is true of foods that are very poor sources of the nutrient. Conversely, foods that are consumed in very varied amounts and that are rich sources of the nutrient in question tell us much about the resulting individual differences. This suggests rank-ordering the foods, say by the absolute magnitude of their standard deviations for the nutrient in question, and concentrating the sampling on those foods that are high on the list.

Energy, Survival, and Fitness

In the discussion above of relationships among fitness variables, I noted that the number of infants and yearlings a female baboon produces in her lifetime is closely correlated with the length of her reproductive period: those that live longer produce more offspring. Exactly these three variables—numbers of infants and yearlings produced and length of reproductive period—are the ones best predicted by energy shortfalls. Specifically, energy shortfall accounted for 96% of the variance of females' fecundity (number of liveborn infants) and 81% of their reproductive success (number of surviving juveniles).

These results are consistent with the hypothesis that the energy shortfall of baboons' diets is the major dietary factor affecting their fitness. This influence appears to be exerted primarily through effects on survival: differences in energy shortfall were sufficient to predict which individuals survived to adulthood and to account for 92% of the variance in the females' subsequent survival; that is, in the duration of their reproductive life span. Survival, in turn, appears to be the key variable through which differences in fitness are mediated. In short, those whose energy intakes are closer to their optima live longer and thus produce more offspring.

Furthermore, relatively small differences in diet during infancy presage large differences in lifetime fitness. The coefficient of variation in energy shortfalls (males and females combined) was just 21%. For protein surplus, the coefficient of variation was 14%. Yet lifetime fitness ranged from zero to four surviving offspring—twice the average requirement for a population to sustain itself—and as indicated above, these differences in reproductive success primarily reflect differences in survival. Charles Darwin would not have been surprised at this result. In a letter of September 5, 1857, to Asa Gray (quoted in Simpson 1982), Darwin wrote, "What a trifling difference must often determine which shall survive, and which perish!"

One final note about the consequences of nutrient intake levels. A common belief (e.g., FNB 1980, 10–11) is that nutrient intakes above the required amounts provide no benefits. In chapter 5, I stated a different assumption, that "for every component *except one,* intake variations within the range of adequate intakes . . . are assumed for the purpose of this model to have no significant effects on fitness." The assumed exceptional diet component— variously energy, protein, or harvesting time—formed the basis for the objective function of each optimal diet model. The relationships demonstrated in this chapter between several measures of biological fitness and dietary intakes of energy provide evidence for just such incremental benefits. Beyond that, the relationship of the fitness measures to protein consumption suggests that fitness may be simultaneously limited by at least two nutrients. As I mentioned earlier, there is no a priori reason to assume that limiting factors are unique.

Small-Sample Biases

I am troubled, as the reader may well be, by the small number of subjects on which my regressions are based: far fewer than ideal (e.g., Kerlinger and Pedhazur 1973). This means that my estimates of regression statistics, R^2 and the regression coefficients, are biased estimates. The R^2 values have been adjusted to reflect my small samples, and only adjusted R^2 values (R^2_{adj}) are given above. The regression coefficients may be over- or underestimated. As more predictor variables are incorporated into models, these biases probably become progressively greater.

Although no confidence can be placed in the exact numeric values given here for regression statistics, I believe that the basic patterns described above are real. First, as I mentioned above, energy and protein—two diet components that on biological grounds are suspected of being important limiting factors— appear repeatedly as outstanding predictors of various fitness indicators. Second, the models presented above predicted relatively uncorrelated measures. Of the twenty-eight pairwise correlations between the eight continuous fitness indicators that I used (all except survival to age six in table 8.8), twenty-four were not significant at the 0.05 level (table 8.6). Finally, I believe coincidence does not account for the fact that multiple regression models using energy shortfall, protein surplus, or both as predictor variables successfully predicted all and only those fitness indicators that primarily reflect survivorship, not fecundity (table 8.8).

Diets and Play

Predicting Play Bout Lengths

As I indicated earlier, play bout length was selected, a priori, as the main candidate for a shorter-term physical fitness surrogate for biological fitness. How well can play bout length at ages 30–70 weeks be predicted from dietary information at that age?

On the assumption that play bout length might be energy-limited—a contemporary version of the old hypothesis that play represents an expenditure of excess energy (Beach 1945)—I regressed mean play bout length of the male and female yearlings on energy variables (variables $r, \ldots, x$ in table 8.4). Although "excess energy" was never clearly defined in the early play literature, perhaps the most straightforward interpretation is that it is energy in excess of minimal requirement, which, expressed as a proportion, is the energy surplus, variable JDR in table 8.4. By itself, energy surplus is worthless as a predictor of individual differences in mean play bout length: $R^2_{adj} = 0.07$, and the regression is not significant. Indeed, none of the energy-based variables I checked showed any significant correlation or regression with play bout length, either alone or in any linear combination. These results are not consistent with the hypothesis that play, or at least play bout length, is energy-limited.

Can it be that play bout length is better predicted from the protein content of the yearlings' diets? No. When just the protein variables were used in a stepwise multilinear regression, no statistically significant predictor of play bout duration was found. If, however, we allow combinations of protein and energy variables, we obtain good predictors of play bout length. (Coefficients for this and all other play models discussed below are given in table 8.11). Unfortunately, these models all have at least four predictor variables, and in view of my small sample sizes, such models should be considered only suggestive of possible relations. The smallest significant model (model 32) contains four nutrient variables and accounts for 82% of the variance in play duration. The best five-variable model found is model 33, for which $R^2_{adj} = 0.94$, indicating a trivial opportunity for improvement in predictability. With six nutrient variables in the model, all based on protein or energy intakes, play bout lengths become even more highly predictable (model 34, $R^2_{adj} = 0.99$). Clearly, these models suggest that individual differences in mean play bout length at 30–70 weeks are closely related to what the baboons eat at that age.

Predicting Play Frequencies

The frequency and duration of play bouts are uncorrelated (table 8.6): these are independent attributes of social play in yearling baboons. Judging by the

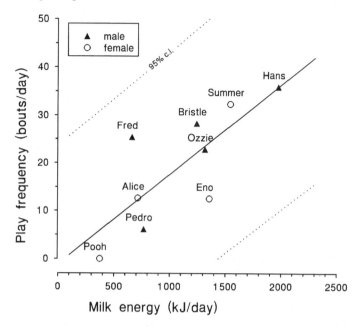

Fig. 8.14. Play frequency (bouts per day, y) versus energy obtained from milk (kilojoules per day, x). Regression: $N = 9$, $R^2_{adj} = 0.54$, $p \le 0.05$, $y = -1.13 + 0.0187x$. Data from table 8.4.

correlation and regression coefficients, the best one-variable non-play predictor of play bout frequencies is the amount of energy baboons derive from their mothers' milk (model 35, table 8.11), which accounts for just over half the variance among yearlings in play frequencies (fig. 8.14). This model indicates that on average each kilojoule per day of milk energy at 30–70 weeks of age resulted in (or at least was correlated with) an additional 0.0187 play bouts per day; put another way, each 53.34 kJ of milk energy translates into a one-bout difference in play per day. The frequency of play also has correlation coefficients that are nearly as high with several other energy and protein variables, especially daily energy intake, energy shortfall and surplus, and protein surplus (table 8.6).

When all non-play predictor variables were included in a stepwise multilinear regression analysis, a six-variable model (model 38) was found that accounts for most of the variance among subjects in play frequencies ($R^2_{adj} = 0.98$). When the regressions were limited to nutrient-based variables (that is, protein and energy variables but not REBUFF, FOOD1, or FOOD2), another series of significant models was obtained. For example, linear combinations of four of these protein- or energy-based variables (model 36) accounted for much of the variance in play frequencies ($R^2_{adj} = 0.90$). A five-variable model (model 37) accounted for virtually all the variance: $R^2_{adj} > 0.99$.

So individual differences in play bout frequencies, though independent of play bout durations, are also predictable from energy- and protein-based characteristics of the baboons' diets.

Predicting Play Time

One non-play variable alone, protein surplus (fig. 8.15), was enough to account for nearly half the variance between individuals in the amount of time per day that they devoted to social play (model 39 in table 8.11, for which $R^2_{adj} = 0.43$). Model 39 indicates that on average each percentage point difference in protein intake in excess of requirement (so-called protein surplus) translates into a difference of 0.1053 minutes of play per day. In other words, an additional minute per day of play represents an additional 9.5 percentage points of surplus protein above requirement levels. Various models containing five, six, or seven variables (e.g., models 40–43, table 8.11) predicted play time with R^2_{adj} values exceeding 0.90.

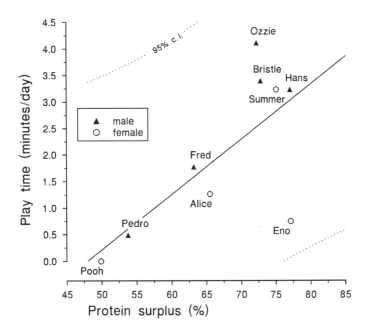

Fig. 8.15. Play time (minutes per day, y) versus protein surplus (percentage of above requirement, x). Regression: $N = 9$, $R^2_{adj} = 0.43$, $p \leq 0.05$, $y = -5.06 + 0.105x$. Data from table 8.4.

If I limit the predictors of play time to nutrient-based variables, the one-variable model based on protein surplus is discussed above, those that the program MAXR generated with two to six variables are not statistically significant, but a seven-variable model (model 43, table 8.11) accounts for virtually all the variance: $R^2_{\text{adj}} \approx 1.00$.

In sum, all three components of social play are highly predictable from linear models, not only those derived from combinations of relational and nutrient-based variables but also those based solely on protein and energy.

Does Social Play Predict Lifetime Fitness?

Baboon diets at 30–70 weeks of age accurately predict virtually all my measures of lifetime fitness and all three components of social play. Does social play, in turn, predict fitness?

When taken one by one, my three play variables—mean play bout length, bout frequency, and play time—showed only one significant correlation with a fitness variable: play frequency was negatively correlated with age at menarche (table 8.6). This one significant correlation out of twenty-one cases (seven fitness variables × three play variables) is just the number of pseudosignificant outcomes expected by chance at the $p \leq 0.05$ level, even if all variables are uncorrelated, so this isolated case should not be taken too seriously. Furthermore, as I noted above, this regression is based on just three females, and the regression depends almost entirely on one.

What about using the three play components in multilinear combinations? Twelve of the twenty-one cases are limited to three females each. For example, there are three females whose age at menarche I know and whose play behavior I sampled; for these, multivariate models cannot be used for lack of degrees of freedom. However, for three of the fitness variables—reproductive period of a female and the numbers of infants and yearlings she produced—two-variable models can be used. For each of these three fitness measures, I checked all of the nine possible two-variable linear combinations of play variables. None provided a significant multilinear model for any of these three fitness measures.

In sum, aside from the one apparently pseudosignificant case, none of the three play variables I used, either singly or in any linear combination, provided a statistically significant predictive model of female lifetime fitness. Play, while highly predictable from diet, is not a satisfactory surrogate for biological fitness.

9

This I know . . . that to the human body it makes a great difference whether the bread be fine or coarse, of wheat with or without the hull, whether mixed with much or little water, strongly kneaded or scarcely at all, baked or raw. . . . For by every one of these things, a man is affected and changed this way or that, and the whole of his life is subjected to them. . . . Nothing else, then, can be more important or more necessary to know than these things.

—Hippocrates, 4th century B.C.

Why Be Choosy?

In this chapter I consider the results of the baboons' selectivity and show, in several specific cases, the biochemical consequences for Amboseli baboons of fine-grained food discrimination. Then I compare the diet of Amboseli baboons with the diet of vervet monkeys in the same area and with the diets of baboons living elsewhere.

The Wheat and the Chaff

No mobile animal eats a random assortment of available foods. Each is selective to some degree. Even a minute, sessile organism like *Stentor* can exhibit considerable feeding selectivity (Schaeffer 1910; Rapport 1980).

Animals exhibit food selection on several hierarchically arranged levels: by choosing the habitat and area in which to live; by selecting some parts of their home range as feeding areas in preference to others; by selecting some food species and some individuals of those species within each area; and finally, by selecting among the parts of their prey, whether plants or animals. Such selectivity among foods may depend on size, hardness, degree of ripeness, or other characteristics.

Certain potential foods are rejected altogether. Some of these rejected foods may be literally inedible, given the animal's anatomy, such as foods that are too large and too tough. Others may be so hazardous to consume or may yield so little in the way of nutrients as to be worthless or deleterious even in combination with other available foods. In some cases a potential food may be rejected for purely historical reasons, such as unfamiliarity combined with dietary conservatism.

After selecting food items, foragers may exhibit further selectivity. By various types of manipulation (in primates, often literally, that is, by hand), they

311

separate out the more edible and nutritious parts of the plant and discard materials, such as husks and sheaths, that are lower in nutrients, higher in toxins, or simply too tough. Even after ingestion, selective processes occur, including selective digestion, absorption, and detoxification. Thus one solution to the packaging problem—the inexorable combination of costs and benefits in every food—is to "untie the package." This can never be more than a partial solution, however, because of the inability of herbivores to feed at the biochemical level, consuming only nutrients and avoiding all toxins.

What do animals accomplish by this selectivity? In chapters 4 and 6, I indicated the contribution of each food to the total foraging time and total diet of weanling baboons and showed that their diet is heavily skewed toward certain foods. That skewing is not merely a reflection of what is available to the baboons. To the contrary, they are highly selective feeders.

The question remains, Is such selective foraging adaptive? The strongest answer but the most difficult to provide is to show just how close the actual diets come to the fitness-maximizing optimum. That is the tack followed in chapters 5 and 7. The simplest way to answer the question would be to show that the animals' diets deviate from a random selection of the available foods and that the deviations are in the direction of a more nutritious or otherwise more beneficial dietary intake. However, although I did not gather quantitative data on the relative abundance of various plants and plant parts in Amboseli, the actual array of foods eaten by the baboons was to me so obviously nonrandom that to demonstrate that their diet is selective seems pointless. After all, these animals are neither lawn mowers, chewing up everything in their pathway, nor statisticians, taking random samples.

Much more revealing than demonstrating nonrandomness would be to show that the foods the animals eat provide a better diet than those they seldom eat or completely reject but that are, in some sense, "close" to these. The closeness may be spatial, such as different parts of a plant, it may be taxonomic, or it may involve some physical attribute, such as size or degree of ripeness. Particularly at this level of fine-grained selectivity, an animal's abilities to detect, discriminate, and manipulate are crucial and may be pushed to their limits.

Indeed, the longer and more closely I watched the baboons as they fed, the more impressed I was by their acute selectivity in foraging. In some cases the rationale behind the selectivity was obvious. For example, the calyx capsule around *Solanum coagulans* fruit has numerous long, sharp spines and was discarded. Similarly, once the thorny capsule surrounding the fruits of devil's thorn hardened, the baboons stopped feeding on them; in this case, for an animal the size of a baboon, there apparently is no easy way to extricate the fruit. The lateral meristem of the tall, tough grass *Sporobolus consimilis* is much more tender and succulent than the tissues peripheral and distal to it and the en-

Fig. 9.1. *Sporobolus consimilis* grass.

veloping sheath. The baboons fed on the meristem and discarded the rest (fig. 9.1).

In other cases, however, there was no obvious physical reason for the selectivity, and I suspected that chemical factors were involved. The baboons ate the fruits of *Solanum coagulans,* which are somewhat nourishing (e.g., 5.5% protein and 16% soluble carbohydrate, according to S. Altmann, Post, and Klein 1987), and the fruits apparently are harmless, but as indicated in chapter 4, the baboons did not eat the similar and related but toxic fruit of *Solanum incanum.* After digging up corms of various sedges and grasses, including *Sporobolus rangei,* the baboons removed not only the attached roots and dead leaf blades but even the fine sheath that surrounds each corm. When picking fruits, for example, from the shrubs *Azima tetracantha, Withania somnifera,* or *Salvadora persica,* they showed a decided preference for the ripe ones. After they picked the flowers of *Trianthema ceratosepala,* they carefully removed the pericarp (fruit wall) and adhering floral parts and swallowed only the fruit, although the wall of this fruit is neither hard nor thorny. Similarly, the baboons ate the pulpy endocarp of *Capparis tomentosa* fruit but discarded the surrounding pericarp. After a young baboon has picked a green pod from an umbrella tree, it extracts the seeds from the pod, which it discards, then one by one removes the seed coats and discards those too. Much of this fine-grained separation is not literally manipulation. Rather, it is done in the mouth, apparently (judging by the movements one can see) by the joint action of lips, tongue, and teeth.

For six of the foods mentioned above, I have analyses of nutrients or toxins in both the discarded portion and the consumed part. These examples demonstrate what the baboons can accomplish through fine-grained selectivity. They also suggest some of the characteristics of baboons that facilitate their selective omnivory, a theme I will take up again in the next chapter.

Seeds of Umbrella Trees

When the young baboons in my study fed on green pods of umbrella trees (*Acacia tortilis*), they exhibited a five-stage selectivity. First, certain individual umbrella trees were used in all three years of the study (1974, 1975, 1976), whereas other trees with abundant pods at comparable stages of ripeness were bypassed. I suspect that these trees are polymorphic for a toxin, trypsin inhibitor. (Polymorphisms in toxins have been reported for a number of species of forage plants, and in some cases these have been related to selective foraging; e.g., Cooper-Driver and Swain 1976; Corkhill 1952; Jones 1966; Glander 1978; Pollard 1992 and references therein.) Second, some pods were picked, then discarded uneaten, or were ignored altogether. Third, immature baboons and adult females removed the seeds and discarded the pods. (Adult males sometimes crushed the entire pod, seeds and all, then discarded the fibrous residue.) Fourth, only some of the seeds were removed from any one pod. Many of the seeds that were left behind were very small (near the distal end of the pod) or moldy. Finally, the seed coats were removed, as described above.

I have data on the nutritional consequences of two of these five steps in the selective process: discarding umbrella tree pods in favor of the seeds, and removing the seed coats. Table 9.1 shows the unit weight and percentage composition of green pods with seeds removed, of seeds without their seed coats, of seed coats, and by calculation from these, of seeds with coats and of pods with their enclosed seeds and seed coats. From these data one can calculate the amount of nutrient and toxin per pod, per seed coat, and per seed, and thus the consequences of rejecting the pods and seed coats.

For every hundred green umbrella tree seeds consumed—which represents about thirteen minutes of harvesting (table 6.1)—baboons will have discarded 150 mg protein in the seed coat and 664 mg protein in the pod. But in the process they will have enriched the protein content of their food to 160% of the value of whole pods. In the case of nonfibrous carbohydrates, the baboons have thrown away the best part, namely the seed coat, which in both absolute and relative terms is richer in carbohydrates than the seed they eat. However, the pod is even lower in carbohydrates, so that in eating the naked seed the baboons get a food that is 111% of the carbohydrate concentration of whole pods. As for fiber, removing the seed from the pod eliminates 2.47 g of fiber for every hundred seeds consumed; next, removing the seed coat eliminates a further 575 mg. The naked seed that remains contains only 220 mg of fiber—less than 7% of the fiber that the baboons would have gotten if they had consumed whole umbrella tree pods, as do the gazelles that forage on them beneath the trees (fig. 4.2). Lipids are negligible in all three pod components. The energy values therefore depend almost entirely on the protein and carbohydrate levels. The discarded seed coats, thanks to their high levels of carbohydrate, have higher energy densities (7.0 kJ/g) than the naked seeds (5.0 kJ/g), whereas in the discarded empty pods it is about the same. Overall, the energy density of the naked seeds is 96% of the value for whole pods, probably a negligible difference. In sum, the primary nutritional consequence of discarding the pods and seed coats is to increase the protein concentration and decrease fiber, in favor of the high-protein, low-fiber seeds.

The toughness of the seed coats in *A. tortilis* and particularly of the pods may limit the availability of their nutrients for small animals that cannot readily crush these materials. Perhaps more important than fiber is reduction of trypsin inhibitor, which I had analyzed separately in the seed and seed coat. (So far as I know, the only other analysis of trypsin inhibitor in legume seed coats is by Ambe and Sohonie 1956 for germinating seeds of field beans and double beans. According to Miller 1967, alkaloids are concentrated in the seed coats of some *Crotalaria* species.) For every hundred green umbrella tree seeds consumed, baboons take in 148.4 trypsin-inhibitor units, TIU's (= 4.52 g/seed × 32.84 TIU/g), but the one hundred seed coats, in which trypsin inhibitor is more concentrated (52.9 TIU/g), contain 241.0 TIU. Thus, rejecting seed coats reduces the intake of trypsin inhibitor by 62%. Pods averaged 1.54 g, but on average 6.72 seeds were consumed per pod (table 6.3), so the weight of discarded pod per hundred seeds (ignoring the few unextracted seeds) is 22.90 g. The amount of trypsin inhibitor in pods is unknown, but if we assume that the concentration is the same as in the seed coats, these 22.9 g of pod contain 1,210.5 TIU. This means that by rejecting both seed coats and pods, young baboons reduce their intake of antitrypsin by 91%, which amounts to 1,451 trypsin-inhibitor units in seed coats and pods per hundred seeds consumed. To do so, they give up 151 mg protein in the coats and 664 mg protein in the pod. Once physiologists determine the quantitative effect of antitrypsin on protein digestibility (grams of protein lost per trypsin-inhibitor unit), the balance sheet can be completed: net protein gained by removing seeds from pods, then seed coats from seeds.

Trypsin inhibitor is also present in many domestic legumes (Liener 1982; Chavon and Kadam 1989). In Ethiopia, eight of us ate partially cooked dried beans (presumably *Phaseolus vulgaris*) that we purchased in the marketplace at Gondar. A few hours later we all had the same response: violent vomiting and diarrhea. Analysis of a sample of these beans showed them to be high in trypsin inhibitor (S. Altmann, Post, and Klein 1987), which would ordinarily be deactivated by adequate cooking. On a number of occasions I have seen Amboseli baboons vomit after eating acacia seeds. Often the green vomitus is retained in the mouth and then swallowed, suggesting that baboons have some way of coping with the toxin if they can only keep their food down long enough. The disulfide bridges in the molecule of one component of trypsin inhibitor (Odani and Ikenaka 1973) probably make this toxin very resistant to digestive enzymes. Perhaps all baboons can do is to secrete sufficient additional trypsin to compensate for the bound component. Janzen (1977) demonstrated that bruchid beetle larvae usually failed to penetrate the seed coats of sixty-three species of nonhost plants, including several neotropical acacias, and he speculated that this was because of their hardness and toxicity.

Corms of *Sporobolus rangei*

When feeding on corms of several species of grasses and sedges, baboons often remove the loose outer layers of dry sheath, which resembles the skin of onions and apparently is the remnant of the bases of old blades (fig. 9.2). They then consume the core of the corm. What do the baboons accomplish by this? We have analyzed the proximate composition of both the core and the sheath of their favorite corms, those of *Sporobolus rangei* (*S. kentrophyllus* in S. Altmann, Post, and Klein 1987; see table 8.2 herein). When they discarded the corm sheaths of *Sporobolus rangei,* the percentage of ash in what they consumed (the corm "cores") was reduced to 65% of what it would have been if the whole corm had been consumed, and fat was 79% as abundant. They probably benefited from the former, since salts are so abundant in Amboseli. They lost as a result of the latter, though negligibly so, since lipids are sparse in this food source in either case. Water, protein, and carbohydrate concentrations were improved, to 115, 111, and 107%, respectively, of whole corm values. Fiber, on the other hand, was reduced to 72% of the value in whole corm. The energy density is essentially unchanged: 11.1 J/mg in whole corms versus 10.5 J/mg in the consumed corm sheaths. (The baboons also discard the adherent dry blade bases, but our nutritional data on this material are incomplete.)

One may question whether the changes indicated here are sufficient to

Fig. 9.2. *Sporobolus rangei* grass.

account in a cost-benefit way for the labor the baboons invested in removing the corm sheaths. Two other factors may be relevant. One is the possibility of important nutrients (or toxins) other than those analyzed here. Another is that the percentage of fiber may be a poor indicator of the toughness of a material, which probably is more dependent on fiber structure than on fiber concentration.

Ovaries of *Trianthema ceratosepala*

At certain times of the year, baboons spend long periods feeding on a succulent prostrate plant, *Trianthema ceratosepala,* that grows extensively in moist depressions near fever trees. They are after just one part of the plant: the fruit (fig. 9.3). One after another, the tiny flower buds are picked and put into the mouth. There the fruit walls are removed, apparently through the combined action of lips, incisors, and tongue, and are allowed to drop from the lips. This liberates the developing ovaries (fruits), which are consumed.

For *Trianthema ceratosepala,* table 9.3 gives the macronutrient composition of fruits, fruit wall, and intact floral buds. The lipid content in the fruits is

Fig. 9.3. *Trianthema ceratosepala.*

higher than in the whole buds but is still low. Protein concentration, however, is considerably higher in the fruit: 241% of the value in the whole buds. The fruit is nearly three times as fibrous, perhaps because of the seeds. Nonfibrous carbohydrates are also higher in the fruit. Since all contributors to energy are higher in fruit than in whole buds, so too is the energy density: 5.95 versus 2.83 J/mg, respectively. Beyond that, ascorbic acid and riboflavin are higher in fruits, with, respectively, 152% and 363% of the values in whole buds. In sum, by removing the calices from *Trianthema ceratosepala* buds, baboons obtain a fruit that has higher concentrations of protein, lipid, fiber, nonfibrous carbohydrates, ascorbic acid, and riboflavin.

Note, as a check on consistency, that the floral bud analysis should give the same results as the sum of the fruit and fruit wall analysis, since the fruits and fruit walls were obtained by partitioning buds. For comparison, both sets of values are given in table 9.3. The primary difference is in water content, 76% versus 81%, which is not surprising since my field notes indicate that one of the two fruit samples, 776/75 in S. Altmann, Post, and Klein (1987), was not as well protected from evaporation as the other. Adjusting the values for fruit plus fruit wall based on 81% moisture brings all these values within 2% of each other.

Fruit of *Withania somnifera*

A preference for the ripe fruit of *Withania somnifera* (fig. 9.4) appears to have definite advantages (table 9.4). Although the ripe fruit is somewhat lower in lipids than green fruit, it is higher in protein, fiber, carbohydrate, and energy. The ascorbic acid content of the ripe fruit (62 mg/100 g) is comparable to that of domestic oranges (50 mg/100 g; Watt and Merrill 1963), and the orange color suggests a high concentration of carotenoids.

During my samples I was usually unable to see whether the *Withania* fruit the baboons ate was ripe. *Withania* fruit of unknown ripeness is among the core foods (WSFX in table 6.2). I assigned it a value intermediate between ripe and unripe fruit, but this value probably underestimates the nutritional contribution of this food because on occasions, primarily outside my samples, when I could see adequately, the baboons appeared to feed selectively on the ripe fruit. The baboons almost always fed on fruit from within inflated calyces (Dave, Patel, and Menon 1985) that had turned from green to beige; limited observations suggest that this change occurs simultaneously with ripening of the berries, changing from green to orange-red.

In 1891 Simmonds reported that *Withania somnifera,* part unspecified, was being used at the Civil Hospital in Alger, Algeria, as a sedative and hypnotic.

Fig. 9.4. *Withania somnifera.*

Kopaczewski (1948) isolated a hypnotic fraction from the fruit juice of the plant. Majumdar (1952) reported the isolation of five alkaloids from the roots, leaves, and stem of the plant: nicotine, somniferine, somniferinine, withanine, withananine, and withaninine. Since that time, a wide variety of other alkaloid compounds have been isolated from nonreproductive parts of this plant, but little other chemical or pharmacological work has been done on its fruit. I have never noticed signs of sedation or intoxication in baboons after they have fed on *W. somnifera* fruit.

Fruit of *Azima tetracantha*

Two of the most common shrubs in the area are *Azima tetracantha* and *Salvadora persica*. Baboons eat the fruit of both and in each case show a decided preference for the ripe fruit. In ripening, berries of *Azima tetracantha* go through four stages, first green ("unripe"), then greenish yellow or greenish white ("semiripe"), then opaque white, then translucent gray-white (both recorded as "ripe"). The baboons preferred the fully ripe, gray-white berries but also ate ones that were opaque white. A few individuals (e.g., female Alto) ate them green. Infants seem to be primarily after the seeds inside the fruit. The berries of *Azima tetracantha* are well protected from the lips of gazelles and other browsers by radiating thorns (fig. 9.5), which have very sharp points

Fig. 9.5. Thorns and fruit of *Azima tetracantha.*

with minute, recurved barbs (visible at 30× magnification). Because baboons have opposable thumbs, however, they can reach between the thorns and pluck the berries.

Table 9.5 gives the composition of unripe, semiripe, and ripe fruit of *Azima tetracantha*.[13] The changes in composition of *Azima* fruit as it ripens are surprisingly small, at least at the level of macronutrients. The same is true for ascorbic acid, riboflavin, and energy value; indeed, the latter two appear to be somewhat lower in the ripe fruit. Perhaps most significant is a 41% increase in nonfibrous carbohydrates. Thus the available data indicate higher carbohydrate levels as the only nutritional advantage to selecting ripe *Azima* fruit, and the baboons' preference for the ripe fruit may be based on its sugar content. On the other hand, the remarkable ability of the unripe fruit to etch stainless steel suggests that the baboons may be responding to compounds that our analyses are not detecting. The seeds of *Azima tetracantha* contain over 9% each of ricinoleic acid and cyclopropenoid fatty acids, in addition to the more common ones (Daulatabad, Desai, Hosamini, and Jamkhandi 1991). The leaves contain a rare glycoside, isorhamnatin 3-O-rutinoside (Williams and Nagarajan 1988). The pharmaceutical significance of these compounds is unknown.

Fruit of *Salvadora persica*

The toothbrush bush, *Salvadora persica,* is common in Amboseli. When the berries ripen, areas in which these bushes are abundant attract many birds, gazelles, and monkeys (both baboons and vervets), and these animals feed on the racemes of fruit (fig. 9.6). The baboons appear to prefer fully ripe (dark red) *Salvadora* berries, but they will eat them semiripe. Indeed, they feed in longer bouts on semiripe fruits than on ripe ones (table 6.1). I believe the explanation is that selective feeding on ripe berries by birds and other animals reduces the number of fully ripe berries per raceme.

For *Salvadora persica,* table 5.1 gives the macronutrient composition of unripe fruit plus flowers (rank 21 in the table), ripe fruit (rank 24), and their mean, as an estimate for semiripe fruit (rank 22). In addition, ripe fruit contains 5.1 mg/100 g ascorbic acid, 0.1 mg/100 g riboflavin, and at least 8.5 μg/100 g folic acid (S. Altmann, Post, and Klein 1987). In the macronutrients, we have an enigma. The baboons prefer the ripe fruit over unripe fruit, yet the latter is higher in protein (6.0% vs. 0.7%), carbohydrates (9.9% vs. 9.5%) and energy (343 vs. 255 kJ/100 g). Lipid values are about the same. Based on the data now available, it seems that in selecting the ripe fruit of this bush, the baboons, vervets, gazelles, and birds are making a mistake! I suspect the animals know something about this fruit that the chemists do not.

The roots and stems of *Salvador persica* are widely used in Africa and the Middle East as chewing sticks and toothbrushes (*miswak*), hence the plant's

Fig. 9.6. *Salvadora persica.*

common name, toothbrush bush or tree. The antiplaque effect of these tooth-brushes has been extensively documented, for example, by Gazi, Lambourne, and Chagla (1987) and by Quinlan, Robson, and Pack (1994). The sticks are reported to have antiplaque, antiperiopathic, anticaries, antibacterial, and antimycotic effects, probably attributable to one or more root contents, which include chlorine, trimethylamine, alkaloid resins, elemental sulfur (rare in plants), and various sulfur compounds (Ezmirly, Cheng, and Wilson 1979; al-Bagieh, Idowu, and Salako 1994). I do not know of any chemical or pharmacological studies of this plant's fruit. The periodontal health of baboons in Amboseli National Park, Kenya, and Awash National Park, Ethiopia, has been described by Phillips-Conroy, Hildebolt, J. Altmann, Jolly, and Muruthi (1993). Isothiocyanate, a cyanogenic glycoside, is found in *Salvadora persica;* this compound has proven antimicrobial qualities (Elvin-Lewis 1982) and may reduce periodontal diseases (Phillips-Conroy, Hildebolt, J. Altmann, Jolly, and Muruthi 1993).

Fruit of *Capparis tomentosa*

Capparis tomentosa is a scandent that grows primarily on acacias in Amboseli, particularly on *Acacia tortilis,* the umbrella tree. When *Capparis* fruits are ripe, they are eaten by vervet monkeys (Klein 1978) and baboons. The baboons eat the sweet, pink, pulpy flesh (endocarp) but discard the husk (pericarp). *C. tomentosa* fruit is reputedly toxic to camels. Cornforth and Henry (1952) therefore tested the fruit for various toxins that had been found in a related plant. Only one was found: L-stachydrine. The yield of L-stachydrine periodide was 2.92 g/1,000 g pulp and 3.18 g/500 g husk; that is, L-stachydrine is more than twice as concentrated in the husks, which the baboons discard. They do, however, sometimes gorge on endocarps, and neither they nor I have exhibited any ill effects from eating them. Very little work has been done on the toxicity of this compound (Massiot and Deloude 1986).

Other Toxins and Hazards

In Amboseli, baboons eat many of the same foods as do vervet monkeys (*Cercopithecus aethiops*). In a study of feeding behavior in Amboseli vervet monkeys, Wrangham and Waterman (1981) showed that in a group of vervets that had ready access to all parts of the two Amboseli acacias, umbrella trees and fever trees, a significant negative correlation was found between food preference (feeding time) and the phenol/tannin content of the food, and they suggest that condensed tannins are acting as major feeding deterrents. The correlations are based on immature leaf, seed, flower, and exudate for each of the two acacia species. The correlation is dominated by one food, the exudate of umbrella trees: vervets spent very little time feeding on this material (about 0.5%), which contains very high quantities of total phenolics (38–56% dry weight) and condensed tannins (23–71% dry weight). Conversely, the vervets, like the baboons, spend much time feeding on the exudate of the other acacia, the fever tree, which has only small amounts of total phenolics (0.09–0.23% dry weight) and condensed tannins (0.23–0.31%).

The correlation across foods of the phenolics and condensed tannins in them versus the time the vervets spent feeding on them is noteworthy and suggests that these secondary compounds may be feeding deterrents. However, these food preferences are also correlated with other characteristics of these foods. In particular, the gum of umbrella trees is very much sparser than that of fever trees, and it has nothing like the latter's extraordinary sugar content (S. Altmann, Post, and Klein 1987). Sugars, which are known to be of considerable importance in mammalian food preferences, probably play a major role in the strong preferences of both vervets and baboons for the gum of the fever tree and, conversely, in the near total neglect by both primate species of umbrella tree gum (Hausfater and Bearce 1976). Similarly, for seeds extracted from green pods, both primate species spend more time per day feeding on seeds of the umbrella tree, and those seeds have more sugars and other carbohydrates (16.3% vs. 6.7% of fresh weight for seeds in their coats: S. Altmann, Post, and

Klein 1987) as well as smaller amounts of phenolics (1.0% vs. 2.5% dry weight) and condensed tannins (1.0% vs. 4.2% dry weight: Wrangham and Waterman 1981, table 10).

However, sugar content alone cannot explain all the other food preferences that occur in both of these primates. For example, both spend more time feeding on the blossoms of umbrella trees than on the blossoms of fever trees, yet the sugar content of blossoms of the preferred species is only about half that of the other (7.2–9.7% vs. 16.3–19.0% fresh weight: S. Altmann, Post, and Klein 1987). On the other hand, the preferred blossoms of umbrella trees have smaller amounts of phenolics (2.3% vs. 3.5%) and condensed tannins (0.8% vs. 2.4% dry weight: Wrangham and Waterman 1981, table 10).

To date, the only other non-nutritive toxins that have been identified in Amboseli plants are solanine in *Solanum incanum* (chapter 4), trypsin inhibitor in acacia seeds and pods, and L-stachydrine in the fruit of *Capparis tomentosa,* all discussed in the preceding section. Some lilies contain phytoestrogens that can adversely affect a mammal that eats them. Dutch women who ate tulip bulbs for survival during World War II were plagued by reproductive problems (Harborne 1977). I have no information on phytoestrogen levels in the several species of lilies eaten by baboons.

Because of the prevalence of grasses in their diet, baboons may be exposed to phytic acid (inositol phosphoric acid), which might interfere with their absorption of calcium (Mellanby 1950). However, we have not seen any indication of rickets in our animals, and it may be that with a steady intake of grass material, the gut adapts and acquires the capacity to hydrolyze phytic acid (Walker 1951) or that for other reasons phytic acid does not impair calcium absorption (Graf and Eaton 1985). Other possible sources of toxins are plants that are abundant in Amboseli and often encountered by the baboons, but that are not eaten by them. Particularly suspicious is *Dicliptera albicaulis,* whose flowers, leaves, and terminal growth are moderately nutritious (S. Altmann, Post, and Klein 1987) but, so far as Muruthi (1988), Post (1982), or I know, are never eaten by the baboons. (The three records in table 4.1 are probably of insects being picked from the plant.) As I mentioned in chapter 4, the abundance of butterflies around the flowers of this plant suggests the presence of a toxic secondary compound. On the other hand, vervets eat the flowers of this plant (Klein 1978). (Postscript: in more recent observations, Amboseli baboons have occasionally eaten the flowers of *Dicliptera,* but never in quantities needed to test the possibility that they include toxic compounds.)

In addition to toxic secondary plant metabolites, each of the essential nutrients, like all other food constituents, is toxic at some level, although in the extreme this level may approach gut capacity. Of particular significance are those nutrients that have a relatively narrow safety margin (ratio of toxic to required intakes) and where very rich sources occur naturally.

Vitamin A is a prime example. Its toxic effects are described in virtually every textbook of nutrition. Eskimos are aware of the toxicity of polar bear liver, which Rodahl and Moore (1943) have attributed to its extraordinary levels of provitamin A: 16,333 IU ($\equiv$ 4,900 μg retinol) per gram wet weight (mean of three samples). Seal liver is nearly as rich (Rodahl and Moore 1943): 13,000 IU ($\equiv$ 3,900 μg retinol) per gram wet weight (one sample) and can cause acute poisoning (Cleland and Southcott 1969). The minimum amount that produces chronic toxic reactions is not well established, but in fifteen reported human cases (Rodriguez and Irwin 1972) the lowest reported intake level that led to chronic hypervitaminosis A was 33,333 IU retinol ($\equiv$ 10,000 μg retinol) taken daily for eight to twelve months. To date, the one analyzed source of vitamin A for baboons, the fruit of *Lycium "europaeum"* (fig. 9.7), contains 1.04 mg beta-carotene per 100 g fresh weight (chapter 5), equivalent to about 174 μg retinol/100 g. Since the fruits of this shrub average 37.4 mg (S. Altmann, Post, and Klein 1987; samples 14/76 and 644/76), a daily intake of over 15,000 fruits would be needed to reach 10,000 μg retinol. Thus if *Lycium* fruit is the baboons' only major source of vitamin A or its precursors, hypervitaminosis seems unlikely.

The foods of baboons present hazards other than toxins. I have mentioned the sharp spines of the calyx of *Solanum coagulans* (fig. 9.8) and the thorns guarding the fruit of *Azima tetracantha*. Umbrella trees have short, recurved thorns. On one occasion an infant fell asleep in its mother's lap as she fed in one of these trees. As the infant relaxed, its head fell back until one of its ears was pricked by a thorn of the tree. The infant screeched. Its mother pulled it to her, and in the process the recurved thorn tore through the edge of the infant's ear. Similar tears on the ears of several other baboons may have been acquired in this manner rather than (as we at first thought) in fights.

When baboons pull up a grass plant, they sometimes brush off the adhering dirt. This behavior doubtless reduces wear on the molar crowns (Washburn and DeVore 1962), though that effect would be more significant in areas with gritty soils rather than the very friable types that predominate in Amboseli, which break down into fine dust. Even so, twenty-year-old females in Amboseli have very worn teeth (J. Altmann, pers. comm.). Fibrous and woody plants present a similar problem. Beyond that, for animals with a limited ability to digest fiber and whose foraging may be limited by gut capacity, large volumes of tough material may be detrimental.

Some hazards are indirect. The grass *Sporobolus rangei* grows in waterless areas with no shade, and long periods of feeding on its corms present the dual problems of dehydration and heat prostration. Conversely, the grass *Cynodon nlemfuensis* and three of the baboons' favorite shrubs, *Withania somnifera*, *Salvadora persica*, and *Azima tetracantha*, grow in areas where dense foliage would provide good hiding places for leopards.

Fig. 9.7. Eno (age 36 weeks) eating *Lycium "europaeum"* fruit.

Fig. 9.8. *Solanum coagulans,* its fruits covered with spiny calices. B. Ozzie (age 18 months) eating *Solanum coagulans* fruit, retracting his lips to avoid the thorns. The dark specks on his whiskers are the tiny fruits of "sticky-fruit," *Commicarpus pedunculosus.*

Fig. 9.9. *Balanites pedicellaris.*

I have dealt here only with the immediate effects of toxins and other hazardous food components. In the next chapter I will consider their effects on
dietary diversity and on foraging behavior.

Still largely unexplored is the possibility that some plants eaten by baboons
are toxic to their parasites and other pathogens, that is, that some plants have
medicinal value, as has been demonstrated or suggested for several other primates (Glander 1994). Phillips-Conroy (1986) has discussed this possibility,
with particular reference to the effects of a saponin, diosgenin, in the fruit of
Balanites aegyptica on bilharzia (schistosomiasis). At Amboseli, baboons eat
the seeds and presumably the fruit of *Balanites pedicellaris* (fig. 9.9), but
bilharzia does not occur there. At Mikumi National Park, Tanzania, yellow baboons eat the fruit, exudates, bark, and seeds of this plant (Norton, Rhine,
Wynn, and Wynn 1987).

A Note on Methods

What is the evidence that the foraging of our subjects is selective, and how can
the sensory basis for their discriminations be discerned? As I mentioned ear-

lier in this chapter, the claim that an animal is foraging selectively is, at a minimum, a claim that it is not feeding at random from the foods available to it. By means of concurrent habitat and foraging samples, Barton, Whiten, Strum, Byrne, and Simpson (1992) demonstrated selective foraging at this level in anubis baboons in the Laikipia area of Kenya.

Evidence for finer discrimination can be obtained if animals feed on one food in preference to another food that is close to it. The two foods may be physically close, as when yearling baboons eat the green seeds of umbrella trees yet reject both the pods and the seed coats, which they must handle to get at the seeds. Or they may discriminate between foods that are structurally similar, either because they are taxonomically close, as when baboons feed on the fruit of *Solanum coagulans* but reject the very similar fruit of *Solanum incanum,* or because they are products of convergent evolution, as seems to be the case for many grasses and sedges that grow intermixed on the African savannahs and appear very similar although they are from different plant families—for example, *Sporobolus rangei* and *Cyperus obtusiflorus.*

At a minimum, selectivity seems to require that the foods be available to the animals at the same time of year, for otherwise feeding more on one food at one time of year than on another at a different time may simply reflect differences in the alternative foods available at each time. Consider, for example, that Amboseli baboons and vervets both spend more time feeding on the blossoms of umbrella trees than on those of the fever trees, as described above. Yet these two species bloom at different times of the year, so these results, however suggestive, are not literally an indication of a preference, since the animals are choosing not between them but between each of them and concurrently available alternatives; similarly for the green seeds of the two species. The possible role of phenols and tannins as feeding deterrents should be investigated experimentally.

The other question is, On what basis do selective foragers make their discriminations? In field studies, the common method for answering that has been to search for components of foods that correlate with particular foods being selected or rejected. Such studies have used univariate correlation or various multivariate methods (see below, in the discussion of other baboons). However, the possibility of false correlates, illustrated above by the example of acacia gums, is a problem in all such studies. Just because some toxin or other detrimental food component is lower, or a nutrient is higher, in the foods on which an animal feeds preferentially does not mean that this toxin, or nutrient, is the basis of the animal's preference. Such correlates of food selection can suggest possible cues, but other, covarying food components may be involved and, indeed, may be the primary basis for the discrimination.

Experimental manipulation of a putative cue, holding other potential cues constant, is the most direct approach, but it is seldom feasible in the field. In some cases laboratory experiments on diet selection can be very suggestive. For example, mammals can be conditioned to only a very few nutrients. Those for which conditioning attempts have repeatedly failed are unlikely to be the basis for dietary selection. Some further comments on food selection cues are included below, in the discussion of other baboon populations.

Another way of treating correlates of food selection is to recognize that they are *consequences* of selective foraging, even if they are not necessarily causes of or cues to that selection. Here the student of naturalistic behavior is on more secure ground. For example, as I noted above, baboons and vervet monkeys in Amboseli eat more fever tree gum than umbrella tree gum, and as a consequence of that selection, whatever its sensory basis, they take in smaller amounts of phenolics and condensed tannins than if they consumed these two gums indiscriminately or in equal amounts. Looking at the dietary consequences of selective foraging is the tack I have taken in this book.

Diets of Amboseli Vervets and Baboons

Comparison of Amboseli baboons with the local vervet monkeys is particularly enlightening (cf. S. Altmann and J. Altmann 1970; Hall 1965; Struhsaker 1967a). They are both members of the same subfamily (Cercopithecinae) and have almost identical repertoires of behavior. They have similar social systems (permanent, multimale, multifemale groups), similar sexual dimorphism (adult males have long canines and weigh about twice as much as adult females), and similar gross morphologies. Their geographic distributions in Africa overlap extensively, and in areas in which both species occur, they have overlapping home ranges. In Amboseli, both baboons and vervets are omnivores. They eat many of the same foods and drinks from the same waterholes. They are, in turn, preyed on by many of the same predators. They sleep in the same species of trees, sometimes in the same grove. They respond to each other's alarm calls to the same mammalian predators, and youngsters of the two species occasionally play together.

The numerous similarities between baboons and vervets reflect their common heritage combined with their common experiences. Conversely, many of their differences probably reflect different solutions to the same set of environmental problems. An alternative hypothesis, that some of the differences reflect adaptations to different characteristics of the habitat (S. Altmann 1974), may be applicable to many of the baboons' adaptations for harvesting certain foods, such as grass corms, that vervets appear physically unable to utilize and to those differences between the species that enable baboons to move with near impunity across open terrain, far from trees. However, the distinction between these alternatives—adaptations to the same or to different problems—may be mere artifacts of how the investigator views the world. The "problems" posed by the environment are not natural units, they are human inventions (Lewontin 1979). So, for example, the size and strength that enable baboons to utilize grass corms can be regarded either as a solution to a special problem, digging, that vervets do not encounter or as an alternative solution to the common problem of finding adequate sources of nutrients during the dry season. The important fact is that baboons, unlike vervets, feed on the corms of grasses and sedges, and that their ability to do so depends on certain physical attributes that vervets do not share.

At the time that my 1975–76 field study was under way on the baboons of Amboseli, David Klein (1978) was studying the diets of the vervet monkeys in the same area, using methods very similar to mine. I shall now indicate some of the dietary similarities and differences between baboons and vervets. Reasons for the differences will be considered in the next chapter.

Surprisingly, vervets spent less time feeding than did baboons. The proportion of daytime devoted to feeding in adult female vervets (Klein 1978, table 2.9) ranged from a low of 23% (June 2–24) to a high of 33% (December 15–January 18), whereas for Amboseli baboons foraging by adult females of Alto's Group took 45% of daytime (Bronikowski and J. Altmann 1996, table 2, mean for 1982–91, extrapolated to eleven-hour day) and feeding by the weanlings I sampled during Klein's study year averaged 36.4% (240.5 minutes out of 660 potential foraging minutes per day, table 6.5, seasonally adjusted diet).

In table 9.6, I have listed the foods eaten by adult female vervets in Amboseli (Klein 1978, table 2.2). Of these fifty-five foods, all but eighteen were also eaten by the young baboons in my samples, and twenty of the fifty-five were among the baboons' core foods, that is, the fifty-two foods ranking highest by amount of foraging time. Furthermore, of twenty-two "major" vervet foods (Klein 1978), eight also were among the top twenty-two baboon foods, judging by foraging time. The foods eaten by both species account for two-thirds of the foods eaten by vervets but only a quarter of the foods eaten by baboons. That is, baboons eat almost everything that vervets do and share many of their preferences, but many baboon foods are not eaten by vervets. The diet of Amboseli baboons includes about 2.5 times as many foods as the diet of Amboseli vervets. Perhaps coincidentally, this is roughly the ratio of body weights in the two species. (In calculating the diet ratio, I took into account that I divide food categories more finely than does Klein, e.g., by degree of ripeness, so that I tend to have 2.2 categories for each corresponding one of his.)

Furthermore, the diets of Amboseli vervet monkeys show a degree of temporal specialization that is apparently greater than that of Amboseli baboons, although the appropriate analyses are lacking. The vervets "tended to concentrate on two or three important foods on a particular day" (Klein 1978, 54). Beyond that, vervets in the group in Klein's study that had both species of acacia trees within their home range "focused their feeding almost entirely on the flowers of *Acacia tortilis* from 15 January to 11 April" (ibid., v). During that period, these flowers constituted at least 85% of the vervets' feeding time (ibid., 63). Unfortunately I do not have at hand the comparable feeding values for baboons during this period, but I doubt they were so concentrated.

Looking at foods eaten by one species but not by the other, particularly major foods, is revealing. (On the other hand, some of the reported differences in rarely eaten items probably are small-sample errors.) Let's begin with the few foods commonly eaten by vervets but seldom or never eaten by baboons. Only the vervets eat the succulent base of new fever tree thorns, and these appear to be fairly nutritious (see S. Altmann, Post, and Klein 1987 for documentation of all statements here about nutritive values). Similarly, the vervets eat the young leaves of both acacias in the area, and these too are nutritious. I have several records of young baboons eating young fever tree leaves, but only once did I see one try an umbrella tree leaf. A watery exudate of umbrella trees that sometimes collects at the junction of branches (fig. 9.10) was used by vervets but not baboons and may have been the primary source of water for some vervet groups (Klein 1978). Vervets eat the cotyledons from sprouting fever tree seeds; so far as I know, yearling baboons do not, and I have only one record of a baboon eating an umbrella tree seedling. Acacia seed pods, even mature ones, are eaten by older baboons (Barton, Whiten, Byrne, and English 1993; Muruthi 1988) but not, so far as I know, by the yearlings.

As I mentioned earlier, young baboons stop feeding on the fruit of devil's thorn once the spiny fruit wall begins to harden, but vervets do not: they break the stiff, short spines in the mouth before mastication. Vervets commonly eat the herb *Amaranthus graecizans,* and it has some moderate nutritive value. So far as I know, baboons do not eat this plant, although it may have been among my unidentified foods. Vervets but not baboons eat the flowers of *Solanum coagulans.*

The list of foods eaten by baboons but not vervets is long: about two hundred foods! Many of these fall into two clear-cut categories. First, baboons, unlike vervets, sometimes eat the meat of snails and of several vertebrates and the eggs of birds (table 4.1; also S. Altmann 1970; Hamilton and Busse 1982; Hausfater 1976; Strum 1983). Second, as already noted, baboons dig up and feed on the corms of several species of grasses and sedges and also the bulbs of lilies. Vervets probably are not strong enough to do this (Struhsaker 1967a), and the baboons' favorite corms, from *Sporobolus rangei,* occur primarily on the open plains, often distant from trees, where vervets seldom go and rarely linger. Finally, some species that baboons utilize for food may have been ab-

Fig. 9.10. Watery exudate in crotch of branches in an umbrella tree, *Acacia tortilis,* being used by vervet monkeys.

sent from the much smaller home ranges of the two vervet groups that Klein studied. Some of these foods occur in open, treeless areas, perhaps beyond the reach of most vervet groups (e.g., *Cyperus obtusiflorus, Ornithogalum donald-sonii, Rhamphicarpa montana, Sporobolus rangei*). Others have very localized distributions in Amboseli and may be utilized by other vervets in whose home ranges they occur (e.g., *Maerua angolensis, Asparagus ?africanus, Aloe* sp., *Opilia campestris, Solanum coagulans*).

Some interesting differences exist between vervets and baboons when they feed on the same foods. Usually, the parts of a plant that are eaten by both primate species are extracted to the same extent; for example, when feeding on the ovaries and young fruits of *Trianthema ceratosepala* both primates remove the outer fruit wall. However, at least two exceptions occur. Vervets eat just the succulent inner portion of *Cynodon nlemfuensis* stolons, whereas baboons eat the inner and outer portions; and when feeding on the open seedheads of this grass, vervets eat only the pliable, white inner portion at the base of the seed-head, whereas baboons eat the entire head. In both cases, and in the ability of vervets to remove thorns from devil's thorn fruit, described earlier, the more fine-grained selectivity of vervets probably reflects their smaller body size, which makes such microextraction practicable.

Diets of Other Baboons

So much for comparison of Amboseli baboons and vervets. What about the diet of other baboon populations and species? The available literature is consistent with the characterization of baboons, presented here, as selective omnivores, searching for the most palatable and nutritious foods available in their habitat. So, for example, *Papio ursinus* in southern Africa (Hall 1963), *P. anubis* in central Kenya (Forthman Quick and Demment 1988; Barton, Whiten, Byrne, and English 1993), and *P. cynocephalus* in Mikumi National Park, Tanzania (Norton, Rhine, Wynn, and Wynn 1987), and Amboseli (this study), each eat many foods that the others do not, for a simple reason: the floras of these three areas are different, and each provides a different array of prime foods from which the baboons select. For example, of the 185 species of food plants fed on by *P. cynocephalus* in Mikumi National Park, only seven (*Balanites aegyptica, Capparis tomentosa, Acacia xanthophloea, Abutilon mauritianum, Sporobolus fimbriatus, S. ioclados,* and *Salvadora persica*) are also sources of food for Amboseli baboons.

Similarly, of thirty-eight baboon plant foods at Chololo, Kenya, that have been identified to species (Barton, Whiten, Byrne, and English 1993), only eight also provide food for Amboseli baboons: *Cynodon dactylon* (if actually *C. nlemfuensis*), *Cynodon plectostachyus, Tragus bertorianus [berteronianus], Amaranthus graezicans [graecizans], Tribulus terrestris, Asparagus africanus, Lycium "europaeum,"* and *Acacia tortilis.* Surprisingly, these authors list as a baboon food the green and ripe fruits of the toxic Sodom apple, *Solanum incanum,* which are not eaten by baboons in Amboseli, though these plants occur commonly there. On the other hand, most of the species and even genera used by Mikumi and Chololo baboons are not known from Amboseli.

Conversely, Nagel (1973) was unable to detect any major difference between adjacent hamadryas baboons (*Papio hamadryas*) and anubis baboons (*P. anubis*) in the plants they eat.

Even within a small area, floristic differences can lead to differences in diet. For example, among both baboons (Muruthi 1988, 1997; Stacey 1986) and vervets (Klein 1978) in Amboseli, between-group differences in diet exist, even for animals of the same age-sex class, and these seem to be largely due to local differences in the plant communities within the home ranges of the groups.

To what can one attribute these similarities and differences? When comparing diet selection in anthropoid primates of the same genus, one has difficulty evaluating the relative roles of individual experience, of experience transmitted from one individual to another, even across generations, and of genetic differences in perception or motor skills. Certainly the ability of many anthropoid primates to adopt diverse new foods, such as cultivars and tourist handouts (e.g., Muruthi 1988, 1997; Fa and Southwick 1988), and to do so even within one generation, indicates that for an intelligent omnivore, major dietary shifts can occur without genetic change. Conversely, among baboons, none of the species differences in size, pelage, limb proportions, dentition, or other anatomical features bear any obvious relation to dietary differences, though subtle correlations may exist. Population and species differences in baboon diets, when they do occur, probably primarily reflect baboons' adaptability at exploiting available resources in whatever habitat they are in, rather than being evolved local adaptations to distinctive features of each habitat.

Dietary profiles for the various species of baboons, based on time spent feeding on various foods (table 9.7), show some major commonalities in this genus, despite the variability. (For reasons given in chapter 4, these time budget differences may be poor indicators of the nutritional and dietary contributions of various foods.) Although predation on vertebrates occurs in most baboon populations and insect eating probably occurs in all, in none is animal matter a major component of the diet. Indeed, only at Gombe do baboons spend more time on animal matter than the Okavango baboons and all other populations spend less than 5% of their time on it. The high mean value at Okavango resulted from an outbreak of Mopane scale insects, during which one group fed on these insects for 72% of their feeding time. At other times insects account for less than 2% of their feeding time. A similar shift in diet occurred in Amboseli baboons one year in which an enormous outbreak of "army worms" (caterpillars) occurred. In short, baboons are opportunistically faunivorous.

As for plant material, reproductive parts (flowers, seeds, fruits) occupy the majority of feeding time in all baboon species and populations except two (yellow baboons at Ruaha, chacma baboons [*Papis ursinus*] in the Drakensbergs). On average, plant reproductive parts (flowers, fruits, seeds) account for 47% of baboon feeding time, and in one case, *P. papio* at Mount Assirik, for 82%. In almost all populations, appreciable time was spent digging up and eating bulbs, corms, and other subterranean plant parts: 25% on average. Leaves account for an additional 18% of baboons' feeding time. Judging by foraging time, almost all species and populations of baboons rely extensively on grasses and trees.

The pattern described above is consistent with baboons' combined arboreal and terrestrial way of life. The observations of Norton, Rhine, Wynn, and Wynn (1987, 115) confirm our own about the feeding niche of baboons: "Eclectic feeding and selectivity are a powerful combination; together they probably go a long way toward explaining the baboons' success over a large part of a continent with diverse ecosystems." Whiten, Byrne, Barton, Waterman, and Henzi (1991) characterize the baboons' foraging strategy as "adaptably broad, yet selective." Indeed, the baboons' eclectic omnivory has been

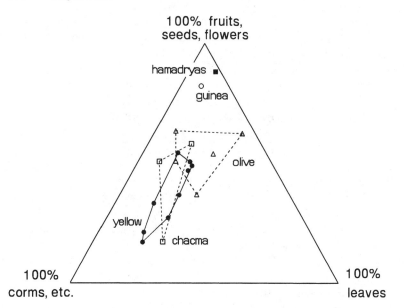

Fig. 9.11. Percentage of feeding time that wild baboons of several populations devote to three classes of foods. Percentages are plotted on three axes, each perpendicular to one of the sides. Because the triangle is equilateral, the sum of the distances from any point within the triangle to each of the three sides is equal to the height of the triangle, that is, to 100%. Foods in the first category, fruits, flowers, and seeds, constitute the sexual reproductive parts of the plant. Leaves, including grass blades and meristem (blade bases), make up the second class. All other foods, both plant and animal, make up the third category, which in almost all baboon populations consists predominantly of underground plant parts: roots, bulbs, tubers, corms, and rhizomes. Data from table 9.7, with one data point plotted for each study. For similar plots based on feeding time and amount of food consumed by various primates, see MacKinnon and MacKinnon (1978) and Chivers and Hladik (1984, 1994); their categories differ somewhat from those used here.

commented on by virtually all who have studied their behavioral ecology. The foundations of baboons' omnivory are the subject of the next chapter.

When the baboons' foods are grouped into three classes and the percentage of time spent on each is plotted on a triangular graph (fig. 9.11), similarities and differences among the species become apparent. For yellow and chacma baboons, the ranges of food mixes are virtually the same. Olive baboons, which generally live in areas of higher rainfall and correspondingly greater plant productivity, tend toward a diet with a smaller component of corms and other underground plant parts. Hamadryas and guinea baboons differ markedly from the others in consuming much larger proportions of seeds and fruits; in both, feeding on these takes at least two-thirds of their feeding time. Yet the habitats of the two study sites are markedly different: desert and montane forest, respectively. I do not know whether the species differences indicated by figure 9.11 represent actual differences in competence rather than just habitat differences in where, for historical reasons, these species now live or where they have been studied; but as I indicated above, I suspect the latter is of greater import.

Noteworthy among field studies of baboon foraging is an ongoing series of projects carried out by past and present members of the Scottish Primate Research Group at the University of St. Andrews (Barton 1989, 1990, 1993; Barton and Whiten 1993, 1994, 1995; Barton, Whiten, Strum, Byrne, and Simpson 1992; Barton, Whiten, Byrne, and English 1993; Whiten, Byrne, Barton, Waterman, and Henzi 1991; Whiten, Byrne, and Henzi 1987; Whiten, Byrne, Waterman, Henzi, and McCullough 1990).

Whiten, Byrne, Waterman, Henzi, and McCullough (1989) studied chemical correlates of diet selection by montane chacma baboons (*Papio ursinus*) in the Drakensberg area of South Africa. By means of discriminant analyses they compared the composition of foods eaten with those nonfoods that were overtly rejected by the baboons or that looked similar but were not consumed (e.g., an uneaten corm in the same genus as a species whose corm is eaten). They refer to the resulting discriminant function, expressed in terms of standardized weighting coefficients, as a "food value score." For food leaves versus nonfood leaves, the discriminant function detected foods without error based on composition. Protein content was the only food component with a positive weighting coefficient; negative weightings were given to fiber, alkaloids, and total phenolics. Lipids, starch, total carbohydrates, and water content did not figure in discriminating leaf foods from nonfoods.

When alkaloids were eliminated from the analysis of leaves, fiber was the only discriminating variable, and it misclassified two nonfoods and one food. That is, the value of a plant component for deciding whether it would be eaten by baboons depends on which other components are included in the analysis. Discriminant analyses were also carried out for underground storage organs (corms, tubers, bulbs), for the swollen bases of shoots, and for inflorescences.

An instructive result came from an analysis of the baboons' winter foods. The baboons eat the ripe but not the unripe fruit of *Cussonia,* yet the discriminant function classified them both as foods. The authors speculate reasonably that the unripe fruit provides a distasteful element that was not included in their analyses and that prevents the fruit from being eaten before it is ripe.

These authors speak of "rules underlying selective foraging," of "nutritional choices underlying [the baboons] selection," and of "rules about payoffs which guides their foraging behavior" (Whiten et al. 1989). However, they indicate that their suggestive terminology "does not imply, of course, that the baboons can directly distinguish chemical composition; presumably, the selective decisions which clearly are made are based on proximal sensations of taste and texture" (ibid., 17).

Selective foraging probably depends on two quite different sets of cues. The first are those that initially lead the animal to avoid some potential foods and eat others. These may involve not only taste and texture but many other factors, including social learning from other members of their group, the long-term effects that eating certain foods have on health (as exemplified by negative conditioning to poisonous ingredients), and inborn specific hungers. Second, there are sensory cues by which an animal detects and recognizes its foods without having to taste them: cues such as the array of colors, odors, and fleshy arils by means of which fruits with animal-dispersed seeds attract to their ripe fruit the animals they depend on, and more subtle cues for recognizing other foods. Whether or not any given array of compounds that enables us to distinguish foods from nonfoods also provides the cues animals use in their food selection, identifying such arrays is valuable. It reveals the dietary results of food selection: its consequences if not its causes.

More recent studies by this group of investigators have focused on a population of anubis baboons (*Papio anubis*) at Chololo, on the eastern edge of the Laikipia Plateau in central Kenya. In one project (Barton, Whiten, Strum, Byrne, and Simpson 1992), Barton carried out systematic monthly samples of food availability combined with focal samples of foraging and ranging behavior. The study documented extraordinary seasonal fluctuations in food availability that resulted, as in Amboseli, from alternating rainy and dry seasons. As a result of the time lag in the plants' responses to rain, peak food biomass occurred in June, near the end of the long rains and about two months after the peak in rainfall. Much of the rainy season increase was due to herb-layer foods (grass blades, herb leaves, shoots, flowers, and fruits). As the herb layer's biomass increased, the extra distance per day that the baboons walked, above the estimated minimum (the average distance between sleeping trees), decreased, and on a log-log scale, this relation was linear.

The parts of the baboons' home range where they spent the most time changed with the seasons: the proportion of time they spent each month in the different habitat zones was significantly correlated with the productivity in each zone of several acacia species (*Acacia etbaica, A. mellifera, A. tortilis*) that are major sources of food for them. Not surprisingly, zone occupancy time was also correlated with the proportion of feeding time the baboons devoted to these species, especially *A. etbaica* ($r = 0.71$). Although the amount of time the baboons spent in various quadrats was negatively correlated with the distance to waterholes—that is, they tended to stay near waterholes—this association fell off during periods of rain, when water was more widely available.

To test the hypothesis that a negative relation exists between the abundance and the quality of food items, Barton et al. calculated correlations for the major food classes between mean monthly food biomass and several phyto-chemical measures: protein, fiber, condensed tannins, and protein/(fiber + condensed tannins). None were significant (Barton, Whiten, Byrne, and English 1993). This analysis leaves open the question whether, in a given month, the digestibility (e.g., crude fiber content) of various potential foods is positively correlated with their abundance, as in Demment and van Soest (1985). This latter is a correlation among simultaneously present foods, not a comparison across seasons.

That last-cited study by Barton et al. provides extensive information on the chemical composition of baboon foods at Chololo. Despite some differences in sites and methods, analysis of foods that are also eaten by Amboseli baboons showed reasonable agreement. As a result of that study at Chololo, of the work of Norton, Rhine, Wynn, and Wynn (1987) at Mikumi National Park, Tanzania, and of ours at Amboseli, Kenya, we now have three baboon research sites at which extensive analyses of diet and diet composition have been carried out. This data bank provides outstanding opportunities for comparative and in-depth analyses of baboon foraging.

10

Les animaux sauvages sont des gourmets.

—H. Gillet, 1964

Baboons are excellent generalists. They combine mental facilities and physical morphology to exploit a wide range of conditions. They are capable of remembering details of the location of food items, and are sensitive to the production cycles of plants and life cycles of animals. Their manual dexterity allows both the separation of nutritious portions of plants from toxic or indigestible portions and the efficient harvest of a wide size range of food items. Baboons can exploit all levels of the habitat from below ground surface to the treetops.

—Montague W. Demment, 1983

How to Be an Eclectic Omnivore

B aboons are among the most successful of primates. This is so by any of a variety of biological criteria for success: wide geographic distribution, large biomass, ability to live in diverse biological communities and, within each, to utilize many local habitats both arboreal and terrestrial, to withstand considerable fluctuations in local conditions, and to deal successfully with predators and competitors (S. Altmann and J. Altmann 1970).

At the heart of the baboons' success is their food-getting ability. Their foraging has three salient characteristics, each of which I have documented in preceding chapters. First, they feed on a great diversity of foods, warranting their designation as omnivores. Reported diets of baboons span several phyla of plants and animals. When feeding on plants, they consume many plant parts: flowers, arils, seeds, gums, roots, tubers, corms, rhizomes, stolons, and so on. Second, they are highly selective, both taxonomically and in terms of plant parts. They consume the fruit of one plant but ignore that of a congener. They feed on some acacia seeds but not others, even from the same pod, then laboriously remove the seed coat from those they consume. Available evidence indicates that their selectivity results in a diet that is of higher nutritive value and less toxic or otherwise harmful. Third, they are flexible and opportunistic. They incorporate new foods into their diet when these become available, for example, as a result of seasonal changes or movements of the baboons into new habitats. This combination of dietary breadth, selectivity, and flexibility means that they are omnivores of a particular kind. They are eclectic omnivores. In the terminology of Schoener (1971), they are generalists, though that term would apply to ensemble feeders as well.

In chapter 8 I described some of the peculiarities of this food niche. In this last chapter, I want to consider several key attributes of baboons that make them so successful at exploiting it. Along the way, I shall touch on several important aspects of these animals that are at present poorly understood and that provide opportunities for research. The focus here is on the capacities of adult baboons—on what young baboons become if they survive. From time to time I shall touch on the extent to which the capabilities of yearling baboons fall short of those of adults. In most cases the sources of their shortcomings will be obvious: smaller size, less experience, and so forth.

As I indicated at the beginning of chapter 4 and documented in the rest of that chapter and in chapters 6 and 8, baboons achieve their high-quality, low-bulk diet by being highly selective omnivores. Baboons appear to have carried to the extreme among cercopithecine primates the trade-off of quantity in favor of quality, independent of habitat quality. Indeed, if it were not for the great effort baboons exert to obtain their food, we might caricature them as simian gourmets.

The success of baboons at exploiting this feeding niche is not attributable to any one trait but rather depends on an entire complex of traits, all of which baboons share, to varying degrees, with other cercopithecine primates. In what follows I shall discuss the implications of several key features of baboons: their size and strength, their sexual dimorphism and fighting ability, their locomotor and manipulative abilities, their cheek pouches, binocular color vision, and keen senses of taste and smell, their memory and intelligence, and their multimale, multifemale social groups. Many of these attributes are directly related to body size in mammals. That adult baboons are among the largest of living primates is therefore significant. Indeed, they are exceeded in body mass only by the great apes (orangs, gorillas, chimpanzees) and humans (Napier and Napier 1967).

Yearling baboons, however, are considerably smaller. That raises the question of how they too survive. I suggest three answers. First, many of them don't, and in some cases the disadvantages of being smaller than the adults may be decisive. Even those that survive are sometimes handicapped by their smaller size. For example, the smooth trunks of fever trees can be difficult for them to climb when very young, and they cannot pull up corms as large as those that adults feed on. Second, during the first year and a half, they fall back on mother's milk to supplement whatever they can glean on their own. Third, they rely on adults for some things they cannot do themselves. Lacking the adults' defensive abilities, the youngsters rely on them for safety. Without the adults' ability to traverse great distances, they ride on their mothers for part of the day's journey. Lacking the adults' years of experience, they rely on them for decisions about where to go, what to eat, and which trees to sleep in. At night they gain heat by snuggling into the mother's ventrum; during the day, on the open plains, they sometimes seek the shade of adults.

Among mammals, annual natality decreases with body size at about the same rate that life expectancy increases. The former scales as $M^{-0.33}$ (Western 1979), the latter as $M^{0.35}$ (Calder 1983). Consequently, lifetime production of offspring, which is approximately the product of annual natality and life expectancy, is essentially independent of body size in mammals (Fleming 1975; Eisenberg 1981). Age at sexual maturity, gestation length, weaning age, and interbirth interval all increase with body size; litter size, where variable, decreases (Clutton-Brock and Harvey 1983). As a result, age at first reproduction and mean generation time have a major impact on the intrinsic rate of natural increase (Cole 1954), which declines with body size in mammals (Bonner 1965; Fenchel 1974).

These reproductive costs associated with major size increases suggest there may be compensating advantages to large size in mammals, which may vary from group to group. Clutton-Brock and Harvey (1983) review the evidence for several. For example, increased size may enable mammals to reduce heat loss, to withstand periods of acute food shortage, to produce larger and stronger offspring, to reduce predation, to capture larger prey (carnivores), to feed on tougher material (herbivores), to travel farther in search of food and water, to digest food more rapidly and thus to subsist on food of lower nutritive value, and to compete successfully for food resources. Might some of these potential advantages of large size accrue to baboons? That is the subject of this chapter. A discussion by Temerin, Wheatley, and Rodman (1984) of the relation between body size and foraging in primates provides a different perspective on this topic.

Heat Conservation

Reduction of heat loss is commonly regarded as a major advantage of large size in endotherms. Even in a mammal's thermoneutral zone, a large part of its daily metabolic expenditure is required to maintain its body temperature. Across species of eutherian mammals, basal metabolism, which is largely the energy flux required for temperature regulation, increases more slowly than body mass, approximately in proportion to body mass to the three-quarter power (Kleiber 1961) or even less rapidly, judging by the much larger samples used by Hayssen and Lacy (1988), who got a slope of 0.696 for eutherian mammals.[7] This implies that at rest a large mammal generates (and loses) less heat per unit body mass than does a small one, and on that ground large size is said to be advantageous.

However, a reduction in heat turnover per unit mass appears to have no intrinsic value, and in fact the total heat turnover in large mammals—evaluated per corpus rather than per kilogram—is on average greater than in small ones. Since lost heat in endotherms is replaced primarily through the catabolism of food constituents, the higher rate of heat loss in large mammals ultimately translates into an increased foraging requirement: energy costs are food costs.

Thus large size is thermally advantageous only if, *as a consequence of larger size,* the capacity for energy intake increases more rapidly than does the increased thermal production that larger size entails. Such an advantage would accrue even if, say, basal rate increased more rapidly than mass, so long as the increase in total metabolism was outstripped by the increase in foraging efficiency. To test this hypothesized advantage of larger size, it would not suffice to show that the diets of larger mammals have on average a higher absolute energy value. They must have, since the absolute heat loss of a larger mammal is greater, all else being equal. What must be asked is, How much of the increment in the energy value of the diet is attributable to larger size, and how does that value compare with the increment in heat loss? How one could answer those questions is not obvious; one possibility is described in the section on strength and competitive ability, below.

Starvation, Dehydration, and Insolation

The lower heat turnover rate (power per unit mass) in larger birds and mammals suggests that larger endotherms should be able to fast longer and thus be less susceptible to periods of acute food shortage. I do not know of any direct evidence for this in mammals, but it seems reasonable. Peters (1983) reviews evidence that larger birds take longer to starve and, by calculating the time required for the metabolism of large and small mammals to consume the energy stores of their tissues, shows that large mammals should be able to fast longer. Conversely, a very small mammal like a shrew must feed almost continuously to maintain its body temperature.

Similarly for water and heat, larger mammals probably can go long periods without water in arid habitats, and because their greater mass gives them greater thermal inertia, they are better able to withstand long periods of intense insolation without suffering from dangerously high body temperatures. For baboons, who often walk long distances through the middle of the day without shade or water, such benefits of large body size probably are of considerable importance. In Amboseli their problems are made more acute by back-radiation from very light-colored soils. Stelzner (1988), in a study of behavioral responses of Amboseli baboons to temperature and insolation, showed that at high temperatures the baboons did not preferentially move to parts of their home range in which shade was more available, but that upon getting into shade, they spent more time resting and less time walking. That is, baboons linger in shade opportunistically.

As is typical of arid regions, nights in Amboseli are chilly. Infant baboons, handicapped by their smaller size, huddle against their mothers.

Turnover Rates and Diets

Although the basal metabolism of placental mammals tends to increase approximately in proportion to body mass to the three-quarter power (Kleiber 1961), the individual values show considerable scatter about the regression line of metabolic rate on body size (McNab 1983). At least three components of mammalian reproduction—gestation length, postnatal growth rate, and young per female-year—appear to be related to metabolic rate independent of mass (McNab 1983). Furthermore, metabolic rate in mammals is correlated with diet, independent of mass. For example, according to McNab (1983), although most placental mammals that have been tested have metabolic rates within 10% of Kleiber's regression value, very low basal rates—less than 60% of the Kleiber value—occur among flying insectivores, ant and termite eaters, arboreal folivores, soil and litter omnivores, and a few desert herbivores.

Although no colobine primate was included in McNab's survey (1983), mammals of the four genera of arboreal folivores that were (*Phascolarctos, Trichosurus, Phalanger,* and *Pseudocheirus*) all have basal rates that are less than 89% of the Kleiber values for animals of their sizes. McNab (1983) has suggested that "the antiarthropod chemical compounds stored in many tree leaves may be a factor reducing basal rates of arboreal folivores. . . . [The] cost to the consumer for detoxification is great enough to reduce the value of a high rate of metabolism." The expectation, then, is that by comparison with colobine primates of the same size, baboons and other cercopithecines have higher metabolic rates independent of body size and, as a consequence, longer gestation periods and more rapid postnatal growth.

Digestive Capacity

Limits on body size in a group of mammals may be set by the abundance of food of the types the animals have the competence to digest efficiently. The gut capacity of mammalian herbivores increases nearly isometrically with body mass (Demment 1983), whereas basal metabolism and total metabolism increase more slowly (Kleiber 1961; Garland 1983), so that the ratio of metabolism to gut capacity (and presumably therefore to food processing capacity) is higher in smaller mammals. Consequently, rates of passage through the gastrointestinal tract in small herbivorous mammals must be higher than in large ones to meet metabolic requirements on the same quality of food. This rapid throughput sets a lower limit on the size for efficient microbial digestion of cellulose (Demment 1983; Clutton-Brock and Harvey 1983).

Thus small herbivores must select food with a relatively high nutritional value, whereas large herbivores can subsist on vegetation of low nutritional value. In fact, most large herbivores have no alternative. The most nutritious parts of plants are less abundant than the high-fiber, low-nutrient components, so that most large herbivores satisfy their large absolute requirements by consuming large volumes of food that has low concentrations of nutrients and high concentrations of fiber (Bell 1970, 1971; Gaulin 1979; Jarman 1972, 1974; Jarman and Sinclair 1979). According to Demment (1983), such material contains larger concentrations of lignins, which are indigestible, and this sets an upper limit on the size range within which a ruminant system is more efficient than a nonruminant one.

A similar restriction may apply to baboons. "My hypothesis," writes Demment (1983), "is that male body size [in baboons] is in part limited by the density of low fiber items. Increased body size would increase intake requirements that would force the utilization of a higher fibre category of plant material. Baboon guts are not designed to efficiently digest these higher fiber foods." That is, the largest baboons should be at a selective disadvantage as a result of being less able to satisfy their greater nutritional requirements.

Demment (1983) further argues that in the middle level of fiber contents, baboons are competing with ungulates, whose presence "exerts a competitive effect that increases the cost of larger body size in baboons." Such competition may be minimal, however, judging by our observations in Amboseli. Grazing ungulates are restricted to the aboveground portions of grass plants, but baboons are not, and I have indicated the importance to baboons of the subterranean corms of perennial grasses and sedges. By feeding on corms, baboons are destroying potential sources of food for grazers; this competition is in the opposite direction from that hypothesized by Demment. Also in contrast to ungulates, Amboseli baboons eat the blades of grasses and sedges only when green, that is, primarily during the rainy season. At that time the new growth of blades is so copious and widespread, and so few of the large migratory ungulates are then within the baboons' home ranges, that competition for this food between baboons and grazers must be very limited. In the long run, however, competition with grazing ungulates may have been sufficient to preclude the evolution of a true grazer among primates, with one exception: geladas (*Theropithecus gelada*), which live high in the Simēn Mountains of Ethiopia, where few ungulates occur (Dunbar and Dunbar 1975; Kawai 1979).

As for browsing ungulates, those primates that subsist primarily on dicot leaves, such as howlers (*Alouatta*) and colobines, forage primarily in trees at heights that are out of reach for ungulates. This is not true of baboons, however. Many of the leaves they eat are from forbs and shrubs that are eaten by various other mammals. The intensity of competition among them is unknown.

Do larger primates, like larger ungulates, eat food of higher fiber content? Although we know the fiber content of too few primate diets to be sure, the proportion of leaves in the diet has been taken as an indicator of the amount of dietary fiber by Clutton-Brock and Harvey (1977). They plotted body mass against proportion of leaves in the diets of numerous primates and claimed that a tendency exists for the proportion of leaves to go up with mass. However, the trend in the plotted data (ibid., fig. 1) appears to be almost entirely because the largest primate, the gorilla (*Gorilla gorilla*), is largely folivorous and the very smallest of primates eat virtually no leaves. At intermediate sizes over several orders of magnitude, the other primates show no obvious trend. Furthermore, orangs (*Pongo pygmaeus*), the second largest primates, are largely frugivorous (Rodman 1979).

Reduced Cost of Locomotion

In the literature on animal energetics, locomotion has repeatedly been said to be less expensive for large animals than for small ones. Do baboons, as the largest of monkeys, obtain benefits from lower energy expenditures for locomotion? As I show below and in a previous publication (S. Altmann 1987), the available literature on locomotion is inadequate to answer that question, either for baboons or for eutherian mammals in general.

Of the various ways the energy cost of locomotion has been defined (S. Altmann 1987), the energy required per unit body mass to move a given distance probably is the most widely used measure. It decreases with body mass, and that is true not only for running mammals but also for flying birds and swimming fishes (Schmidt-Nielsen 1972). "The significance of this relationship," writes Schmidt-Nielsen (1979, 194), "is that, for a running mammal, it is metabolically less expensive, and thus an advantage, to be of large body size." By this criterion, the energy cost of locomotion can be expressed in units of $J\,kg^{-1}\,km^{-1}$.

However, from an ecological standpoint, this measure is not a satisfactory indicator of the impact on the organism of energy expended on locomotion by mammals of various sizes (S. Altmann 1987). A moving animal does not just move a kilogram of body mass. It transports its entire body, whatever that weighs, and must pay the energy cost of doing so. That cost goes up with body size, not down. Yet so does total energy expended on all other activities. Thus the cost of locomotion per unit distance should be measured not in absolute terms but relative to total energy expenditure—or preferably, total nonlocomotor expenditure, because mammals do not all walk the same relative distance, so that locomotion adds a variable increment to metabolic expenditure.

Furthermore, with rare exceptions food is the source of energy for movement, so that any increase with size in energy expended on locomotion translates into an increase in foraging requirements (or possibly, a decrease in food energy deployed to other activities). Thinking in terms of foraging requirements makes clear why the energy costs of locomotion must be measured relative to total energy costs. An absolute energy requirement that an elephant could easily satisfy in a few minutes of foraging might represent a month's labor to a mouse.

How does total locomotor energy change with body size relative to other metabolic expenditures? For a mammal of any given size, the amount of energy devoted to locomotion, expressed as a percentage of nonlocomotor energy, can readily be calculated:

$$(10.1) \quad T(\%) = \frac{100 \left(\begin{array}{c}\text{energy}\\\text{increment}\\\text{per km}\end{array}\right) \left(\begin{array}{c}\text{km moved}\\\text{per day}\end{array}\right)}{\left(\begin{array}{c}\text{total daily}\\\text{metabolism}\end{array}\right) - \left(\begin{array}{c}\text{energy}\\\text{increment}\\\text{per km}\end{array}\right) \left(\begin{array}{c}\text{km moved}\\\text{per day}\end{array}\right)}.$$

The product on the right of the denominator—energy expended on locomotion—is, as shown below, a small fraction of total metabolism, so for our purposes it can be ignored, as I did in the following calculations. So modified, T becomes equivalent to Garland's (1983) *ecological cost of transport:* the percentage of total daily energy expenditure that is devoted to locomotion. (I prefer to call this the *ecological cost of travel* and to reserve the former term for carrying external loads, as primate mothers do with their infants.) Estimated mean values for the variables in equation (10.1) as a function of body size in eutherian mammals were obtained as follows. Energy increment per unit distance $= 10700 \, M^{0.684} \, \text{J}' \, \text{km}^{-1}$ regardless of speed (from $10.7 \, M^{-0.316} \, \text{J}'$ $\text{m}^{-1} \, \text{kg}^{-1}$ in Taylor, Heglund, and Maloiy 1982 multiplied by M). (Here and elsewhere in this section, $M = $ body mass in kilograms and $d = $ day. All other abbreviations are standard SI metric units.) Distance moved per day $= 0.875$ $M^{0.22} \, \text{km} \, d^{-1}$ (Garland 1983 for noncarnivorous, terrestrial mammals). Total energy expended per day is $8.95 \times 10^5 \, M^{0.762} \, \text{J}d^{-1}$ (Nagy 1994).[7]

Using these values, an estimate $\hat{T}$ of this cost can be obtained:

$$(10.2) \quad \hat{T}(\%) = \frac{100 \, (10700 \, M^{0.684}) \, (0.875 \, M^{0.22})}{8.95 \times 10^5 \, M^{0.762}}$$

$$= 1.046 \, M^{0.142}.$$

As can be seen from these values, the distance moved per day by placental mammals scales in proportion to body mass to the 0.22 power, and energy expended per unit distance scales in proportion to body mass to the 0.68 power.

Therefore their product, energy expended per day on locomotion, goes up as $M^{0.90}$; that is, it increases with size more rapidly than total metabolism but less rapidly than body mass. Larger mammals thus expend a larger percentage of their metabolism on locomotion than do smaller ones. Far from being better off, larger mammals are, according to these values, at a disadvantage in terms of the percentage of their available energy that is consumed by locomotion.

Beyond that, the expenditures for locomotion are quite small. For terrestrial eutherian mammals of 100 kg, their daily trek accounts for only 2.1% of their daily energy expenditure, according to equation (10.2). For a 10 g shrew, the results are even more extreme: locomotion accounts for a mere 0.54% of total energy expenditure, and even a 4,000 kg elephant expends only about 2.8% of its energy on locomotion according to equation (10.2). That is, the percentage of total metabolism that eutherian mammals devote to locomotion is greater in larger mammals but is in any case a trivial fraction of the total according to this estimate. For size differences within a species, a level at which natural selection operates, differences in the ecological cost of travel would be minute. If the estimate given in equation (10.2) for T is reliable and is also applicable to the ancestors of extant mammals, conservation of locomotor energy is unlikely to provide an explanation for evolution of large body size in eutherian mammals, including baboons.

Yet on several grounds estimates obtained from equation (10.2) appear to be grossly inaccurate (S. Altmann 1987). Consider two predictions that can be made based on the values given above but that are unreasonable. First, the mean daily travel time budget of mammals (hours per day spent traveling) should equal the mean day-journey length divided by mean travel velocity: hours/day = (km/day) (km/hour)$^{-1}$, assuming independence. Mean day-journey length in noncarnivorous terrestrial mammals has been estimated at 0.875 $M^{0.22}$ km d^{-1} (Garland 1983). For mammals, the mean velocity of walking is 0.33 $M^{0.21}$ m/s, according to Peters (1983), based on Buddenbreck's (1934) regression of walking speed in captive mammals on body length combined with mass-length relationships. Note that these regressions for mean day-journey length and walking velocity scale to essentially the same exponent of mass, which implies that the mean daily travel time budget should be essentially independent of size and very nearly 0.74 hours. Although I do not know of a compilation of travel time budgets in mammals, this value is much too low for every mammal I am familiar with. That is, either mammals walk farther than Garland's regression indicates, or Buddenbreck's regression for walking speeds gives values too high for the average velocity at which animals traverse their home ranges, or both.

Next, consider that total metabolism of eutherian mammals is on average about 3.3 times basal metabolism,[7] and the difference in adults represents primarily energy expenditure beyond that required for maintaining body temperature. For most mammals, locomotion probably is the largest energy-consuming activity, as well as one of the most time-consuming. Yet, according to the estimate above, locomotion results in a negligible energy expenditure compared with total metabolism. If the rest of the animals' time is spent on activities requiring even lower metabolic increments over basal than does locomotion, these activities in sum would not equal the difference between basal and total metabolism.

Let me illustrate. In Amboseli, an average adult female baboon weighing 11 kg (J. Altmann 1980) walks 6.10 km per day (Post 1981) (versus 1.48 predicted by Garland's regression formula). According to Post (1981), this travel accounts for 23.9% of the 11 daylight hours during which they are active, that is, for 2.63 hours per day (compared with the predicted 0.74 hours, above), with an average travel velocity of 2.4 km/hour (compared with 1.97 km/hour according to Buddenbreck's regression). That is, baboons walk longer, faster, and farther than average for mammals of their size—not a surprise. However, according to equation (10.2) but using the above day-journey length of 6.10 km for an adult female baboon, travel accounts for only 6.0% of the female's total energy budget, and this is unreasonably low, as the following calculations show. For a mammal of 11 kg, basal metabolism requires 1,143 kJ/d and total metabolism[7] requires 5,563 kJ/d, so that the energy of activity (nonbasal metabolism) is 4,118 kJ/d. Energy expended by an 11 kg eutherian to move 6.1 km is, from the regression given above for energy of locomotion, 10700 $M^{0.684}$ J/km $\times$ 6.1 km/d $\times$ 10^{-3} kJ/J = 336.5 kJ/d. That is, although locomotion occupies 23.9% of the eleven-hour active period, it accounts for only 8.2% of nonbasal energy. So the remaining 76.1% of the active period must account for the remaining 91.8% of nonbasal energy expenditure. (During the other thirteen hours, primarily nighttime, the animals usually are asleep or dozing and therefore presumably are near or somewhat below basal metabolic rate.) This would require that during the remaining daytime, the energy expenditure rate would have to be about 3.4 times greater than when locomoting! This is surely false: the rest of a female baboon's time budget during the eleven daylight hours consists of feeding (46%), resting (22%), and socializing (7%) (Post 1981).

In short, the available data on energetics and locomotion appear to grossly underestimate the fraction of metabolic energy mammals expended on locomotion. For now we cannot say whether baboons have lower costs of locomotion relative to their total metabolism than do smaller primates.

If, as I strongly suspect, the estimate provided by equation (10.2) greatly underestimates the ecological cost of travel, then either the parameter estimates in the numerator are too small, the denominator is too large, or both. I strongly suspect that both factors in the numerator, energy expended per kilometer and kilometers of travel per day (day-journey length), are underestimates and that in both cases circuitous travel is a major source of errors (S. Altmann 1987). In the laboratory, travel energy is measured during linear locomotion, typically on a treadmill. However, in the wild, appreciable additional energy may be required for repeated changes in speed and direction. In addition, when patterns of locomotion on treadmills and over ground were compared, they were different in cats (Wetzel, Atwater, Wait, and Stuart 1975) and in humans (Nelson, Dillman, Lagasse, and Bickett 1972), and such differences may affect locomotor energetics.

As for day-journey length, I believe that almost all methods that have been used for estimating mammalian day-journey lengths systematically underestimate the true length. An animal's actual route through its home range is more devious than the trajectory of the center of mass of its group, than a straight line connecting successive point samples of its location, than a straight line connecting the midpoints of quadrats through which the animal moves, or than a plot of its route, particularly on a small-scale map.

In Amboseli we have used several methods for estimating day-journey length of baboons. The most accurate method involves counting paces of individuals whose individual-specific mean pace length has been determined (J. Altmann and Samuels 1992). Results of this new method can be compared with those obtained by other methods. During a seventeen-month period in 1983–84, a sample of paces was taken on thirty baboon mothers in Alto's Group and Hook's Group (ibid.). According to the pace-count samples, these females walked 8–10 km/day. By comparison, hourly mapping of the location of these two groups during the same two years was used to determine which 165 m × 165 m quadrat they were in. Distance was then estimated to be the distance from the center of the old quadrat to the center of the new quadrat (Bronikowski and J. Altmann 1996). The average of the distances traveled is 5.72 km/9-hour day (ibid., table 2 data) or, adjusting to a full eleven hours of potential travel, 7.0 km/day. These values suggest that the quadrat-to-quadrat method gives estimates that are 70–88% of those obtained from pace counts.

In another Amboseli study (S. Altmann, unpublished) I simultaneously sampled the day-journey length of a single female by our benchmark method—pace counts combined with individual-specific mean pace length—and mapped her route. This enabled me to estimate her day-journey length at the same time by nine methods, including differences in map scale. Pace counts consistently gave the largest estimates, as expected. The other eight methods gave mean values that were 67–82% of the mean of pace count estimates.

By comparison with many other primates, particularly arboreal species, baboons traverse much of their day-journey length along relatively direct routes; underestimates of day-journey lengths for species with more circuitous pathways probably are subject to even larger underestimates. Thus the magnitude of underestimates that we have found for Amboseli baboons cannot be applied directly to other species. Until accurate data on day-journey length are available for a representative sample of other mammals, we cannot determine how much the biases in methods now in common use for estimating day-journey length affect estimates of the ecological cost of travel as a function of body size (equation 10.2).

Velocity and Foraging Range

If conservation of locomotor energy is not an advantage of large size in mammals, what then? Although the unit-cost of locomotion appears to be little affected by body mass in mammals and to be negligibly small in any case, size affects two other components of locomotion: velocity and aerobic scope. Large mammals walk faster than small ones, and as indicated above, mean velocity scales in proportion of $M^{0.21}$. Furthermore, even though the energy cost per unit distance is independent of speed, small mammals cannot completely compensate for their slow walk by running instead of walking: an upper limit on the rate of sustained activity in a mammal is established by the fastest rate at which it can carry out sustained locomotion aerobically without accumulating lactic acid (Bennett and Ruben 1979). Furthermore, other processes, as yet poorly understood, apparently keep it from moving continuously even at this rate.

Increased walking velocity can be advantageous in several ways. First, for a time-limited animal, increased velocity enables it to cover a given distance in less time. For an animal whose risk rate from predation or other hazards is higher when foraging or migrating, spending less time on these activities can mean lower mortality rates.

Furthermore, increased velocity of walking may be advantageous in enabling animals to cover greater distances in a given amount of time, thereby increasing the foraging area available to them. Although home range is on the average larger in larger mammals, the amount of variability is very high (Harestad and Bunnell 1979; but see also Mace and Harvey 1983), perhaps largely because of great differences in the spatial distribution of resources within habitats. An animal's home range is limited to the area that lies within cruising range of the essential resource with the most limited distribution (S. Altmann 1974). In arid regions, water often is that resource. Consider a water-dependent herbivore under dry season conditions—that is, no plant production taking place—in which the animal commutes outward from a water source, traveling progressively farther as the vegetation near the water is grazed down. According to Pennycuick (1979), herbivores, ruminant and otherwise, cannot deplete their gut contents below a certain point without disrupting their gut flora, so that under these conditions the cruising range (foraging radius) of a water-dependent herbivore depends on the round-trip distance to water that can be traversed before those reserves are exhausted, and that distance depends in turn on the speed and thus the size of the animal.

Pennycuick (1979) argues that "the need for a large foraging radius may well be a potent source of selection for large size. If two similar species are using the same water source during a prolonged drought, the one with the larger foraging radius will last longer, other things being equal." Not necessarily: Pennycuick's rough estimate of the effect of body mass on foraging radius in ungulates is that the radius increases as body mass to the 0.40 power. However, several indicators of food requirements—basal metabolism, total metabolism, and energy expended per kilometer—all increase more rapidly with mass. If, with increased size, food availability increases, say, only as rapidly as the distance walked per day, as approximated by the foraging radius, then with increased size, requirements are growing faster than energy input. The impact of energetics on the evolution of body size depends on all of the energy costs and benefits that a change in body size entails.

Perhaps of greater importance than the size of the area within the foraging radius is the ability of large mammals to get to areas that are too remote from water for small mammals and thereby to utilize otherwise untapped food sources. By the end of the dry season on the savannahs of Africa, forage for grazing mammals is impoverished in the vicinity of the permanent water sources to which they are tied at that time of year.

The greater distances that larger mammalian herbivores can travel enable them to take advantage of three aspects of spatiotemporal heterogeneity in plant communities. First, the individual plants of a species usually are not randomly distributed: in many species they have a patchy distribution. Second, floristic diversity increases with area, so that a mammal that forages over a greater distance encounters a greater number of plant species. Finally, both within and across species of plants, individuals within a community may differ in the time at which they reach each phenological stage: germination, flowering, fruiting, and so forth.

By traversing larger amounts of terrain per day, large mammals are more likely to find local concentrations of food. That will be particularly important if the location of patches is to some extent unpredictable—for example, if the various trees of a species come into fruit asynchronously. Conversely, large mammals, because they walk farther, may be less susceptible to localized seasonal depletions in available food. That is, they take advantage of the law of large numbers and thereby minimize their chance of impoverishment. Kortlandt (1984) has argued that the geographic distribution of chimpanzees is limited to areas of sufficient size, and thereby of sufficient floristic diversity, to guarantee them a year-round supply of fruit. As indicated in chapter 9, floristic diversity also enables animals to take advantage of nutrient complementarity to solve the problems posed by foods that are each deficient in one or more essential nutrients. It also enables them to minimize the impact of toxic secondary compounds by taking advantage of the fact, noted earlier, that the effects of various toxins are often nonadditive (chapter 9). In short, the locomotor advantage of large size in mammals may be conservation not of energy but of time, and that in turn can mean greater safety and more or better food.

Concurrently, however, size alters animals' nutrient requirements and the array of foods available to them, so that the ability of larger mammals to move farther and faster does not necessarily result in comparable increases in available food or in improved proficiency at meeting nutritional needs. Distance and time have different size-dependent meanings to animals. A foraging distance that would satisfy a mouse would leave a gazelle hungry. At present, little information is available about the effects of mammalian size either on nutritional requirements (Peters 1983) or on the adequacy and reliability of available foods to satisfy those requirements. I suggest that in many mammalian groups, including baboons, the latter have increased more rapidly than the former, and that this has been a consequence of the ability of larger mammals to forage over greater distances per day. If so, then increased locomotor velocity has been a factor selecting for large size in mammals.

Baboons fully exploit their capacity to walk great distances. Perhaps the longest recorded mean day journeys among nonhuman primates are those of hamadryas baboons (*Papio hamadryas*), among which bands average 8.6 to just over 13 km per day (Kummer 1968; Sigg and Stolba 1981), apparently a necessity in their sparse desert habitat (S. Altmann 1974). As I said above, yellow baboons in Amboseli walk about 6.1 km per day, four times the average for mammals the size of a baboon female. Similar distances have been reported for other baboon species (S. Altmann and J. Altmann 1970). By contrast, the smaller vervet monkeys in Amboseli averaged only 1.2 km per day (Struhsaker 1967, average of two groups). The home range of a group of baboons in Amboseli is also much larger than that of vervets living in the same area: 24.1 km^2 for a group of about forty baboons (S. Altmann and J. Altmann 1970) versus an average of 0.42 km^2 for four groups of vervets studied the same year (Struhsaker 1967) and averaging twenty-eight vervets per group.

Long-range locomotion in baboons is primarily devoted to moving to, from, and between feeding sites. The considerably greater diversity in the diet of baboons compared with that of vervets, which I described in the preceding chapter, may to a considerable extent reflect the greater floristic diversity in the baboons' much larger home range and longer day journeys. These long treks help reduce seasonal fluctuations in food supply, and that in turn may explain why the vervets of Amboseli, but not the baboons, have a sharply demarcated breeding season. Beyond that, the baboon's long day journeys may buffer them against year-to-year fluctuations in local food supply and thus make their populations less susceptible than those of vervets to the effects of "ecological bottlenecks" brought on by several years of adverse conditions.

Strength and Competitive Ability

I do not know of any direct measurements of the strength of various primates, and the extensive compilation by Peters (1983) of size-correlated traits in mammals does not list strength. However, since strength goes up approximately as the cross-sectional area of muscles, overall strength should scale approximately in proportion to body mass to the two-thirds power.

A major advantage of greater strength is the ability to harvest foods that require considerable force to process before ingestion. For baboons, the ability to dig out and pull up the corms of grass plants is of particular significance during times of the year when more ephemeral but more accessible foods, such as blossoms and fruits, are in short supply. By feeding on grass corms, baboons, like warthogs, defeat a major mechanism of survival and reproduction in savannah grasses. Most of these grasses are perennial: when the rains come, most of their new growth comes not from seeds but from the corms of established plants. New plants often are established through subdivision of old corms and by means of runners (rhizomes and stolons) that emanate from corms. The corms are underground and thus escape the many grazing ungulates on the African plains, but baboons and warthogs can dig them out of the ground. In feeding on corms and runners, they are living off capital rather than interest. At present the impact of corm feeding on populations of grasses and sedges is unknown.

Several observations of circumstances in which the feeding behavior of young baboons was constrained by limits on their physical capabilities illustrate the significance of size and strength. On several occasions, the yearlings I studied were unable to turn over large boluses of elephant dung.

> August 9, 1975. Female Pooh, age 40 weeks, tried to turn over a fresh bolus
> of elephant dung, but it was too heavy for her. (The same observation was made
> on Eno at age 55 weeks.)

In chapter 9, I mentioned that adult males, unlike other baboons, crush the tough green pods of *Acacia tortilis* in their mouths. In all likelihood this difference in feeding technique is due to differences in mandibular crushing strength. Physiological work on the closely related rhesus monkey (*Macaca mulatta*) provides some indication that the unusual jaw strength of adult males depends not only on greater muscle mass but on histological differences in the masseter muscles: those of adult males contain proportionally larger type 2 (fast) fibers and can generate at least 2.5 times as much isometric tetanic tension as the masseters of adult females (Maxwell, Carlson, McNamara, and Faulkner 1979).

One would think that two baboons feeding next to each other on the same parts of plants of the same species would be competing with each other, even if only indirectly, in that each is depleting the other's resources. The following observation on corm feeding indicates that this may not always be so. The largest *Sporobolus cordofanus* that I observed any yearling baboon pulling out of the ground was probably at the limit of the yearling's capability. With plants of that size, the seated youngster's rump came clear of the ground as it pulled, so that leg muscles were also involved, and the yearling then sometimes fell over backward when the grass plant finally came out. Yet such plants were at or below the lower limit of the size range on which the yearling's mother, seated next to it, was feeding.

Until baboons are at least six months old, they are very ineffective at digging, and for some time after that they appear to be handicapped by inadequate strength.

> September 16, 1975. I noted that Dotty, age 116 weeks, seemed to be testing grass plants for "digability." She walked from plant to plant, briefly tugging on each one at less than her full strength.

Underground foods such as grass corms and the larvae of dung beetles, both of which baboons eat seasonally in considerable quantities, are obtained by small infants from other individuals, particularly their mothers (J. Altmann 1980) and their "godfathers" (Stein 1984), who tolerate them and allow them to pick up scraps as material is unearthed and bitten into. Yet food is not handed to infants. They are sometimes tolerated but never fed.

An infant's short arms and torso put it at a disadvantage when feeding on the berries of thorny *Azima tetracantha* bushes. It can reach few berries from the ground, and when those are exhausted it must gingerly climb onto the bush to get at those higher up.

Although baboons are primarily herbivorous, they occasionally kill and eat various animals, both vertebrate and invertebrate (S. Altmann and J. Altmann 1970; Hausfater 1976; Hamilton and Busse 1982; Strum 1983). Considerable strength may be required to hold down and kill a vervet monkey or a young gazelle, or to crack open a mollusc shell. We have seen an infant gazelle escape after being caught by a baboon because the baboon could not hold it down long enough to kill it.

Does large size give baboons an advantage in interspecific competition for food? Direct competition in the form of confrontations occurs seldom between baboons and the other animals that specialize in some of the baboons' favorite foods, and these few encounters, although they have not been systematically studied, seem to have negligible impact on food intake. Baboons readily displace vervets, which eat many of the same foods, but usually this seems to cause only a brief delay in the vervets' ongoing activities. Kori bustards (*Ardeotis kori*), like baboons, eat large quantities of fever tree gum, and as I noted already, warthogs feed extensively on grass corms; but again, the rare confrontations we have observed between baboons and these animals suggest that interference competition plays a negligible role in the foraging of any of them.

What about indirect competition? For example, to what extent does foraging by baboons deplete the food supply and thereby the food intake of the vervets that live within their home ranges? To answer that question, we need information we do not now have: How do food intake rate and food density in the overlap areas affect each species? One could then determine whether the loss to vervets from baboon foraging was greater than the loss to baboons from vervet foraging. Beyond that, to substantiate a claim that size confers a competitive advantage on baboons, one would have to show that some of this difference in foraging can be attributed to the difference in body size.

Vervet monkeys—even the adults, which are about the size of small juvenile baboons—do not feed on grass corms (Struhsaker 1967; Klein 1978), apparently, as suggested in chapter 9, because they lack the strength to dig them out of the ground. The result is a major dietary difference between the two species. The size difference between baboons and vervets may be the most important one. I do not know of anything baboons do that vervets could not do if they were as large as baboons. Although yearling baboons are about the same body length as adult vervets, they are stockier and presumably stronger. Because corms are eaten by Amboseli baboons but not by the weaker vervets, a lower bound on the increment in nutrients and food energy that results from the greater strength of baboons independent of body length can be obtained by measuring the nutrient composition of the grass and sedge corms they eat as yearlings, when they are at the transition stage in terms of adequate strength to dig up corms. Corms contribute 6.1% of the proteins the yearlings get from non-milk foods, 2.7% of the lipids, 4.6% of the carbohydrates, 5.9% of the minerals, and 5.1% of the energy. Doing so takes 23.6% of the yearling's non-nursing time (from values in table 6.5). Adult females in the adjacent Hook's Group, studied by Muruthi (1988, table 16), devoted 30.3% of their feeding time to corms.

Binocular Color Vision

Several other traits of baboons, not obviously related to their size, contribute to their success. Their visual capabilities are extremely important: "For us and probably for a number of Old World primates as well, color dominates the daytime visual world" (Gruber 1978). Current understanding of color vision in primates has recently been summarized by Jacobs (1996). Regarding anthropoids, Jacobs writes:

> Direct studies of color vision show that there are several modal patterns of color vision among groupings of primates: (i) Old World monkeys, apes, and humans

all enjoy trichromatic color vision, although the former two groups do not seem prone to the polymorphic variations in color vision that are characteristic of people; (ii) most species of New World monkeys are highly polymorphic, with individual animals having any of several types of dichromatic or trichromatic color vision.

In addition to color perception, baboons, like the other anthropoid primates, have overlapping visual fields and thus binocular vision, with its attendant advantage of stereoscopic perception.

> The elaboration of the visual mechanism in diurnal primates is an adaptation to specialized arboreal life in which a high premium is placed on the ability to judge distance accurately in jumping and leaping and on the finer discriminatory functions required in searching for and manipulating food. Visual judgment of spatial relationships depends on two factors, overlapping of the visual fields and the presence of a cone-type retina with a *fovea centralis*. (Napier and Napier 1967, 18)

The evolution of binocular vision was not without its drawbacks, however, particularly reduction in lateral vision. Hershkovitz (1977) has suggested an interesting consequence of this reduction for primate sociality:

> Evolution of binocularity among primates may be correlated with increasing sociability, particularly among diurnal species. As binocularity advanced and the visual field narrowed to the front, the individual presumably became increasingly dependent on conspecifics for sighting enemies outside his own line of vision and for sounding alerts. The survival value of such interdependence would tend to strengthen mating and family bonds. This in turn would provide the basis for the evolution of a complex, enduring society with defense as one of its functions.

Binocular color vision enables baboons and other anthropoids to detect food at a distance and to make fine discriminations according to degree of maturity, ripeness, and so forth. This ability is particularly important to baboons when they feed on fruit. Baboons eat several species of fruit and, in every case, appear to prefer them fully ripe. This is not surprising: fruits and fruit eaters (primarily birds and mammals) have coevolved. Immature seeds will not germinate. The ripening of animal-dispersed fruits occurs at about the time the seed becomes viable. Ripening of animal-dispersed fruits typically involves production of sugars and other animal nutrients, decreases in toxins that protect immature seeds, and production of various animal attractants, including odors, flavors, and colors (McKey 1975; Van der Pijl 1969). Animals eat the fruit, digest the aril of the fruit, then later deposit the seed, complete with fertilizer, often at some distance from the parent plant. Polyak (1957) suggested that color vision in vertebrates coevolved with colored fruits, and Snodderly (1978) has studied the relation of color vision in primates to the spectral characteristics of their native foods.

Although baboons, like other fruit-eating primates, generally prefer fully ripened fruits, competition among various fruit-eating birds and mammals, combined with shortages of other food sources, puts ripe fruit at a premium and sometimes pushes baboons into eating much fruit that is not fully ripe. "Baboons . . . tend to harvest immature fruits which are bitter to our taste and rejected by birds" (Hamilton, Buskirk, and Buskirk 1978).

The sense of smell and taste are also important in food selection, though almost nothing is known about species differences among primates in the acuity of these senses. "Before ingestion, [food] is, if unusual, passed before the nares to ascertain palatability" (Hill 1966, 159). Scott (1963) has suggested that both dental and olfactory factors account for the prominence of the muzzle in baboons. "The skull of the baboon demonstrates . . . that a long muzzle and frontally-facing [optic] orbits are not incompatible" (Napier and Napier 1967, 18).

Manipulation

Selective feeding includes two processes, selection among foods and food manipulation. The functional significance of such behavior—that it increases the concentration of nutrients and reduces the impact on an herbivore of the admixture of nutrients with various deleterious materials—selects for proficiency in manipulation.

In contrast with most groups of mammals, anthropoid primates have relatively few anatomical restrictions on what foods they can obtain. Baboons, like other cercopithecines, are agile. They can move readily in the trees, on shrubs, and on the ground. This locomotor skill, combined with a hand suited for digging, plucking, holding, pulling, and breaking and for complex, fine manipulation, enables them to get at concentrated food sources wherever they occur (S. Altmann and J. Altmann 1970).

A few foods, such as soft berries and young grass shoots, are plucked from the plant and eaten "as is," but most baboon foods are processed in various ways, manually or orally, after being picked. In some forms of processing ("dissociation") distinct fractions are produced, one of which is discarded, as in separating wheat from chaff. In others ("fragmentation") there is only mechanical breakdown into smaller particles, as in chewing, and no selection is involved. In baboons, manual processing of food involves almost entirely dissociation, whereas the mouth parts (teeth, lips, and tongue) are used for both dissociation and fragmentation.

The methods baboons use for harvesting the tiny fruits of *Trianthema ceratosepala*, described in the last chapter, are a good illustration of their food-processing capabilities. This is a skill that apparently improves with age. On one occasion when yearling Eno and her mother Este were near each other and both were feeding on these fruits, I compared their intake rates. Although Eno kept fairly busy, Este picked the fruit about four times faster.

Equally impressive is the handling of green acacia pods by adult females and immature individuals, mentioned in chapter 4. Each pod in turn is picked with the hand, then put into the lips. It passes across the lips, sometimes unaided by the hands, apparently by the joint action of lips, teeth, and tongue. As it does so, the seeds are removed from the pod, which then drops from the mouth. The seeds are then stored in the buccal pouches. Meanwhile, the baboon is busy with its hands, picking another pod. Later, while the baboons are resting in the shady lower branches of the tree, the seeds are brought from the buccal pouches back into the oral cavity and the seed coats are removed, again apparently by the joint action of lips, incisors, and tongue; the hands are not involved. One by one, the seed coats fall from the lips. The naked seeds are then chewed and swallowed.

Dissociative manipulation is a selective process that requires time and energy. What does the animal get in return? Basically, an increased concentration of nutrients or a decreased concentration of useless or toxic materials, or both. Because already picked material is discarded in the process, the animal cannot thereby increase its absolute intake of nutrients. However, manipulation often increases the *concentration* of nutrients in food. Another function of food manipulation is to remove tough materials, such as hulls and corm sheaths—materials that are largely indigestible and that compete with more nutritious food components for limited gut capacity. Finally, some parts of food plants are hazardous, such as thorns, toxins, and adherent soil that wears down the teeth, and as documented in the last chapter, dissociative manipulation can reduce the concentration of such materials. In the words of an old song, preingestion food manipulation "accentuates the positive and eliminates the negative."

Much has been written about the importance of eye-hand coordination in primates (e.g., Watt 1925). With few exceptions, primarily certain soft fruits and leaves, the foods of baboons and other cercopithecines are processed manually before being eaten. Opposition of the thumb facilitates both picking of food and its subsequent manipulation.

> The opposable thumb enables baboons to be highly selective in their utilization of some plant food parts. Most other animals utilizing the same food plants cannot be as selective as baboons. For example, baboons crop most grasses at or below the soil surface. Often only the bases of grass shoots are eaten. By comparison, buffalo and warthogs grazing the same areas rip away masses of the tops of these same grasses but do not take the bases which are favored by baboons. Manual dexterity allows enhanced food value for selected items. The use of *Cyperus* bulbs and *A[cacia] giraffe* seeds . . . shows that relatively protein-rich

parts are selected. To obtain the protein-rich components of these plants requires extensive processing, which baboons accomplish by manual dexterity. (Hamilton, Buskirk, and Buskirk 1978)

The opposable thumb of baboons and many other anthropoids probably permits more effective manipulation than other primates are capable of (Bishop 1962). The usefulness of opposable digits is beautifully illustrated by the way baboons feed on the fruit of the thorny *Azima tetracantha,* an abundant and widespread shrub of African savannahs. The thorns of this plant are long, very sharp, armed with microscopic recurved barbs, and so oriented as to provide maximum protection of the fruit from the mouths of browsers. As mentioned in chapter 9, baboons bypass this defense system by reaching between the thorns with their prehensile fingers to pluck the fruit.

By contrast, "The colobus [monkey, *Colobus angolensis*], with its reduced thumb, apparently has difficulty in grasping small or single food items. Feeding was done primarily by bringing the mouth to the food or breaking off a piece of a branch and bringing it to the mouth" (Moreno-Black 1978). Similarly, in describing the foraging behavior of New World howler monkeys, Milton (1980, 81) wrote as follows: "*Gustavia superba* produces a large fruit with hard seeds bound into a tough matrix that howlers, with their lack of manual dexterity, might find difficult to handle."

Like most primates, baboons are agile—surprisingly so for such large and primarily terrestrial primates. In this, young baboons have an advantage over the adults. Their flexibility and small size enable them to clamber through openings in thorny umbrella tree branches that adults must go around. Their speed and agility make them quite skillful at capturing grasshoppers, despite the insects' evasive jumps. Usually the youngsters capture a grasshopper by a series of quick swats at it on the ground or grass, but occasionally they will catch it in midair as it jumps away.

Skill at catching grasshoppers improves with age. The yearlings' capture rate per attempt was 22.2% of swats, and they averaged 4.5 swats per captured grasshopper; juveniles two and three years old had a success rate of 31.4%, and each captured grasshopper took on average 3.1 attempts. Unfortunately I have not sampled this behavior in adults, but my impression is that they less often go after a grasshopper, and when they do so they make fewer attempts before quitting if unsuccessful.

Cheek Pouches

Cheek pouches occur in all cercopithecine primates but in no other primates (Murray 1970; Schultz 1969). Their primary function is short-term storage of small food items when harvesting rates exceed the rates of preingestion oral processing, as illustrated above by the way baboons feed on the green seeds of umbrella trees. Cheek pouches are of considerable importance to an animal that feeds on patchy foods and finds choice items. By storing such material in their cheek pouches, they can reduce competition with conspecifics and exposure to excess insolation or other adverse environmental conditions at the food sites. Not surprisingly, according to Murray the smallest food pouches, relative to body size, occur in geladas, *Theropithecus gelada,* whose diet is different from that of all other primates; consisting primarily of grass (96.9% of feeding records in Bole Valley, Ethiopia; Dunbar and Dunbar 1974) and changing seasonally, with individual values of 79–99% of diet mass during the rains and 35–81% during the dry season (Iwamoto 1979) in the Simēn Mountains, Ethiopia. On such a diet, buccal pouches are seldom of much use.

Intelligence

Baboons appear to be extraordinary clever animals, as evinced not only by their ability to cope with complex social relationships but also by their foraging. They search a very large area containing many different habitats, in which they feed opportunistically on a wide variety of foods that are patchily distributed in both time and space. In so doing, they somehow manage to find many highly nutritious foods and still avoid poisoning themselves on toxic plants. All of this seems to require a considerable sophistication at information processing.

Intelligence in animals is a curious thing. It is the topic of much research and many measurements. Its neurological basis and its evolution have been extensively studied (Jerison 1973). Yet no one knows what it is; or rather, several people purport to know, but they don't agree with each other. (For examples, see Parker and Gibson 1979; Thomas 1980, 1982; and various chapters in Hoage and Goldman 1986 and Weiskrantz 1985.) Whatever intelligence is or is defined to be, it is most highly developed in humans. This is not an empirical fact but a constraint on any acceptable definition of intelligence.

For my present purpose, the following definition from Alan Kamil (1994) is sufficient. Kamil defines animal intelligence as

> those processes by which animals obtain and retain information about their environments and use that information to make behavioral decisions. [This broad definition] includes all processes that are involved in any situation where animals change their behavior on the basis of experience. It encompasses the processes studied with traditional methods, such as operant and classical conditioning. It also includes processes such as memory and selective attention. . . . It includes processes involved in complex learning of all sorts, including that demonstrated in social situations. It also includes the study of more "specialized" learning, such as song learning and imprinting.

So defined, animal intelligence seems to include any behavior more complex than a simple signal-response relationship, such as suggested by the ethologists' concept of a releaser.

Two major sources of selection for intelligence in primates have been suggested (McGrew 1981): first, the requirements for complex social processes (Alexander 1979; Beck 1982; Byrne 1994, 1996; Byrne and Whiten 1988; Cheney and Seyfarth 1990; Humphrey 1976; Jolly 1966; Kummer 1982), and second, the requirements for foraging (Clutton-Brock and Harvey 1980; Milton 1981; Parker and Gibson 1977, 1979; Kortlandt 1984). Cheney and Seyfarth (1990) succinctly put the case for the former view:

> Although ecological factors have undoubtedly contributed to the evolution of primate intelligence, nonhuman primates do not appear to manipulate objects in their physical environment to solve ecological problems with as much sophistication as they manipulate each other to solve social problems.

However, to the extent that feeding and foraging are socially mediated processes and in turn constrain social behavior, socializing and foraging are not mutually exclusive. Cheney and Seyfarth's book provides a fascinating survey of social cognition and intelligence in the vervets that have been the focus of their long-term research in Amboseli and of other primates. Some social aspects of primate foraging will be discussed in the next section.

Five components of foraging by baboons and other primates suggest a considerable amount of information processing: foraging for very widely dispersed food patches, the apparent use of "cognitive maps" as they navigate over very large home ranges, extractive foraging, fine-grain discrimination among potential foods, and maintenance of a large yet highly select repertoire of foods despite marked seasonal changes.

Dispersed patches. Patchiness presents foragers with complex spatial problems, and Milton hypothesized that

> the extreme diversity of plant foods in tropical forests and the manner in which they are distributed in space and time have been a major selective force in the development of advanced cerebral complexity in certain higher primates. . . . All else being equal, those primate species dependent on the most hyperdispersed and patchy foods should show greater evidence of mental development than primates eating more uniform dietary resources. (1981, 537, 539)

Fruits, for example, are usually more patchily distributed than leaves, and in primates (Clutton-Brock and Harvey 1980), bats (Eisenberg and Wilson 1978; Pirlot and Pottier 1977), and rodents (Mace, Harvey, and Clutton-Brock 1981) the extent of cerebral development is greater in more frugivorous species. Eisenberg and Wilson (1978), as well as Clutton-Brock and Harvey (1980), suggest that the greater spatial complexity of patchily distributed foods has selected for large brain size. However, there are alternative explanations, such as greater energy turnover in frugivores: brain size may be metabolically limited (Armstrong 1982). As for the implications of brain size for intelligence, no good evidence exists that this or any other morphologically identifiable characteristic of the brain can be used as a reliable indicator. McPhail, in his extensive critique of the available literature, put it succinctly:

> It is not clear that the emergence of neocortex carries any implications relevant to the evolution of intelligence. . . . There is no necessity to assume differences in intellect between the various mammalian groups. . . . It appears that there is no compulsion to assume variations in intellectual capacity throughout (non-human) vertebrates in general. (1982, 288–89)

Extractive foraging. The term "extractive foraging" refers to feeding on food "encased in a shell or . . . imbedded in a solid matrix such as the earth" (Parker and Gibson 1979; King 1986). The significant aspect of such behavior is that during the initial phases of extractive feeding the food is out of sight and in many cases can also be assumed to be "out of smell" and not otherwise directly perceived. So extractive foraging appears to be goal-directed behavior toward a goal that cannot be perceived. It seems to transcend any simple stimulus-response explanation and to require some cognitive capacity (Menzel and Wyers 1981). This impression is enhanced if the behavior by which the goal is achieved is not a fixed routine of motor patterns but, instead, the animal adjusts its behavior to current circumstances, with the consummatory act (eating the embedded food) as the only common denominator.

Extractive foraging is widespread among primates. For example, orangs open hard-shelled fruits and nuts with their teeth (Galdikas 1982; Rijksen 1978), gorillas dig holes to obtain underground bamboo shoots (Casimir 1975), many terrestrial primates turn over stones to get at arthropods (e.g., chacma baboons; Hamilton, Buskirk, and Buskirk 1978b; yellow baboons, pers. obs.), saddle-backed tamarins remove the seed and pulp from the tough exocarp of *Celtis* fruits (Terborgh 1983), hamadryas baboons dig drinking holes in sandy riverbeds (Kummer 1968), and baboons along the coast of South Africa extract molluscs from their shells (Hall 1963). Terborgh (1983) indicates the percentage of hidden prey eaten by five species of primates at Manu National Park, Peru. The aye-aye of Madagascar may be the only primate for which extractive foraging is the primary mode of feeding: they scoop out insect larvae from holes in trees, using a specialized claw on the third digit (Petter 1967). To date, cebus monkeys and chimpanzees are the only primates that are reported to use tools in the wild for extractive foraging. Cebus monkeys use rocks to open oysters (Hernandez-Comacho and Cooper 1976). Chimps crack open very hard nuts using rock hammers and long anvils (Boesch and Boesch 1983, 1984) and extract termites and ants by "fishing" for them with twigs or grass probes (Goodall 1963; McGrew and Rogers 1983; Nishida and Hiraiwa 1982; Uehara 1982).

The baboons of Amboseli include a wide variety of hidden foods in their diet. These include the pulp inside the pods of *Capparis tomentosa* and *Maerua* sp.; the underground corms and blade bases of several species of grasses and sedges; the lateral meristem of *Sporobolus consimilis;* the subterranean larvae of dung beetles and butterflies; the calyx-enclosed fruit of *Withania somnifera;* various seeds, sprouts, and insects from within and under dung, bark, and rocks; gum on the trunk of young fever trees (where it is concealed by the low canopy); the fruit inside the thorny shell of the devil's thorn plant; the ovaries of *Trianthema ceratosepala;* seeds extracted from the pods of several legumes; the nut of *Balanites glabra* inside its hard shell; lily bulbs; terrestrial gastropods; bird eggs; and the flesh of various birds and mammals.

Extractive feeding is not restricted to primates, so if such behavior is taken as a hallmark of intelligence, one must concede intelligence to many other animals as well. Numerous examples of extractive feeding are available, not only among birds and other mammals, but among a wide variety of invertebrates, including leeches, mosquitoes, and various predators on molluscs, such as other molluscs that drill through their shells and starfish that pry them open. Many other examples can be found in any textbook on invertebrate biology. Some of these examples of extractive feeding are known to be stereotyped responses to simple sign stimuli. McGrew (1981) claims that while some nonprimates practice extractive foraging, they do so "only in a limited and stereotyped way compared to the opportunistic, omnivorous monkeys and apes." Although this claim may be correct, evidence available at present on the sensory, motor, and mental capacities utilized by primates and other animals during extractive foraging is so fragmentary that no claim for special extraction talents by primates can be substantiated.

Cognitive maps. Baboons, like any mobile predators, must navigate through their home range, getting at its resources while avoiding its hazards. The ability of baboons to locate key resources over a very large home range is remarkable. The area involved is so large that months may go by before some parts of their range are revisited.

On what basis do they select their routes through these large areas? Post (1982) showed that the time baboons spend in various vegetation zones, however scattered through their home range, can be largely predicted from the density of a small number of major foods: two grasses (*Sporobolus rangei* and *S. marginatus,* primarily *S. cordofanus*), fever trees, and the shrub *Azima tetracantha.* Beyond that, the relative abundance of "patches" of these vegetation zones within different areas of the home range is in turn a significant predictor of the overall pattern of area occupancy.

Of course this correlation does not tell us how baboons locate these resources. I am convinced that baboons are not just relying on their perceptions of the moment but are integrating those perceptions with a large fund of stored information from their experience in the area. By doing so they can navigate from any part of their home range to a goal in another part, even though that goal is completely out of sight, and can do so without repeating their previous approach. Furthermore, I believe this cognitive ability enables them to greatly minimize searches through areas whose resources they have recently depleted. Elsewhere we have referred to this store of spatial information as a "mental map" (S. Altmann and J. Altmann 1970), but the term "cognitive map" avoids any implication of conscious processes.

That the baboons' foraging is goal directed sometimes seems apparent. Within a short time of finding ripe berries on isolated bushes of *Lycium "europaeum"* in the core of their home range, the baboons of Alto's Group would make a long trek to the base of Naarbala Hill, a place where these bushes were concentrated and the baboons gorged on the ripe fruit, but where they seldom went at other times. When, in the morning, Alto's Group moved northwest away from sleeping grove 23, we knew they would make a long trek to the edge of Lake Amboseli and feed for a long time on gum from the regenerating fever trees there.

Similarly, Boesch and Boesch (1984) provide strong evidence for cognitive maps in wild chimpanzees: they use stones to crack the nuts of *Panda oleosa,* and in doing so

> they seem to remember the location of stones and to chose the stones so as
> to keep the transport distance minimal. The chimpanzees seem to possess an
> Euclidean space, which allows them to somehow measure and remember distances; to compare several such distances so as to choose the stone with the
> shortest distance to a goal tree; to correctly locate a new stone location with
> reference to different trees; and to change their reference point so as to measure
> the distance to each *Panda* tree from any stone location.

The remarkable collective decision-making and signaling processes that are involved in movements of hamadryas baboons within their home range have been analyzed in a brilliant field study by Sigg and Stolba (1981). Clearly, we are just beginning to reveal the range of mental capacities of anthropoid primates.

I mentioned two other activities of baboons that suggest extensive information processing. The first is fine-grained discrimination among potential foods, even those close to each other physically (e.g., the seed vs. the seed coat of acacias), taxonomically (e.g., *Solanum dubium* vs. *Solanum incanum*), or as a result of convergent evolution, such as some of the halophytic grasses and sedges. Yet, although finicky, the baboons maintain a very large repertoire of foods throughout the year, despite major seasonal changes in the array of foods available to them. Much of this book documents these aspects of baboon foraging.

Terrestriality, Sociability, and Omnivory

Two genera of large-bodied cercopithecines, *Papio* (baboons) in Africa and *Macaca* (macaques) in Asia, stand out for their mastery of terrestriality and their extensive adaptive radiations. For baboons, full exploitation of African savannah habitats involves not only an ability to walk great distances, often without shade or water in the heat of the day, but also to do so with relative immunity to predators, even in areas remote from trees. The large home range and long day journeys of baboons increase their susceptibility to predation in several ways. In the course of each day's travels, the baboons' movements carry them into a wide variety of habitats. Unlike vervets, baboons stride unhesitantly into the open from the cover of the woodland. Consequently there are large portions of the open grassland in the home range of a baboon group, far from any tree, where no vervets go, and other places that vervets use only in rapid transitions between patches of woodland or thicket (Hall 1965). The long day journey of baboons not only increases the diversity and number of potential predators they are exposed to, it also means they spend much of each day in open grassland—on the ground, away from trees, conspicuous and vulnerable. Beyond that, because baboons repeatedly move from one area to another, they have little time to check each new area for hidden predators. Except when approaching areas of very dense foliage, they seldom pause for long before moving from one habitat to the next.

Group life contributes to the safety of baboons in several ways beyond the general antipredator advantage of living in proximity to other potential prey individuals (S. Altmann 1974; Hamilton 1971). The immunity of baboons to most predators most of the time apparently results from a constellation of closely related characteristics: not only their body size but also their considerable strength, stamina, and fighting ability. On one occasion several adult male baboons wrested off a python—leaving the males exhausted (J. Altmann, pers. comm.). A weaker animal might have succumbed. An adult male baboon is about twice the size of a female and is armed with long canine teeth that are honed on a specialized premolar (Ankel-Simons 1983). For a predator to approach a group of baboons without being seen is very difficult, and once spotted, the predator usually is driven away by the baboons' barking, canine displays, and mobbing behavior. Although this extreme form of antipredator behavior occurs rarely, it probably conditions predators against baboons, who would be formidable animals to attack. Furthermore, I once saw the adult male baboons of a group drive a leopard off its prey, a vervet monkey, which the males then ate. On the other hand, baboons are not immune to predation, and we have several records of adult male baboons killed by leopards.

Washburn and DeVore (1962) suggested that the large size, fighting ability, and sexual dimorphism of baboons may result primarily from sexual selection via competition among males for mates, and thus that the mating system of baboons selects for traits of adult males that secondarily protect them and other members of their group from predators. Unfortunately, we do not know how much of the sexual dimorphism of baboons results from differential mating success and how much from differential mortality entailed by predator repulsion. The former probably predominates, however. Hausfater (1975) demonstrated that within a year about half of the variance among Amboseli baboon males in their access to estrous females at the time of maximum fertility was accounted for by the males' dominance ranks, that is, by their fighting ability vis-à-vis the other males in their group. The lifetime effect of this rank-correlated difference in mating success depends on the still unknown long-term pattern of time-in-rank and rank-to-rank transitions (Saunders and Hausfater 1978). Other factors affecting males' lifetime reproductive success include initial rank upon entering a group, duration of group residency, year-to-year fluctuations in the number of conceptions in their group, and development of mating preferences by females (J. Altmann, Hausfater, and S. Altmann 1988).

Every member of a baboon group benefits from the presence of the adult males, not because males specifically defend them but because when it comes to predators, adult male baboons are literally repulsive. If a predator attacks and the baboons flee, very young baboons often will be picked up either by their mothers or by other members of the group. Adult males sometimes flee less rapidly and thus become interposed between the predator and the other group members. At other times, males flee with the group.

Predator detection, while less spectacular than mobbing or threatening a predator, may have a larger impact on survivorship. Youngsters and immigrants in a group can learn about local hazardous areas from the place-specific alerting responses of other baboons in their group, and young baboons apparently have to learn which of the many animals in their home range are dangerous. In the course of a day's activities, each baboon glances up occasionally, monitoring its environment. Although baboons do not literally "post sentries," as was once thought (Hall 1960), individuals at rest repeatedly look about them, sometimes from the vantage point of a tree or other promontory. Because the baboons of a group are not all facing the same direction and their glances are not synchronized, it is very difficult for a predator to approach (or be approached) without some members of the group seeing it and alerting the others. Thus each baboon is part of a "vast sentient web" (Galton 1871, quoted in Hamilton 1971).

The relation between the baboons' omnivory and their reliance on the safety of their group has been vividly described by Hamilton, Buskirk, and Buskirk (1978):

> Baboon-style omnivory (i.e., time-consuming processing of various plant foods and the taking of many small animal prey items) is favored by a multimale social organization. The advantage of sociability in enhancing time available for feeding activities has been reviewed by Powell (1974) and Brown (1975). Predation protection offered by the troop enables baboons to develop large social units capable of moving through their home range relatively free from predation. Omnivory enables low-ranking individuals to reduce competition with higher-ranking individuals while still living in close proximity to them, thereby gaining protection from predation. Predation protection in turn enhances the value of omnivory, because baboons can effectively compete with more specialized species by increasing processing time of potential plant foods, thus avoiding plants' antiherbivore adaptations. Predation protection also allows baboons to concentrate on selective foraging for prolonged intervals. More vulnerable species are restricted to more localized spaces and less diverse environments, where predation evasion on the basis of individual escape behavior is effective. Sociality may also allow animals to focus their attention on foraging sites and to move farther from cover. Many baboon foods require time-consuming processing activities, and the predation-buffered social group provides this time.

Hamilton et al. continue:

> The hierarchical arrangement of baboon society limits the rate at which small patches of resources are utilized. A single fruiting fig tree may attract and hold the movement of the troop, while low-ranking individuals are forced to peripheral, less desirable foods. In this circumstance the potential to utilize alternative foods effectively is particularly advantageous to low-ranking individuals. Thus, the interaction of a troop with patchy resources favors the development of omnivory, especially by relatively low-ranking individuals.

The advantages to each individual of a group's antipredator behavior may be sufficient to offset the disadvantage of foraging competition from conspecifics. Such competition is reduced by dietary differences between members of a group, and I have described several attributes of baboons that contribute to this diversity.

Furthermore, some apparently competitive interactions may not be what they seem. Consider the following. When a baboon is feeding, it is sometimes interrupted by another individual. These interruptions are sometimes accompanied by agonistic behavior on the part of one or both and so appear to be a form of food competition. Yet a study by Shopland (1987) of twelve immature baboons from the same group I studied provides contrary evidence of several sorts. Dispersion of food and proximity of neighboring baboons had little or no effect on the frequency or intensity of such interruptions, and a juvenile's mean number of neighbors within five meters did not differ between clumped and uniformly distributed foods. Measures of food quality (energy content, protein content, yield of food units per bout, yield rate, and the amount of time needed to process foods) provided no correlation of the quality of various foods with the rate of interruption of feeders or the probability that the interrupter would usurp the victim's food patch. Indeed, in three-quarters of all interruptions, the interrupter did not usurp the victim's patch.

Furthermore, the timing of these interruptions was not what would be expected if interruptions were a form of food competition (S. Altmann and Shopland in prep.). Interruptions did not occur at or near the optimal time, in terms of maximizing the interrupter's food gain (or its victim's losses). Indeed, for most of the foods examined, interruption times were indistinguishable from random (time-independent) distributions. Furthermore, even when the victim's food patch was usurped, interrupters fared worse, in terms of their subsequent food harvest from that patch, than they would have done by finding their own patch. While these interruptions may still involve an element of competition, in that interrupters have lowered the relative foraging success of their competitors, interruptions among juveniles may have a different significance: they may be agonistic contests, used in establishing and asserting relative rank in an age class undergoing transitions in dominance relationships. Whether direct competition at food patches is more important for adults than for juveniles remains to be seen.

On the other hand, baboons appear to obtain at least four types of information about food from other members of their group: what foods to eat, where in the home range to obtain them, which patches are particularly rich, and what new or newly available foods are currently present. For none of these do we yet have adequate studies, despite their potential importance. The following descriptions will indicate some of the contexts of such information transfer.

Baboons watch each other and thus see when and on what foods the other members of their group are feeding. No specialized food signals are required, only the feeding activities themselves—what Darwin (1896) referred to as serviceable associated habits. Baboons also sniff each other's muzzles, almost always when the sniffed individual is chewing (S. Altmann and J. Altmann 1970), and in that way they probably can tell what their companions have found to eat. An infant baboon, clinging to its mother's ventrum as she feeds, will often pick up scraps of food that drop from her hands or lips (J. Altmann 1980). In these ways information about foods is transmitted from one individual to another and from one generation to the next.

In chapter 1, I mentioned Whitehead's (1985) study of infant howler monkeys (*Alouatta palliata*) in Costa Rica. He showed that "a form of socially dependent learning, like observational learning, governs the ingestion of new and mature leaves; a learning process independent of social influences, like trial-and-error learning, governs feeding on fruits." Infant howlers rarely if ever ate species of leaves before seeing their mothers do so. On the other hand, they fed on various fruits and fruitlike objects both during and outside the context of the group's feeding. Leaves, far more than ripe fruits, are likely to contain toxic secondary compounds.

Perhaps we should not be surprised that primates learn something of what to eat from other members of their group. Laboratory rats do so, even when they do not see their partners feeding, perhaps by using olfactory cues from food clinging to the animals' fur or from the animals' breath (Galef and Beck 1990; Galef and Stein 1985). More astonishing are demonstrations that laboratory rats can learn what to eat and what not from flavor compounds in their mother's milk (Galef 1973; Galef and Clark 1971; Galef and Henderson 1972; Wuensch 1978) and even, prenatally, from compounds in the mother's bloodstream that cross the placental barrier or that occur in amniotic fluid (Smotherman 1982)!

Can primates learn from others what foods *not* to eat? Fairbanks (1975) was unable to demonstrate communication about distasteful foods either in captive pigtailed macaques (*Macaca nemestrina*) or in free-ranging red spider monkeys (*Ateles geoffroyi*). In Whitehead's study, no instance of an adult's removing a plant part from an infant's mouth was observed, nor have I seen any such behavior in baboons. On the other hand, Fletemeyer (1978) observed adult male chacma baboons (*Papio ursinus*) threatening subadults and juveniles away from oranges into which an anesthetizing drug (Cyanalin, Parke-Davis) had been injected.

On a larger scale, baboons learn from other members of their group about the location of resources within their home range. They move several kilometers each day through a habitat containing dozens of local plant zones (Post 1978). As they move, infants ride on or follow their mothers, juveniles go wherever the group goes, and the adults, in ways we do not understand, somehow make collective decisions about where to go and when. How well do they do? We cannot say, except to point out the obvious: that much of the time, they find enough to eat. At present, however, there does not exist an adequate predictive model for the optimal way to exploit a mosaic of resources, so little can be said about just how adaptive the baboons' pattern of home range utilization is. Because of considerable long-term stability in their home range occupancy pattern, with relatively gradual shifting under changing ecological conditions, I believe that much of their detailed knowledge about the distribution of resources within their home range and many of the decisions about group movements rest with the older adult females, rather than the adult males, who from time to time move into a new group. On the other hand, such immigrant males can bring with them a knowledge of nearby areas into which a group may extend its range.

Baboons converge on large, concentrated food sources found by any member of their group, apparently as a result of "conspecific cueing" (Kiester 1979). If, for example, one individual begins feeding rapidly and continuously on the berries of *Azima tetracantha,* other members of the group gradually converge on the area, even though the group had been scattered over several hundred meters. Because fruit on all the bushes in one local area tends to mature at roughly the same time, the baboons benefit from their ability to respond in this way. If the patches are large enough, the total gain in food from such convergences exceeds the total loss to the finders that results from resource depletion of the patches by their companions, so that in the long run everyone benefits. Because of the very patchy, localized nature of many savannah plants, this conspecific cueing may be of considerable importance to baboons.

At smaller food patches and ones that require some time to harvest, baboons may reduce competition with other members of their group by increasing interindividual distances. On the open plains, baboons reduce competition over corms by spacing themselves in a long, irregular row. Such a rank formation is the unique configuration that provides a separate foraging swath for each individual while simultaneously minimizing the distance from each individual to its nearest neighbor (S. Altmann 1974).

The observations presented here only hint at the numerous relations between social processes and subsistence activities. Much has been written about how ways of making a living have constrained the evolution of primate social systems and about how group foraging leads to competition. Largely overlooked are the ways social processes facilitate subsistence activities. Of these, none can be more important than socially transmitted information about what to eat and what not. "As social animals, omnivores can expand the range of environments in which they successfully avoid toxins and select balanced diets" (Galef and Beck 1991). In all probability, socially transmitted, pooled information about food resources is a major advantage to group life for baboons as well as a key element in their adaptability to different habitats and to temporal and spatial variations with each habitat. Because of the aridity of most baboon habitats, such environmental fluctuations are often of considerable magnitude.

Baboons and Savannahs

The baboons' mastery of terrestriality while still retaining the ability to exploit the arboreal habitat has enabled them to inhabit nearly all of the terrestrial communities of Africa, from equatorial rain forest to desert steppe. In the absence of human intervention, their distribution appears to be limited locally only by lack of permanent water or by the most severe montane conditions. Of particular importance for the success of baboons is the fact that their arboreal-terrestrial complex of traits enables them to exploit savannahs, that is, habitats with some trees, often scattered, and in which grasses form the primary ground cover. This is exactly the habitat most common in Africa (Keay 1959). Baboons are widespread and numerous largely because they are successful at exploiting the predominant habitat of one of the largest continents.

They are as sick that surfeit with too much as they that starve with nothing.

—William Shakespeare, *The Merchant of Venice*, 1.1.5

Primate Nutrient Requirements and Toxin Limits

The State of the Art

I must apologize for the very existence of this appendix. In it I pose a seemingly simple question: What is the minimum amount of each nutrient that a mammal must consume in order to survive and reproduce? And similarly for toxins, for which maxima pertain. These values are needed to write constraint equations for the optimal diet models and to evaluate the adequacy of the baboons' actual diets. When I began this research project, I assumed that the answer to my questions was to be found in the literature on mammalian nutrition—indeed, even on the back of my box of breakfast cereal! All I would need to do would be to scale the requirements to baboon size and present the results in a summary table.

My assumption that mammalian nutritional requirements are well established, while naive in retrospect, was not unreasonable. After all, many prestigious committees, both national and international (e.g., FNB 1980, 1989; FAO/WHO 1962, 1965, 1973; WHO 1974), have met repeatedly to convert the latest information on mammalian nutrition into recommendations about our own diet. "Recommended Dietary Allowances" (1989), published by the National Academy of Sciences (USA), is now in its tenth edition! There are, in addition, several reviews of nutritional requirements and deficiencies in nonhuman primates (Ausman and Gallina 1979; Day 1944, 1962; Greenberg 1970; Harris 1970; Hayes 1979; Kerr 1972; NAS 1978; Nicolosi and Hunt 1979; Wixson and Griffith 1986) and even in baboons, the subjects of my study (Dunbar MS; Hummer 1970). Surely, thought I, the experts must by now be refining the third or fourth significant digit in the requirements!

Alas, this is not the case. For proteins, Dorothy Calloway (1975) tersely summarized the extent of our ignorance: "It seems likely that we do not know accurately the human requirement for dietary protein even under short-term laboratory conditions." Yet proteins are among the most intensively investigated classes of nutrients. For every other nutrient except iron, the situation is no better: at present we cannot estimate, for any given level of dietary intake of a nutrient, the likelihood of deficiency symptoms. Yet that is exactly what we must do in order to make a rational dietary recommendation for a nutrient: we must know the consequences of not meeting that recommendation (Hegsted 1972). I shall return to this issue below.

I found it impossible to obtain from the literature a clear-cut answer to the pivotal questions. What is the minimum daily intake of nutrient below which the fitness of an organism is appreciably reduced? What nutrients limit fitness? For toxins, what is the maximum intake above which fitness is compromised? There are several reasons for this inadequacy of the literature:

1. Biological fitness is hardly ever the criterion for establishing nutritional requirements or recommendations. At the heart of this failure lies a pervasive inadequacy of the medical sciences: they are focused on recognizing and treating the sick but have little to say about degrees of health and their causes:

> For almost any nutrient we might mention there is most likely a minimum intake necessary for life, an optimum intake compatible with excellent health and, finally an excessive level that will cause injury. It is comparatively easy to define the minimum intake and the level that will cause injury. It is usually very difficult to define an optimum intake, however, because we are not so clear on what constitutes optimum health or how to measure it. (Stare 1958)

> It was also recognized that the proposed [nutrient] levels might not represent optimal intakes, which remain undefined in the absence of criteria of optimal health. (FAO/WHO 1973, 10)

Instead, the search is for dietary minima, for which a wide variety of indirect measures and criteria are used, including the minimum amount known to be consumed by healthy individuals, the amount below which deficiency symptoms occur and above which they are eliminated, the minimum amount of the nutrient necessary to keep the body's input and output of the nutrient in balance, and in young animals, the minimum amount necessary to maintain "good" growth, which often means maximum growth, a point to which I shall return. Not surprisingly, these various criteria for minima do not always yield the same values.

2. Individuals of the same species and age may differ in their nutritional requirements. Some of these differences are genetic (Durnin et al. 1976; NAS 1975). Some are ontogenetic for example, the body can adapt to lower levels of some nutrients. Unfortunately, few attempts have been made to measure the variability in nutritional requirements. "Apart from the data upon iron losses in women about the only other nutrient where some estimate of this variance is available is for protein. . . . Unfortunately, I do not believe that anyone has ever done an experiment designed to estimate the real variation in requirements of any nutrient in man" (Hegsted 1972).

In establishing recommended daily allowances (RDA's) for human nutrition, the Committee on Dietary Allowances (Food and Nutrition Board, U.S. National Academy of Sciences [NAS]) tries to take such variability into account: the RDA's are "adequate to meet the known nutritional needs of practically all healthy persons" (NAS 1980). Although the RDA's are an attempt to specify something like the mean plus two standard deviations of requirements in the population (Hegsted 1972), the proportion of individuals in a sampled population whose intakes fall below the recommended intake for a particular nutrient is not equivalent to the expected proportion that are deficient for that nutrient. At specified nutrient intake levels, calculating the proportion of a sample population that is "at risk," or the likelihood of a deficiency in individuals of known intake, requires a knowledge of the statistical distribution of nutrient requirements. Beaton

(1974), Hegsted (1972), and Sukhatme (1972, 1975) have provided rare and enlightening discussion of this and other population aspects of nutritional requirements. Their basic approach, that of calculating for each nutrient the probability of a deficiency as a function of intake level, and similarly for toxins, would have been used in this study if adequate information had been available. The upper and lower lines specified in my equations (5.1), and drawn for the example shown in figure 5.1, can be thought of as approximations to the outer limits of such probability functions, which might in some cases be sharp, and in others more gradual.

3. None of the reviews I have consulted has covered all of the relevant literature. A computerized search, carried out for me by the National Library of Medicine, turned up numerous publications on primate nutritional requirements that have not been cited in any of the reviews mentioned above. I estimate that a thorough review of the literature on primate nutritional requirements would take at least a year.

4. There are many sources of errors in the available studies of nutritional requirements (e.g., Munro 1985; Munro and Crim 1988), but they are seldom mentioned, much less estimated. "A major defect in the collecting and processing of dietary data lies in our inability to make precise or even approximate statements concerning the validity and reliability of the various procedures in current use" (Young and Trulson 1960; see also Hegsted 1972 and Woolf 1954). Some of these errors come from surprising sources. For example, at some laboratories of nutritional chemistry, the accuracy and reproducibility of quantitative analyses of nutrients, on which virtually all research in nutrition rests, leave much to be desired. This problem is not unique to laboratories of nutrition; it is widespread in analytic chemistry (Hunter 1980). (For some simple field methods of checking the results of nutrient analyses, see S. Altmann 1979.)

5. For most nutrients, the appropriate parametric studies of requirements have not been carried out. Recent nutritional research has tended to bypass this level of research in favor of the more fashionable subcellular levels, so that we know more about, say, the biochemical reactions vitamin K is involved in than about the amount of it we require.

Is Bigger Always Better?

For most nutrients, requirements are stated in terms of amount per unit body mass. For immature individuals, these requirements are usually based on a search for the amounts necessary to produce "good" or "satisfactory" growth (e.g., FNB 1980; FAO/WHO 1965), not on the amount below which survival or reproduction is compromised.

> Protein intake compatible with a satisfactory rate of growth has been accepted as the basis for estimating protein requirements in infancy. It must therefore be emphasized again that values thus derived are not minimum requirements but are probably optimal levels. (Gopalan and Narasinga Rao 1979)

> The dualism between the physiological minimum and the optimum has exercised all committees since then [Sherman's recommendations of 1933, and rightly so because these figures are of great economic and political importance. (Truswell 1976)

In practice, amino acid "requirements" for infants and children are predicated neither on satisfactory growth nor on optimal growth, whatever those terms mean, but primarily on maximum growth, that is, "on the least amounts compatible with maximum growth" (Munro and Crim 1988). The idea that bigger is better is widespread in the nutrition literature. For example:

> Without some milk it is difficult to supply the amounts of calcium required for high rates of growth and superior adult stature, with which so many other qualities of value are associated. (Leitch 1964; 302)

Yet there are reasons for believing that optimal body size is well below the maximum attainable by dietary manipulation. Slonaker (1931) and McCay, Maynard, Sperling, and Barnes (1939) published the first systematic demonstrations of the effectiveness of dietary restriction—below ad lib saturation levels but above the level of malnutrition—at retarding growth and sexual maturity while extending life span. In 1955 Silberberg and Silberberg reviewed a growing body of evidence that dietary restriction at levels between satiation and malnutrition prolonged the life span of laboratory rats, *Rattus norvegicus.*

> As far as the length of life is concerned, there thus seems to exist an optimum of the duration of underfeeding and apparently also of the age at which food intake is restricted.

In the years since then, the effects of restricted diets on albino rats have been studied by a few investigators (e.g., Berg 1960; Berg and Simms 1960; Ross 1961, 1969). Experimental and epidemiological evidence reviewed by Murray and Murray (1977) suggests that, within limits, "undernutrition" in humans and other animals decreases susceptibility to infection with viruses, malaria, and some bacteria, probably those preferring an intracellular environment. They suggest that Western nutritional standards are set too high for optimal defense against viral infection. Similarly, Stunkard (1983), Guigoz and Munro (1985), Everitt and Porter (1976), and Young (1979) review a growing body of evidence from a variety of mammals (primarily laboratory rodents) and invertebrates that indicates greater life span and lower susceptibility to disease in animals on diets below the levels that lead to maximum growth rate. Clearly, many processes other than total body growth are affected by dietary restrictions. Ross (1969) summarized these changes in laboratory rats as follows:

> Rapid growth rates, structural or biochemical, are not commensurate with prolonged life span and reduced risk to age-associated diseases. Indeed, the dietary regimen which evoked the greatest rate of change with age was most detrimental and such animals had the shortest life expectancy and the greatest incidence and severity of [age-related] diseases.

Miller and Payne (1968) criticized Ross (1961) for feeding his rats a constant diet, that is, for not adjusting the diet for age and sex differences in requirements, but that criticism does not apply to Ross's other experiments (e.g., 1969) and to others in the literature in which the diet was adjusted for age. Like Ross, Miller and Payne obtained the longest life expectancy in rats by giving them a protein-restricted diet that resulted in less than maximum rates of growth, namely, a diet adequate to support maximum growth only for the first 120 days, then essentially to maintain body mass thereafter.

At present, the long-term consequences for fitness of diets that are between nutritionally minimal and growth maximizing are equivocal even in albino rats, the mammals for which the most extensive data are available. While extension of age, reduced susceptibility to age-related maladies, and reduced infertility for animals on restricted diets all augur well, the time to first reproduction in such animals is greater (e.g., Berg 1960), yet this parameter of life histories has been shown to have a particularly strong effect on the intrinsic rate of increase in a population (Cole 1954). Beyond that, rats on restricted diets had smaller litters and higher infant death rates and eating the young was common (Berg 1960). Finally, studies of diet-altered growth in the standard laboratory rat may not generalize to primates. Because the epiphyses of laboratory rats do not ossify, their skeleton continues to grow throughout life (Strong 1926). The physiology of growth in these rodents must be quite different from that in mammals in which the epiphyses ossify.

The restricted diet studies all have been conducted outside the perspective of evolutionary biology, although they address what are basically questions about adaptation. The comparatively moderate conditions in a standard laboratory setting allow the survival of animals with deviant traits that, if present in animals in the wild or in a laboratory setting that more closely approximated natural conditions, would lead to strong counterselection. One finds hints of this in the literature. McCay, Maynard, Sperling, and Barnes (1939) noted that although their rats with diet-retarded growth lived longer than controls, more of them perished when the heating system failed!

Richard Lee apparently shares my dissatisfaction with what is now known about the functional consequences of diet-induced differences in growth rates and body size, and the common assumption that bigger means better. In his book on the !Kung people of the Kalahari desert, Lee (1979; 291) writes:

It may be true, as Truswell and Hansen suggest, that under traditional hunting and gathering conditions, !Kung do not reach their maximum genetic potential (1976:191), but implicit in this statement is the assumption that *bigger* is somehow *better.* For hunting, precisely the opposite may be the case.

Lee provides some evidence that beyond age thirty-five, shorter men have greater hunting success. Similarly, Frisancho, Sanchez, Pallardel, and Yanez (1973) have shown that among Quechua natives of Peru, short subjects (stature less than 160 cm) were better adapted to high altitudes than tall ones, as measured by maximum oxygen uptake per unit body mass.

Martorell and Ho (in press; cited in Martorell 1985) reviewed the literature on anthropometric indicators and the risk of infection. They concluded that no evidence exists for associating size with a greater incidence of infection, nor have nutrition interventions been shown to reduce the incidence of infection. "In a nutshell, nutrition seems to have little to do with who gets sick." However, malnutrition can affect the severity of infections.

Stini (1975) estimated that the low mean body mass of adult male humans in Columbia (60 kg) compared with those in the United States (70 kg) results in an efficiency difference equivalent to about one-third of the daily nutrition requirement. This suggests that one of the major adaptive values for a growth rate that depends in part on available food is that it enables organisms to adapt to an unpredictable and changeable food supply. By growing rapidly only when conditions are particularly favorable, an animal takes advantage of good conditions. However, growth at any given age increases subsequent dietary requirements: it is a commitment. Conversely, slow growth when conditions are poor curtails feeding requirements. For mammals, however, this system is limited. Although many young mammals, including monkeys, exhibit "catch-up growth" when returned to a condition of ready access to abundant, nutritious food (Kerr 1972; Rutenberg and Coelho 1988), dietary restrictions imposed during certain critical periods in early development lead to permanent reduction in adult size (McCance 1976). It is interesting that these animals showed no physiological signs of malnutrition; they behaved as if they were genetically small (ibid.).

In sum, the available information on diet-induced growth retardation provides suggestive but equivocal evidence that fitness is maximized at some dietary intake level below that leading to maximum growth and maximum adult size. Whether altered growth and size are causally necessary for the effect or are only secondary consequences of the dietary restriction is at present unknown. So too are differences between protein-induced retardation and retardation induced by other growth-limiting nutrients.

Body Mass

(Note: Much of the literature in biology confuses mass with weight. Weight is a measure of force and is therefore in units of kilogram meters per second squared, that is, in newtons, whereas mass is measured in kilograms. Thus the weight of an object varies with altitude and látitude and approaches zero in outer space; its mass does not. I have freely used literature indicating the "weights" of animals when clearly only a mislabeling was involved.)

The body masses of the animals in my study are not known. Average first-year growth rate of Amboseli baboons was estimated by J. Altmann (1980) to be 5–6 g/day and birth mass, 0.775 kg. On this basis, my first approximation of growth rate (equation (A9.4)) is 6 g/day, which leads to an estimate of mean body mass (in kilograms) of $0.775 + .042w$, where w is age in weeks. That value, in turn, was used to obtain the first approximations of milk intakes (table 6.6), which were used in some of my other calculations. However, a subsequent estimate (J. Altmann and Alberts 1987), based on larger samples, suggests that the first estimate of growth rate probably is too large. In Alto's Group, growth averaged 4.5 g/day. That estimate is close to the growth rate of 4.7 g/day obtained by Nicolson (1982) for anubis baboons at Kekopey Ranch near Gilgil, Kenya. Thus my second approximation of the regression of body mass on age is

(A1.1)   mean body mass (kg) $= 0.775 + 0.0315w,$

where w is age in weeks. (A mean birth mass of 0.710 may be more accurate, according to J. Altmann and Alberts [1987].) Not only is this growth rate below that of laboratory-reared baboons fed on relatively rich laboratory diets (table A1.1), it is also below the growth rates, during the first four months, of 5.2 and 6.7 g/day in a nursing laboratory baboon male and female, respectively, whose mothers were being fed the lowest protein diet (7.2% of dry mass) on which they maintain their body mass (Buss and Reed 1970). However, some loss of mass during lactation, of fat deposited during pregnancy, is to be expected, judging by human data (Widdowson 1976). At the growth rate indicated by my formula above, an average baboon at 35 weeks would weigh 1,878 g; at 65 weeks, 2,823 g.

Individual mass differences among my subjects, some of which were apparent in the field but which I cannot estimate, were ignored. However, one individual for which mean body mass is not a good estimate is female Pooh, who was very much smaller than the others. I do not know the slowest viable growth rate for baboons, but I estimate it to be about 3 g/day:

$$\text{minimum body mass (kg)} = 0.775 + 0.021w,$$

where again, w is age in weeks. This results in an animal that, at one year of age, weighs 72% of the average for Amboseli baboons. Female Pooh weighed considerably less than that (estimated by eye). She had a short, scraggly coat and impaired locomotion. I was astonished that she lived as long as she did: she was killed by an adult male at age twenty-four months (J. Altmann 1980). I can only speculate that her susceptibility to such rare, lethal aggression was increased by her retarded growth, malformed appearance, and odd behavior. I doubt that she would have bred successfully if she had lived.

Diet and Growth

While this estimate of minimal body-mass values involves an element of guesswork, there are reasons for believing that in Amboseli the mean growth rate in our wild-feeding study groups is not far from the minimum. At the time of my study the population was nearly stationary—births were just keeping up with deaths—and the age distribution was nearly stable (J. Altmann, G. Hausfater, and S. Altmann 1985). Baboons in an Amboseli group that regularly scavenges from tourists and from the lodge garbage dumps are conspicuously larger than wild-feeding individuals. In a study using isotope-labeled water, we found that the difference was due to body mass and fat rather than body length. The garbage-feeding females were 50% heavier and averaged 23% body fat; the wild feeding females, just 2% (J. Altmann, Schoeller, S. Altmann, Muruthi, and Sapolsky 1993). Thus, body sizes in Amboseli baboons, and perhaps also their population size, probably are food limited.

That baboon growth rates are sensitive to dietary intake can readily be demonstrated experimentally. Baboon growth rates during the first year of life have been measured in a number of studies (J. Altmann and Alberts 1986; Berchelman, Vice and Klein 1971; Castracane, Copeland, Reyes, and Kuehl 1986; Glassman, Coelho, Cavey, and Bramblett 1984; Lewis, Berhand, Masoro, McGill, Carey, and McMahon 1983; McMahon et al. 1976; Moore and Cummins 1979; Nicolson 1982; Roberts, Cole, and Coward 1985; Rutenberg and Coelho 1988 (and Coelho references therein); Snow 1967; Vice, Britton, Kalter, and Ratner 1966; Vice and Kalter 1971). Although several of these reports provide or refer to the nutrient composition of the diets that were used, only two, other than ours, provide information on intake quantities, namely, the studies by Vice et al. (1966) and by Lewis et al. (1983). Lewis et al. fed infant baboons on commercial human formula (SIMLAC) on one of two dilutions, 3.84 kJ/g or 2.05 kJ/g, for the first eighteen months of life. Those on the high-energy diet consumed more energy per unit of body mass and, not surprisingly, gained body mass more rapidly. After recovery from the usual postnatal loss, the baboons' energy intake and mass increased essentially linearly with age in both sexes on each diet. For example, males on the rich diet grew at about 12.7 g/day, and their dietary intake grew at about 40.3 kJ/day. Note that this represents a steady *decrease* in efficiency of production in that, with age, progressively larger total food intakes are required to obtain the same amount of growth, presumably because of the ever increasing energy requirement just to maintain the body. For example, at 4 weeks of age on the rich diet, infant males took in 956.3 kJ/day and gained 12.7 g/day. That is, they gained 13.3 mg for every kilojoule consumed at 4 weeks, but by the time that they were 16 weeks old, they took in 1,422.56 kJ/day and still were growing at 12.7 g/day, that is, they gained only 8.9 g/kJ.

For the animals on the high-energy diet, neither the growth rate nor the food increment rate of these early weeks is likely to be sustainable. For example, by age three years, at the growth rate indicated above and a 930 g y-intercept, a male would have a body mass of 14.9 kg. According to Snow (1967, fig. 7), laboratory-reared baboons of that age average 9 kg (albeit probably not on such a continuously rich diet). Increases in food consumption would have to be even more extreme than the growth rate. By the age of three years, energy intake would be 49.6 MJ/day, which even with a food source as concentrated as the high-energy formula used in this experiment would come to 12.9 kg of food per day! Even at the mean age of animals in my study, 47.5 weeks, male baboons with the growth rate of the high-energy diet animals in the study of Lewis et al. would weigh 5,150 g (more than twice what the wild-feeding Amboseli baboons actually weigh, but comparable to the size of garbage feeders there), and their diet would be an enormous 18,867 kJ/day (vs. 1,894 kJ/day for the wild-feeding Amboseli animals). Thus, for such baboons, body-mass production (12.7 g/day) would be 1.0 mg of body mass per kilojoule of food, whereas Amboseli baboons are gaining 2.4 mg/kJ. While Amboseli baboons may be more efficient by this criterion, it is more likely that under laboratory conditions, growth efficiency does not continue to decline as rapidly as predicted by the data of Lewis et al. from the first third of a year.

The faster growth rate on the high-energy diet (Lewis et al.) indicates that growth in those baboons on the low-energy diet was food-limited. But must it therefore have been *energy*-limited? I think not. The difference in the energy density of the two formula preparations was produced by dilution with water, and so these two diets differ not only in their energy content but in all other nutrients, of which any one (or any combination) could be responsible for the observed differences in growth. For the same reason, we cannot say that the smaller size of Amboseli baboons is a result of energy-limited growth, for though the mean energy density of their diet (4.5 kJ/g, mean of four age-class means, table 7.9) is higher than even the high-energy formula in the study by Lewis et al., the diets of these animals differ in many other ways, and so indeed do many other aspects of their lives, doubtless including energy expenditure. Although I subscribe to the hypothesis that body growth of Amboseli baboons is primarily food-limited, the available studies on baboon growth do not resolve the question of the active ingredients limiting growth.

The only other baboon growth study that includes quantitative information on dietary intake is that by Vice et al. (1966). However, dietary data are provided only for the first fourteen days of life, a period in which intake and growth are not typical of later stages of infancy. As in the study by Lewis et al., baboons grew faster on a SIMLAC-based formula of higher energy content, but again the difference in the diets was the result of differences in dilution, so every nutrient is potentially involved in the observed effects. As in the study by Lewis et al., one can conclude only that growth of the baboons on the more dilute diet was food-limited.

Nutrient Requirements and Toxic Limits

I now present the estimates I used for requirements and toxic limits for each of the macronutrient classes: proteins, carbohydrates, fiber, minerals, lipids, water, and energy, as well as total consumption capacity. While a few vitamin analyses have been carried out on some primate foods in Amboseli (Altmann, Post, and Klein 1987), our coverage of these micronutrients is not yet adequate to use them in calculating optimal diets.

Protein Requirements

Mammals have no requirement for any specific protein except eight amino acids that they use but cannot synthesize and that are essential in the diet, plus an additional and larger amino nitrogen component for synthesis of nonessential amino acids (FAO/WHO 1973; Rose and Wixom 1955). Two additional amino acids, cysteine and histidine, are regarded as required during infancy in many mammals, possibly excepting humans (Holt and Snyderman 1965), because internal synthesis is not fast enough for maximum growth. As I have indicated above, however, this is a criterion of dubious value. In order to be utilized, the essential amino acids must be consumed within a relatively short time of each other (Block and Mitchell 1946; Albanese and Orto 1973).

To date I have no data on the amino acid composition of any Amboseli plants and therefore shall consider only the total requirement for total amino nitrogen. For convenience, this requirement can be broken down into two components, the requirement for maintenance and the requirement for growth.

One way to estimate the minimum requirement for total amino nitrogen in nongrowing individuals (or the maintenance requirement of growing individuals) is to measure the asymptotic level of nitrogen excretion, in feces, urine, sweat, and so forth, of nongrowing individuals (adults) kept on an essentially nitrogen-free diet (e.g., FAO/WHO 1973). In human subjects on a protein-free diet total daily nitrogen loss from all sources totaled 54 mg N per kg body mass (references in FNB 1980, 39–40). At a conversion rate of 6.25N, this amount of nitrogen would be contained in 0.34 g of protein. For several reasons, however, consumption of just 0.34 g protein per kg body mass per day is inadequate to maintain a human in nitrogen equilibrium—though whether nitrogen equilibrium is the best measure of amino nitrogen requirements is controversial (Carpenter 1994).

The quantity of dietary protein needed to maintain a human subject's nitrogen balance varies from one food to another: we can utilize protein more efficiently from some sources than from others. Efficiency of protein utilization depends on several charac-teristics of foods, including their amino acid balance and fiber content (Mitchell 1964; Nicol and Phillips 1976b) and their digestibility (Block and Mitchell 1946).

Efficiency of protein utilization also depends on the dietary backgrounds of the in-dividuals (Alison 1951; Dubos 1980) in ways that probably are important for Amboseli baboons. In a series of studies, Nicol and Phillips (1976a, 1976b, 1978) demonstrated that moderately active young Nigerian men who customarily ate low-protein diets were more efficient at utilizing dietary proteins than the better-fed Americans and Europeans who have been the subjects in many previous protein utilization studies. For example, the University of California students studied by Calloway and Margen (1971) had a net protein utilization (NPU) of 75% whereas the Nigerians utilized 90% in one study (Nicol and Phillips 1976b, table 5) and 84% in another (Nicol and Phillips 1978, table 3). Some of the physiological processes that are involved in adaptation to low pro-tein intakes have been described by Stephen and Waterlow (1968) and Waterlow (1964). Egg protein is known for its fairly high digestibility, so that, as expected, the proportion of dietary nitrogen retained by the Nigerian subjects was lower when the sources were more typical, primarily vegetable diets composed of local foods based on sorghum, rice, or cassava. The average NPU for eight such diets with no amino acid supplements was 67% (Nicol and Phillips 1978, table 3 and 284), a value indicating greater efficiency than expected by applying the recommendations of FAO/WHO (1973). On these diets, the amount of dietary protein these Nigerian men required to maintain nitrogen balance would be 0.55 g/kg body mass per day (Nicol and Phillips 1978, 284), which, at an average mass of 56.84 kg (SE 0.48; data for eight unsupplemented diets in their table 3) comes to 31.3 g protein per day. By way of comparison, the "allowance for the mixed proteins of the United States diet" is 0.8 g per kg of body mass per day (NRC 1980), and FAO/WHO (1973) allows 0.73 g/kg/day.

Amboseli baboons, like the Nigerian subjects, probably are accustomed to a low-protein diet and have comparably greater protein utilization efficiencies. Further-more, the traditional low-protein nearly vegetarian diets of these Nigerians seem to re-semble the diets of Amboseli baboons more closely than do the much richer customary diets of the Western subjects in most comparable studies. In this regard it is interesting that the protein in the eight unsupplemented diets in the Nicol and Phillips experiments (1978, table 3) had an average "true digestibility" of 89%, and the average for the diets of eight species of herbivorous placental mammals was 87% (data in Robbins 1983, table 13.2). I have therefore assumed that on average their foods have the same NPU value (0.67) and have scaled the results above to Amboseli baboons. In scaling I as-sumed, as did Kleiber (1945), that nutrient requirements for placental mammals are in proportion to basal metabolic rate and thus in proportion to body mass to the three-quarter power.

This scaling procedure is consistent with what little is known about the effects of body size on protein metabolism. Brody, Proctor, and Ashworth (1934) have shown that the rate of endogenous nitrogen secretion is $kM^{0.72}$, where M is body mass in kilograms and k is a constant. Munro (1976) has commented:

> It is important to recognize that the intensity of protein metabolism varies
> in proportion to the body size of the mammal. This is well known in energy
> metabolism, where basal energy output is related to three-fourths power of body-

mass. Most parameters of protein metabolism follow the same general relationship, e.g., requirements for threonine and methionine per kg body-mass of the mature rat that are about four times that of man.

Similarly, Miller and Payne (1963) wrote: "The term M [nitrogen required for maintenance] in equation (2) has been evaluated by a number of authorities and shown to be related to kilograms body weight [mass] (W) to the power of 0.73, in the same way as basal metabolism."

Kleiber's exponent of 3/4, which I took as a first approximation, is higher than the value obtained by Hayssen and Lacy (1985) based on a much larger sample,[7] but the difference at baboon size probably is small compared with other potential sources of error. In particular, little work has been done on the scaling of various nutrients across species, and the scaling of metabolic rate within species, for individuals of different sizes and ages, may scale differently than it does across species.

Taking the scaling coefficient to be 3/4, protein requirement is $kM^{3/4}$, where k is a constant of proportionality and M is body mass in kilograms. As indicated above, body mass of the subjects in the studies of Nicol and Phillips averaged 56.84 kg, hence 31.3 g protein/day $= k\,56.84^{3/4}$. Solving for k we get 1.51 g protein per $M^{3/4}$ per day. Therefore the estimated daily requirement for vegetable protein (NPU $= 0.67$) for a nongrowing baboon (or other placental mammal) of mass M kg is 1.51 $M^{3/4}$ g. If we assume an average of 16% nitrogen in proteins, this requirement is equivalent to 0.24 $M^{3/4}$ g dietary nitrogen per day, which is below all of Robbins's tabulated values (Robbins 1983, table 8.4) of nitrogen intakes at which various placental mammals achieve nitrogen balance, including one primate, rhesus monkeys, at 0.68 $M^{3/4}$ g/day. I suspect that most such values were obtained from experimental animals kept on relatively high-protein diets. A persistent problem in estimating nutrient requirements is that the nutrient intake that is required to achieve any specified level of functioning, such as nitrogen balance or a particular growth rate, depends in part on recent intake levels.

To this maintenance requirement, we must add sufficient protein for minimal viable growth. As a first approximation, I estimated this latter value as follows. Consider a baboon growing at the estimated minimum viable rate, 3 g/day. Approximately 17% of its body is protein.[14] If we assume, as does FNB (1980), that the efficiency of protein utilization for growth is the same as for maintenance, then, at a net protein utilization value of 0.67, as in Nicol and Phillips (1978, table 3, eight unsupplemented diets), a daily increment of 3 g of body mass (510 mg of protein) requires 761 mg dietary protein per day. Similarly, to maintain the average daily gain of 4.5 g body mass would represent 765 mg protein and would require 1.14 g dietary protein (NPU 0.67) per day. More generally, the protein requirement for growth is 0.254 g/day for each gram of growth.

We can now combine the infant baboons' protein requirement for maintenance with their requirement for minimal growth:

$$\text{Protein requirement (g/day)} = 1.51\,M^{3/4} + 0.254\,(\Delta m),$$

where M is body mass in kilograms and Δm is growth rate in grams per day. To illustrate, at the mean sample age of my subjects, 47.5 weeks and so with an estimated mean mass of 2.27 kg, an infant baboon would require 3.9 g of dietary protein (NPU 0.67). This comes to about 1.7 g protein per kg each day. Similarly, at the average growth rate of 4.5 g/day, the protein requirements range from 1.6 g kg^{-1} day^{-1} for 65-week-old infants to 1.9 g kg^{-1} day^{-1} for 35-week-old infants.

Is this estimate of the protein requirement consistent with the limited data available on primate protein requirements? It is well below the 5.0 g kg^{-1} day^{-1} of "high quality protein" (i.e., of correct amino acid ratios) suggested by Nicolosi and Hunt (1979) for growing baboons, but the studies on which they base their recommendations are very limited, and none represent the state of the art (Rand, Seriskav, and Young 1977). Samonds and Hegsted (1973) found that the amount of protein needed to maintain infant cebus monkeys was 2.0 g kg^{-1} day^{-1} for lactalbumin, 2.8 for wheat gluten + lysine, and 8.7 for wheat gluten alone; this emphasizes the significance of the quality of the protein and of the dietary mix. Robbins and Gavan (1966) indicate an average 2.6 g/kg daily intake for adult rhesus who on average were nearly in nitrogen equilibrium, but with a range of nitrogen retention from −31% to +28% of this value. (The sources of this individual variability are unknown.)

Daily protein intake of 2 g protein per kg body mass was inadequate to maintain body mass in young, nonpregnant female rhesus monkeys (Riopelle, Hill, Li, Wolf, Seibold, and Smith 1974) and so, one would think, would be too low for infant rhesus or infant baboons. However, a closer examination of the results in Riopelle et al. (1974) suggests an alternative explanation. In their study, individuals on diets providing 2 g protein and others on diets providing 1 g protein per kg body mass per day dropped to about 85% of their initial body mass; after about 130 days, they began gaining again. (Those on just 0.5% protein continued to lose mass.) Yet unprovisioned but healthy and vigorous wild baboons grow at slower rates than do laboratory-raised animals (J. Altmann and Alberts 1987; J. Altmann and S. Altmann 1977; Nicolson 1982). Probably the same is true of other species. Judging by baboons in Amboseli, a body mass considerably below that of a typical lab monkey is quite consistent with good health (see also J. Altmann, Schoeller, S. Altmann, Muruthi, and Sapolsky 1993). Furthermore, the increase in mass beyond 130 days in the study of Riopelle et al. (1974) suggests accommodation to the low-protein diets. So too does the fact that by 160 days, the plasma protein and plasma albumin concentrations of the animals on the 2 g and 1 g protein diets were virtually identical with the controls on a diet of 4 g protein per kg per day. The animals remained in "apparent general good health."

In a study by Kerr, Allen, Scheffler, and Walsman (1970), rhesus monkeys fed ad lib from age one to seven months on a formula with half the protein of normal formula but made isocaloric by addition of lactose grew as rapidly as the controls on full-strength formula and appeared behaviorally normal. This low-protein diet provided about 1.3 g protein per kg of body mass per day, equivalent to about 1.6 $M^{3/4}$, that is, less than my estimated requirement.

None of these studies are conclusive; and the tendency of animals and humans to adapt over time to low-protein diets has largely been ignored. Therefore, in lieu of better evidence, I will use the requirement levels that I have indicated above.

Protein Optimum

Very large protein intake rates can be harmful (see below). Before toxic limits are reached, however, what are the functional consequences of taking in more protein than the lowest amount required for nitrogen balance (or in the case of infants, for nitrogen balance plus growth)? In particular, is there an optimal protein intake above the minimal requirement?

The Food and Nutrition Board (FNB 1980, 51) maintains that an intake of protein above the recommended daily allowance has no effect except in special circumstances: "There is no compelling evidence to show that such higher intakes are either beneficial or harmful, except in the special case of very small premature infants." Similarly, the FAO/WHO (1973, 9) committee says that "there is no demonstrated harmful or beneficial effect of [protein] intakes well above the probable requirement."

By contrast, Slonaker (1931a, 1931b), in perhaps the first systematic demonstration of optimal nutrient intakes, showed that rats on a diet with 14% protein were more active, had a longer reproductive life span, produced more and larger litters, and had a greater life expectancy than rats on diets with higher or lower protein concentrations. Several other studies have shown benefits of high-protein diets. Perhaps best documented are protein's antitoxin effects. While most of the studies on this topic have used synthetic compounds, such as pesticides, some have used naturally occurring toxins. A protein-rich diet protected rats from the toxic alkaloids produced by ergot and from emetine (Wilson and Eds 1950). The sulfur-containing amino acids protected rats from the toxic alkaloid in tansy ragwort *Senecio jacobaea* (Cheeke and Garman 1974). In chicks, methyl-donating amino acids provided protection from tannic acid (Armanious, Britton, and Fuller 1973; Fuller, Chang, and Potter 1967; Rayudu, Kardivel, Vohra, and Kratzer 1970). In mice, bee venom was more toxic to animals with a protein deficiency and those on a protein-rich diet than those on intermediate diets (Benton, Morse, and Gunnison 1964). This situation is reminiscent of the results of very thorough studies by Boyd and colleagues on effects of protein intake on pesticide resistance, in which an optimum protein intake was found (26% casein in the feed) above and below which pesticide tolerance was lower (Krijnen and Boyd 1970). Such antitoxin effects may be particularly important to animals subjected to large quantities of toxic secondary compounds from plants. For a few synthetic compounds (e.g., carbon tetrachloride and heptachlor), toxicity is lower in animals on low-protein diets. This reverse effect may result from compounds that are not toxic themselves but have a toxic metabolite (Basu 1981).

Taken together, these studies of toxin resistance and those reviewed in the section above on growth rates suggest that for wild mammals there are optimal levels of dietary protein above the amounts required for nitrogen equilibrium. Dietary protein levels probably also affect other components of fitness, such as disease resistance (Faulk and Chandra 1981; Guigoz and Munro 1985), but with few exceptions, such as those by Saxton, Boon, and Furth (1944) on leukemia and those by Ross and Bras (1971) on adenomas, few quantitative studies have been carried out and present data are inadequate for establishing optimal dietary requirements. To date, the quantitative relationship between protein intake and biological fitness has not been determined for any wild animal. In one version of the optimal diets that are presented in chapter 5, I consider a forager that maximizes protein intake while nonetheless remaining below the toxic limit.

Protein Limit

A surfeit of protein may lead to calcium imbalance (Anand and Links-Wiler 1974). However, within limits, the adverse effects of high-protein diets on calcium metabolism can be counteracted by higher intakes of calcium (Anand and Links-Wiler 1974) or phosphorus (Hegsted et al. 1981; Schuette and Links-Wiler 1982; Yuen and Draper 1983). No safe maximum is known for protein intake, apparently because of internal homeostatic mechanisms (Anderson 1988). For the first time, the 1989 edition of the U.S. Recommended Dietary Allowances (FNB 1989) included a recommended maximum: that protein intake not exceed twice the RDA of 0.8 g/kg body mass, but without providing good evidence for the selected value (Carpenter 1994). I have taken the baboons' consumption capacity as setting the upper limit for their protein intake.

The relationship mentioned above between protein toxicity and calcium illustrates a common phenomenon that I believe is of major importance: in many cases, food components interact in such a way that requirements and tolerances for a food component are not fixed values but depend on the intake levels of other food components. Over a hundred such interactions are known for nutrients alone (Wise 1980). For non-nutrient toxins too, many interactions with other food components have been studied. Tannin from leaves, for example, forms relatively indigestible complexes with leaf proteins. Oxalic acid binds with calcium ions to form calcium oxalate, an insoluble precipitate. Many other examples are given by Freeland and Janzen (1974), by Reese (1979), and by Oltersdorf, Mittenburger, and Cremer (1977).

Analytically, such interactions are not a problem in the calculation of optimal diets: they would merely require additional constraints in equation (5.1) for the tolerated or required combinations of components. The problem is empirical: interaction effects are not yet sufficiently well known to use them in establishing nutrient requirements.

Fat Requirement

Several essential fatty acids are now known for primates (FNB 1989), and depending on their representation among fats, these set a lower limit on the viable dietary content of lipids—"fats" in colloquial English, though lipids include more than fats. For humans, FNB (1980, 33) recommended a minimum of 15 g fat intake per day for human adults. If we assume that this is for a 50 kg individual and that the body-mass scaling coefficient is 3/4, the requirement for mammals of other sizes can be written:

$$\text{Fat requirement (g/day)} = 0.8\ M^{3/4}.$$

For example, a baboon at the mean sample age and mass, as above, would require 1.48 g dietary fat per day.

Carbohydrate Requirement

Carbohydrates are a major source of food energy, but proteins and fats provide alternative sources. No carbohydrates are required in the diet. The carbohydrate minimum has therefore been set at zero.

Carbohydrate Limit

No carbohydrate toxic limit is known. The upper limit has been set at consumption capacity.

Fiber Requirement

The disadvantages of low-fiber Western diets have been the subject of considerable research (Burkitt 1982). One study is particularly relevant to our work. Sly et al. (1980) produced deficiency symptoms (chronic inflammation near the iliocecal valve) in four of six anubis baboons on a fiber-free diet but not on a similar one containing 0.70 g fiber per 100 g air-dried food. I have therefore taken half of this value, 0.35 g per 100 g dry food, calculated on the basis of oven-dried material, as the minimum.

Fiber Limit

Although large quantities of fiber can reduce the ability of the gut to absorb minerals (Hallfrish, Powell, Carafelli, Reiser, and Prattner 1987 and references therein), no toxic limit is known. I have set the maximum at consumption capacity.

Energy Requirement

The requirement for dietary energy in a homeotherm depends on activity level, growth rate, ambient temperature, and other factors. Methods currently used to establish human requirements for energy are inadequate: several studies have shown average energy intakes in Third World countries much lower than the requirements proposed by WHO and FAO (Waterlow 1981). The situation is like that just reviewed for proteins.

Unfortunately, we do not know either the actual or the minimum viable energy expenditure level of Amboseli baboons. Several ways of estimating the minimum have been considered. First, the regression of total mammalian metabolism on body mass, while so far based only on studies of small mammals (typically adults), suggests a value of 800 $M^{0.71}$ kJ/day (Garland 1983), which comes to 1,432 kJ/day for the average baboon (2.27 kg) in my samples. This regression is based on actual metabolism, not minimum, and so indicates only an upper bound on the minimum. Conversely, a lower bound is established by basal metabolism, 354 $M^{3/4}$ kJ/day (Hemmingsen 1960), which is 654.7 kJ/d for my average sampled weanling baboon (d = day).

Second, large Old World monkey infants in captivity take in 200–300 kcal (= 837–1255 kJ) per kilogram each day (NAS 1978; Nicolosi and Hunt 1979). It is difficult to know what to do with this figure, however: as pointed out above, laboratory baboons grow considerably faster and to a larger adult size than those in Amboseli, yet they probably get much less exercise.

The method I decided to use is this. Suppose a baboon exerted only as much energy as needed to maintain its basal metabolism plus the amount required to stay with its group, to perform other vital activities, and to grow. The cost of basal metabolism is, as indicated, 354 $M^{3/4}$ kJ/d. Mean distance traversed per day by Amboseli baboons, based on average adult female values, is 6.1 km (Post 1981),[15] and energy expended by mammals on locomotion is 10,700 $M^{0.684}$ J km^{-1}, regardless of speed (10.7 $M^{-0.316}$ J m^{-1} kg^{-1} from Taylor, Heglund, and Maloiy (1982) multiplied by M and by 1,000 m/km), and so baboons expend 65.3 $M^{0.684}$ kJ per day on locomotion. Finally, growth costs about 21 kJ/g (Payne and Waterlow 1971), so at the estimated minimum viable rate of 3 g/d, growth requires 63 kJ/day and at the mean growth rate (4.5 g/day) requires 94.5 kJ/day. The total at mean growth is $354\ M^{3/4} + 65.3\ M^{0.684} + 94.5$ kJ/d. Since there is, within the range of body masses that I am considering, always less than a 1% difference in the sum between using the exponent 0.684 for the second term and using 0.75, I have, for convenience used the latter value. The difference, for a yearling, adds about 6% of the cost of locomotor energy as an allowance for other vital activities. Energy requirements at levels of minimal activity and at average or minimal growth can now be stated:

$$\text{Energy requirement, average growth (kJ/day)} = 419.3\ M^{3/4} + 94.5;$$

$$\text{Energy requirement, minimal growth (kJ/day)} = 419.3\ M^{3/4} + 63.$$

For example, at the mean sample age of 47.5 weeks, with an average mass of 2.2727 kg (i.e., after average growth), this comes to 870.6 kJ/day. We note that this represents a daily intake that is roughly 30–46% of the 837–1,255 kJ/kg that is taken in by captive primates of this size fed ad lib (citations above).

Energy Limit

No energy maximum is known. The energy maximum was therefore determined from a consumption capacity amount of the most energy-rich food available.

Mineral Requirement

Requirements for six minerals (sodium, potassium, chlorine, calcium, magnesium, and phosphorus), in percentages, make up the bulk of the dietary requirement for cations and anions. All others are required in amounts of not more than 100 ppm in the diet (Nicolosi and Hunt 1979). The dietary minima for these six minerals total 1.73% of the diet (ibid., table IV). I therefore took 1.75% as the minimum dietary requirement for total minerals, and of course this assumes that all the minerals are present in just the right proportions. Nicolosi and Hunt's RDA's for the five cations listed above are shown in table A1.2.

Mineral Limits

How much of the ideal mineral mixture, referred to above, could be eaten with impunity? The lowest safety margin I have found in my initial search is for magnesium: it has a cathartic effect on humans at about 3 to 5 g of magnesium salts per day (FNB 1980, 134), the lower value of which, at an assumed body mass of 50 kg, comes to 60 mg kg^{-1} day^{-1}, a value that is 7.5 times the human daily allowance of 8.3 mg kg^{-1} day^{-1}. (FNB [1980, 135] gives 50 mg/day for a 6 kg human infant.) Assuming the same safety margin for baboons, the ideal mineral mixture would become dangerous at $7.5 \times 1.75\% = 13.1\%$ of the diet; that value was used to establish a total mineral constraint for diets 1–5 in chapter 5. Of course, if the proportion of magnesium is higher or lower than in the ideal mix, the limit on total mineral intake will be correspondingly different.

After I calculated diets 1–5, I found an extensive survey of mammalian cation tolerances (NRC 1980). Values from that publication, shown in table A1.2, were used to establish cation constraints on diets 6–10 (chapter 5) and on the semimonthly individual diets (chapter 7).

Water Requirement

The water requirement of baboons in Amboseli is unknown. An upper limit on water requirements is the water turnover rate of wild animals or of captive animals offered water ad lib. In mammals, this turnover is a curvilinear function of body mass (Robbins 1983):

$$\text{Water requirement: turnover rate (g/day)} = 120 \, M^{0.84} \, (r^2 = 0.96),$$

where M is body mass in kilograms. For the animals in my study, this gives an average daily water intake of 238.9 g, or roughly 11% of body mass per day. Similar results were obtained by Adolph (1949; 100 $M^{0.88}$), Richmond et al. (1962; 120 $M^{0.80}$), and Nicol (1978; 100 $M^{0.80}$).

Second, according to FNB (1980; 168), the water requirement depends on the activity level and is 1.5 ml/kcal, presumably of dietary intake. That value is 50% higher than found for pigtailed macaques, *Macaca nemestrina:* 1 ml (1 g) per kcal (4.184 kJ) of gross energy (Pace, Hanson, Rahlmann, Barnstein, and Cannon 1964; Kerr 1972). If, as above, we take the energy requirement to be 419.3 $M^{3/4}$ + 63 kJ at the minimum growth rate and 419.3 $M^{3/4}$ + 94.5 kJ/day at the average growth rate, then the water requirement, based on the pigtailed monkey value, is:

$$\text{Water requirement: metabolic, at minimal growth (g/day)} = 100.2 \, M^{3/4} + 15.1;$$

$$\text{Water requirement: metabolic, at average growth (g/day)} = 100.2 \, M^{3/4} + 22.6.$$

In the latter case, this amounts to an average of 8% of body mass per day.

Third, the amount of obligatory water required for renal clearance depends on the diet, primarily of proteins and salts, whose end-products constitute all but about 15% of the osmolarity of the urine. Each gram of dietary protein results in the production of about 0.03 g of urea, which contributes 5 milliosmols (mosM), and each gram of sodium chloride contributes 34 mosM. The kidneys of humans and presumably also baboons cannot concentrate the urine to more than 1,200 mosM/liter (Davidson, Passmore, and Brock 1972). Consequently there is an obligatory water excretion by the kidneys, which in turn establishes an obligatory water intake. Allowing 15% contribution to osmolarity from other sources, the requirement can be written as:

$$\text{Water requirement: renal clearance (g/d)} = \frac{5 \, p + 34 \, s + 0.15 \, (5 \, p + 34 \, s)}{1.2}$$

$$= 4.79 \, p + 32.58 \, s,$$

where p is the number of grams of protein consumed per day and s is the number of grams of sodium chloride. I have assumed that the other salts were the only other significant contributors to urine osmolarity and also—because I do not know what fraction of the total ash content of the baboon foods was sodium chloride—they had the same impact, per gram, on urine osmolarity as does sodium chloride. I have therefore estimated the obligatory water excretion and its consequent effects on water requirements from

$$\text{Water requirement: renal clearance (g/day)} = 4.79\,p + 32.58\,a,$$

where a is the number of grams of ash in the daily diet and p is as before the number of grams of protein. The amount of water baboons lose by other means, such as evaporation from lungs and epidermis and excretion from the gastrointestinal tract, is unknown.

In my study, the amount of water that the baboons obtained from their food could be estimated, but the amount they obtained by drinking could not; in the absence of evidence of acute dehydration, I can only assume that any water deficiency in the baboons' foods was made up by drinking. Water requirements will be useful primarily in indicating the magnitude of this deficit that must be satisfied by drinking.

Water Limit

In sufficiently large quantities, water is toxic (Wynn and Rob 1954). However, in healthy animals, input/output regulation precludes water intoxication (Andersson, Leskell, and Rungren 1982). Kerr, Allen, Scheffler, and Waisman (1970) suggest that excess water may have been the cause of some deaths of infant rhesus monkeys, age one to seven months, taking in near normal quantities of nutrients by consuming very large quantities of very dilute formulas, in a no-choice situation. The foods available to Amboseli baboons are far too low in water content to present any water intoxication problem, and I have no record of the amount of water they drink. For computational purposes, the water limit was set at the consumption limit, which just means that the water contained in food never reaches toxic levels.

Consumption Capacity

To avoid the possibility of calculating an optimal diet whose volume exceeded the consumption capacity of the animals, I have estimated the maximum amount of material a baboon can eat per day. Because no values for primate consumption capacities could be found in the literature, I turned to the literature on the size and throughput rate of the gastrointestinal tract. The amount of food an animal can consume in a day is limited by two factors, the instantaneous content of the gastrointestinal tract ("gut capacity," Demment 1983) and the mean length of time food takes to pass through the tract ("retention time"). Consumption capacity is maximized only if gut capacity is maximized and retention time is minimized. (Of course, by other criteria, neither of these conditions may be ideal, e.g., retention time may have an optimal value above and below which absorption rate is lower.) Neither gut capacity nor retention time is a constant. Periodic intake of large quantities of food leads to enlargement of the gastrointestinal tract (Fabry 1967). Nonetheless, there must be some upper limit on its size beyond which it cannot be further enlarged. Similarly, retention time is not a constant: it depends on many characteristics of the diet, such as the amount of fiber and fat, but it too must have a limit.

Unfortunately, I have not been able to find maximum values for primate gut capacity or minimal retention times. Milton (1984) provides unique data on food passage time in several primate species (see below). A lower bound on instantaneous gut capacity can be established from the volume of a dissected gastrointestinal tract immersed in and filled with water, without distension. These volumes have been measured for eight species of Old World monkeys of known body mass (Chivers and Hladik 1980), to which I have added another (from Post 1982) by assuming a gut-content specific gravity of 1.0. Where more than one sample was available for a species, I used the mean values of body mass and gut volume for that species. Linear regression of body mass against the logarithm of the volume of gut (stomach + small intestine + cecum + colon) against the logarithm of body mass gives the following relationship:

$$V = 220\ M^{0.83}\ (r = 0.80),$$

where V is the volume capacity of the gut in cm^3 and M is body mass in kilograms. At an assumed specific gravity of 1.0 for gut contents, V is also mass capacity of the gut in grams. For a baboon at mean sample age 47.5 weeks and therefore with a mean body mass of 2.27 kg, instantaneous gut capacity, according to the regression above, is 435 g, about 19% of body mass. If we assume that an animal can fill its gut to capacity at least once a day, consumption capacity is at least that great.

Are the values of V for various ages consistent with the actual intakes of the baboons in this study? No. In figure A1.1, I show the mean daily mass of food consumed by the baboons at various ages and the corresponding values of V. As can be seen in the figure, the actual intakes of the baboons were often well above the putative maximum, V. The greatest discrepancy, female Eno at 30–40 weeks of age, is slightly more than double the value for V_{35}. This suggests either that baboons can put more into their guts than the values obtained from the data in Chivers and Hladik (supra) or that baboons can fill their guts more than once a day, or approximately, that the mean passage time of food through their guts is less than eleven hours, the number of potential feeding hours per day. I have therefore based my estimate of consumption capacity on the baboons' actual intakes, as follows. The values in figure A1.1 do not indicate a systematic increase in mass consumption over the age range 30–70 weeks. Indeed, as the graph shows, there was a slight decrease with age in the mean. Furthermore, there was little consistency over time in the intakes of individuals. (Perhaps both of these results indicate the confounding effects of seasonal shifts in diet, as discussed in chapter 8.) I have therefore treated the twenty-eight mass intakes, each for a different individual × age block (table 7.8), as independent samples from a common population of mass intakes, and tentatively I have adopted the mean plus three standard deviations of these intakes, namely 1,002.4 g/day, as the consumption capacity of Amboseli baboons at 30–70 weeks of age. If intakes are normally distributed, as appears to be the case (fig. A1.2), the mean plus three standard deviations would include 99.87% of the population.

Milton (1984) has published food passage times (first appearance in feces of markers in food) for ten species of New World monkeys, three species of apes, and humans, but no Old World monkeys. These should provide lower bound estimates of mean retention times. Values ranged from 3.5 hours for cebus monkeys to 36–38 hours for the apes and were largely a function of body mass. If the same relationship of mass to passage time applies to baboons as to these primates, yearling baboons should have a mean passage time of about 3–5 hours, a result consistent with my hypothesis that they can eat more in a day than their gut capacity.

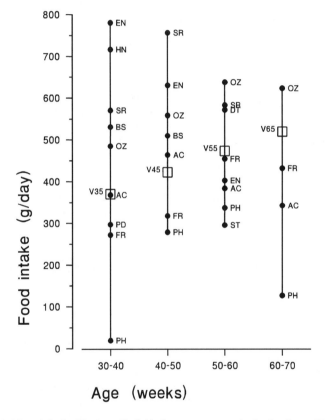

Fig. A1.1. Mean daily food intakes of individuals versus gut capacity. Intakes from table 6.1. Volume capacities of the guts (V's) calculated at the midpoint of each age block (ages 35, . . . , 65 weeks). Volume capacities (cubic centimeters) converted to mass capacities (grams) by assuming a diet with a specific gravity of 1.0.

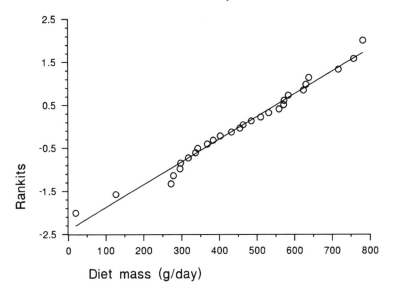

Fig. A1.2. Rankit plot of mean daily diet masses. Based on 28 diets of 11 yearlings in whichever of the four age blocks each was sampled (table 7.8). Regression: $N = 28$, $R^2_{adj} = 0.98$, $p \leq 0.001$, $y = -2.41 + 0.0053x$. A linear rankit plot implies a normally distributed frequency distribution (e.g., Sokal and Rohlf 1981).

Appendix 2

Alternative Definitions of Feeding Bouts

In our fieldwork on foraging by Amboseli baboons, David Post, Jennifer Shopland, and I each used somewhat different definitions of a feeding bout. More than trivial semantics is involved: criteria for when a feeding bout begins and ends affect its putative length. The beginning of feeding bouts is behaviorally more distinct, so our definitions differ little for that.

For Post, a feeding bout was demarcated as follows:

> Feeding began when the subject first made contact with any part of a food plant, excluding contact with the plant as a locomotor substrate. Feeding bouts terminated when the subject either (a) moved more than one full stride, even if it was carrying part of the plant in its hand or mouth, or (b) stopped looking at the food plant. By this definition, a switch to a new food type in the absence of either of these conditions was not sufficient for the bout to be considered terminated; thus a single feeding bout could include more than one food type. (Post 1981)

In this definition, the criterion for bout onsets is virtually identical with mine. For bout terminations, the criterion of at least one full stride serves the same purpose as my two-second criterion: to bridge the gap between foods eaten in volleys. In most situations, little difference in bout lengths would result from using one or the other criterion. The stride of an adult baboon takes about one second. I found the full stride criterion to be more difficult to use in practice, particularly when the monkeys were in complex locomotor situations (e.g., when clambering about in a fever tree to get at gum droplets) and to be irrelevant to the nursing situation. Contrary to the definition of Post, I continued timing a bout whenever food was carried away in the hand or mouth, for otherwise some feeding would take place outside any defined feeding bout. However, such food carrying was sufficiently rare in Amboseli baboons that I doubt it accounts for very much of the difference in our results. His criterion (b), that the animal stopped looking at the food plant, is roughly equivalent to my criterion that they broke contact with it: in most situations the two events occurred almost simultaneously.

However, one appreciable difference occurs in our definitions of a feeding bout. By my criterion, a complete switch to another food always initiates a new bout, regardless of whether the animal is still in contact with the same plant. So, for example, if a monkey switches from green fruit of a *Salvadora persica* bush to the leaves of that bush or to ripe fruit on that bush, I record the onset of a new bout, but this would not be done by the criteria of Post, who in those circumstances would record a bout that included more than one food. (Note that such a multifood bout is different from my "overlapping bouts" in which an animal feeds simultaneously on two or more foods, which was not precluded by my criterion of a *complete* switch to another food but nonetheless occurred very rarely, as indicated in the text.)

For two reasons, recording a new bout whenever the animal switches completely to a new food is preferable. First and foremost, the nutritional content of single-food bouts will be more homogenous, in a statistical sense, than will multifood bouts, thereby leading to more accurate estimates of nutrient intake levels. Second, if food-specific time budgets are calculated only from the single-food bouts, as in Post et al. (1980), then the sum of these time budgets will be different from the total feeding time budgets, since the latter also include the multifood bouts. That procedure only confuses an analysis of the contribution of each food to the total diet or to the total time budget. The one disadvantage of switching whenever the animals do is that the resulting bout length distributions are sensitive to one's criteria for a "new" food, since particularly near the beginning of such a research project, one will have limited information on which items show significant nutritional differences. Fortunately, the major nutritional changes that occur in fleshy fruits when they ripen are usually accompanied by color changes.

Jennifer Shopland carried out another field study of feeding behavior in Amboseli baboons that utilized a definition of feeding bouts nearly the same as mine, yet different in one small respect. In Shopland's samples, a feeding bout is

> an interval of time that begins when a baboon touches any part of a food plant, a potential prey animal, or an edible product of these (e.g., fruit, egg). This defining contact does not include incidental contact while moving past an object or contact with a plant as a support or locomotor substrate, but it does include grasping plant parts to bring a food item within reach. The feeding bout ends when (1) the baboon has been out of contact with the food for 2 s or longer, (2) the baboon does not break contact but begins using the food plant solely as a support or locomotor substrate for more than 2 s, or (3) the baboon switches to another food type, regardless of whether 2 s have elapsed after contact with the previous food. Thus, a feeding bout never involves more than one food type. Moreover, a bout does not end if the baboon shifts its location but continues to feed on the same food type without a 2-s break. By these criteria feeding bouts occurred during which no food was consumed ("empty" bouts)—for example, if a baboon pulled a branch toward itself, examined the fruit on it, and then released the branch without plucking any fruit—but such bouts were rare. (Altmann and Shopland in prep.)

Shopland's criteria and mine differ in only one respect, item (1) in her criteria for bout endings: she included the terminal two seconds after the animal broke contact with the food plant, but I did not. This was a consequence of a difference in the timing equipment we used. With my cumulative timer, it was easy to press the "read" button the instant the animal terminated contact, then ignore the reading if within two seconds the animal made contact with another plant of the same species, whereas with the event recorder that Shopland used, the time was recorded automatically at the instant the appropriate button was pressed. Two seconds seem like a trivial difference until one considers that, by our criteria for bouts and for new foods, almost all feeding bouts lasted less than sixty seconds. However, the difference will be appreciable only for a few foods, such as grass corms, in which the bouts are short and the animals frequently go from one patch to another in less than two seconds.

A similar definition of feeding bouts was used by Barton and Whiten (1993, 1994) in their studies of feeding in olive baboons at Laikipia Plateau, Kenya: "A feeding bout began when the animal began processing a food item (including digging, in the case of underground items), and finished when the animal ceased orientation towards, or handling it, or, in the case of 'travel-feeding' (Whiten et al. 1987), when the animal took two complete paces without harvesting an item." For adult female baboons in Amboseli, paces average about one second each (J. Altmann, pers. comm.).

Appendix 3

Survival Analysis of Bout Lengths

Survival Functions

The beginning and ending of any bout of activity can be regarded as the birth and death of the bout. Between these two, the elapsed time since the beginning of a bout can be regarded as the age of the bout, and the duration of the bout can be likened to a lifetime. Consequently, analytic techniques developed in demography for the study of survivorship (Gross and Clark 1975; Lee 1980; Mann, Schafer, and Singpurwalla 1974) and in engineering for the study of equipment failures (e.g., Kaufmann 1977) can be applied to activity bout length distributions (Bressers, Meelis, Haccou, and Kruk 1991; Fagen and Young 1978).

Survival phenomena—activity bouts or whatever—can be described by means of three related functions:

1. *Survivorship function S (t)*. Let the bout length be represented by a random variable T. The survival function $S(t)$ is the probability that a bout lasts for at least time t ($t > 0$), that is, $S(t) = \text{pr}\{T \geq t\}$, which is the complement of the cumulative distribution function.[16]

2. *Death density function f (t)*. Consider the probability $\text{pr}\{t \leq T < t + \Delta t\}$ that a bout will terminate during the "age" interval $(t, t + \Delta t)$. The death density function is $f(t) = \lim_{\Delta t \to 0} \text{pr}\{t \leq T < t + \Delta t\}/\Delta t$. It is sometimes called the unconditional failure rate.

3. *Failure rate h(t)*. The probability that a bout ends in the interval $(t, t + \Delta t)$, given that it was under way at t, is the conditional failure probability. For $\Delta t = 1$ (typically, one year), this is the familiar age-specific mortality q_x of life tables. Conditional failure probabilities are defined not at particular ages but over age intervals. To obtain a measure of the instantaneous failure rate at any age, we define the instantaneous failure rate $h(t)$ as the ratio of the conditional failure probability to its interval length as the latter goes to zero, that is, $h(t) = \lim_{\Delta t \to 0} \text{pr}\{t \leq T < t + \Delta t \mid T \geq t\}/\Delta t$. The failure rate is also called the conditional hazard rate, the force of mortality, and the instantaneous death rate. The quantity $\Delta t \cdot h(t)$ is the proportion of events of age t that fail during the next instant t to $t + \Delta t$.

These three survival functions are related as follows: (i) $f(t) = -S'(t)$; (ii) $h(t) = f(t)/S(t)$; and (iii) $H(t) = -\ln S(t)$, where $H(t)$ is the cumulative hazard function $H(t) = \int_0^t (0,t) f(u)du$ and $\ln x = \log_e x$. Because we can convert from one survival function to another, the choice between them can be made on practical grounds. The survivorship function $S(t)$ has the advantage of familiarity: it is widely used in ecological work, and has occasionally been used in ethological studies (e.g., Nelson 1964; Wiepkema 1968; Delius 1969; Slater 1978; Machlis 1974; Hailman 1974; Fagen and Young 1978).

When plotted logarithmically, the survivorship function has a second advantage. Perhaps the simplest model of the processes generating a bout length distribution is that bout terminations take place at a constant rate, independent of how long the bout has been under way, that is, $h(t)$ is constant, as in a Poisson process. If so, then $S(t)$ has a negative exponential distribution, which becomes a straight line if we plot the ln $S(t)$ or log $S(t)$ against age. (The choice of logarithm base alters only the slope.) This "lack of memory" is unique to the negative exponential distribution (Feller 1957, 413) and it therefore provides the most obvious null hypothesis for a set of survivorship data. Beyond that, a negative exponential distribution of "giving up times" would be contrary to expectation if animals were to end each feeding bout when they have exhausted the food supply in a local patch or reduced it to a certain level—unless, perchance, the patch sizes themselves have a negative exponential distribution. Conversely, if the log survival distribution is not linear, its shape often suggests alternatives to a constant failure rate, and in this respect it is preferable to the death density function. (One alternative to a constant failure rate, the Weibull distribution, will be described below.)

Another advantage of log survivorship plots is this. For some purposes, failure rates $h(t)$ might seem like the function of choice, but because failure rate estimates for long "lifetimes" are susceptible to small-sample fluctuations, particularly if the estimation intervals are kept small enough to be useful, cumulative hazard rates $H(t)$ are often used instead. However, since $H(t) = -\ln S(t)$, a log survivorship plot serves the same purpose. For all these reasons I used the log survivorship function extensively in my analyses of feeding bout length distributions; in some situations (analyses of "censoring") I inverted its plot in order to obtain a cumulative hazard rate. A disadvantage of log survivorship plots is that logarithmic scaling overemphasizes the right end of the distribution, where sample size is smallest. As a result, the right end of such graphs exaggerates small-sample and integer-value fluctuations.

The Weibull Distribution

The Weibull distribution applies to survival processes in which the hazard rate may increase with elapsed time, decrease, or remain constant, so long as it is a power function of time. The Weibull distribution is therefore a generalization of the exponential and several other survival distributions. For this and other reasons, it is often the distribution of choice in survival analysis (Mann et al. 1974; Gross and Clark 1975; Kaufmann 1977; Barlow and Proschan 1975; David 1974; Pinder, Weiner, and Smith 1978). The Weibull distribution has two parameters, a scale parameter α and a shape parameter β. (A third parameter γ is included if an initial period with no "deaths" occurs.) The method that I used for estimating α and β is described below. If $\beta = 1$, then the failure rate $h(t)$ is constant and the survival function $S(t)$ is negative exponential; for $\beta > 1$, the failure rate is increasing with time and the density function is a unimodal asymmetric curve; and for $\beta < 1$, the failure rate is decreasing (Kaufmann 1977, fig. 6.3; Martz

1987). The scale parameter β is a characteristic time-to-failure, since it is the 100 $(1 - e^{-1}) = 63.2$th percentile of the distribution for any value of the shape parameter (Nelson 1972). Alternatively, the mean duration is $\alpha\Gamma (1 + 1/\beta)$, where $\Gamma(\cdot)$ is the gamma function.

The survivorship function of the Weibull distribution is $S(t) = \exp[-(t/\alpha)^{\beta}]$. The failure function is $h(t) = \beta t^{\beta-1}/\alpha^{\beta}$. The cumulative failure function is $H(t) = t/\alpha)^{\beta}$. The density function is $f(t) = \exp[-(t/\alpha)^{\beta}] \beta t^{\beta-1}/\alpha^{\beta}$. Age-specific mortality rates of the Weibull, which are needed for generating expected values from the model, can be obtained (Pinder, Wiener, and Smith 1978) from $1 - \exp[(t_x/\alpha)^{\beta} - (t_{x+1}/\alpha)^{\beta}]$.

Be forewarned that notation for Weibull parameters is not standardized in the literature, although α, β, and γ, used here, are common (e.g., Lawless 1982; Martz 1987). They correspond respectively with δ, β, and μ of Mann, Schafer, and Singpurwalla (1974, 184), with η, β, and γ (ibid., 128), with $1/\lambda$, β, and t_0 of Kaufmann (1977), with $\lambda^{-1/\gamma}$, γ, and δ of Gross and Clark (1975), with b, c, and ϵ_1 in Pinder et al. (1978), and with $1/\lambda$, γ, and G ("guarantee time") in Lee (1980).

Estimating Weibull Parameters

When I set out to estimate the two parameters α and β of the Weibull distribution, I was unable to find a computer program that would provide estimates of the parameters from randomly censored data. (Since then, at least one such program has been published, SAS Version 5, 1985, but it appears not to have solved the problems in parameter estimation that are discussed below.) I therefore used the following method. First, consider the survival function of the Weibull distribution:

$$S(t) = \exp[-(t/\alpha)^{\beta}].$$

If we twice take the logarithm of the inverse of both sides of the equation, we get a linear function of the logarithm of t (Nelson 1972):

(A3.1) $$\ln \ln S(t)^{-1} = \beta \ln t - \beta \ln \alpha.$$

Consequently, for a random sample of intervals from a Weibull-distributed population, a plot of $\ln \ln S(t)^{-1}$ against $\ln t$, or equivalently of $\ln S(t)^{-1}$ against t on log-log graph paper, will give a straight line except for small-sample fluctuations and errors of measurement. I therefore carried out least-squares linear regressions of $\ln \ln \hat{S}(t)^{-1}$ against $\ln t$, where $\hat{S}(t)$ is a product-limit estimate of $S(t)$. The slope $\hat{\beta}$ of the least-square regression estimates the Weibull parameter β, and since the y-intercept $\hat{a}$ is $-\beta \ln \alpha$ (my equation 4.1, right side), α can be estimated from $\alpha = \exp[-\hat{a}/\beta]$.

This method of estimating Weibull parameters tends to be inaccurate because the transformation of $S(t)$ greatly exaggerates small differences on the left side of the distribution, and this effect is more marked when the sample size is small and the duration of bouts is short. Because of inaccuracies in timing very short feeding bouts, such effects are common.

The effects are illustrated in figures A3.1 and A3.2, both of which show data for the yearlings' grasshopper chases. During these chases, a yearling repeatedly swatted at the

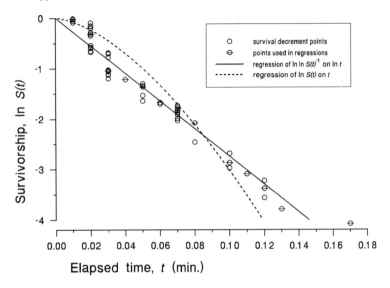

Fig. A3.1. Log-survivorship graph for grasshopper chases. This figure and figure A3.2 show two linear regressions fitted to the same 13 points (marked −). The first regression (solid line in both graphs) regresses ln $S(t)$ on t; the second (dashed line in both) regresses ln ln $S(t)^{-1}$ on ln t.

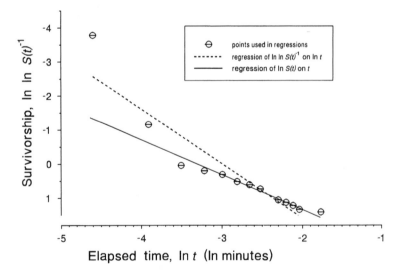

Fig. A3.2. Weibull graph of grasshopper chases. See figure A3.1 for details.

grasshopper on the ground as it flew from place to place. If, after a swat, it failed to capture the grasshopper, the chance that it would try again appears to be independent of how many attempts it had already made to capture this grasshopper, as judged by the linearity of the log-survivorship distribution of these sequences of swats. That the log-survivorship distribution is essentially linear is apparent both from inspection of figure A3.1 and from linear regression through the origin ($t = S(t) = 0$) of log survivorship against time, shown as a solid line in the figures. Even when just the thirteen points indicated are used, the regression ($\hat{S}(t) = -26.94\ t$) is highly significant ($p \leq 0.001$) and accounts for almost all of the variance in bout survivorship ($R^2_{adj} = 0.97$). This linearity suggests time-independent bout terminations. Similarly, when $\ln \ln \hat{S}(t)^{-1}$ is regressed against $\ln t$ (fig. A3.2), using the same thirteen data points, the results are highly significant ($p \leq 0.001$) and account for most of the variance ($R^2_{adj} = 0.85$). However, the regression line has a slope (the Weibull β parameter) of 1.6, which differs more than any other in table 4.6 from the value (1.0) that, under this log-log-inverse transformation, would indicate time-independent terminations.

Furthermore, when the regression line so obtained—the dashed line in both figures—is plotted on the log-survivorship graph (figure A3.1), it clearly does not fit the data as well as does the other regression. The high slope of the dashed line in figure A3.2 is due largely to the effect of the leftmost point, at $t = 0.01$ minute, whose vertical deviation is exaggerated by the log-log-inverse transformation. Without this point, the regression slope is much closer to unity ($1.085 \pm$ SE 0.090) but still significantly different from it ($p = 0.001$), and it accounts for more of the variance ($R^2_{adj} = 0.93$). Restricting the range of regression points to $\hat{S}(t) > 0.01$ reduces but does not completely eliminate this effect. Further evidence that the method produces only rough estimates of the Weibull parameters is that the apparently linear log-survivorship distribution for *Azima tetracantha* leaves, shown in figure 3.4, has a Weibull β parameter estimate (1.039) exceeding unity by more than a standard error (table 4.6), whereas the distribution for the green leaves of *Cynodon nlemfuensis,* which appears to be a sample from a monotone increasing function (fig. 3.13) has a β parameter estimate of 1.011—that is, for a distribution that is linear or very slightly decreasing. Clearly, these Weibull parameter estimates must be viewed with caution. Note that the log-survivorship function itself is not immune to such effects: the right sides of such distributions exaggerate the effects of small-sample and integer-increment fluctuations.

Appendix 4

Observational Censoring of Feeding Bouts

Bout Length Distributions

In chapter 3 I ask, If a baboon's feeding behavior is being sampled, does the likelihood that it will go out of sight depend on the food it is eating at the time, and does that probability change during the course of a feeding bout? These questions were answered in a sample of six representative baboon foods by calculating the "interruption rate" that results from such censoring as a function of elapsed time in the feeding bout. I proceeded as follows:

Let n_i = the number of bouts still under observation i time units after the bout began, c_i = the number of bouts in which the individual went out of sight during bout age interval $(i, i + 1)$, and d_i = the number of bouts that were observed to terminate during a small bout-age interval $(i, i + 1)$ or that I terminated because the sample ended. On the assumption that out-of-sight interruptions (censorings) occur randomly from a uniform distribution during this short interval, the number of bouts at risk of interruption was taken as $r_i = n_i - 1/2 \, d_i$, and therefore the probability of interruption during time interval $(i, i + 1)$ for bouts that lasted at least i time units was estimated as $q_i = c_i/r_i$. This probability is the conditional failure probability described in appendix 3.

Calculations were carried out by means of the survival analysis program (product-limit estimates) of BMDP-77 (Brown 1977). Time units were hundredths of a minute. Except for the interchange of censoring and observed termination, the computation is the same as for obtaining the failure rate resulting from observed terminations, and I therefore interchanged c_i and d_i of the BMDP program by defining them as above. Only bouts whose beginning was observed were included in these calculations, in order to avoid uncertainty about the "age" of a bout at any point in time. The proportions of bouts that were excluded by this criterion from the analysis of each food were always small. (If censoring was known a priori to be independent of bout age, left-censored bouts for which the termination was observed could be time-reversed and included in the sample.) Statistical tests—nominally for equality of survivorship functions but, as a result of the interchange, for equality of the interruption rate function—were also carried out by the BMPD program, using the methods of Breslow (1970) and Mantel/Cox (Mantel 1966). For ease in graphing, censoring frequencies are plotted as the proportion of bouts still under observation (*not* censored) at any specified elapsed time in the bout ("bout age"). These functions are plotted on a logarithmic scale; consequently "random" (bout-age independent) censorship would be represented by a linear graph.

Six foods (table 3.3) were selected for censoring analysis because they represent the diversity of foods eaten by the baboons and because adequate bout sample sizes are available for each. That these foods differ in their censoring rates can be shown in several ways, graphical (figs. 3.5, 3.7–3.11) and statistical (table 3.3). Both methods of statistical testing that were used (Breslow and Mantel/Cox) indicate the miniscule probability (≤ 0.0001) that the six interruption rate distributions could have come from a single population. Similarly, the mean rate of censoring (interruptions per minute of observation on a food) differed from food to food (table 3.3). Consequently, the observed bout lengths are systematically biased, not only against long bouts but also against those foods whose consumption is most likely to result in the animal's disappearance from my view.

For most of these six foods, interruptions were sufficiently uncommon that bout-age-specific interruption rates cannot be estimated. For the few with adequate samples of interruptions (e.g., *Salvadora persica* fruit and umbrella tree blossoms picked from trees, figs. 3.7 and 3.10, respectively), the logarithmic plots of cumulative interruptions (the open triangles and the right-hand ordinate of each figure) are essentially linear.

The results of this pilot analysis of six foods convinced me that the effects of censoring would have to be analyzed separately for each food. I proceeded as follows. First, note that those bouts I terminated when the scheduled sample periods ended can also be considered censored. Because sample durations were scheduled in advance, such censoring will take place at times that are independent of the elapsed time in bouts, and at a rate that depends not only on the food but also on the choice of sample length (twenty minutes in this study) relative to bout lengths. Censoring rates were recalculated, including both types of censoring combined. (Because the probability of sample-end censoring, like disappearance censoring, was essentially independent of bout age, so too is the combination, albeit at a higher rate.) Censoring failure rate functions were than calculated and graphed in the same manner for all of the other foods in the samples on weanlings that occurred at least twenty times and for aggregates of foods (table 4.1) for those that were observed less often. (Whenever one food contributed more than 95% of the sample in an aggregate class, only the aggregate distribution was analyzed.) Inspection of the cumulative failure functions showed only two censoring distributions that, aside from small-sample and integer-value fluctuations, appeared to be nonlinear, namely, the fruits of *Azima tetracantha* and *Tribulus terrestris* (figs. 3.14, 3.15). Even in those two cases, the number of censored bouts was quite small and the nonlinearity appeared primarily on the right side of the distribution, where sample size becomes progressively smaller.

Means of Bout Length Distributions

In chapter 6 I discuss the mean lengths of feeding bouts as well as the distributions themselves, and in that case too I faced the problem that, as a result of observational censoring, the observed bout length distributions were systematically biased. The product-limit method (Kaplan and Meier 1958; Meier 1975) was used to obtain maximum likelihood estimates of the bout length distributions and their means despite these out-of-sight periods.

Product-limit estimates of means and their standard errors were calculated and the log-survivorship distributions were plotted for every food that was eaten at least twenty times in my samples. Those that occurred less often (table 4.1, col. a) were combined with other similar foods into aggregate classes, each with at least twenty cases, and product-limit estimates were carried out on these aggregate classes. Each such food is assumed to have the bout length distribution of its assigned class. Table 7.1 lists the primary aggregate classes, their component foods, and their sample sizes for product-limit estimates. As indicated above, left-censored bouts were not used for product-limit estimates, so the sample sizes for these estimates (tables 4.1 and 7.1) are typically less than the total number of observed bouts.

For each food, column d of table 4.1 indicates its primary category, that is, that in which its data were used for calculating the product-limit estimate of its bout length and (in parentheses) any other product-limit category in which its data were used. Where the primary category is the same as the food's row number (left of table), it was the only food in that category. So, for example, food 10 (*Acacia tortilis* seeds on the ground) had a sufficiently large sample to warrant its own product-limit estimate (number 10), but it also contributed to aggregate category DD, used to estimate the mean bout lengths of foods 11 and 12, for each of which sample size was inadequate.

Appendix 5

Composition of Foods

For each of the fifty-two *core foods* (see chapter 6), table 5.1 indicates the macronutrient composition, that is, the amount of water, ash, lipids, proteins, fiber, and other carbohydrates, each expressed as a percentage (g per 100 g of fresh food), and also the energetic value of the food (kJ per 100 g of fresh food). In using these nutrient classes, one must bear in mind the somewhat imperfect relations between the material actually determined in standard chemical analyses and the corresponding nutrient classes (see, e.g., Merrill and Watt 1973). For example, protein content is commonly estimated by analyzing a sample for nitrogen content, then multiplying the result by a conversion factor of 6.25, but this method assumes that the protein in the sample is 16% N (because $0.16^{-1} = 6.25$) and that non-protein nitrogen compounds are removed or absent. For human foods of plant origin, nitrogen/protein conversion factors range from 5.18 to 6.31 (Jones 1941; Merrill and Watt 1973). Milton and Dintzis (1981) provide conversion factors for a variety of primate plant foods of the neotropical forests. Comparable values for African plants are not available at present.

The sources of the macronutrient compositions are indicated in the last column of table 5.1. Most come from tables in S. Altmann, Post, and Klein (1987), in which we report on nutrients and toxins in wild foods eaten by baboons and vervets in Amboseli. In that publication, we provide quantitative data on macronutrient composition (water, ash, fat, protein, fiber, and other carbohydrates), fourteen minerals, four vitamins or provitamins (folic acid, ascorbic acid, riboflavin, beta-carotene) and a toxin (trypsin inhibitor). In total, we report on 864 analyses, carried out on ninety-one foods from thirty-nine species of plants. In collecting plant samples for these analyses, we tried to approximate as closely as possible the foods of Amboseli baboons and vervet monkeys. We collected all of these samples in Amboseli during the phenophase in which they were being eaten by baboons and vervets, from within their range, and we attempted to obtain representative samples of exactly what these primates eat, paying attention to plant part, size, color, degree of ripeness, shear zones, rejected material (e.g., rind, calyx, seed, insect-infested material), and so forth. Further details of the samples and of the methods of preservation are available from the authors.

Where only a single sample is available and is believed to be representative of what I saw yearling baboons eat, the source is indicated in table 5.1 as a single sample code, for example, M331/75. In several cases, a food for which no chemical analysis is available was assigned the composition of a similar food; this is indicated in table 5.1 by the four-letter code of the assigned composition. For example, single, dry seeds of fever trees picked up from the ground (FTDG) are assumed as a first approximation to have the same mean composition as dry fever tree seeds in pods on the ground (indicated by FTDR in the source column), although they might differ, as a result of, for example, isolation and ambient moisture.

In some cases we sampled the same food more than once. On the assumption that these repeated samples are drawn at random from the array of material that is consumed by the baboons, their normalized unweighted means were used, that is, for each of the six chemical components of a macronutrient analysis, an unweighted mean was calculated and when necessary (because of missing analyses) these values were then divided by their sum then multiplied by 100 to give percentage by mass. These unweighted means are indicated in table 5.1 by the notation $\hat{x}(\cdot)$

For pooled categories, intake-weighted means were used in most cases; these are indicated by $w(\cdot)$. For example, for corms of unidentified grasses, I averaged the composition of all identified grasses, each weighted by the weanling baboons' intake in grams per day of that grass, then normalized the results to get percentages. In actual practice I could, of course, use only those grasses whose composition and intake rates are known, and similarly for other pooled classes. The materials actually used are indicated in the table. The intake rates are given in table 6.1. For one pooled category, *Withania somnifera* fruit, ripeness unknown (WSFX), I used unweighted means because I do not think I have adequate data to establish weighting coefficients.

What is the rationale for the $w(\cdot)$'s, that is, for weighting by grams consumed per day (instead of by frequency of consumption, time spent in consumption, grams per minute of feeding on that food, or whatever)? The mean daily intake (g/day) of any food component i from food j is calculated by multiplying the mean daily intake (g/day) of that food ($A_j B_j C_j$ in equation 3.1) by the mean fraction of that component in the food (d_{ij} = grams of i per gram of food j), and so the composition represented by D_{ij} must be characteristic of the average consumed gram of that food. Thus if several foods are pooled statistically, the composition of the pool should reflect the relative mass of the foods that make up the corresponding dietary pool.

For three foods, the macronutrient composition was obtained from the literature. These are milk, for which I used the composition of baboon milk published by Buss (1968); unidentified insects, for which I used the value given by Leung (1968) for grasshoppers; and dung-beetle larvae (species unknown), for which I used the value in Leung (1968) for caterpillars. Finally, for one food, the fruit of *Capparis tomentosa,* no analysis is available, and I have arbitrarily assigned it the composition of the fruit of *Azima tetracantha.* Because relatively little *Capparis* fruit was included in the baboons' diets, this should not greatly affect estimation of their total intake.

The energy values given in table 5.1 were estimated from the proportions of lipids, proteins, and total carbohydrates (fiber plus others), which are the primary sources of metabolic energy, by means of updated *Atwater factors* (Atwater 1902; Atwater and Bryant 1900a, 1900b; Maynard 1944; Merrill and Watt 1973). These factors lead to reasonably accurate estimates of the energy values of human foods (Merrill and Watt 1973; Southgate and Drunin 1970) and are widely used for this purpose by the U.S. Department of Agriculture and various international health organizations. For estimating physiologically available energy, they are preferable to the raw results of bomb calorimetry, because some components of the diet (e.g., lignins) are not digested or metabo-

lized by mammals, and thus the gross energy content of food is greater than its metabolically available energy. Crudely put, animals, unlike bomb calorimeters, do not excrete ashes.

The physiological fuel value of each of the baboons' foods was estimated from
$lL + pP + cC$, where L, P, and C are the grams of lipids, proteins, and total carbohydrates in the food, and l, p, and c are the respective Atwater factors (physiological energy per gram of ingested constituent). The most recent revision of the Atwater factors,
albeit based on human foods and human digestibility coefficients, was published by
Merrill and Watt (1973). Based on their monograph and assuming that their terms "calories" and "Calories" always mean *kilocalories,* I have adopted Atwater factors of 8.37,
3.40, and 3.92 kcal of physiological energy per ingested gram of plant lipids, proteins,
and total carbohydrates, respectively, each then multiplied by 4.184 to give kilojoules
per gram.

For energy from lipids, the value of 8.37 kcal/g is the only one given by Merrill and
Watt (1973, table 13) for foods of plant origin, and is based on a heat of combustion of
9.30 kcal/g and an assumed coefficient of digestibility of 90%.

For energy from plant proteins in human foods, the tabulated conversion factors
(Merrill and Watt 1973, table 13) are variable and are based on cultivars that are often
extensively processed and cooked before being eaten. As an approximation, I have used
a value of 3.40 kcal/g, which is the unweighted mean for sixteen plant foods in their
table, selected to emphasize foods that are least processed and including one fruit value,
eleven grain values, three legume/nut values, and one other vegetable value.

For energy from carbohydrates of plant origin, I have used a value of 3.92 kcal per
gram of carbohydrate including fiber, based on the same sixteen foods. Some error in
the carbohydrate energy factor, of unknown magnitude, results from the fact that many
of the human foods were cooked before their energy values were determined (Merrill
and Watt 1973) and from the high fiber content of some baboon foods. The ability of humans to digest fiber has not been extensively studied, but is appreciably higher than
commonly assumed: 54% for neutral-detergent fiber from bran, including celluloses
and hemicelluloses (Milton and Demment 1988), and up to 100% for the cellulose in
cabbage (Ehle, Robertson, and Van Soest 1982). In chimpanzees, the proportion of the
gut devoted to fermentation, the hindgut, is longer than in humans (52% vs. 17–20%);
correspondingly, they digest even more of the bran fiber, 71%, on comparably fibrous
diets (Milton and Demment 1988). Rhesus monkeys (*Macaca mulatta*) digest 38.8%
and 47.0% of the crude fiber in two laboratory diets (Knapka and Morin 1979). Doubtless baboons too can digest a considerable portion of the fiber in their diet, but at present the amount they utilize is unknown.

As indicated in the text, I have assumed that baboons can digest the polymerized
sugars in fever tree gum.

The energy conversion factors given in the publications of Atwater (e.g., Atwater and
Bryant 1900) and reevaluated by Merrill and Watt (1973) apply to *carbohydrates by difference,* meaning the difference between 100 and the sum of the moisture, lipid, ash, and
crude protein, and thus including fiber (Atwater and Bryant 1900; Merrill and Watt
1973). (Currently, when some laboratories report carbohydrates by difference, they are
excluding fiber, a fact that initially escaped my attention and led to a lot of recalculat-
ing!) Southgate and Durnin (1970) have argued in favor of basing the carbohydrate-

to-energy conversion factor on fiber-free carbohydrate, but if an appreciable amount of fiber is digested, as seems to be the case (Saunders 1978; Cummings 1984; Ehle, Robertson, and Van Soest 1982; Van Soest 1994), that procedure would be ill advised unless, in addition, a fiber-to-energy factor is included.

On my core food list (table 6.1) foods of animal origin include milk, various unidentified insects, and the larvae of dung beetles. For milk I have used the energy factors for the components of human milk given by Southgate and Barrett (1966): 8.74, 4.39, and 3.83 kcal/g for lipids, proteins, and carbohydrates, respectively. For the insect material, I have used the values given by Merrill and Watt (1973, table 13) for meat and fish: 9.02, 4.27, and (ibid., table 7) 3.82 kcal/g for lipids, carbohydrates, and proteins, respectively. As with plant materials, these values were multiplied by 4.184 to convert kilocalories to kilojoules.

In sum, the physiological energy values that I used, in kJ per 100 g of ingested material, for lipids, proteins, and carbohydrates-plus-fiber, respectively, are as follows: 35.02, 14.23, and 16.40 for plant foods; 36.57, 18.37, and 16.02 for milk; 37.74, 17.87, and 15.98 for insect material. Stacey (1986) used different conversion factors and therefore shows different energy values for Amboseli baboon foods.

Appendix 6

Michael Altmann

Maximizing Rates

The task of constructing an optimal diet is a variant on a classic linear programming problem (Altmann and Wagner 1978; Dantzig 1963). Typically, costs, nutrients, toxins, gut capacity, etc., are assumed to impose linear constraints on the amounts of each food in the diet. After setting up these constraints one minimizes or maximizes some dietary component, say, its cost or its energy content, that is a linear function of the amounts of each of the foods in the diet. As just stated, the optimal diet problem has linear constraints and a linear objective function and can therefore readily be solved numerically. Recently, however, several authors have suggested that animals might be expected to choose a diet that maximizes their average energy intake *rate,* say, per second of feeding or foraging. This modification makes the objective function a nonlinear function of food amounts, but a simple transformation can remove the nonlinearity and make the problem amenable to standard linear programming techniques, such as the simplex method.

Making the Problem Linear

To fix notation, let x_j be the amount (in grams) of the jth food in the proposed diet. We assume that all constraints are linear in the x_j's and so they may be expressed as

$$\sum_{j=1}^{n} a_{i,j} x_j \geq b_i \qquad i = 1, 2, \ldots, m.$$

Let c_j be the energy intake rate (in joules/sec) for the jth food. Then the mass-weighted mean energy intake rate is

$$\sum_{j=1}^{n} c_j x_j / \sum_{j=1}^{n} x_j .$$

When this appendix was written, Michael Altmann was at the Courant Institute of Mathematical Sciences, New York City. He is currently at the Division of Health Computer Sciences, University of Minnesota, Minneapolis.

More generally, we might consider something like the time-weighted mean rate. If d_j is the time (in seconds) it takes the animal to eat one gram of the jth food, then the time-averaged caloric rate is

$$\sum_{j=1}^{n} c_j x_j / \sum_{j=1}^{n} d_j x_j .$$

This is the general form of the objective function I shall consider. The task is now to find the values of the x_j's that

(A6.1) $$\text{maximize} \sum_{j=1}^{n} c_j x_j / \sum_{j=1}^{n} d_j x_j$$

(A6.2) $$\text{subject to} \sum_{j=1}^{n} a_{i,j} x_j \geq b_i \qquad i = 1, 2, \ldots, m,$$

where the c_j's, d_j's, $a_{i,j}$'s and b_i's are known constants. Clearly the objective function does not make sense if $\sum_{j=1}^{n} d_j x_j = 0$, so for the rest of this discussion I shall assume that the $a_{i,j}$'s are such that for any set of x_j's satisfying equation (A6.2) (this is called the feasible set), we have $\sum_{j=1}^{n} d_j x_j \neq 0$. From the fact that the feasible set is necessarily connected it follows that $\sum_{j=1}^{n} d_j x_j$ is of one sign. Without loss of generality we may assume it is always positive.

The problem, as stated, is nonlinear and therefore not directly amenable to standard linear optimization techniques. However, a simple change of variables linearizes the problem. Consider the following change of variables (sometimes known as a change to homogeneous coordinates):

$$\beta = 1 / \sum_{j=1}^{n} d_j x_j$$

$$y_j = x_j \beta \qquad j = 1, 2, \ldots, n.$$

It is not hard to check that this maps the set where $\sum_{j=1}^{n} d_j x_j > 0$ in a one-to-one fashion onto the region defined by $\sum_{j=1}^{n} d_j y_j = 1$ and $\beta > 0$. The inverse mapping is given by

(A6.3) $$x_j = y_j / \beta \qquad j = 1, 2, \ldots, n.$$

Substituting this into equations (A6.1) and (A6.2), we find that our problem is changed into

$$\text{maximize} \sum_{j=1}^{n} c_j y_j$$

$$\text{subject to } \sum_{j=1}^{n} a_{i,j} y_j \geq b_i \beta \qquad i = 1, 2, \ldots, m,$$

$$\sum_{j=1}^{n} d_j y_j = 1$$

$$\beta > 0.$$

Or bringing the independent variables (including β) to one side,

(A6.4)
$$\text{maximize } \sum_{j=1}^{n} c_j y_j$$

(A6.5)
$$\text{subject to } \sum_{j=1}^{n} a_{i,j} y_j - b_i \beta \geq 0 \qquad i = 1, 2, \ldots, m,$$

$$\sum_{j=1}^{n} d_j y_j = 1$$

$$\beta > 0.$$

This is now a standard linear optimization problem and can be solved by the simplex method using a standard package. Note that in comparison with equations (A6.1) and (A6.2), the system described in equations (A6.4) and (A6.5) has one additional independent variable, β, and two additional constraints, one of which is an equality constraint. Once the optimal solution to equations (A6.4) and (A6.5) has been found, the optimal solution to the original problem (the optimal diet) may be recovered via equation (A6.3). Also, the optimal value of the objective function is the same in the transformed problem and the original problem.

If a standard linear programming package is used to solve the transformed problem, the user probably will be provided with information about various auxilliary quantities. We now consider the relations of several of these to their counterparts in the untransformed original problem.

Slack Values

For each inequality constraint, there is an associated variable whose value is simply the difference between the two sides of the inequality. Thus, the ith slack variable for the transformed problem, s_i, satisfies the equation

$$\sum_{j=1}^{n} a_{i,j} y_j - b_i \beta + s_i = 0.$$

Rearranging this we see

$$\sum_{j=1}^{n} a_{i,j}x_j + s_i/\beta = b_i,$$

so s_i/β are the slack values for the original nonlinear problem in the sense that they are the differences between the two sides of the inequalities that appear in the nonlinear problem.

Reduced Costs

At the optimal diet there is a reduced cost variable for each food. The value of the reduced cost variable is the amount by which the energy intake rate for that food would need to be increased before the food would be included in the optimal diet (and so is zero if the food is already in the optimal diet). The change to homogeneous coordinates does not affect the value of the c_j's or whether a food is included in the optimal diet, so the reduced cost variables are the same for the nonlinear problem.

Cost Coefficient Ranges

At the optimal diet the cost coefficient range for each food (joules per gram) gives the range of energy density values for that food that would leave the optimal diet unchanged. As with the reduced costs, these will be the same for the nonlinear problem.

Marginal Costs

For a linear programming problem the marginal cost of the ith equation is defined as the rate of change of the optimal value of the objective function as the right-hand side of the ith equation is varied. To examine the marginal costs of the first m constraints in equation (A6.5), let e_i denote the ith right-hand side and let g (a, b, c, d, e) denote the optimal value of the system:

$$\text{maximize} \sum_{j=1}^{n} c_j y_j$$

$$\text{subject to} \sum_{j=1}^{n} a_{i,j}y_j - b_i\beta \geq e_i$$

$$\sum_{j=1}^{n} d_j y_j = 1$$

$$\beta > 0$$

and let $y*$ (a, b, c, d, e) and $\beta*$ (a, b, c, d, e) denote the values of y and β at the maximum.

By definition, the ith marginal cost for our linear transformed problem is

$$N_i = \frac{\partial g}{\partial e_i}\bigg|_{e_i = 0}.$$

It seems natural to define the ith marginal costs for the nonlinear problem as the rate of change of the optimum as the right-hand side of ith constraint equation is changed. Therefore let $f(a, b, c, d, \epsilon)$ denote the maximum of

$$\sum_{j=1}^{n} c_j x_j / \sum_{j=1}^{n} d_j x_j$$

subject to

$$\sum_{j=1}^{n} a_{ij} x_j \geq b_j + \epsilon_j$$

and let $y**(\epsilon)$ and $\beta**(\epsilon)$ denote the values of y and β at the maximum. Note that $y*(\vec{0}) = y**(\vec{0})$ and $\beta*(\vec{0}) = \beta**(\vec{0})$.
The ith nonlinear marginal value may defined as

$$M_i = \frac{\partial f}{\partial b_i}\bigg|_{\epsilon_i = 0}.$$

Assuming that this really is a useful definition, and that our software package provides us with the marginal values for the transformed system, we need a way to go from M_i to N_i.

From the point of view of the transformed system, M_i represents the change with respect to coefficient of the $m + 1$st variable in the ith constraint because b_i is the coefficient on β in the ith equation.
According to Gass (1984, equation 3.7 on p. 220 and equation 3.1 on p. 218) the dependence on a constraint coefficient is given by

$$y_j^{**}(0, \ldots, 0, \epsilon_i, 0, \ldots, 0) = y_j^{**}(\vec{0}) + \frac{q_{j,i} \epsilon_i \beta^*(\vec{0})}{1 + q_{m+1,i} \epsilon_i}$$

(A6.6)

$$\beta^{**}(0, \ldots, 0, \epsilon_i, 0, \ldots, 0) \frac{\beta^*(\vec{0})}{1 + q_{m+1,i} \epsilon_i},$$

and the dependence on a right-hand side is given by

(A6.7) $$y_j^*(0, \ldots, 0, e_i, 0, \ldots, 0) = y_i^*(\vec{0}) + q_{j,i} e_i.$$

In these formulas, $q_{j,i}$ are the entries in the inverse of the matrix of active columns. (Gass writes b instead of q, but this might be confusing here because we already have the variables b_i in use.) The exact meaning of $q_{j,i}$ need not concern us, we simply observe that equations A6.6 and A6.7 imply that

$$\frac{\partial y_j^{**}}{\partial \epsilon_i}\bigg|_{\epsilon_i = 0} = q_{j,i}\beta^*(\bar{0})$$

and

$$\frac{\partial y_j^*}{\partial e_i}\bigg|_{e_i = 0} = q_{j,i}\ .$$

Therefore

$$M_i = \frac{\partial f}{\partial \epsilon_i}\bigg|_{\epsilon_i = 0} = \frac{\partial}{\partial \epsilon_i}\bigg|_{\epsilon_i = 0} \sum_j c_j y_j^{**} = \beta^*(\bar{0})\sum_j c_j q_{j,i}$$

and

$$N_i = \frac{\partial g}{\partial e_i}\bigg|_{e_i = 0} = \frac{\partial}{\partial e_i}\bigg|_{e_i = 0} \sum_j c_j y_j^* = \sum_j c_j q_{j,i}\ .$$

From this we see that the marginal values for the the nonlinear system are obtained by scaling the marginal values of the linear system by $1/\beta^*(\bar{0})$.

Right-Hand-Side Ranges

The marginal value tells us how the optimal solution changes as the right-hand side of the inequalities varies but this change is smooth only over a particular range of values for the right-hand side. The range for the ith equation is called the ith RHR. In our particular case that translates into the range for e_i over which $g(a, b, c, d, e)$ varies smoothly. For the nonlinear problem we can make a similar definition. Let the ith RHR for the nonlinear system be defined as the range of values for ϵ_i over which $f(a, b, c, d, \epsilon)$ varies smoothly. It is not hard to see that the ith nonlinear RHR is the range for ϵ_i over which equation (A6.6) is valid, and that the ith linear RHR (which is what we obtain from our standard package) is the range for e_i over which equation (A6.7) is valid. Looking at equations (A6.6) and (A6.7) we see that y^* varies linearly in e_i but nonlinearly in ϵ_i, so the range over which equation (A6.6) is valid is unlikely to correspond in any nice way to the range over which equations (A6.7) is valid. In fact, Gass gives the ranges for equations (A6.6) and (A6.7) as follows. According to Gass (equation 3.10 on 221), the range for equation (A6.6) is

$$(A6.8) \quad \max_{j,j \neq i} \frac{y_j^*(\bar{0})}{(q_{j,i}\beta^* - q_{m+1,i}y_j^*(\bar{0}))} < 0$$

$$\leq -\epsilon_i \leq \min_{j,j \neq i} \frac{y_j^*(\bar{0})}{(q_{j,i}\beta^* - q_{m+1,i}y_j^*(\bar{0})) > 0}\ .$$

Note: here the inequality in the fraction indicates that the max or min should be taken only over those values of j for which the inequality holds. Again, this is Gass's notation. Similarly, Gass (218) shows that the range for equations (A6.7) is

$$\max_{j,j \neq i} - \frac{y_j^*(\vec{0})}{q_{j,i} > 0} \leq e_i \leq \min_{j,j \neq i} - \frac{y_j^*(\vec{0})}{q_{j,i} < 0} .$$

These ranges are different in a nontrivial way: there is no way to go from one to the other via scaling. This means that if we really want to know the RHR for the nonlinear problem we must determine the range specified in equation (A6.8). If the linear programming package allows one to do sensitivity analysis for the equation coefficients (which is what the b_i's are for the linear system), then there should be no trouble: we simply ask for the range corresponding to the coefficient b_i of β.

Appendix 7

Procedure for Maximizing Rates

In appendix 6 Michael Altmann provides a solution to the problem of maximizing an objective function that is a rate in a linear optimization problem. Based on his method, I used the following procedure for maximizing intake rates of energy and protein, as specified by diets 4, 5, 9, and 10. The procedure is here described in terms of objective function G_4' for diet 4 but is applicable mutatis mutandis to the three other rate functions. See chapter 5 and appendix 6 for notation.

For simplicity, let $\beta = 1/\Phi = (\Sigma F_a)^{-1}$ and let $c_j = e_j B_j C_j = $ energy intake rate (J/minute) when feeding on the jth food. First, multiply the numerator and denominator of the objective function by β and relabel each βF_j (the proportion of food j in the diet) as y_j. This gives

(A7.1)
$$G_4' = \sum c_a y_a.$$

Note that equation (A4.1) is a linear function of the y's and so it can be maximized in the usual way.

Second, scale all constraints by β: multiply each constraint equation by β and relabel each βF_j as y_j. For example, in diets 6 to 10, feeding time is limited to 436 minutes per day, as expressed by $\Sigma t_a F_a \geq 436$, where t_i is the time required to eat each gram of the ith food. After rescaling, this becomes

(A7.2)
$$\sum t_a y_a \leq 436\beta.$$

Then, to satisfy the input requirements of LINDO, each constraint equation is rearranged so that only constant terms appear on the right side. Call this the β-form of the constraint equations. Thus equation (A7.2) in β-form is

(A7.3)
$$\sum t_a y_a - 436\beta \leq 0.$$

Now use standard linear optimization methods to solve this β-scaled form of the rate maximization problem: find those values of the y's and of β (respectively $y_1^*, \ldots, y_n^*$ and β^*) that maximize $\Sigma c_a y_a$, subject to $\Sigma y_a = 1$ and to all previous constraints, in β-form. In addition to these constraints, all variables (β and the y's) must be non-negative; this can be achieved by multiplying the corresponding equations by -1 if necessary.

Finally, to convert the solution of the β-scaled problem to that for the original problem, rescale the computer output as follows. For each food, divide the output value (y_j^*) by β^* to get F_j^*, the amount of the jth food in the optimal diet. Delete the right-hand-side allowances (or see appendix 6). Divide each slack value by β^*. When dealing with marginal values, adopt the following simplified notation: write y^* for $y^*(\vec{0})$ and β^* for $\beta^*(\vec{0})$. Multiply each marginal value by β^* and interpret it as an instantaneous rate of change in the objective function value. Do not alter the objective function value, the reduced costs, or the cost coefficient ranges (leeway).

Justifications for these rescaling procedures are as follows (cf. appendix 6). Because the original objective function was multiplied by β/β, the β-scaled form of the objective function is equivalent to the original. That is, $\Sigma\, c_a y_a^*$ in the β-scaled problem is the maximized value of G_4', the mass-weighted mean energy harvest rate of the optimal diet.

As for the slack values, in a typical β-scaled constraint such as that for harvesting time, shown in equation (5.10), with the optimal values β^* and the y^*'s, the slack value s satisfies $\Sigma\, t_a y_a^* - 436\beta^* + s = 0$. Solving for s and recalling that $y_j^* = \beta^* F_j^*$, we get $s = \beta^*(436 - \Sigma\, t_a F_a^*)$. In the original harvesting time constraint, if S is the slack value, it satisfies $\Sigma\, t_a F_a^* - 436\,S = 0$, and so $S = 436 - \Sigma\, t_a F_a^*$. Thus $S = s/\beta^*$. That is, the slack value of the β-scaled problem must be divided by β^*.

Now look at the marginal values. Suppose that in the β-scaled problem, the feeding time constraint, in β-form, has a marginal value m. This means that at the optimum, if the right-hand side of equation (A7.3) were increased by one unit, thereby changing it into

$$(A7.4) \qquad\qquad \Sigma\, t_a y_a^* - 436\beta^* \geq 1,$$

then a new optimal diet could be obtained with a mean energy harvest rate that would be m kJ/minute higher than that of the current optimal diet (assuming that we have not gone beyond the allowable change in the right-hand side). To see what this corresponds to in the original problem, divide equation (A4.4) by β^*. This gives $\Sigma\, t_a F_a^* \geq 436 + 1/\beta^*$. That is, at the optimum, the objective function value changes at the rate of m units per $1/\beta^*$ change in the right-hand side of the corresponding constraint in the original problem, or equivalently, the objective function values changes at a rate of $\beta^* m$ per unit change in the right-hand side of the corresponding original constraint. Because β^* itself is changing, these marginal values are instantaneous rates of change, not changes valid within bounds set by right-hand-side allowances, and so no such allowances are given in my tables.

Since the β-scaling procedure did not involve any change in the cost coefficients (e.g., for energy-rate maximization, the energy yield rate of each food), no rescaling is needed for the reduced costs or the cost coefficient ranges (leeways).

Computer outputs for the four diets herein that involve rate maximization (diets 4, 5, 9, and 10) have been rescaled as indicated above (tables 5.7, 5.8, 5.12, 5.15).

By including the constraint $\Sigma\, y_a = 1$, we ensured that the y_i's would be the *proportions* of the various foods in the optimal diet. In rate maximization problems, that constraint is always binding, its marginal value is always the same as the objective function's value at the optimum, and the slack value of equality constraints is always zero. Thus this constraint is uninformative and has not been included in the tables.

Appendix 8

Seasonal Adjustments

Mean Annual Intake

If the baboons' diet did not change through the year, then even with an uneven distribution of sampling over the year, their observed diet would be representative of their annual intake. However, because of the marked seasonality of Amboseli, the mean observed diet (table 6.1) is affected by the particular seasonal distribution of sample times in my study. The consumption values of the observed diet do not completely reflect the baboons' actual mean of daily intakes over the year because they include more data from times of the year when I sampled more. For that reason these values could not be directly compared with or replicated by values from another study with a different sample-time distribution. How can we convert such season-biased samples of mean intakes into estimates of what, on average, was actually consumed? The method I used is as follows.

Consider such a biased annual mean intake value E_j for any food j, as in table 6.1. It can be written as the product of three factors that were used to calculate it: $E_j = g_j S/T$, where S = potential foraging minutes per day = 660, $g_j = B_j C_j L_j n_j$ is the total number of grams of food j consumed during my samples, $T = \Sigma_x L_x n_x$ (where x indexes bouts on all feeding and nonfeeding bouts) is the total product-limit sample minutes in the study year, and B_j, C_j, L_j, and n_j are as in equations (3.1) and (3.2). Thus

$$(A8.1) \qquad\qquad g_j = \frac{E_j T}{S}.$$

Now, all g_j grams of food j must have been consumed while food j was "in season," by definition (see below), and so the *in-seasonal* daily consumption e_j for the jth food is $e_j = g_j S/T_j$ g per day, where T_j is the number of sample-minutes when food j was in season. (The value of each T_j was obtained by identifying all semimonths in which the jth food was in season, indicated by p's and h's in table A8.2, then summing my corresponding in-sight sample time, table A8.3.) Thus,

$$(A8.2) \qquad\qquad g_j = \frac{e_j T_j}{S}.$$

Substituting the expression of g_j given in equation (A8.2) into equation (A8.1) and solving for e_j, we get $e_j = E_j T/T_j$, or

418

(A8.3)
$$e_j = \frac{E_j}{t_j},$$

where $t_j = T_j/T$ is the proportion of my sample time (table A8.3) when food j was in season (table A8.2).

By averaging the in-season consumption rate e_j with the out-of-season rate (zero), each weighted by the length of the corresponding time period, we obtain an estimate $\hat{E}_j$ of the actual annual mean daily consumption, in grams per day, for food j, namely, $\hat{E}_j = e_j f_j$, where f_j is the fraction of a year that food j was in season (table A8.1), and so

(A8.4)
$$\hat{E}_j = \frac{E_j f_j}{t_j}.$$

As I will explain presently, the fraction of the year during which a food was in season was obtained from $f_j = (2p_j + h_j)/48$, where p_j is the number of semimonths in which the jth food is "present" (available on all days) and h_j is the number in which it is "half-present" (available for half the days, on average). Thus

(A8.5)
$$\hat{E}_j = \frac{E_j(2p_j + h_j)}{48t_j}.$$

For each core food, table A8.1 gives my estimate ($\hat{E}$) of the baboons' actual (seasonally adjusted) annual mean intake, in grams per day, my estimate (e) of their daily intake of each core food during the part of the year that it was in season, the correction factors f/t, and the total numbers of the p and h values for each.

Values that I used for the adjustment factors f_j/t_j are not perfect. Some discrepancy results because the tabulated values of the f's and t's (table A8.1) are based solely on the 1975–76 data, not the 1974 data, whereas the E_j values (table 6.1) are based on both. However, the 1974 samples accounted for less than 5% of the total (Dotty and Striper only, table 3.1), so the effects should be small. More significantly, small samples in some semimonths lead to underestimates of the values of the f's. The overall effect of these small-sample omissions is that the $\hat{E}$ values tend to be underestimates.

The Seasons of Foods

The fraction of a year that each core food was available (f_j, above) was obtained as follows. I divided each month into two semimonths, the first including the first fifteen days of the month, the second containing the remainder (variously fourteen, fifteen, or sixteen days). I used this level of resolution because the array of foods that are available in Amboseli changes from one fortnight to the next. Then I determined the frequency with which each food was consumed in each semimonth in the 1975–76 samples. The 1974 samples were not included because I do not have phenophase data for that year.

For any semimonth in which a food was eaten at all by any yearling in the study group, I considered that the food was available to all yearlings of the study group throughout that semimonth. An exception was made for any such semimonth if it was immediately preceded or succeeded by a semimonth in which that food was not eaten; in such semimonths I considered that the food was available for only half the semimonth, on the grounds that these semimonths were at the beginning or end of a season for that food and that on average the onset or termination of availability for such foods would occur halfway through the interval. (Such semimonths will be referred to below as times when the food was "half-present," whereas during other in-season times, the food will be referred to as "present.") Each h represents availability for half a semimonth, that is, a quarter-month, of which there are forty-eight in a year; each p represents two quarter-months. Thus, the fraction of a year during during which a given food j is available is $f_j = (2p_j + h_j)/48$, where p_j is the number of semimonths when food j is present, as defined above, and h_j is the number when it is only half-present.

Table A8.2 indicates the semimonths during which each core food was present or half-present. At the same time, this table provides a concise summary of seasonal shifts in the array of foods that are available to the baboons and it documents my claim that this "menu" changes from one fortnight to the next.

This table includes two types of interpolation that I made in the data, to handle semimonths with very small samples. First, only a very small sample was obtained during the second semimonth of July (14.2 minutes), and so as a result of small-sample errors many foods are scored as being not present at that time and, consequently, as being only half-present during the preceding and succeeding semimonths. I therefore interpolated values for this semimonth by considering that a food was eaten then if it was eaten in the immediately preceding and following semimonths. These interpolated values are indicated in table A8.2 by $\hat{p}$'s, with a circumflex. Second, although no nursing was observed during my samples in the second half of May, I am quite sure this gap is just a small-sample error: sample time during that semimonth was 182.8 minutes. I have therefore indicated milk as available (present) during that time.

The reader may be bothered, as I am, that availability is here being judged by use. I must leave it for future research to determine whether foods that are sometimes present in the baboons' home range outside the period during which they are eaten are equivalent (in terms of ripeness, and so forth) to those that the baboons eat. For example, green pods were present on fever trees from about October 27 to February 7 (fig. 1.5) but were not eaten by any yearlings during my October or January samples. Umbrella tree flowers were still present in late March but were not eaten by yearlings then. Several examples are mentioned in chapter 4, and figure 1.5 provides other examples. Perhaps these are just small-sample artifacts. Yet my impression is that baboons sometimes fail to find or to utilize some sources of familiar foods, that a delay sometimes occurs between when foods become available and when the baboons begin to feed on them, and that these failures and delays are not always due to the presence of other, preferred foods. To the extent that this is so, utilization underestimates availability. It cannot overestimate availability, however: if a food is eaten, it is available.

In-Season, Peak-Season, and Maximum Consumption

For chapter 8, I computed optimal diets for each semimonth. For that purpose, I needed estimates of the maximum amount of every food that a baboon could consume during those parts of the year when the food was available. I proceeded as follows.

During the part of a year when a given food j is "in season," its mean daily intake, e_j, is, as indicated in equation (A8.3), $e_j = E_j/t_j$. However, e_j is an average not only of semimonths during which the food was available every day ("present") but also those during which the food was coming into season or going out, during which, on average, it was available only half the days ("half-present"). So, first, what was the yearlings' mean (not maximum) daily consumption during semimonths when the food was available every day—what one might call "peak-season intake"? On the assumption of a uniform mean intake on all days when a food is available, mean daily intake of a food during a semimonth in which it is half-present would be only half that for a semimonth during which it is always present, so that the mean intake of food (g/day) during the entire period of availability (i.e., both p's and h's in table A8.2) can also be written as

$$(A8.6) \qquad e_j = \frac{h_j v_j/2 + p_j v_j}{h_j + p_j} \, ,$$

where v_j is the mean daily intake of food j just during semimonths when food j is presumed to be present every day. Combining equations (A8.3) and (A8.6), then solving for v_j we get

$$(A8.7) \qquad v_j = \frac{E_j(h_j + p_j)}{t_j(h_j/2 + p_j)} \, .$$

For any semimonth when food j is only half-present, mean daily intake would average half as much, $v_j/2$. Values of v_j for the fifty-two core foods are given in table A5.1.

As I indicated, the v_j's are means. What about the maxima? For convenience, the derivation of equation (A8.7) was worded in terms of mean daily intake, but the same argument applies to the mean plus three standard deviations or other percentiles of the distribution. I therefore estimated maximum daily intake during a semimonth when food j was present every day by substituting s_j (the greater of the annual mean plus three standard deviations of daily intakes or the highest observed value in a semimonth, as in chapter 5) for E_j in equation (A8.7):

$$(A8.8) \qquad w_j = \frac{s_j(h_j + p_j)}{t_j(h_j/2 + p_j)} \, .$$

Values for the w_j's are given in table A8.1. They were used in calculating semimonthly optimal diets for each infant to establish constraints on the amount of each food that could be called for. For a semimonth in which food j was only half-present, maximum daily intake was limited to $w_j/2$. I believe this method of estimating intake maxima is conservative, in that use of intervals shorter than ten weeks doubtless would have turned up periods with still higher value than those used to obtain the s_j values.

Appendix 9

Estimation of Individual Milk Intakes

The procedure I used to estimate the yearlings' milk intakes is as follows.

Step 1

For each yearling i during each ten-week age block j, I estimated the time budget T_{ij} (suckle) of diurnal suckling, in minutes of suckling per "day" (0700 h to 1800 h), from

$$(A9.1) \qquad \hat{T}_{ij} \text{ (suckle)} = \frac{\hat{L}_{ij} \text{ (suckle) } n_{ij} \text{ (suckle) } S}{s_{ij}k},$$

where $\hat{L}_{ij}$ (suckle) is the product-limit estimate of the mean duration in minutes of suckling bouts of yearling i during age block j (calculated as in chapter 4), n_{ij} (suckle) is the number of suckling bouts (censored or complete) in my samples of yearling i during age block j, $S = 660$ is the number of minutes between 0700 h and 1800 h, and s_{ij} is the number of minutes of in-sight sampling time (table 3.1) on yearling i during age block j. Finally, $k = 1.116$ is obtained from the "product-limit sample time"—that is, the sum ($\sum_x L_x n_x = 22,299.84$ minutes) of all activity-specific mean bout lengths (L_i) times bout frequencies (n_i), as described in chapter 3, and including all foods (9,384.23 minutes, table 4.1) and also the nonfeeding bouts (12,915.61 minutes), and calculated with data for all yearlings at all ages pooled—divided by total actual in-sight time (19,978.74 minutes, table 3.1). (Because the product-limit estimates of bout lengths approximate the true values, the product-limit sample time should approximate the scheduled sample time including both the actual [in-sight] sample time and the time when the subjects were out of sight. Indeed, it does: as indicated in chapter 3, the subjects were out of sight for 9.24% of the time, that is, scheduled sample time was 10.2% greater than in-sight time, and product-limit sample time is 11.6% greater.)

This method of calculating an activity-specific time budget is the same as that used in chapter 3 and 4, equation (3.2)—that is, mean minutes of nursing per daytime minute equals mean minutes per nursing bout times mean number of bouts per daytime minute—except for the denominator of equation (6.1). The product $s_{ij}k$ approximates the summation over all activities of the numerator term of equation (3.2), which is what was used as the denominator of equation (3.2). On the assumption that in the large-sample case the ratio (approximated by k) of the summation over all activities of the numerator to the in-sight time is essentially constant across individual $\times$ age-block classes, $s_{ij}k$ will be asymptomatically equivalent to the sum of the feeding and nonfeeding times, such as is shown for milk in the numerator of equation (6.1). I used $s_{ij}k$ as the denominator rather than the summation over all activities in order to avoid the gargantuan task of calculating a separate product-limit estimate of the mean for each of the 277 foods plus nonfeeding bouts for each of eleven yearlings in each of four age blocks. (The normalizing factor $k = 1.116$ is used again in chapter 7 for estimating individual-specific bout rates and time budgets.)

Table 6.6 shows the mean diurnal suckling bout lengths (product-limit estimates), suckling bout frequencies and rates (expressed for convenience as bouts per 100 minutes), and suckling time-budget estimates (step 1), the latter both as numbers of minutes per daytime and as percentages of the daytime, 0700 h to 1800 h. (The amount of nocturnal suckling time is unknown.) The methods used for calculating standard errors of rates and products are given in chapter 3.

Step 2

In each age block, the unweighted mean diurnal suckling time of the yearlings was calculated (table 6.8); then a least-squares linear regression of these four means against age was carried out. One very small sample (14.2 minutes without any suckling, Pooh at 30–40 weeks of age) was not included in the calculations. The x-intercept of the regression line was taken as an estimate of the mean age at which suckling would cease. In carrying out the regression, the mean daily suckling time in each age block was taken to be at the age midpoint of that block.

The results of step 2 are that the estimated mean age of last suckling is 85.08 weeks. This is the value of x when $y = 0$ in the least-squares regression equation

$$(A9.2) \qquad\qquad y = 148.08 - 1.74w,$$

where y = minutes of suckling per daytime and w = age in weeks ($w = 30, \ldots, 70$). The high regression coefficient of this equation ($r^2 = 0.93$) indicates that almost all of the variances with age in mean suckling times is accounted for by a linear decline of 1.74 minutes of suckling per week, that is, about one-quarter minute per day.

For comparison, Wasser and Wasser (1995) studied rates of infant development in three groups of yellow baboons at Mikumi National Park, Tanzania, during the first three months of development. Sampling was done during 9.5 daylight hours. Their linear regression ($r = 0.95$) of percentage of daytime spent in nipple contact against age is:

$$z = 61.06 - 0.227d,$$

where z is percentage of time and d is age in days. If we assume that this percentage holds on average over eleven daylight hours, as in my study, and convert age to weeks, as in equation (A9.3), this becomes:

$$y = 403.0 - 10.49w,$$

with y in minutes per day on the nipple and w = age in weeks ($w = 0, \ldots, 13$). Barring population or methodological differences, these two equations for y suggest a marked tapering off in the rate of decline in nipple time after about thirty weeks of age, and nursing time appears to be a linear function of age only if a restricted age span is sampled. A curvilinear function is suggested by the composite data that are presented in figure 6.8.

Step 3

I assumed that mean daily milk consumption per unit of body mass fell off linearly from 210 ml per kg body mass per day at age 17.5 weeks—the last value given by Buss and Voss (1971)—to the estimated zero point of suckling, 85.08 according to the results of step 2. This resulted in the relationship

$$(A9.3) \qquad m/M = 264.38 - 3.11w,$$

where m/M is daily milk consumption (in milliliters) per kilogram of body mass and w is age in weeks. I then interpolated to the age-block midpoints, 35, . . . , 65 weeks of age, to obtain mean daily milk intakes per unit body of 155.6, 124.5, 93.5, and 62.4 ml kg^{-1}d^{-1}, respectively. This decrease in milk intake with age is consistent with Nicolson's (1982) study, in which the quantity of milk that could be expressed manually from trapped, sedated, wild female anubis baboons after oxytocin injection averaged 4 ml for females with infants under 200 days ($N = 4$), whereas the average amount obtained from females with infants 201–400 days ($N = 7$) was only 2.8 ml.

Equation (A9.3) is not equivalent to extrapolating from Buss and Voss's data from the first eighteen weeks of baboon life. Values in their figure 1b (using at each age the midpoint of average male and average female values) result in a linear regression of $m/M = 262.6 - 1.59\ w$ ($R^2_{adj} = 0.50$, $p \leq 0.01$), which if extrapolated to zero intake predicts an age of last nursing at more than three years—essentially twice our value (step 2, above). A more rapid decline in mass-specific milk intake at older ages is to be expected as infants progressively increase the amount of solid food they themselves get.

Step 4

Within each ten-week age block, my estimates of individual daily milk intakes (ml/24 hr) were based on time on the nipple during 0700 h to 1800 h. In doing so, I assumed (1) that throughout each age block, the mean dirunal suckling time budget corresponded with mean daily (24 hr) milk intake per unit body mass (step 3) at the midpoint of the age block; and (2) that there were no significant differences in the yearlings' masses in a given age class, so that mean mass at the midpoint of each age class could be used for each of them.

For milk intake calculations, mean body masses M in kilograms were estimated on the assumption of a birth mass of 0.775 kg and a mean daily mass increase of 6 g (J. Altmann 1980):

$$(A9.4) \qquad M = 0.775 + 0.042w \qquad \text{(first approximation)},$$

where w is age in weeks. This gave 2.245, 2.665, 3.085, and 3.505 kg at the ten-week age intervals 35, . . . , 65 respectively (first approximation).[17]

Mean daily milk consumption at each age was obtained from the product of these pairs of numbers, that is, this age-specific body masses and the milk intakes per unit mass, obtained in step 3. Finally, I assumed (3) that a given baboon's daily milk intake at any age differed from the mean intake for infants of that age by the same percentage as its diurnal nursing time differed from mean nursing time for that age. For example, if in a given age block a yearling spent, say, only 95% of the average amount of daytime suckling, it was assumed to get 95% of the average amount of milk during that age block. (Note that I am not assuming milk was obtained only during the daytime.) Thus the mean daily milk intake of an infant could be estimated at any age. For ease of computation, body mass and milk-intake rate per minute of suckling were assumed uniform for each yearling throughout each ten-week period and were computed for the midpoint of each interval.

Milk-intake estimates that made use of equation (A9.4), the first approximation of infant growth rate, are used extensively in chapter 6. Estimates based on a mean growth rate of 4.5 g/day (second approximation, for which see appendix 1) are discussed further along in chapter 6.

Appendix 10

Calculating Nutrient Intakes of Individuals at Specified Ages

When calculating the nutrient intake of an individual at a specified age, I took feeding bout rate and bout length values for each core food from the most specific (most narrowly partitioned) level—by individual × age, just by age, or all data pooled—for which significant differences were demonstrated for that food. That criterion led to the following procedural rules in carrying out the calculations.

For bout lengths, use table 7.3A–D, as appropriate for the individual's age. In the table, if the row for that food shows significant individual differences, that is, if the food is listed in table 7.3 at that age, use the individual's value in that row. If not significant, then use that food's age-specific pooled mean, table 7.2, if those values are significantly different, that is, if one is given in table 7.2; otherwise, use that food's overall pooled mean bout length (table 4.1). For bout rate, use table 7.6A–D, as appropriate for the yearling's age. If, in the table, the row for that food shows significant individual differences at that age, use the individual value in that row. If not significant, then use the row's pooled mean if significant age differences are indicated; otherwise use the overall pooled mean (table 7.6E).

Let me illustrate this procedure by showing the computation of the amount of water that Alice at 30–40 weeks got from unidentified grass corms (GRCX, food 2 in the time-ranked lists). The value, 0.794 g/day, was obtained from the product of the following numbers:

L_2	Bout length	0.41 minutes of feeding on this food per bout	table 7.2
R_2	Bout rate	0.1191 bouts per potential foraging minute	table 7.6
S	Potential foraging time	660 minutes of potential foraging time per day	chapter 3
B_2	Unit intake rate	3.80 food units per minute of feeding on this food	table 6.1
C_2	Unit mass	0.0235 grams per fresh food unit	table 6.4
D_2	Water content	0.276 grams water per gram fresh food	table 5.1

Because the intake rate of milk is not known, it was treated differently, as described in chapter 6. Table 6.6 gives milk intake values for each yearling during each age block in which it was sampled. Those values, given in milliliters per twenty-four hours, were converted to mass per twenty-four hours by multiplying them by the specific gravity of baboon milk, 1.027 g/ml (Busse 1968).

Alice's total daily water intake, 254.5 g, was obtained by summing the yield of water from the fifty-one non-milk core sources, adding 9.5% of that for the non-core foods, then adding the water obtained from milk (PCMX) to get the total intake of water from all foods.

The standard errors of these intakes were obtained as follows. The daily intake K_i of the ith food component is the sum $K_{i1} + \ldots + K_{in}$ of the intakes of that component from milk, from each of the other core foods, and from non-core foods. As indicated in chapter 3, the standard error of the sum of random variables is approximately the square root of the sum of the squared standard errors of each, that is, $S(K_i) = [S^2(K_{i1}) + \ldots + S^2(K_{in})]^{1/2}$. For the intake of component i from the jth core food, the standard error $S(K_{ij})$ was obtained from the standard errors of the six factors L_j, R_j, S, B_j, C_j, and D_{ij} via equations (3.3) and (3.4). Then the standard error for the contribution of the non-core foods was obtained as follows. As indicated above, the total contribution K_{ir} of any food component i made by the non-core foods is assumed to be 9.5% of the total contribution of the fifty-one non-milk core foods, that is,

$$K_{ir} = 0.095 \sum_{j=2}^{52} K_{ij} = \sum_{j=2}^{52} 0.095 K_{ij}$$

where the summations are over the fifty-one non-milk core foods. For each of the terms K_{ij}, note that $S(0.095K_{ij}) = 0.095S(K_{ij})$, that is, the standard error of a constant times a random variable equals the constant times the standard error of the variable (chapter 3). Thus $S^2(0.095K_{ij}) = 0.095^2 S^2(K_{ij})$, or more generally, the squared standard error of a constant times a random variable equals the square of the constant times the squared standard error of the variable. So for the non-core foods,

$$S^2(K_{ir}) = 0.095^2 \sum_{j=2}^{52} S^2(K_{ij}).$$

We can now write the standard error of food component i, with terms for milk, non-milk core foods, and non-core foods, as follows:

$$S(K_i) = \left[S^2(K_{i1}) + \sum_{j=2}^{52} S^2(K_{ij}) + 0.095^2 \sum_{j=2}^{52} S^2(K_{ij}) \right]^{1/2}$$

$$= \left[S^2(K_{i1}) + 1.00903 \sum_{j=2}^{52} S^2(K_{ij}) \right]^{1/2}.$$

Tables

Table 3.1 Minutes of Actual Sampling (in-Sight Time) by Individual and Age

Name	Abbre-viation	30–40 Weeks (block 1)	40–50 Weeks (block 2)	50–60 Weeks (block 3)	60–70 Weeks (block 4)	Total
Alice	AC	504.12	975.78	394.65	1001.01	2875.56
Bristle	BS	810.31	1035.01	0	0	1845.32
Dotty	DT	0	0	405.43	0	405.43
Eno	EN	582.08	794.31	1212.82	0	2589.21
Fred	FR	653.81	1012.91	203.72	964.82	2835.26
Hans	HN	1214.14	0	0	0	1214.14
Ozzie	OZ	694.83	995.15	202.03	1030.07	2922.08
Pedro	PD	646.91	0	0	0	646.91
Pooh	PH	14.22	839.87	718.74	397.43	1970.26
Striper	ST	800.05	790.43	595.88	0	2186.36
Summer	SR	0	0	488.21	0	488.21
Total		5920.47	6443.46	4221.48	3393.33	19978.74

Note: Figures do not include time spent nursing after scheduled sample endings, even when observed and recorded.

Table 3.2 The Subjects

Name	Abbreviation	Sex	Birth Date	Mother	Maternal Style[a]
Alice	AC	F	1 Jan 75	Alto	
Bristle	BS	M	17 Aug 75	Brush	Restrictive
Dotty	DT	F	21 Jun 73	Alto	
Eno	EN	F	22 Apr 75	Este	
Fred	FR	M	1 Jan 75	Fem	
Hans	HN	M	15 Oct 75	Handle	Restrictive
Ozzie	OZ	M	24 Dec 74	Oval	
Pedro	PD	M	9 Jul 75	Preg	Laissez-faire
Pooh	PH	F	31 Oct 74	Plum	
Striper	ST	F	5 Jul 73	Mom	
Summer	SR	F	7 Jul 75	Scar	Restrictive

[a] From J. Altmann 1980.

Table 3.3 Censoring Analysis, Six Test Foods

PLE Category and Food Code[a]	G SPF	M FTG	B SKC	8 TOBS	7 TOBR	90 CDLU
PLE sample size	228	1334	430	95	182	518
Bout length (minutes), PLE mean ± SE	0.71 ± 0.079	0.37 ± 0.036	0.74 ± 0.058	≥1.85 ± 0.317	0.08 ± 0.016	0.45 ± 0.043
Percentage of bouts right-censored	13.2%	5.4%	3.7%	12.6%	0%	4.8%
Number of right-censorings						
Disappearances	24	50	1	5	0	14
Sample endings	6	22	15	7	0	11
Minutes before right-censoring, PLE mean ± SE	≥2.85 ± 0.371	≥3.66 ± 0.197	≥5.68 ± 0.286	9.37 ± 1.283	≥0.70 ± 0	≥4.92 ± 0.258
Right-censorings per minute of feeding[b]	≤0.35	≤0.27	≤0.18	0.11	≤1.4	≤0.20
Minutes before disappearing, PLE mean ± SE	≥3.26 ± 0.415	≥4.08 ± 0.129	≥6.47 ± 0.015	≥11.70 ± 1.075	≥0.70 ± 0	≥5.41 ± 0.198
Disappearances per minute of feeding[b]	≤0.31	≤0.25	≤0.15	≤0.085	≤1.43	≤0.18
Minutes of observed feeding	135.20	456.23	305.30	145.83	14.38	217.04
Disappearance per minute of observed feeding	0.18	0.11	0.0033	0.034	0	0.065
Right-censorings per minute of observed feeding	0.22	0.16	0.052	0.082	0	0.12

Note: Based on bouts with observed onsets. PLE = product-limit estimate; SE = standard error of mean.

[a] SPF = fruit of toothbrush-bush (*Salvadora persica*); FTG = fever tree gum from trees; SKC = corms of *Sporobolus rangei*; TOBS = umbrella tree blossoms from trees; TOBR = umbrella tree blossoms from ground; CDLU = green leaves of *Cynodon nlemfuensis*.

[b] Inverse of number above. Assumes negligible within-food variance in rate (Welsh, Peterson, and Altmann 1988).

Table 4.1 Statistical Characteristics of Feeding Bouts

Food	Number of Bouts (a)	Species Total (b)	Number of Complete Bouts (c)	PLE Categories (d)	PLE Sample Size (e)	PLE Mean Bout Length (min) ± SE (f)	PLE Feeding Time (min) (g)	Percentage of Feeding Time (h)
				A. Plants and Plant Products				
AA *Asparagus ?africanus*								
1. XX unknown material	1	1		ZZ	961	0.20 ± 0.03	0.20	0.002
AB *Abutilon* sp.		10						
2. BS blossoms on plant	4			BB	421	0.67* ± 0.13	2.68	0.03
3. FX fruit, condition unknown	2			FF	1721	0.51 ± 0.05	1.02	0.01
4. PU pods, green	1			PD(XX)	21	0.75 ± 0.21	0.75	0.01
5. PX pods, condition unknown	1			PD(XX)	21	0.75 ± 0.21	0.75	0.01
6. XX unknown material	2			ZZ	961	0.20 ± 0.03	0.40	0
TO *Acacia tortilis*, umbrella tree		913						
7. BR blossoms on ground	185		182	7(BB)	182	0.08 ± 0.02	14.80	0.16
8. BS blossoms on tree	112		83	8(BB)	95	1.85* ± 0.32	207.20	2.21
9. DD seeds from dung	6			C(DD)	22	0.72 ± 0.25	4.32	0.05
10. DG seeds on ground	79		75	10(DD)	75	0.20 ± 0.03	15.80	0.17
11. DU' seeds, green	1			DD	3349	0.32 ± 0.06	0.32	0
12. DX' seeds, condition unknown	1			DD	3349	0.32 ± 0.06	0.32	0
13. EU seedlings	1			PT(XX)	29	0.51 ± 0.13	0.51	0.01
14. GX gum, condition unknown	2			UU	1716	0.40 ± 0.04	0.80	0.01
15. LU leaves, green	1			OO(LL)	486	0.35 ± 0.03	0.35	0.00
16. DR seeds from dry pods	118		90	16(DD)	100	0.72 ± 0.12	84.96	0.91
17. DS seeds from semidry pods	9			DD	3349	0.32 ± 0.06	2.88	0.03
18. DU seeds from green pods	385		247	18(DD)	304	1.27 ± 0.11	488.95	5.21
19. DX seeds from pods, condition unknown	6			DD	3349	0.32 ± 0.06	1.92	0.02
20. WR wood, dead	4			W(WW,XX)	114	0.50 ± 0.12	2.00	0.02
21. WX wood, condition unknown	3			W(WW,XX)	114	0.50 ± 0.12	1.50	0.02
FT *Acacia xanthophloea*, fever tree		3396						
22. BR blossoms on ground	5			BB	421	0.67 ± 0.13	3.35	0.04
23. BS blossoms on tree	80		66	23(BB)	72	0.88 ± 0.14	70.40	0.75
24. DG seeds on ground	795		754	24(DD)	773	0.26 ± 0.03	206.70	2.20

Table 4.1 (*Continued*)

Food	Number of Bouts (a)	Species Total (b)	Number of Complete Bouts (c)	PLE Categories (d)	PLE Sample Size (e)	PLE Mean Bout Length (min) ± SE (f)	PLE Feeding Time (min) (g)	Percentage of Feeding Time (h)
25. DP seeds on tree	2			DD	3349	0.32 ± 0.06	0.64	0.01
26. DX seeds, condition unknown	5			DD	3349	0.32 ± 0.06	1.60	0.02
27. GR gum from tree, clear to light amber	10			M(UU)	1334	0.37 ± 0.04	3.70	0.04
28. GX gum from tree	1493			M(UU)	1334	0.37 ± 0.04	552.41	5.89
29. GU gum from tree, amber to brown-black	1			M(UU)	1334	0.37 ± 0.04	0.37	0.004
30. HG gum on ground	287		240	30(N,UU)	258	0.50 ± 0.06	143.50	1.53
31. HP gum on other plant	40		35	31(N,UU)	37	0.62 ± 0.17	24.80	0.26
32. HR gum not from tree, amber to brown-black	6			N(UU)	380	0.49 ± 0.06	2.94	0.03
33. HS gum from ground, amber	1			N(UU)	380	0.49 ± 0.06	0.49	0.01
34. HW gum from wood	53		48	34(N,UU)	50	0.46 ± 0.12	24.38	0.26
35. HX gum from other sources	30		29	35(N,UU)	29	0.23 ± 0.09	6.90	0.07
36. LU leaves, green	13			R(OO,LL)	25	0.68 ± 0.24	8.84	0.09
37. LX leaves, condition unknown	12			R(OO,LL)	25	0.68 ± 0.24	8.16	0.09
38. DR seeds from dry pods	435		401	38(DD)	414	0.38 ± 0.04	165.30	1.76
39. DU seeds from green pods	74		55	39(DD)	64	1.03 ± 0.18	76.22	0.81
40. DX seeds from pods, condition unknown	14			DD	3349	0.32 ± 0.06	4.48	0.05
41. QU thorns, green	1			PT(XX)	29	0.51 ± 0.13	0.51	0.01
42. RU roots	1			PT(YY)	29	0.51 ± 0.13	0.51	0.01
43. TX stalks, condition unknown	2			Y(XX)	173	0.41 ± 0.04	0.82	0.01
44. WR wood, dead	12			W(WW,XX)	114	0.50 ± 0.12	6.00	0.06
45. WU wood, live	5			W(WW,XX)	114	0.50 ± 0.12	2.50	0.03
46. WX wood, condition unknown	19			W(WW,XX)	114	0.50 ± 0.12	9.50	0.10
AC *Acacia* sp. (*tortilis* or *xanthophloea*)		17						
47. DD seeds from dung	17			C(DD)	22	0.72 ± 0.25	12.24	0.13
FU Agaricales (mushrooms), other		6						
48. EX whole sporophores	5			PT(XX)	29	0.51 ± 0.13	2.55	0.03
49. TX stalk of sporophores	1			Y(YY)	173	0.41 ± 0.04	0.41	0.004

				Code	Mean ± SD	N		
AL *Aloe* sp.	1							
50. XX unknown material		1		ZZ	0.20 ± 0.03	961	0.20	0.002
AH *Atriplex* sp. nr. *halimus*	1							
51. LU leaves, green		1		OO(LL)	0.35 ± 0.03	486	0.35	0.004
AT *Azima tetracantha*	590							
52. DR seeds, dry		1		DD	0.32 ± 0.06	3349	0.32	0.003
53. EU seedlings		4		PT(XX)	0.51 ± 0.13	29	2.04	0.02
54. FR fruit, ripe		218	168	54(E, FF)	0.42 ± 0.06	185	91.56	0.98
55. FS fruit, semiripe		7		E(FF)	0.34 ± 0.04	470	2.38	0.03
56. FU fruit, green		90	75	56(E, FF)	0.35 ± 0.04	83	31.50	0.34
57. FX fruit, condition unknown		215	183	57(E, FF)	0.23* ± 0.03	195	49.45	0.53
58. LR leaves, yellow		3		O(OO, LL)	0.41 ± 0.09	24	1.23	0.01
59. LU leaves, green		17		O(OO, LL)	0.41 ± 0.09	24	6.97	0.07
60. LX leaves, condition unknown		5		O(OO, LL)	0.41 ± 0.09	24	2.05	0.02
61. WR wood, dead		1		W(WW, XX)	0.50 ± 0.12	114	0.50	0.01
62. WU wood, live, from sprout		2		W(WW, XX)	0.50 ± 0.12	114	1.00	0.01
63. WX wood, condition unknown		5		W(WW, XX)	0.50 ± 0.12	114	2.50	0.03
64. XX unknown material		22	20	64(ZZ)	0.15 ± 0.03	20	3.30	0.04
BS *Battarraea stevenii*, a mushroom	3							
65. TX stalks		1		Y(YY)	0.41 ± 0.04	173	0.41	0.004
66. XX unknown material		2		ZZ	0.20 ± 0.03	961	0.40	0.004
CT *Capparis tomentosa*	61							
67. BX blossoms, condition unknown		7		BB	0.67* ± 1.33	421	4.69	0.05
68. FR fruit, ripe		11		F(FF)	0.62 ± 0.13	30	6.82	0.07
69. FS fruit, semiripe		2		F(FF)	0.62 ± 0.13	30	1.24	0.01
70. FU fruit, green		5		F(FF)	0.62 ± 0.13	30	3.10	0.03
71. FX fruit, condition unknown		19		F(FF)	0.62 ± 0.13	30	11.78	0.13
72. LX leaves, condition unknown		2		OO(LL)	0.35 ± 0.03	486	0.70	0.01
73. PR pods, ripe		3		PD(XX)	0.75 ± 0.21	21	2.25	0.02
74. PX pods, condition unknown		5		PD(XX)	0.75 ± 0.21	21	3.75	0.04
75. WX wood, condition unknown		5		W(WW, XX)	0.50 ± 0.12	114	2.50	0.03
76. XX unknown material		2		ZZ	0.20 ± 0.03	961	0.40	0.004
CH *Chenopodium opulifolium*	4							
77. LX leaves, condition unknown		1		OO(LL)	0.35 ± 0.03	486	0.35	0.004
78. XX unknown material		3		ZZ	0.20 ± 0.03	961	0.60	0.01
CB *Chlorophytum* sp. nr. *bakeri*	21							
79. CX bulbs, condition unknown		2		PT(XX)	0.51 ± 0.13	29	1.02	0.01
80. LX leaves		19		OO(LL)	0.35 ± 0.03	486	6.65	0.07

Table 4.1 (*Continued*)

Food	Number of Bouts (a)	Species Total (b)	Number of Complete Bouts (c)	PLE Categories (d)	PLE Sample Size (e)	PLE Mean Bout Length (min) ± SE (f)	PLE Feeding Time (min) (g)	Percentage of Feeding Time (h)
SF *Commicarpus pedunculosus*, sticky-fruit plant		73						
81. FS fruit, semiripe	1			J(FF)	73	0.23* ± 0.05	0.23	0.002
82. FX fruit, condition unknown	72			J(FF)	73	0.23* ± 0.05	16.56	0.18
CD *Cynodon dactylon*, Bermuda grass*		705						
83. CR corms of dry plant	1			CC	3052	0.53* ± 0.04	0.53	0.01
84. CU corms of green plant	1			CC	3052	0.53* ± 0.04	0.53	0.01
85. CX corms, condition unknown	12			CC	3052	0.53* ± 0.04	6.36	0.07
86. DU seedheads, closed	16			D(DD)	44	0.54 ± 0.07	8.64	0.09
87. DX seedheads, condition unknown	18			DD	3349	0.32 ± 0.06	5.76	0.06
88. LR leaves, dry	1			NN(LL)	8	0.18 ± 0.10	0.18	0.002
89. LS leaves, semidry	1			NN(LL)	8	0.18 ± 0.10	0.18	0.002
90. LU leaves, green	546		493	90(GG,LL)	518	0.45 ± 0.04	245.70	2.62
91. LX leaves, condition unknown	3			NN(LL)	8	0.18 ± 0.10	0.54	0.01
92. SR stolons, dry	1			U(RS,XX)	88	0.32 ± 0.05	0.32	0.003
93. SS stolons, semidry	1			U(RS,XX)	88	0.32 ± 0.05	0.32	0.003
94. SU stolons, green	35		34	94(U,RS,XX)	35	0.26 ± 0.04	9.10	0.10
95. SX stolons, condition unknown	51		50	95(U,RS,XX)	51	0.36 ± 0.08	18.36	0.20
96. TU blade bases	2			Y(YY)	173	0.41 ± 0.04	0.82	0.01
97. XX unknown material	16			ZZ	961	0.20 ± 0.03	3.20	0.03
CP *Cynodon plectostachyus*		12						
98. CR corms of dry plant	1			CC	3052	0.53* ± 0.04	0.53	0.01
99. CX corms, condition unknown	7			CC	3052	0.53* ± 0.04	3.71	0.04
100. DX seeds	1			DD	3349	0.32 ± 0.06	0.32	0.003
101. LU leaves, green	1			GG(LL)	2495	0.38 ± 0.04	0.38	0.004
102. XX unknown material	2			ZZ	961	0.20 ± 0.03	0.40	0.004
CI *Cyperus immensus*		2						
103. XX unknown material	2			ZZ	961	0.20 ± 0.03	0.40	0.004
CL *Cyperus laevigatus*		4						
104. TX stalks, condition unknown	1			Y(YY)	173	0.41 ± 0.04	0.41	0.004
105. XX unknown material	3			ZZ	961	0.20 ± 0.03	0.60	0.01
CO *Cyperus obtusiflorus*		29						
106. CX corms, condition unknown	28		27	106(CC)	28	0.77 ± 0.16	21.56	0.23

						Mean ± SE		
107. XX unknown material	1			ZZ	961	0.20 ± 0.03	0.20	0.002
CV *Cyperus* sp.		4						
108. BS flowers from plant	1			BB	421	0.67* ± 1.33	0.67	0.01
109. XX unknown material	3			ZZ	961	0.20 ± 0.03	0.60	0.01
DB *Dactyloctenium bogdanii*		8						
110. CX corms, condition unknown	1			CC	3052	0.53* ± 0.04	0.53	0.01
111. DX seeds, condition unknown	2			DD	3349	0.32 ± 0.06	0.64	0.01
112. SX stolons/rhizomes, condition unknown	1			RS(XX)	120	0.30 ± 0.04	0.30	0.003
113. TU blade bases, green	4			Y(YY)	173	0.41 ± 0.41	1.64	0.02
DP *Dasysphaera prostrata*, hairy-ball plant*		12						
114. XX unknown material	12			ZZ	961	0.20 ± 0.03	2.40	0.03
DI *Dicliptera albicaulis*		3						
115. XX unknown material	3			ZZ	961	0.20 ± 0.03	0.60	0.01
GR Gramineae (grasses) or Cyperaceae (sedges), unspecified		4749						
116. CU corms of green grass	5			CC	3052	0.53* ± 0.04	2.65	0.03
117. CX corms, condition unknown	2539		2272	117(CC)	2379	0.47* ± 0.04	1193.33	12.72
118. DU seedheads, green, closed	25		24	118(D,DD)	25	0.53 ± 0.09	13.25	0.14
119. DX seeds	33		31	119(DD)	32	0.23 ± 0.04	7.59	0.08
120. EU seedlings	2			PT(XX)	29	0.51 ± 0.13	1.02	0.01
121. LU leaves, green	1958		1836	121(GG,LL)	1886	0.36 ± 0.04	704.88	7.51
122. LX leaves, condition unknown	1			NN(LL)	8	0.18 ± 0.10	0.18	0.002
123. SU stolons/rhizomes, green	5			RS(XX)	120	0.30 ± 0.04	1.50	0.02
124. SX stolons/rhizomes, condition unknown	22		21	124(RS,XX)	22	0.22 ± 0.06	4.84	0.05
125. TR blade bases, dry	4			Y(YY)	173	0.41 ± 0.04	1.64	0.02
126. TU blade bases, fresh	137		133	126(Y)	136	0.38 ± 0.04	52.06	0.55
127. TX blade bases, condition unknown	3			Y(YY)	173	0.41 ± 0.04	1.23	0.01
128. XX unknown material	15			ZZ	961	0.20 ± 0.03	3.00	0.03
KO *Kochia indica*		2						
129. LU leaves, green	1			OO(LL)	486	0.35 ± 0.03	0.35	0.004
130. XX unknown material	1			ZZ	961	0.20 ± 0.03	0.20	0.002
LC *Leucas stricta*		2						
131. XX unknown material	2			ZZ	961	0.20 ± 0.03	0.40	0.004
LS *Ludwigia stolonifera*		11						
132. LX leaves, condition unknown	2			OO(LL)	486	0.35 ± 0.03	0.70	0.01
133. XX unknown material	9			ZZ	961	0.20 ± 0.03	1.80	0.02

Table 4.1 (Continued)

Food	Number of Bouts (a)	Species Total (b)	Number of Complete Bouts (c)	PLE Categories (d)	PLE Sample Size (e)	PLE Mean Bout Length (min) ± SE (f)	PLE Feeding Time (min) (g)	Percentage of Feeding Time (h)
TF *Lycium* "*europaeum*," trumpet-flower plant		195						
134. BS blossoms on plant	30			A(BB)	31	0.41 ± 0.07	12.30	0.13
135. BX blossoms, condition unknown	1			A(BB)	31	0.41 ± 0.07	0.41	0.004
136. FR fruit, ripe	30		30	136(H,FF)	30	0.24 ± 0.06	7.20	0.08
137. FX fruit, condition unknown	2			H(FF)	32	0.24 ± 0.05	0.48	0.01
138. LR leaves, dry	4			OO(LL)	486	0.35 ± 0.03	1.40	0.01
139. LU leaves, green	16			Q(OO,LL)	109	0.36 ± 0.05	5.76	0.06
140. LX leaves, condition unknown	97			Q(OO,LL)	109	0.36 ± 0.05	34.92	0.37
141. WR wood, dead	1			W(WW,XX)	114	0.50 ± 0.12	0.50	0.01
142. XX unknown material	14			ZZ	961	0.20 ± 0.03	2.80	0.03
MX *Maerua ?angolensis*		27						
143. BS blossoms on plant	5			BB	421	0.67 ± 0.13	3.35	0.04
144. FR fruit, ripe	2			FF	1721	0.51 ± 0.05	1.02	0.01
145. FX fruit, condition unknown	13			FF	1721	0.51 ± 0.05	6.63	0.07
146. LU leaves, green	1			OO(LL)	486	0.35 ± 0.03	0.35	0.004
147. PX pods, condition unknown	6			PD(XX)	21	0.75 ± 0.21	4.50	0.05
OC *Opilia campestris*		1						
148. XX unknown material	1			ZZ	961	0.20 ± 0.03	0.20	0.002
OD *Ornithogalum donaldsonii**		9						
149. BX flowers, condition unknown	1			BB	421	0.67* ± 0.13	0.67	0.01
150. CX bulbs, condition unknown	1			PT(XX)	29	0.51 ± 0.13	0.51	0.01
151. LX leaves, condition unknown	3			OO(LL)	486	0.35 ± 0.03	1.05	0.01
152. TX stalks, condition unknown	3			Y(YY)	173	0.41 ± 0.04	1.23	0.01
153. XX unknown material	1			ZZ	961	0.20 ± 0.03	0.20	0.002
PJ *Psilolemma jaegeri**		12						
154. CX corms, condition unknown	1			CC	3052	0.53 ± 0.04	0.53	0.01
155. LU leaves, green	6			GG(LL)	2495	0.38 ± 0.04	2.28	0.02
156. LX leaves, condition unknown	1			NN(LL)	8	0.18 ± 0.10	0.18	0.002
157. SX stolons/rhizomes, condition unknown	4			RS(XX)	120	0.30 ± 0.04	1.20	0.01

RM *Rhamphicarpa montana*		44						
158. BU blossom buds	3			BB	421	0.67* ± 0.13	2.01	0.02
159. BX blossoms, condition unknown	1			BB	421	0.67* ± 0.13	0.67	0.01
160. XX unknown material	40		39	160(ZZ)	40	0.22 ± 0.05	8.80	0.09
SP *Salvadora persica*, toothbrush bush		550						
161. EU seedlings	2			RS(XX)	120	0.30 ± 0.04	0.60	0.01
162. FR fruit, ripe	55		39	162(G,FF)	47	0.95 ± 0.17	52.25	0.56
163. FS fruit, semiripe	63		35	163(G,FF)	42	1.05 ± 0.19	66.15	0.70
164. FU fruit, green	142		112	164(G,FF)	124	0.47* ± 0.07	66.74	0.71
165. FX fruit, condition unknown	18			G(FF)	228	0.71 ± 0.08	12.78	0.14
166. LR leaves, dry	3			OO(LL)	486	0.35 ± 0.03	1.05	0.01
167. LU leaves, green	94		79	167(P,OO,LL)	83	0.35 ± 0.04	32.90	0.35
168. LX leaves, condition unknown	101		74	168(P,OO,LL)	82	0.25 ± 0.04	25.25	0.27
169. WR wood, dead	9			X(WW,XX)	59	0.44 ± 0.06	3.96	0.04
170. WU wood, live	13			X(WW,XX)	59	0.44 ± 0.06	5.72	0.06
171. WX wood, condition unknown	42		38	171(X,WW,XX)	39	0.40 ± 0.07	16.80	0.18
172. XX unknown material	8			ZZ	961	0.20 ± 0.03	1.60	0.02
SH *Sericocomopsis hildebrandtii**		6						
173. DX seeds, condition unknown	1			DD	3349	0.32 ± 0.06	0.32	0.003
174. LX leaves, condition unknown	1			OO(LL)	486	0.35 ± 0.03	0.35	0.004
175. XX unknown material	4			ZZ	961	0.20 ± 0.03	0.80	0.01
SV *Setaria verticillata*		21						
176. DU seedheads, green	2			D(DD)	44	0.54 ± 0.07	1.08	0.01
177. DX seedheads, condition unknown	1			DD	3349	0.32 ± 0.06	0.32	0.002
178. LU leaves, green	7			GG(LL)	2495	0.38 ± 0.04	2.66	0.03
179. TU blade bases, green	11			Y(YY)	173	0.41 ± 0.04	4.51	0.05
SB *Solanum dubium**		21						
180. FU fruit, green	1			K(FF)	20	0.55 ± 0.12	0.55	0.01
181. FX fruit, condition unknown	20			K(FF)	20	0.55 ± 0.12	11.00	0.12
SI *Solanum incanum*		3						
182. FX fruit, condition unknown	1			FF	1721	0.51 ± 0.05	0.51	0.01
183. XX unknown material	2			ZZ	961	0.20 ± 0.03	0.40	0.004
SA *Sporobolus africanus**		5						
184. CX corms, condition unknown	2			CC	3052	0.53* ± 0.04	1.06	0.01
185. DX seeds, condition unknown	1			DD	3349	0.32 ± 0.06	0.32	0.003
186. LU leaves, green	2			GG(LL)	2495	0.38 ± 0.04	0.76	0.01
SC *Sporobolus consimilis*		615						
187. CX corms, condition unknown	8			CC	3052	0.53* ± 0.04	4.24	0.05
188. LU leaves, green	13			GG(LL)	2495	0.38 ± 0.04	4.94	0.05
189. LX leaves, condition unknown	1			NN(LL)	8	0.18 ± 0.10	0.18	0.002

Table 4.1 (Continued)

Food	Number of Bouts (a)	Species Total (b)	Number of Complete Bouts (c)	PLE Categories (d)	PLE Sample Size (e)	PLE Mean Bout Length (min) ± SE (f)	PLE Feeding Time (min) (g)	Percentage of Feeding Time (h)
190. TU blade bases, green	13			V(YY)	521	0.54 ± 0.04	7.02	0.07
191. TX blade bases, condition unknown	561		474	191(V,YY)	509	0.53 ± 0.04	297.33	3.17
192. XX unknown material	19			ZZ	961	0.20 ± 0.03	3.80	0.04
SO *Sporobolus ioclados*								
193. CX corms, condition unknown	5	231		CC	3052	0.53* ± 0.04	2.65	0.03
SM *Sporobolus cordofanus*								
194. CX corms, condition unknown	176		168	194(CC)	172	0.71 ± 0.07	124.96	1.33
195. DU seedheads, green	2			D(DD)	44	0.54 ± 0.07	1.08	0.01
196. DX seeds, condition unknown	1			DD	3349	0.32 ± 0.06	0.32	0.003
197. LU leaves, green	45		43	197(GG,LL)	45	0.55 ± 0.08	24.75	0.26
198. TU blade bases, green	1			Y(YY)	173	0.41 ± 0.04	0.41	0.004
199. XX unknown material	6			ZZ	961	0.20 ± 0.03	1.20	0.01
SK *Sporobolus "kentrophyllus"*** (*S. rangei*)		459						
200. CS corms, semigreen plant	2			B(CC)	430	0.74 ± 0.06	1.48	0.02
201. CU corms, green plant	2			B(CC)	430	0.74 ± 0.06	1.48	0.02
202. CX corms, condition unknown	440			B(CC)	430	0.74 ± 0.06	325.60	3.47
203. LU leaves, green	10			GG(LL)	2495	0.38 ± 0.04	3.80	0.04
204. TU blade bases, green	1			Y(YY)	173	0.41 ± 0.04	0.41	0.004
205. TX blade bases, condition unknown	1			Y(YY)	173	0.41 ± 0.04	0.41	0.004
206. XX unknown material	3			ZZ	961	0.20 ± 0.03	0.60	0.01
SS *Sporobolus spicatus**		8						
207. CX corms, condition unknown	1			CC	3052	0.53* ± 0.08	0.53	0.01
208. LU leaves, green	7			GG(LL)	2495	0.38 ± 0.04	2.66	0.03
SU *Suaeda monoica*		60						
209. LU leaves, green	56		43	209(OO,LL)	46	0.25 ± 0.03	14.00	0.15
210. LX leaves, condition unknown	3			OO(LL)	486	0.35 ± 0.03	1.05	0.01
211. XX unknown material	1			ZZ	961	0.20 ± 0.03	0.20	0.002
TC *Trianthema ceratosepala*		431						
212. BU blossom buds (ovaries)	3			BB	421	0.67* ± 0.13	2.01	0.02
213. FU fruit, green	8			L(FF)	422	0.68 ± 0.07	5.44	0.06

214. FX fruit, condition unknown	414			L(FF)	422	0.68 ± 0.07	281.52	3.00
215. XX unknown material	6			ZZ	961	0.20 ± 0.03	1.20	0.01
TT *Tribulus terrestris*, devil's thorn			425					
216. FX fruit, condition unknown	417	396		216(FF)	409	0.46 ± 0.06	191.82	2.04
217. LU leaves, green	7			OO(LL)	486	0.35 ± 0.03	2.45	0.03
218. XX unknown material	1			ZZ	961	0.20 ± 0.03	0.20	0.002
WS *Withania somnifera*			83					
219. BU blossom buds	5	36		BB	421	0.67* ± 0.13	3.35	0.04
220. FR fruit, ripe	1			I(FF)	48	0.47 ± 0.06	0.47	0.01
221. FU fruit, green	10			I(FF)	48	0.47 ± 0.06	4.70	0.05
222. FX fruit, condition unknown	47			222(I,FF)	38	0.48 ± 0.07	22.56	0.24
223. LR leaves, dry	1			OO(LL)	486	0.35 ± 0.03	0.35	0.004
224. LU leaves, green	5			OO(LL)	486	0.35 ± 0.03	1.75	0.02
225. LX leaves	2			OO(LL)	486	0.35 ± 0.03	0.70	0.01
226. WX wood, condition unknown	2			W(WW,XX)	114	0.50 ± 0.12	1.00	0.01
227. XX unknown material	10			ZZ	961	0.20 ± 0.03	2.00	0.02
XX Plant, unidentified			2456					
228. BS blossoms on plant	1			BB	421	0.67* ± 0.13	0.67	0.01
229. BU blossom buds	4			BB	421	0.67* ± 0.13	2.68	0.03
230. BX blossoms, condition unknown	2			BB	421	0.67* ± 0.13	1.34	0.01
231. CX tubers of monocot	3			PT(XX)	29	0.51 ± 0.13	1.53	0.02
232. DD seeds from dung	8			DD	3349	0.32 ± 0.06	2.56	0.03
233. DG seeds from ground	14			DD	3349	0.32 ± 0.06	4.48	0.05
234. DP seeds from pod	1			DD	3349	0.32 ± 0.06	0.32	0.003
235. DX seeds, source unknown	23	23		235(DD)	23	0.22 ± 0.09	5.06	0.05
236. FX fruit, condition unknown	9			FF	1721	0.51 ± 0.05	4.59	0.05
237. LR leaves, dry	1			OO(LL)	486	0.35 ± 0.03	0.35	0.004
238. LU leaves, green	16			OO(LL)	486	0.35 ± 0.03	5.60	0.06
239. LX leaves, condition unknown	41	37		239(OO,LL)	40	0.33* ± 0.06	13.53	0.14
240. PX pods, condition unknown	5			PD(XX)	21	0.75 ± 0.21	3.75	0.04
241. RX roots, condition unknown	9			PT(XX)	29	0.51 ± 0.13	4.59	0.05
242. TX petioles, condition unknown	2			Y(YY)	173	0.41 ± 0.04	0.82	0.01
243. WR wood, dead	32	28		243(W,WW,XX)	28	0.26 ± 0.07	8.32	0.09
244. WU wood, green	2			W(WW,XX)	114	0.50 ± 0.12	1.00	0.01
245. WX wood, condition unknown	32	29		245(W,WW,XX)	29	0.46 ± 0.16	14.72	0.16
246. XD material from dung	769	760		246(DD)	764	0.06 ± 0.03	46.14	0.49
247. XG material from the ground	664			DD	3349	0.32 ± 0.06	212.48	2.26

Table 4.1 (*Continued*)

Food	Number of Bouts (a)	Species Total (b)	Number of Complete Bouts (c)	PLE Categories (d)	PLE Sample Size (e)	PLE Mean Bout Length (min) ± SE (f)	PLE Feeding Time (min) (g)	Percentage of Feeding Time (h)
248. XP from plants	12			ZZ	961	0.20 ± 0.03	2.40	0.03
249. XW from wood (arthropods?)	277			ZZ	961	0.20 ± 0.03	55.40	0.59
250. XX unknown material	529			ZZ	961	0.20 ± 0.03	105.80	1.13
B. Animals and Animal Products								
IN Arthropoda, other		87						
251. KU larvae	13			S(MM)	230	0.25 ± 0.05	3.25	0.03
252. KX meat	74		73	252(S,MM)	74	0.29 ± 0.05	21.46	0.23
BI Aves (birds), other		3						
253. KX meat	1			T(MM)	37	0.65* ± 0.28	0.65	0.01
254. NX eggshells	1			T(MM)	37	0.65* ± 0.28	0.65	0.01
255. OX nests, torn open	1			T(MM)	37	0.65* ± 0.28	0.65	0.01
VT *Cercopithecus aethiops*, vervet monkey		3						
256. KX meat	3			T(MM)	37	0.65* ± 0.28	1.95	0.02
CG *Cichladusa guttata*, spotted morning warbler		1						
257. NX eggshells	1			T(MM)	37	0.65* ± 0.28	0.65	0.01
BE Coleoptera (beetles), other		17						
258. KX meat	17			S(MM)	230	0.25 ± 0.05	4.25	0.05
WI *Connochaetes taurinus*, gnu		1						
259. AX dung	1			T(MM)	37	0.65* ± 0.28	0.65	0.01
GS *Galago senegalensis*, bushbaby		1						
260. KX meat	1			T(MM)	37	0.65* ± 0.28	0.65	0.01
GA Gastropoda (snails)		12						
261. KX meat	8			T(MM)	37	0.65* ± 0.28	5.20	0.06
262. ZX shell, empty	4			T(MM)	37	0.65* ± 0.28	2.60	0.03
GG *Gazella granti*, Grant's gazelle		1						
263. KX meat	1			T(MM)	37	0.65* ± 0.28	0.65	0.01
HY Hymenoptera (ants)		12						
264. KX bodies	12			S(MM)	230	0.25 ± 0.05	3.00	0.03
PO *Hystrix galeata*, porcupine		1						
265. QX quills	1			T(MM)	37	0.65* ± 0.28	0.65	0.01

Code / food	a	b	c	d	e	f	g	h
HR *Lepus capensis*, Cape hare	5							
266. KX meat	5			T(MM)	37	0.65* ± 0.28	3.25	0.03
GH Aeriidae (grasshoppers)	180		175					
267. KX bodies	180		175	267(MM)	178	0.04 ± 0.01	7.20	0.08
SN Ophidia (snakes), unidentified	1							
268. KX meat	1			T(MM)	37	0.65* ± 0.28	0.65	0.01
PC *Papio cynocephalus*, yellow baboon	1353		1170					
269. JX ejaculate	3			T(MM)	37	0.65* ± 0.28	1.95	0.02
270. MX milk	1350		1170	270	1202	1.65 ± 0.17	2227.50	23.74
DU Scarabaeidae (dung beetles)	116		113					
271. KU larvae	115		113	271(S,MM)	114	0.17 ± 0.05	19.55	0.21
272. ZX exoskeletons	1			S(MM)	230	0.25 ± 0.05	0.25	0.003
MO *Varanus niloticus*, monitor lizard	1							
273. OX eggs	1			T(MM)	37	0.65* ± 0.28	0.65	0.01
XX Unidentified mammals	6							
274. AX dung	6			T(MM)	37	0.65* ± 0.28	3.90	0.04

C. Inorganic Materials

Code / food	a	b	c	d	e	f	g	h
EA Earth	2							
275. EX	2			XX	345	0.44 ± 0.06	0.88	0.01
WA Water	350		350					
276. VG from ground	315		303	276(VV)	306	0.15 ± 0.02	47.25	0.50
277. VP drops on plants	35		32	277(VV)	32	0.27 ± 0.04	9.45	0.10
Totals	18460	18460	16516				9384.23 min	100.00%

Notes:

Column c: Complete bouts are those that were seen from beginning to end. All others are called censored. Values are given for all foods for which individual product-limit estimates (PLE's) were calculated, that is, for which a number is given in column d.

Column d: PLE = product-limit estimate. Each food's *primary category*; the category from which its PLE mean is taken is indicated as follows. If a food was eaten fewer than twenty times, its bout-length data were pooled with data from the most similar foods, whether or not they were eaten at least twenty times, to make up an aggregate primary food class of $n \geq 20$. These aggregate primary food classes are indicated by letter codes not in parentheses. One PLE mean was calculated for each aggregate class and was used for every food in that class that did not have a PLE of its own (see below). A separate PLE mean was calculated for every food that was eaten at least twenty times unless that food contributed more than 95% of the sample in an aggregate primary food class (in which case it was given the class value) or was one of the following three categories of unidentified foods: XXXG, XXXW, XXXX. If an individual PLE was calculated for a food, a number (the food's ID number) appears in column d. *Secondary categories*, shown in parentheses, are other aggregate classes to which the food's sample contributed.

Column e: Number of bouts with observed beginnings. Size of sample used to obtain mean, column f.

Column f: Asterisks indicate values that may be underestimates because the largest bout was censored.

Column g: $(g) = (a)(f)$

Column h: $(h) = 100(g)/\Sigma(g)$, where $\Sigma(g) = 9384.23$ minutes.

Table 4.2 Additional Wild Foods Eaten by Amboseli Baboons

Species	Family	Parts Eaten	Reference/Observer
		A. Plants	
Abutilon prob. *grandiflorum*	Malvaceae	Flower buds	Raphael Mututua, pers. comm.
Achyranthes aspera	Amaranthaceae	Ripe fruit	Stuart Altmann
Agaricus prob. *bukavuensis*	Agaricaceae	Fruiting bodies	David Post, pers. comm.
Amaranthus graecizans	Amaranthaceae	Leaves?	Stuart Altmann
Balanites glabra	Balanitaceae	Seeds from impala dung	Susan Alberts, pers. comm.
Battarraea stevenii and other mushrooms	Tumostomatales	Fruiting body	Stuart Altmann
Brachiaria pubifolia	Gramineae	Green seedheads	Stuart Altmann
Cadaba farinosa	Capparaceae	(Seeds from?) pods	Stuart Altmann
Cassia italica	Leguminosae	Seeds?	Ronald Noë, pers. comm.
Chenopodium opulifolium	Chenopodiaceae	Fruit	Joan Silk, pers. comm.
Cistanche tubulosa	Orabanchaceae	Tubers	David Post, pers. comm.
Commelina africana	Commelinaceae	Leaves	Joan Silk, pers. comm.
Commicarpus plumbagineus	Nyctaginaceae	Fruit	Joan Silk, pers. comm.
Cordia gharaf	Boragineae	Leaves	Post 1982
Cucumis prophetarium	Cucurbitaceae	Tubers	Stuart Altmann
Cucumis sp.	Cucurbitaceae	Ripe fruit	Stuart Altmann
Cyperus bulbosa	Cyperaceae	Corms	Stuart Altmann; Joan Silk, pers. comm.
		Leaves	Joan Silk, pers. comm
		Seedheads	Joan Silk, pers. comm.
Cyperus immensus	Cyperaceae	Leaf bases	Post, Hausfater, and McCuskey 1980
		Leaves	Joan Silk, pers. comm.
Cyperus laevigatus	Cyperaceae	Corms	Joan Silk, pers. comm.
Cyperus obtusiflorus	Cyperaceae	Leaves	Joan Silk, pers. comm.
		Seedheads	Joan Silk, pers. comm.
Cypholepsis yemenica		Seedheads	Joan Silk, pers. comm.
Delonix alata	Leguminosae	(Seeds from?) pods	Susan Alberts, pers. comm.
Dicliptera albicaulis	Acanthaceae	Flowers	Stuart Altmann
Drakebrockmania somalensis	Gramineae	Green leaves	Stuart Altmann; Joan Silk, pers. comm.
		Seedheads	Joan Silk, pers. comm.
		Corms	Joan Silk, pers. comm.
Erucastrum arabicum	Cruciferae	Pods, green and brown	David Post, Peter Stacey, Joan Silk, pers. comm., Stuart Altmann
		Leaves	Joan Silk, pers. comm.
Justicia uncinulata	Acanthaceae	Leaves?	Stuart Altmann
Kedrostis giref	Cucurbitaceae	Fruit	Stuart Altmann
Lessertia pauciflora	Leguminosae	Seeds from pods	Stuart Altmann
Ludwigia stolonifera	Onograceae	Stolons	David Post, pers. comm.
		Meristems	Joan Silk, pers. comm.
Maerua crassifolia	Capparaceae	Arils?	Stuart Altmann
Opilia campestris	Opiliaceae	Fruit	Peter Stacey, pers. comm.
Pennisetum mezianum	Gramineae	Green leaves?	David Post, pers. comm.
Pennisetum straminium	Gramineae	Leaves; seedheads	Joan Silk, pers. comm.
Phoenix reclinata	Palmae	Fruit	Jeanne Altmann, pers. comm.

Table 4.2   (*Continued*)

Species	Family	Parts Eaten	Reference/Observer
Pycreus mundtii	Cyperaceae	Leaves	David Post, pers. comm.
Rhamphicarpa montana	Scrophulariaceae	Leaves	Joan Silk, pers. comm.
Sarcophyte piriei Hutch.	Balanophoraceae	Tubers	Stuart Altmann
Solanum setaceum	Solanoceae	Green and ripe fruit	Stuart Altmann
Sphaeranthus suavolens	Compositae	Stems and leaves	David Post, pers. comm.
Sporobolus fimbriatus	Gramineae	Leaves	Joan Silk, pers. comm.
Sporobolus ioclados	Gramineae	Leaves	Joan Silk, pers. comm.
Sporobolus rangei	Gramineae	Seedheads	Joan Silk, pers. comm.
Syzygium guineensis	Mortaceae	Ripe fruit	Stuart Altmann
Tragus berteronianus	Graminaceae	Seedheads; pedicel cores	Stuart Altmann
Viscium hilderbrandtii	Viscaceae	Fruit? flowers?	Stuart Altmann
Water lily	Nymphaeaceae	Petioles	Stuart Altmann

B. Animals

Species	Family	Parts Eaten	Reference/Observer
"Army worms"	Lepidoptera	Larvae	Stuart Altmann
Belenois prob. *aurota*	Pieridae	Bodies	Stuart Altmann
Cichladusa guttata, spotted morning warbler	Turdidae	Eggs	Stuart Altmann
Crytacanthacris tatarica and other grasshoppers	Acrididae	Bodies	Stuart Altmann
Francolinus leucocepus, yellow-throated spurfowl	Phasianidae	Chicks	Stuart Altmann
Numido meleagris, Helmeted guineafowl	Phasianidae	Eggs and chicks	Stuart Altmann
Odontotermes sp., termites	Termitidae	Alates	Stuart Altmann
Pterocles exustus, chestnut-bellied sandgrouse	Pteroclididae	Eggs	Stuart Altmann

Table 4.3 Food Taxa, Rank-Ordered by Bout Frequency

Rank	Code	Taxon	Fre-quency	CUM. %
1	FT	*Acacia xanthophloea* Benth. Mimosaceae (fever tree)	3409	21.31
2	CD	*Cynodon dactylon* (L.) Pers. Gramineae (common star grass, Bermuda grass)	2506	36.97
3	SK	*Sporobolus "kentrophyllus"* (K. Schum.) W. D. Clayton Gramineae	2133	50.30
4	PC	*Papio cynocephalus* Cercopithecidae (yellow baboon)	1353	58.76
5	SM	*Sporobolus ioclados* (Trin.) Nees Gramineae	1024	65.16
6	TO	*Acacia tortilis* (Forsk.) Hayne Mimosaceae (umbrella tree)	917	70.89
7	SC	*Sporobolus consimilis* Fresen. Gramineae (reed-grass)	836	76.12
8	AT	*Azima tetracantha* Lam. Salvadoraceae	590	79.80
9	SP	*Salvadora persica* L. Salvadoraceae (toothbrush bush)	550	83.24
10	TC	*Trianthema ceratosepala* Volkens & Irmscher Aizoaceae	431	85.94
11	TT	*Tribulus terrestris* L. Zygophyllaceae (devil's thorn)	425	88.59
12	WA	Water	350	90.78
13	TF	*Lycium "europaeum"* L. Solanaceae ("trumpet-flower plant")	195	92.00
14	GH	Acrididae (grasshoppers, incl. *Crytacanthacris tatarica* L.)	180	93.12
15	CO	*Cyperus obtusiflorus* Vahl Cyperaceae	133	93.96
16	DU	Scarabaeidae (dung beetles)	116	94.68
17	IN	Arthropoda, other	87	95.22
18	WS	*Withania somnifera* (L.) Dunal Solanaceae	83	95.74
19	SF	*Commicarpus pedunculosus* (A. Rich.) Cuf. Nyctaginaceae ("sticky-fruit")	73	96.20
20	CT	*Capparis tomentosa* Lam. Capparaceae	61	96.58
21	SU	*Suaeda monoica* J. F. Gmel. Chenopodiaceae	60	96.96
22	SV	*Setaria verticillata* (L.) Beauv. Gramineae (burr grass)	47	97.25
23	CP	*Cynodon plectostachyus* (K. Schum.) Pilg. Gramineae (giant dog-tooth grass; Naivasha star-grass)	46	97.54
24	RM	*Rhamphicarpa montana* N. E. Br. Scrophulariaceae	44	97.81
25	PJ	*Psilolemma jaegeri* (Pilg.) S. M. Phillips Gramineae	39	98.06
26	SS	*Sporobolus spicatus* (Vahl) Kunth Gramineae (spike-grass, spike dropseed)	33	98.26
27	MX	*Maerua ?angolensis* DC. Capparaceae	27	98.43
28	SO	*Sporobolus cordofanus* (Steud.) Coss. Gramineae	24	98.58
29	CB	*Chlorophytum* sp. nr. *bakeri* Poelln. Liliaceae	21	98.71
30	SB	*Solanum dubium* Fresen. Solanaceae	21	98.84
31	SA	*Sporobolus africanus* (Poir.) Robyns & Tournay Gramineae	19	99.06
32	BE	Coleoptera (beetles), other	17	99.07
33	DB	*Dactyloctenium bogdanii* S. M. Phillips Gramineae (creeping crowfoot?)	14	99.16
34	DP	*Dasysphaera prostrata* (Gilg) Cavaco Amaranthaceae ("hairy-ball plant")	12	99.23
35	GA	Gastropoda (snails, etc.)	12	99.31
36	HY	Hymenoptera (ants, etc.)	12	99.38
37	LS	*Ludwigia stolonifera* (Guill. & Perr.) Raven Onagraceae	11	99.45
38	AB	*Abutilon* sp. Malvaceae	10	99.51
39	OD	*Ornithogalum donaldsonii* (Rendle) Greenway Liliaceae	9	99.57
40	CL	*Cyperus laevigatus* L. Cyperaceae	6	99.61
41	CV	*Cyperus* sp. Cyperaceae	6	99.64
42	FU	Agaricales (mushrooms), other	6	99.68
43	SH	*Sericocomopsis hildebrandtii* Schinz Amaranthaceae	6	99.72
44	HR	*Lepus capensis* Leporidae (Cape hare)	5	99.75

Table 4.3 (*Continued*)

Rank	Code	Taxon	Fre-quency	CUM. %
45	CH	*Chenopodium opulifolium* Koch & Ziz. Chenopodiaceae	4	99.77
46	BI	Aves (bird), other	3	99.79
47	BS	*Battarraea stevenii* (Liboschutz) Fr. Gasteromycetes (mushroom)	3	99.81
48	CI	*Cyperus immensus* C. B. Cl. Cyperaceae	3	99.83
49	DI	*Dicliptera albicaulis* (S. Moore) S. Moore Acanthaceae	3	99.85
50	SI	*Solanum incanum* L. Solanaceae (Sodom apple)	3	99.87
51	VT	*Cercopithecus aethiops* Cercopithecidae (vervet monkey)	3	99.89
52	EA	Earth	2	99.90
53	KO	*Kochia indica* Wight Chenopodiaceae	2	99.91
54	LC	*Leucas stricta* Benth. Labiatae	2	99.92
55	AA	*Asparagus ?africanus* Lam. Liliaceae (asparagus)	1	99.93
56	AH	*Atriplex* sp. nr. *halimus* L. Chenopodiaceae	1	99.94
57	AL	*Aloe* sp. Liliaceae	1	99.94
58	CG	*Cichladusa guttata* Turdidae (spotted morning warbler)	1	99.95
59	GG	*Gazella granti* Bovidae (Grant's gazelle)	1	99.96
60	GS	*Galago senegalensis* Galagidae (bushbaby)	1	99.96
61	MO	*Varanus niloticus* Varanidae (monitor lizard)	1	99.97
62	OC	*Opilia campestris* Engl. Opiliaceae	1	99.97
63	PO	*Hystrix galeata* Hystricidae (porcupine)	1	99.98
64	SN	Ophidia (snakes), unidentified	1	99.99
65	WI	*Connochaetes taurinus* Bovidae (blue wildebeest)	1	100
		Feeding bouts with taxon identified: 15,998		
		Feeding bouts with taxon unidentified: 2,462		
		Total number of feeding bouts: 18,460		

Note: For the purposes of this table, 4749 bouts in which the food taxon was indicated only as grass/sedge have been apportioned among 15 identified grass/sedge species in proportion to their observed bout frequencies. Similarly 17 bouts in which the food taxon was given as *acacia* have been apportioned among the two acacias that occur in the area. Cumulative percentages (CUM. %) include only bouts of identified taxa. Data from table 4.1. Common names in quotation marks are our inventions; others are from the literature.

*Some lilies that I recorded as *Ornithogalum donaldsonii* may have been *O. ecklonii,* which also occurs in Amboseli. Several species have recently been separated from *Cynodon dactylon* (see FTEA Gramineae, part 2). Material gathered around a waterhole in the Amboseli study area during 1980 and typical of material formerly identified as *C. dactylon* was identified at the Kenya Herbarium (formerly the East African Herbarium) as *C. nlemfuensis. Dasysphaera prostrata* is now a synonym of *Volkensinia prostrata. Solanum dubium* is now a synonym of *S. coagulans.* Plants that I referred to as *Sericocomopsis hildebrandtii* probably are *Achyranthes aspera* L. var. *pubescens. Psilolemma jaegeri* (syn. *Odyssea jaegeri* and *Diplachne jaegeri*) may be confused in the field with *Odyssea paucinervis.* Only the latter has stolons, which the baboons sometimes ate. As of 1986, no *S. africanus* is known from Amboseli, nor is it likely that this high-altitude grass occurs there (pers. comm., C. Kabuye, Kenya Herbarium). Amboseli grasses originally identified at the East African Herbarium as *Sporobolus africanus* probably are *S. spicatus.* If so, my records for these two species should be pooled. The Amboseli grass commonly referred to as *Sporobolus kentrophyllus* has been misidentified. According to *The Flora of Tropical East Africa,* part 2, p. 369, *S. kentrophyllus* has "tufts often connected by stolens" and "leaf-blades flat, rarely rolled." However, the grass in question has no stolons, and the leaf-blades are almost all convoluted. Amboseli specimens examined at Kew Herbarium were identified by Cope as *S. rangei* Pilger.

Table 4.4 Food Types (Plant Parts, etc.)

Food Type and Subsumed Foods	Number of Bouts (1)	Number of Complete Bouts (2)	PLE categories (3)	PLE Sample Size (4)	PLE Mean Bout Length (min.) ± SE (5)	PLE Feeding Time (min.) (6)	Percentage of Feeding Time[c] (7)
A. Plant Products							
1. Seeds (mostly acacias, grasses, and sedges): D in column 3. XXXD, XXXG	3530	3223	DD	3349	0.32 ± 0.06	1377.65[b]	14.7
2. Corms of grasses and sedges: C in column 3 except ODCX, CBCX, XXCX	3232	2920	CC	3052	0.53 ± 0.04	1692.26[b]	18.0
3. Leaves: L in column 3	3132	2877	LL	2982	0.37 ± 0.04	1173.76[b]	12.5
3a. Green leaves of grasses and sedges: CDLU, CPLU, GRLU, PILU, SALU, SCLU, SKLU, SMLU, SSLU, SVLU	(2595)	(2416)	GG(LL)	2495	0.38 ± 0.04	986.10[a]	(10.5)
3b. Nongreen leaves of grasses and sedges: CDL-, CLL-, COL-, CIL-, CPL-, DBL-, GRL-, PIL-, SAL-, SCL-, SKL-, SML-, SOL-, SSL-, SVL-, where hyphens mean not U	(8)	(8)	NN(LL)	8	0.18 ± 0.10	1.44[a]	(0.02)
3c. Leaves other than those of grasses/sedges: L in column 3 except codes included in no. 3a or 3b	(536)	(460)	OO(LL)	486	0.35 ± 0.03	187.60[a]	(2.0)
4. Gums (always of acacias): G or H in column 3	1923	1621	UU	1716	0.40 ± 0.04	760.29[b]	8.1
5. Fruits: F in column 3	1895	1613	FF	1721	0.51 ± 0.05	952.05[b]	10.1
6. Petioles, pedicels, blade bases: T in column 3	748	654	YY	694	0.51 ± 0.04	371.58[b]	4.0
7. Flowers, flower buds: B in column 3	450	400	BB	421	0.67 ± 0.13	333.25[b]	3.6
8. Wood, cambium, bark: W in column 3	189	169	WW(XX)	173	0.49 ± 0.09	80.02[b]	0.85

9. Rhizomes, stolons: S in column 3	120	RS(XX)	117	120	0.30 ± 0.04	35.94[b]	0.38
10. Other plant parts (bulbs, roots, thorns, whole plants, etc.): XXCX, FTQU, XXRX, ODCX, CBCX, ATEU, GREU, TOEU, SPEU, FUEX, FTRU	31	PT(XX)	28	29	0.51 ± 0.13	15.39[b]	0.16
11. Pods: ABPU, ABPX, CTPX, MXPX, XXPX, CTPR	21	PD(XX)	21	21	0.75 ± 0.21	15.75[b]	0.17
B. Animal Products							
12. Milk: PCMX	1350	270	1170	1202	1.65 ± 0.17	2227.50[a]	23.7
13. Meat and other animal products: K, J, O, Q, N, A, or Z in column 3	451	MM	434	445	0.20 ± 0.05	84.31[b]	0.89
C. Inorganic Materials							
14. Water: V in column 3	350	VV	335	338	0.16 ± 0.02	56.70[b]	0.60
15. Soil: EAEX	2	(XX)	345	345	0.44 ± 0.06	0.88[a]	0.009
16. Food type unknown: all **XX and all XXX* except XXXD and XXXG	1036	ZZ	932	961	0.20 ± 0.03	206.90[b]	2.2
Totals	18,460		16,514	17,506		9384.23	100.00

Note: The sixteen numbered food types are exclusive and exhaustive. The four-letter codes for the foods included in each type are explained in table 6.2. See table 4.1 for further information on column headings. Parenthetical values are subtotals not included in the column totals.

[a] Feeding time obtained from one BMDP computer program run on all included bouts.

[b] Feeding time is sum of feeding times of the component foods.

[c] See table 6.5 for seasonally adjusted values.

Table 4.5 Foods Rank-Ordered by Bout Length

Rank	Number in Table 4.1	Food	Mean Bout Length (Minutes)
1	8	Umbrella tree blossoms on tree	1.85
2	270	Milk	1.65
3	18	Umbrella tree seeds from green pods	1.27
4	163	*Salvadora persica* fruit, semiripe	1.05
5	39	Fever tree seeds from green pods	1.03
6	162	*Salvadora persica* fruit, ripe	0.95
7	23	Fever tree blossoms on tree	0.88
8	106	*Cyperus obtusiflorus* corms	0.77
9–14	4, 5, 73, 74, 147, 240	Pods of various species	0.75
15–17	200–202	Corms of *Sporobolus rangei*	0.74
19	16	Umbrella tree seeds from dry pods	0.72
18, 20	9, 47	Acacia tree seeds from dung	0.72

Note: Only the twenty foods with the longest bout lengths are shown. Data from table 4.1.

Table 4.6 Weibull Distribution Parameters for Feeding Bout Durations of Primary Food Classes

Class	Correlation Coefficient	$m =$ y − intercept	SE(m)	$\alpha =$ exp[− m/β]	$\beta =$ slope	SE(β)	d.f.
			A. Food Classes				
A	.978	0.871	.097	0.435	1.045	.054	17*
ATFR.54	.972	1.058	.055	0.388	1.116	.038	49
ATFU.56	.997	1.242	.025	0.362	1.221	.016	34
ATFX.57	.971	1.788	.079	0.255	1.309	.048	46
ATXX.64	.981	2.859	.198	0.152	1.516	.090	11
B	.989	0.317	.025	0.768	1.202	.021	67
BB	.904	0.559	.063	0.448	0.696	.045	53
C	.981	0.604	.087	0.551	1.013	.054	14*
CDLU.90	.978	0.813	.035	0.448	1.011	.029	56*
CDSU.94	.985	1.906	.120	0.275	1.476	.064	16
CDSX.95	.976	1.353	.094	0.358	1.318	.060	24
COCX.106	.992	0.336	.036	0.717	1.010	.030	18*
D	.997	0.724	.025	0.581	1.333	.019	32
DD	.931	0.903	.039	0.267	0.683	.033	66
DUKU.271	.991	1.385	.027	0.122	0.658	.016	30
E	.968	1.299	.054	0.345	1.220	.037	71
F	.977	0.553	.064	0.543	0.905	.043	22
FF	.949	0.628	.041	0.530	0.990	.035	87*
FTBS.23	.998	0.213	.012	0.770	0.815	.008	47
FTDG.24	.979	1.209	.028	0.253	0.879	.020	83
FTDR.38	.982	1.060	.034	0.404	1.170	.026	74
FTDU.39	.965	0.177	.059	0.846	1.060	.049	35
FTHG.30	.974	0.749	.035	0.455	0.952	.025	77
FTHP.31	.977	0.604	.058	0.507	0.888	.039	24
FTHW.34	.966	0.986	.080	0.363	0.972	.047	31
FTHX.35	.954	1.643	.169	0.154	0.878	.069	16
G	.977	0.459	.033	0.659	1.099	.026	89
GG	.970	0.853	.027	0.386	0.896	.025	85*
GHKX.267	.931	4.787	.555	0.050	1.085	.090	10*
GRCX.117	.986	0.689	.020	0.470	0.912	.018	70
GRDU.118	.993	0.677	.043	0.551	1.134	.030	19
GRDX.119	.994	1.724	.056	0.234	1.188	.031	17
GRLU.121	.958	0.898	.039	0.376	0.918	.034	64
GRSX.124	.985	2.179	.138	0.198	1.344	.067	12
GRTU.126	.982	1.241	.052	0.420	1.429	.040	48
H	.973	1.434	.101	0.207	0.911	.051	18
I	.997	0.984	.030	0.493	1.393	.021	27
INKX.252	.973	1.094	.064	0.276	0.850	.035	32
J	.985	1.628	.064	0.208	1.037	.039	21*
K	.982	0.627	.089	0.536	1.004	.056	12
L	.970	0.415	.042	0.676	1.059	.035	57
M	.965	0.917	.038	0.387	0.966	.032	70*
N	.957	0.736	.040	0.455	0.934	.029	94
NN	.981	2.443	.291	0.105	1.083	.108	4*
O	.996	0.965	.045	0.395	1.039	.027	13
OO	.991	1.120	.032	0.354	1.079	.022	41
PCMX.270	.895	−0.147	.066	1.299	0.713	.041	75
PD	.953	0.365	.116	0.664	0.891	.073	15
PT	.984	0.768	.061	0.438	0.931	.037	21
Q	.986	1.164	.042	0.353	1.119	.027	50

Table 4.6 (*Continued*)

Class	Correlation Coefficient	$m =$ y −intercept	SE(m)	$\alpha =$ exp[$- m/\beta$]	$\beta =$ slope	SE (β)	d.f.
R	.995	0.998	.049	0.422	1.156	.031	15
RMXX.160	.941	1.977	.215	0.222	1.312	.115	17
RS	.979	1.596	.075	0.319	1.395	.047	38
S	.947	1.284	.064	0.232	0.878	.042	51
SMCX.194	.974	0.406	.035	0.726	1.266	.031	91
SMLU.197	.996	0.723	.029	0.558	1.241	.020	28
SPFR.162	.987	0.168	.042	0.864	1.152	.034	29
SPFS.163	.986	0.020	.041	0.982	1.088	.035	28
SPFU.164	.982	0.898	.042	0.456	1.142	.030	54
SPLU.167	.992	1.323	.046	0.378	1.361	.029	36
SPLX.168	.992	1.452	.043	0.237	1.007	.022	35*
SPWX.171	.982	1.013	.076	0.398	1.100	.042	25
SULU.209	.991	1.694	.073	0.270	1.292	.037	23
T	.991	0.859	.046	0.383	0.895	.024	25
TFFR.136	.975	1.377	.099	0.207	0.873	.050	16
TOBR.7	.929	3.313	.337	0.097	1.418	.151	14
TOBS.8	.977	−0.275	.034	1.393	0.830	.024	59
TODG.10	.986	1.659	.063	0.195	1.014	.032	29*
TOPR.16	.978	0.489	.044	0.652	1.144	.032	57
TOPU.18	.997	−0.246	.011	1.249	1.105	.010	74
TTFX.216	.964	0.809	.033	0.439	0.982	.029	86*
U	.980	1.523	.074	0.330	1.374	.049	33
UU	.960	0.824	.036	0.411	0.927	.030	80
V	.991	0.662	.024	0.575	1.196	.022	54
W	.982	0.844	.034	0.352	0.809	.022	53
WAVG.276	.981	2.380	.083	0.168	1.336	.042	39
WAVP.277	.993	1.557	.064	0.287	1.247	.036	17
X	.988	0.949	.052	0.450	1.189	.032	34
XXDX.235	.943	0.950	.104	0.135	0.474	.050	11
XXLX.239	.987	1.024	.054	0.304	0.860	.030	22
XXWR.243	.973	1.595	.131	0.226	1.071	.061	17
XXWX.245	.949	0.886	.124	0.330	0.798	.064	17
XXXD.246	.808	1.904	.115	0.050	0.636	.067	48
Y	.990	1.030	.030	0.435	1.237	.024	56
ZZ	.947	1.432	.044	0.187	0.855	.031	86
B. Nonfood Classes							
RTP[a]	.982	1.907	.042	0.095	0.811	.026	36
OS[b]	.902	0.265	.070	0.763	0.980	.054	74*
ISNF[c]	.937	0.353	.048	0.664	0.861	.039	69
Contact[d]	.833	0.362	.066	0.539	0.585	.042	87

Note: Includes all primary food classes (except CC, XX, 191, and 222), as defined in table 7.1, and (in part B) several nonfood classes. Foods marked with an asterisk have β values within one standard error of unity and therefore may have exponential survivorship distributions; d.f. $= n - 2$, where n is the number of product-limit-estimated survival decrement values (limited to $n \leq 100$) that were used in the Weibull regressions. See table 7.1 for sample sizes of the product-limit estimates.

[a] Rough-and-tumble play.

[b] Out of sight.

[c] In sight, not feeding.

[d] Physical contact of yearling subject with those at least two years old.

Table 4.7 Time Budgets of Food Taxa, Rank-Ordered

Rank	Taxon	Minutes per Day	Percentage of Feeding Time	Cumulative Percentage
1	*Papio cynocephalus* (milk)	69.71	25.10	25.1
2	*Acacia xanthophloea*	41.67	15.00	40.1
3	*Sporobolus rangei*	28.35	10.21	50.3
4	*Sporobolus consimilis*	26.97	9.71	60.0
5	*Acacia tortilis*	25.99	9.36	69.4
6	*Cynodon nlemfuensis*	25.53	9.19	78.6
7	*Sporobolus cordofanus*	12.97	4.67	83.2
8	*Trianthema ceratosepala*	9.07	3.27	86.5
9	*Salvadora persica*	8.94	3.22	89.7
10	*Azima tetracantha*	6.09	2.19	91.9
11	*Tribulus terrestris*	6.08	2.19	94.1
12	*Lycium "europaeum"*	2.06	0.74	94.9
13	*Cyperus obtusiflorus*	1.85	0.67	95.5
14	Water	1.77	0.64	96.2
15	*Capparis tomentosa*	1.16	0.42	96.6
16	*Withania somnifera*	1.15	0.42	97.0
17	Arthropoda, other	0.77	0.28	97.3
18	*Setaria verticillata*	0.73	0.26	97.5
19	Scarabaeidae	0.62	0.22	97.8
20	*Commicarpus plumbagineus*	0.53	0.19	97.9
21	*Maerua* sp.	0.50	0.18	98.1
22	*Suaeda monoica*	0.48	0.17	98.3
23	*Chlorophytum* sp. nr. *bakeri*	0.45	0.16	98.5
24	*Solanum dubium*	0.36	0.13	98.6
25	*Rhamphicarpa montana*	0.36	0.13	98.7
26	*Psilolemma jaegeri*	0.36	0.13	98.8
27	*Sporobolus spicatus*	0.27	0.097	98.9
28	*Dactyloctenium bogdanii*	0.26	0.095	99.0
29–65	All others	0.26	<0.095	100.0
Totals		277.74	100.00	

Note: Time spent on unidentified grasses and sedges has been prorated among the fifteen identified grasses/sedges. Time spent on unidentified acacias (seeds) has been prorated between the two acacias in the area. Percentages are of time spent on foods of known taxa (i.e., taxon code not XX); that is, time spent on food of unknown taxa has in effect been prorated among all known taxa. Data from table 4.1.

Table 5.1 Proximate Composition of Core Foods

Time Budget Rank	Food Code	Water	Ash	Lipid	Protein	Fiber	Carbohydrate	Energy	Sources
1	PCMX	86.0	0.3	4.9	1.6	0.0	7.1	322.33	Buss 1968 for days 36–279, converted from ml to g
2	GRCX	27.6	7.3	1.5	9.5	16.2	37.9	1074.96	w{all available grass/sedge corms: SKCX, SMCX, COCX}
3	GRLU	69.7	3.7	2.2	6.4	7.9	10.1	463.32	w{all available grass/sedge leaves: CDLU, SMLU}
4	FTGX	21.4	6.1	1.1	1.2	7.3	62.9	1206.88	$\bar{x}${14b, 15b, 1894/74, 2101/74}
5	TODU	67.9	1.6	0.7	5.3	5.0	19.6	503.37	$\bar{x}${38B, M331/75}
6	SKCX	29.3	4.1	1.5	10.3	16.5	38.5	1101.10	$\bar{x}${16B, M324/75}
7	SCTX	78.6	2.9	0.9	4.6	4.7	8.4	311.82	$\bar{x}${M97/75, 243/76}
8	TCFX	62.7	3.3	2.7	6.5	14.6	10.3	595.41	$\bar{x}${1896/74, 776/75}
9	CDLU	70.0	3.6	2.2	6.3	7.9	10.0	460.25	$\bar{x}${8B, 8Bsup, 1893/74}
10	XXXG	10.5	3.8	3.9	19.1	20.6	42.1	1436.65	w{FTDG, TODG}[a]
11	TOBS	80.2	1.2	2.0	4.1	3.7	8.9	335.02	$\bar{x}${13B, M55/75, 12/76}
12	FTDG	9.7	3.9	4.0	18.7	21.3	42.6	1454.14	FTDR
13	TTFX	82.3	2.2	0.6	3.9	3.4	7.7	258.55	24B[b]
14	FTDR	9.7	3.9	4.0	18.7	21.3	42.6	1454.14	$\bar{x}${20B-2, 21B, 22B, 1899/74}
15	FTHG	21.4	6.1	1.1	1.2	7.3	62.9	1206.88	FTDG
16	SMCX	24.0	10.0	1.2	9.6	16.9	38.2	1082.27	$\bar{x}${3B, M330/75}
17	ATFR	75.2	2.9	2.7	4.9	3.5	10.9	400.44	$\bar{x}${9B, 777/75, 1895/74}
18	TODR	14.3	3.2	3.6	20.8	17.7	40.4	1374.90	721/75
19	FTDU	75.3	2.0	2.6	10.7	2.7	6.7	397.47	50/76
20	FTBS	74.6	1.6	1.0	1.1	3.8	17.7	403.27	$\bar{x}${5B, 720/75}
21	SPFU	73.4	5.8	0.8	6.0	4.1	9.9	343.00	1898/74
22	SPFS	78.9	3.5	0.9	3.4	3.8	9.7	301.30	$\bar{x}${1898/74, 722/75}[c]
23	XXXW	21.4	6.1	1.1	1.2	7.3	62.9	1206.88	FTGX[d]
24	SPFR	84.3	1.2	0.9	0.7	3.5	9.5	254.68	722/75
25	GRTU	71.3	3.1	3.0	8.3	6.7	7.6	457.69	8B[c]
26	ATFX	74.7	3.0	2.8	5.2	3.7	10.6	406.57	w{ATFR, ATFU}
27	XXXD	10.5	3.8	3.9	19.1	20.6	42.1	1436.65	XXXG[a]
28	TFLX	85.4	3.5	0.7	4.5	1.9	3.9	183.67	$\bar{x}${M53/75, 359/75, 644/76}
29	SPLU	72.1	6.5	0.7	4.9	3.0	12.9	355.00	$\bar{x}${1897/74, 1B}
30	ATFU	71.7	4.0	3.8	7.1	5.1	8.4	455.51	$\bar{x}${M325/75, 34B}
31	SPLX	72.1	6.5	0.7	4.9	3.0	12.9	355.00	SPLU
32	FTHP	21.4	6.1	1.1	1.2	7.3	62.9	1206.88	FTGX
33	SMLU	56.2	8.7	2.2	8.5	8.6	15.9	599.80	$\bar{x}${M101/75, M102/75}
34	FTHW	21.4	6.1	1.1	1.2	7.3	62.9	1206.88	FTGX

35	WSFX	68.9	1.1	3.5	5.8	10.3	10.5	546.22	$\bar{x}\{\bar{x}\{36B, 2100/74\}, \bar{x}\{645/76, 35B\}\}$ [c]
36	COCX	46.5	4.4	3.4	3.8	9.5	32.7	865.22	$\bar{x}\{25B, M86/75\}$
37	INKX	62.7	1.2	3.8	26.8	2.4	3.1	710.22	Leung 1968, no. 1123: grasshopper spp.
38	DUKU	81.1	1.4	2.7	10.6	2.8	1.4	358.44	Leung 1968, no. 1096: caterpillar spp.
39	CDSX	87.1	0.7	0.3	3.8	2.9	5.1	195.78	531/76
40	SPWX	0	0	0	0	0	0	0	0 [f]
41	SFFX	57.4	4.3	9.6	13.0	4.9	10.8	778.66	30B
42	TODG	14.3	3.2	3.6	20.8	17.7	40.4	1374.90	TODR
43	TOBR	80.2	1.2	2.0	4.1	3.7	8.9	335.02	TOBS [c]
44	XXWX	0	0	0	0	0	0	0	0 [f]
45	SULU	81.5	4.8	0.8	3.5	1.6	7.8	231.98	$\bar{x}\{2096/74, M322/75\}$
46	XXLX	78.8	5.0	0.7	4.7	2.5	8.4	270.16	$\bar{x}\{$available non-grass leaves: SPLX, TFLX$\}$
47	GRDU	75.8	1.6	0.9	4.9	6.9	9.9	376.77	$w'\{SMDU, CDDU\}$ [g]
48	SPFX	78.4	3.7	0.9	3.6	3.8	9.7	304.15	$w\{SPFU, SPFS, SPFR\}$
49	TFBS	80.3	1.7	1.0	5.5	4.4	7.2	303.53	364/76
50	ACDD	10.5	3.8	3.9	19.1	20.6	42.1	1436.65	XXXG [h]
51	CTFX	75.2	2.9	2.7	4.9	3.5	10.9	400.44	ATFR [c]
52	SBFX	71.2	1.2	1.4	5.5	5.0	15.7	466.77	33B

Note: Food codes as in table 5.2. Values given as percentage (by mass) of fresh food, except for energy, given as kJ/100 g fresh food. Sources identified only by their reference numbers (148, 1894/74, etc.) are from Altmann, Post, and Klein 1987. Symbols: $w\{a, b, \ldots\}$ = weighted mean compositions of a, $b, \ldots$ using intakes in g/day (table 6.1) as weighting coefficients. $\bar{x}\{a, b, \ldots\}$ = unweighted mean composition of $a, b, \ldots$. Foods are ranked by time budgets, as in table 6.1.

[a] Unidentified items picked from dung and from ground, assumed to be primarily seeds of fever and umbrella trees, in same ratio as when species of such consumed seeds could be identified.

[b] Condition of sample M56/75 not recorded, but looks too dry.

[c] Better data needed.

[d] Most items picked from wood probably were fever tree gum droplets.

[e] Assumes equal proportions of unripe and ripe fruit.

[f] Assumes no digestible material ingested even if wood gnawed.

[g] Weighted by minutes per day because grams intake per day not known.

[h] Ratio of fever and umbrella tree seeds taken from ungulate dung assumed same as from identifiable ground sources.

Table 5.2 Diet 1: Maximum Energy, Macronutrient Constraints

Food-Related Measures

Time Budget Rank	Food Code	Intake (g/day)	Reduced Cost (MJ/g food)	Energy Density[a] (MJ/g food)	Energy Leeway (MJ/g food)	
					Allowable Increase	Allowable Decrease
1	PCMX	735.4	0	0.003	0.011	0.0005
2	GRCX	0	0.004	0.011	0.004	Infinity
3	GRLU	0	0.004	0.005	0.004	Infinity
4	FTGX	200.9	0	0.012	0.001	0
5	TODU	0	0.001	0.005	0.001	Infinity
6	SKCX	0	0.001	0.011	0.001	Infinity
7	SCTX	0	0.004	0.003	0.004	Infinity
8	TCFX	0	0.003	0.006	0.003	Infinity
9	CDLU	0	0.004	0.005	0.004	Infinity
10	XXXG	0	0.0001	0.014	0.0001	Infinity
11	TOBS	0	0.002	0.003	0.002	Infinity
12	FTDG	66.1	0	0.015	0.014	0
13	TTFX	0	0.003	0.003	0.003	Infinity
14	FTDR	0	0	0.015	0	Infinity
15	FTHG	0	0	0.012	0	Infinity
16	SMCX	0	0.007	0.011	0.007	Infinity
17	ATFR	0	0.003	0.004	0.003	Infinity
18	TODR	0	0.0003	0.014	0.003	Infinity
19	FTDU	0	0.004	0.004	0.004	Infinity
20	FTBS	0	0	0.004	0.001	Infinity
21	SPFU	0	0.007	0.003	0.007	Infinity
22	SPFS	0	0.004	0.003	0.004	Infinity
23	XXXW	0	0	0.012	0	Infinity
24	SPFR	0	0.001	0.003	0.001	Infinity
25	GRTU	0	0.004	0.005	0.004	Infinity
26	ATFX	0	0.003	0.004	0.003	Infinity
27	XXXD	0	0.0001	0.014	0.0001	Infinity
28	TFLX	0	0.005	0.002	0.005	Infinity
29	SPLU	0	0.007	0.004	0.007	Infinity
30	ATFU	0	0.004	0.005	0.004	Infinity
31	SPLX	0	0.007	0.004	0.007	Infinity
32	FTHP	0	0	0.012	0	Infinity
33	SMLU	0	0.009	0.006	0.009	Infinity
34	FTHW	0	0	0.12	0	Infinity
35	WSFX	0	0.0004	0.005	0.0004	Infinity
36	COCX	0	0.001	0.009	0.001	Infinity
37	INKX	0	0.004	0.007	0.004	Infinity
38	DUKU	0	0.003	0.004	0.003	Infinity
39	CDSX	0	0.002	0.002	0.002	Infinity
41	SFFX	0	0.003	0.008	0.003	Infinity
42	TODG	0	0.0004	0.014	0.0004	Infinity
43	TOBR	0	0.002	0.003	0.002	Infinity
45	SULU	0	0.006	0.002	0.006	Infinity
46	XXLX	0	0.006	0.003	0.006	Infinity
47	GRDU	0	0.002	0.004	0.002	Infinity
48	SPFX	0	0.004	0.003	0.004	Infinity

Table 5.2 *(Continued)*

Food-Related Measures

Time Budget Rank	Food Code	Intake (g/day)	Reduced Cost (MJ/g food)	Energy Density[a] (MJ/g food)	Energy Leeway (MJ/g food) Allowable Increase	Energy Leeway (MJ/g food) Allowable Decrease
49	TFBS	0	0.003	0.003	0.003	Infinity
50	ACDD	0	0.0001	0.014	0.0001	Infinity
51	CTFX	0	0.003	0.004	0.003	Infinity
52	SBFX	0	0.001	0.005	0.001	Infinity

Constraint-Related Measures

Constraint, Units	Slack (units/day)	Marginal Value (MJ/unit)	Current RHS (units/day)	Right-Hand-Side Ranges (units/day) Allowable Increase	Right-Hand-Side Ranges (units/day) Allowable Decrease
Energy minimum, MJ/Day	4.88	0	0.87	4.88	Infinity
Protein minimum, g/day	22.60	0	3.94	22.60	Infinity
Consumption capacity, g/day	0	0.0057	1002.4	Infinity	650.7
Water minimum 1, g/day	442.63	0	239.20	442.63	Infinity
Water minimum 2, g/day	0	−0.005	0	78.88	386.31
Ash maximum, %	114.27	0	0[b]	Infinity	114.27
Ash minimum, %	0	−0.06	0[b]	1.82	5.05
Fiber minimum, g/day	27.96	0	0[b]	27.95	Infinity
Lipid minimum, g/day	39.41	0	1.48	39.41	Infinity

Note: Only foods 1, 4, and 12 are prescribed for this diet. Food numbers and codes as in table 6.2. Objective function value, 5.75 MJ per day.

[a] Values for these current coefficients (energy density) are given with greater precision in table 5.1.

[b] Because of program requirements, these constraint equations were entered with zero on the right-hand side. To illustrate, consider the ash minimum: 1.73% of the diet mass. This can be written $\Sigma\, a_i F_i \geq 0.0173\, \Sigma\, F_i$, where a_i is the ash density (proportion) of food i and F_i is its mass in the diet. Rearranged, this becomes $\Sigma\, (a_i - 0.173)\, F_i \geq 0$; the ash minimum was entered in that form. At the allowable increase in the right-hand side, 1.82 g/day as shown above for diet 1, total ash would become $\Sigma\, a_i F_i \geq 0.0173\, \Sigma\, + 1.82$. At a diet mass of 1002.4 g/day, the additional 1.82 g/day would add 0.18% ash.

Table 5.3 Nutrient Requirements and Toxin Limits for Proximate Nutrients

Nutrient	Requirement at Minimal Growth[a]	Requirement at Average Growth[b]	Age-Specific Requirement, Average Growth					Toxic Limits
			47.55[c]	35	45	55	65 Weeks	
Water (1)	$120M^{0.84}$	→|	239.2	203.7	232.0	259.7	286.9	G
Water (2)	$4.79p + 32.58a$						|→	G
Water (3)	$100.2M^{3/4} + 15.1$	$100.2M^{3/4} + 22.6$	208.1	183.3	203.1	222.3	240.8	C
Minerals	0.0173Φ	→|				|→	0.131Φ	0.131Φ
Lipid	$0.8M^{3/4}$	→|	1.48	1.28	1.44	1.59	1.74	G
Protein	$1.51M^{3/4} + 0.762$	$1.51M^{3/4} + 1.14$	3.94	3.56	3.86	4.19	4.43	G
Fiber	$3.5 \times 10^{-3}(\Phi - w)$						|→	G
Carbohydrate	0								G
Energy	$419.3M^{3/4} + 63$	$419.3M^{3/4} + 94.5$	871	767.0	850.0	930.0	1007.6	$\max\{e_i\}G$

Note: Requirements and limits in g/day or kJ/day. Arrows indicate repeat of values at their left.

Variables: M = body mass in kg; Φ = mean daily grams of food (fresh mass) in the diet; w = mean daily grams of water in the diet; p = mean daily grams of protein in the diet; a = mean daily grams of "ash" (minerals) in the diet; G = daily consumption capacity (estimated as 1002.4 g/day); e_i = energy density (kJ/g) of the ith food.

[a] Growth at 3 g body mass per day.

[b] Growth at 4.5 g body mass per day.

[c] Mean age of subjects when sampled.

Table 5.4 Terminology of Linear Optimization

Terminology of Economics	Terminology Used Here
Objective	Objective, currency
Objective function	Objective function
Decision variable	Amount to be eaten of each food
Cost coefficient	Energy density[a] (concentration) of foods
Cost coefficient range	Energy leeway[a] of foods
Reduced cost	Reduced cost
Slack value	Slack value, safety margin
Binding constraint	Binding constraint, limiting factor
Marginal value, dual price, shadow price	Marginal value, sensitivity of objective function, leverage, exchange rate
Right-hand-side value	Minimum or maximum, requirement or tolerance
Degenerate	Degenerate
Unique	Unique

Definitions:

Objective: Food component, assumed commensurate with fitness, that will be maximized or minimized in the optimal diet.

Objective function: Linear function relating the objective component to the amount of each food in the diet.

Decision variable: Amount that the objective component's density in a food could increase or decrease without altering the composition of the optimal diet.

Reduced cost: Amount that the objective component's density in a potential food would need to increase in order for it to be included in the optimal diet.

Slack: Difference between the amount of a food component and the animal's requirement or tolerance for it.

Binding constraint: Food component that, in the optimal diet, is at the animal's requirement or tolerance.

Marginal value: Amount by which the optimal value of the objective function would improve (increase or decrease, as appropriate) per unit change in a specified constraint.

Requirement; tolerance: Long-term intake of a nutrient or toxin (or exposure to any other food component) beyond which the animal becomes ill or in other ways has its fitness compromised.

Degenerate: Optimal diet problem whose solution includes more binding constraints than there are foods in the optimal diet.

Unique: Diet such that no other would provide this much (this little) of the objective and still satisfy all the constraints.

[a] Similarly for protein density, harvesting time, or other food component that is to be maximized or minimized.

Table 5.5 Diet 2: Maximum Protein, Macronutrient Constraints

		Food-Related Measures		Protein Leeway (g protein/g food)	
Time Budget Rank	Food Code	Intake (g/day)	Protein Density (g protein/g food)	Allowable Increase	Allowable Decrease
28	TFLX	279.3	0.045	0.011	0.005
38	DUKU	314.6	0.106	0.023	0.002
39	CDSX	408.4	0.038	0.005	0.004

	Constraint-Related Measures			Right-Hand-Side Ranges (units/day)	
Constraint, Units	Slack (units/day)	Marginal Value (g protein/ g food)	Current RHS (units/day)	Allowable Increase	Allowable Decrease
Energy minimum, MJ/day	1.57	0	0.87	1.57	Infinity
Protein minimum, g/day	57.50	0	3.94	57.50	Infinity
Consumption capacity, g/day	0	0.061	1002.4	Infinity	644.6
Water minimum 1, g/day	610.26	0	239.20	610.26	Infinity
Water minimum 2, g/day	0	−0.18	0	117.37	203.14
Ash maximum, %	114.27	0	0	Infinity	114.27
Ash minimum, %	0	−5.86	0	3.41	4.75
Fiber minimum, g/day	25.37	0	0	25.37	Infinity
Lipid minimum, g/day	10.20	0	1.48	10.20	Infinity

Note: For brevity in the tables for diets 2–10, food-related measures are shown only for prescribed foods. Reduced costs for prescribed foods are always zero, and so are omitted. Other conventions as in table 5.5. Objective function value, 61.4 g protein per day.

Table 5.6 Diet 3: Minimum Feeding Time, Macronutrient Constraints

		Food-Related Measures		Time Leeway (minutes/g food)	
Time Budget Rank	Food Code	Intake (g/day)	Time Density (minutes/g food)	Allowable Increase	Allowable Decrease
1	PCMX	257.7	0.208	0.063	0.344
15	FTHG	82.0	0.427	0.062	0.375

	Constraint-Related Measures			Right-Hand-Side Ranges (units/day)	
Constraint, Units	Slack (units/day)	Marginal Value (minutes/ unit)	Current RHS (units/day)	Allowable Increase	Allowable Decrease
Energy minimum, MJ/day	0.95	0	0.87	0.95	Infinity
Protein minimum, g/day	1.17	0	3.94	1.17	Infinity
Consumption capacity, g/day	662.7	0	1002.4	Infinity	662.7
Water minimum 1, g/day	0	−0.37	239.20	466.56	54.69
Water minimum 2, g/day	26.73	0	0	26.73	Infinity
Ash maximum, %	38.73	0	0	Infinity	38.73
Ash minimum, %	0	−7.90	0	0.63	3.89
Fiber minimum, g/day	5.74	0	0	5.74	Infinity
Lipid minimum, g/day	12.05	0	1.48	12.05	Infinity

Note: Conventions as in table 5.5. Objective function value, 88.7 minutes per day.

Table 5.7 Diet 4: Maximum Energy Intake Rate, Macronutrient Constraints

Food-Related Measures

Time Budget Rank	Food Code	Intake (g/day)	Energy Rate kJ/minute	Energy Rate Leeway (kJ/minute)	
				Allowable Increase	Allowable Decrease
1	PCMX	729.1	15.47	12.77	14.55
15	FTHG	273.3	28.24	Infinity	7.6
$\beta = 1002.4^{-1}$			0	0	Infinity

Constraint-Related Measures

Constraint, Units	Slack (units/day)	Marginal Value (kJ/minute/unit)
Energy minimum, MJ/day	4.78	0
Protein minimum, g/day	11.01	0
Consumption capacity, g/day	0	0
Water minimum 1, g/day	446.3	0
Water minimum 2, g/day	0	−0.005
Ash maximum, %	112.5	0
Ash minimum, %	1.82	0
Fiber minimum, g/day	19.13	0
Lipid minimum, g/day	37.25	0

Note: Conventions as in table 5.5. Objective function value, 18.95 kJ energy per minute of feeding.

Table 5.8 Diet 5: Maximum Protein Intake Rate, Macronutrient Constraints

Food-Related Measures

Time Budget Rank	Food Code	Intake (g/day)	Protein Rate (g/minute)	Protein Rate Leeway (g protein/minute)	
				Allowable Increase	Allowable Decrease
1	PCMX	384.8	0.077	0.0008	0.078
17	ATFR	617.6	0.117	Infinity	0.0004
$\beta = 1002.4^{-1}$			0	0	Infinity

Constraint-Related Measures

Constraint, Unit	Slack (units/day)	Marginal Value (g/minute/unit)
Energy minimum, MJ/day	2.84	0
Protein minimum, g/day	32.47	0
Consumption capacity, g/day	0	0
Water minimum 1, g/day	556.2	0
Water minimum 2, g/day	0	−0.00004
Ash maximum, %	112.3	0
Ash minimum, %	2.02	0
Fiber minimum, g/day	21.00	0
Lipid minimum, g/day	34.05	0

Note: Conventions as in table 5.5. Objective function value, 101.3 mg protein per minute of feeding.

Table 5.9 Diet 6: Maximum Energy, Extended Constraints

Food-Related Measures

Time Budget Rank	Food Code	Intake (g/day)	Energy Density (MJ/g food)	Energy Leeway (MJ/g food)	
				Allowable Increase	Allowable Decrease
1	PCMX	667.7	0.003	0.0003	0.0002
3	GRLU*	108.7	0.005	Infinity	0.001
4	FTGX*	66.7	0.012	Infinity	0.009
6	SKCX	0.2	0.011	0.002	0.0001
9	CDLU*	23.3	0.005	Infinity	0.0006
10	XXXG*	7.3	0.014	Infinity	0.010
12	FTDG*	8.2	0.015	Infinity	0.009
14	FTDR*	1.6	0.015	Infinity	0.005
15	FTHG*	24.2	0.012	Infinity	0.009
16	SMCX*	10.4	0.011	Infinity	0.005
17	ATFR*	27.2	0.004	Infinity	0.0007
18	TODR*	3.0	0.014	Infinity	0.007
20	FTBS*	13.7	0.004	Infinity	0.0002
21	SPFU	9.4	0.003	0.0002	0.0004
26	ATFX*	10.4	0.004	Infinity	0.0007
27	XXXD*	0.9	0.014	Infinity	0.006
30	ATFU*	4.3	0.005	Infinity	0.0008
32	FTHP*	0.7	0.012	Infinity	0.005
34	FTHW*	8.2	0.012	Infinity	0.009
36	COCX*	0.4	0.009	Infinity	0.003
37	INKX*	0.6	0.007	Infinity	0.002
41	SFFX*	3.0	0.008	Infinity	0.004
42	TODG*	1.4	0.014	Infinity	0.010
50	ACDD*	0.1	0.014	Infinity	0.010
52	SBFX*	0.8	0.005	Infinity	0.0003

Constraint-Related Measures

Constraint, Unit	Slack (units/day)	Marginal Value (MJ/unit)	Current RHS (units/day)	Right-Hand-Side Ranges (units/day)	
				Allowable Increase	Allowable Decrease
Energy minimum, g/day	3.82	0	0.87	3.82	Infinity
Protein minimum, g/day	25.19	0	3.94	25.19	Infinity
Consumption capacity, g/day	0	0.003	1002.4	6.16	33.7
Ash maximum, %	114.27	0	0	Infinity	114.27
Ash minimum, %	0	−0.003	0	0.24	0.50
Fiber minimum, g/day	26.17	0	0	26.17	Infinity
Lipid minimum, g/day	37.97	0	1.48	37.97	Infinity
Time maximum, minutes/day	0	0.0008	436.0	27.11	2.00

Note: For brevity in the tables for diets 6–10, food-related measures are shown only for pre-scribed foods. Also, constraint information for seasonal availability, for individual minerals, and for $\Sigma y = 1$ is not shown. Asterisks indicate foods with zero seasonal slack. Other conventions as in table 5.5. Objective function value, 4.69 MJ per day.

Table 5.10 Diet 7: Maximum Protein, Extended Constraints

Food-Related Measures

Time Budget Rank	Food Code	Intake (g/day)	Protein Density (g protein/g food)	Protein Leeway (g protein/g food)	
				Allowable Increase	Allowable Decrease
1	PCMX	630.4	0.016	0.009	0.001
3	GRLU*	108.7	0.064	Infinity	0.045
4	FTGX	42.4	0.012	0.004	0
9	CDLU*	23.3	0.063	Infinity	0.013
10	XXXG*	7.3	0.191	Infinity	0.121
11	TOBS	35.3	0.041	0.001	0.004
12	FTDG*	8.2	0.187	Infinity	0.035
15	FTHG*	24.2	0.012	Infinity	0.007
16	SMCX*	10.4	0.096	Infinity	0.004
17	ATFR*	27.2	0.049	Infinity	0.033
18	TODR*	3.0	0.208	Infinity	0.020
21	SPFU*	18.5	0.060	Infinity	0.043
22	SPFS*	33.5	0.034	Infinity	0.009
26	ATFX*	10.4	0.052	Infinity	0.033
28	TFLX*	5.3	0.045	Infinity	0.009
30	ATFU*	4.3	0.071	Infinity	0.036
37	INKX*	0.6	0.268	Infinity	0.171
41	SFFX*	3.0	0.130	Infinity	0.109
42	TODG*	1.4	0.208	Infinity	0.156
48	SPFX*	5.1	0.036	Infinity	0.012
50	ACDD*	0.1	0.191	Infinity	0.099

Constraint-Related Measures

Constraint, Unit	Slack (units/day)	Marginal Value (g protein/ day)	Current RHS (units/day)	Right-Hand-Side Ranges (units/day)	
				Allowable Increase	Allowable Decrease
Energy minimum, g/day	3.48	0	0.87	3.48	Infinity
Protein minimum, g/day	27.04	0	3.94	27.04	Infinity
Consumption capacity, g/day	0	0.002	1002.4	67.7	34.1
Ash maximum, %	114.27	0	0	Infinity	114.27
Ash minimum, %	0	−0.36	0	1.28	1.57
Fiber minimum, g/day	25.85	0	0	25.85	Infinity
Lipid minimum, g/day	36.68	0	1.48	36.68	Infinity
Time maximum, minutes/day	0	0.04	436.00	10.20	20.24

Note: Conventions as in tables 5.5 and 5.9. Objective function value, 31.0 g protein per day.

Table 5.11 Diet 8: Minimum Feeding Time, Extended Constraints

Food-Related Measures

Time Budget Rank	Food Code	Intake (g/day)	Time Density (minutes/g food)	Time Leeway (minutes/g food)	
				Allowable Increase	Allowable Decrease
1	PCMX	119.2	0.208	0.012	0.110
4	FTGX	5.9	0.585	0	0.071
15	FTHG*	24.2	0.427	0.158	Infinity
17	ATFR	20.4	0.421	0.021	0.043
41	SFFX*	3.0	0.651	0.346	Infinity
42	TODG*	1.4	1.240	0.395	Infinity

Constraint-Related Measures

Constraint, Unit	Slack (units/day)	Marginal Value (minutes/ unit)	Current RHS (units/day)	Right-Hand-Side Ranges (units/day)	
				Allowable Increase	Allowable Decrease
Energy minimum, g/day	0	−41.63	0.87	0.32	0.09
Protein minimum, g/day	0	−5.07	3.94	0.30	0.89
Consumption capacity, g/day	828.3	0	1002.4	Infinity	828.3
Ash maximum, %	19.84	0	0	Infinity	19.84
Ash minimum, %	0	−0.49	0	0.59	0.86
Fiber minimum, g/day	3.18	0	0	3.18	Infinity
Lipid minimum, g/day	5.58	0	1.48	5.58	Infinity
Time maximum, minutes/day	385.15	0	436.00	Infinity	385.15

Note: Conventions as in tables 5.5 and 5.9. Objective function value, 50.9 minutes per day.

Table 5.12 Diet 9: Maximum Energy Intake Rate, Extended Constraints

Food-Related Measures

Time Budget Rank	Food Code	Intake (g/day)	Energy Rate (kJ/minute)	Energy Rate Leeway (kJ/minute)	
				Allowable Increase	Allowable Decrease
4	FTGX*	66.7	20.638	1.801	2.217
10	XXXG*	7.3	8.421	16.532	3.817
15	FTHG*	24.2	28.241	4.966	10.747
18	TODR	2.5	3.197	2.063	0.949
34	FTHW*	8.2	20.638	14.582	1.066
37	INKX*	0.6	3.387	210.456	5.217
41	SFFX*	3.0	11.964	40.204	2.244
42	TODG*	1.4	11.091	88.360	7.894
50	ACDD*	0.1	6.522	1877.659	1.899
$\beta = 113.96^{-1}$			0	120.170	473.642

Constraint-Related Measures

Constraint, Unit	Slack (units/day)	Marginal Value (kJ/unit)
Energy minimum, g/day	0.51	0
Protein minimum, g/day	0	−0.74
Consumption capacity, g/day	88.85	0
Ash maximum, %	8.34	0
Ash minimum, %	4.65	−0
Fiber minimum, %	9.26	0
Lipid minimum, g/day	0.34	0
Time maximum, minutes/day	353.60	0

Note: Conventions as in tables 5.5 and 5.9. Objective function value, 20.65 kJ per minute of feeding.

Table 5.13 Energy Yield Rates of All Core Foods in Diet 9 and All Other Core Foods of Comparable Yield Rates

| Energy Rate Rank | Included Foods | | | Excluded Foods | | |
	Time Budget Rank	Food Code	Energy (kJ/minute)	Time Budget Rank	Food Code	Energy (kJ/minute)
1	15	FTHG	28.2			
2	4	FTGX	20.6			
3	34	FTHW	20.6			
4				1	PCMX-2	15.5
5	41	SFFR	12.0			
6	42	TODG	11.0			
7				17	ATFR	9.5
8	10	XXXG	8.4			
9				3	GRLU	8.3
10				26	ATFX	8.1
11	50	ACDD	6.5			
12				21	SPFU	5.2
13				48	SPFX	4.6
14				22	SPFS	4.5
15				12	FTDG	4.1
16				16	SMCX	4.0
17				11	TOBS	4.0
18				24	SPFR	3.8
19				9	CDLU	3.7
20	37	INKX	3.4			
21				43	TOBR	3.3
22	18	TODR	3.2			

Note: Food codes as in table 6.2.

Table 5.14 Protein Yield Rates of Core Foods

Protein Rate Rank	Included Foods			Excluded Foods		
	Time Budget Rank	Food Code	Protein (g/minute of feeding)	Time Budget Rank	Food Code	Protein (g/minute of feeding)
1	42	TODG	0.17			
2	37	INKX	0.13			
3				17	ATFR	0.12
4				3	GRLU	0.12
5	10	XXXG	0.11			
6				26	ATFX	0.10
7				21	SPFU	0.09
8	50	ACDD	0.09			
9				1	PCMX	0.08
10				30	ATFU	0.08
11				22	SPFS	0.05
12				28	TFLX	0.05
13				11	TOBS	0.05
14				12	FTDG	0.05
15	18	TODR	0.05			

Note: Rank-ordered, including all foods in diet 9 whose energy yield rate (table 5.13) is exceeded by at least one excluded non-milk core food.

Table 5.15 Diet 10: Maximum Protein Intake Rate, Extended Constraints

Food-Related Measures

Time Budget Rank	Food Code	Intake (g/day)	Protein Rate (g protein/minute)	Protein Rate Leeway (g protein/minute)	
				Allowable Increase	Allowable Decrease
3	GRLU*	108.7	0.115	0.017	0.016
10	XXXG*	7.3	0.112	0.248	0.059
12	FTDG	6.1	0.053	0.026	0.008
17	ATFR*	27.2	0.117	0.066	0.023
26	ATFX*	10.4	0.104	0.436	0.008
37	INKX*	0.6	0.128	3.158	0.044
42	TODG*	1.4	0.168	1.326	0.112
50	ACDD*	0.1	0.087	28.177	0.033
$\beta = 161.655^{-1}$			0	1.803	1.756

Constraint-Related Measures

Constraint, Unit	Slack (units/day)	Marginal Value (g/minute/unit)
Energy minimum, g/day	0	−0.03
Protein minimum, g/day	7.86	−0
Consumption capacity, g/day	840.7	0
Ash maximum, %	15.49	0
Ash minimum, %	2.94	−0
Fiber minimum, %	12.79	0
Lipid minimum, g/day	2.54	0
Time maximum, minutes/day	321.94	0

Note: Conventions as in tables 5.5 and 5.9. Objective function value, 112.6 mg protein per minute of feeding.

Table 5.16 Constituents and Constraints in Diets 6–10

	Diet				
	6	7	8	9	10
Availability constraints					
1. F_1 = Milk maximum	+	+	+	0	0
2. F_2 = GRCX maximum	0	0	0	0	0
3. F_3 = GRLU maximum	max	max	0	0	max
4. F_4 = FTGX maximum	max	[+]	[+]	max	0
5. F_5 = TODU maximum	0	0	0	0	0
6. F_6 = SKCX maximum	+	0	0	0	0
7. F_7 = SCTX maximum	0	0	0	0	0
8. F_8 = TCFX maximum	0	0	0	0	0
9. F_9 = CDLU maximum	max	max	0	0	0
10. F_{10} = XXXG maximum	max	max	0	max	max
11. F_{11} = TOBS maximum	0	[+]	0	0	0
12. F_{12} = FTDG maximum	max	max	0	0	+
13. F_{13} = TTFX maximum	0	0	0	0	0
14. F_{14} = FTDR maximum	max	0	0	0	0
15. F_{15} = FTHG maximum	max	max	max	max	0
16. F_{16} = SMCX maximum	max	max	0	0	0
17. F_{17} = ATFR maximum	max	max	+	0	max
18. F_{18} = TODR maximum	max	max	0	+	0
19. F_{19} = FTDU maximum	0	0	0	0	0
20. F_{20} = FTBS maximum	max	0	0	0	0
21. F_{21} = SPFU maximum	+	max	0	0	0
22. F_{22} = SPFS maximum	0	max	0	0	0
23. F_{23} = XXXW maximum	0	0	0	0	0
24. F_{24} = SPFR maximum	0	0	0	0	0
25. F_{25} = GRTU maximum	0	0	0	0	0
26. F_{26} = ATFX maximum	max	max	0	0	max
27. F_{27} = XXXD maximum	max	0	0	0	0
28. F_{28} = TFLX maximum	0	max	0	0	0
29. F_{29} = SPLU maximum	0	0	0	0	0
30. F_{30} = ATFU maximum	max	max	0	0	0
31. F_{31} = SPLX maximum	0	0	0	0	0
32. F_{32} = FTHP maximum	max	0	0	0	0
33. F_{33} = SMLU maximum	0	0	0	0	0
34. F_{34} = FTHW maximum	max	[0]	[0]	max	0
35. F_{35} = WSFX maximum	0	0	0	0	0
36. F_{36} = COCX maximum	max	0	0	0	0
37. F_{37} = INKX maximum	max	max	0	max	max
38. F_{38} = DUKU maximum	0	0	0	0	0
39. F_{39} = CDSX maximum	0	0	0	0	0
(F_{40} = SPWX) maximum			(excluded a priori)		
40. F_{41} = SFFX maximum	max	max	max	max	0
41. F_{42} = TODG maximum	max	max	max	max	max
42. F_{43} = TOBR maximum	0	0	0	0	0
(F_{44} = XXXW) maximum			(excluded a priori)		
43. F_{45} = SULU maximum	0	0	0	0	0
44. F_{46} = XXLX maximum	0	0	0	0	0
45. F_{47} = GRDU maximum	0	0	0	0	0
46. F_{48} = SPFX maximum	0	max	0	0	0

Table 5.16 (*Continued*)

	Diet				
	6	7	8	9	10
47. F_{49} = TFBS maximum	0	0	0	0	0
48. F_{50} = ACDD maximum	max	max	0	max	max
49. F_{51} = CTFX maximum	0	0	0	0	0
50. F_{52} = SBFX maximum	max	0	0	0	0
Other constraints					
51. Energy minimum	•	•	min	•	min
52. Protein minimum	•	•	min	min	•
53. Consumption maximum	max	max	•	•	•
54. Ash maximum	•	•	•	•	•
55. Ash minimum	min	min	min	•	•
56. Fiber minimum	•	•	•	•	•
57. Lipid minimum	•	•	•	•	•
58. Feeding time maximum	max	max	•	•	•
59–72. All cation maxima	•	•	•	•	•

Note: max or min = binding constraint; [   ] = alternative food in a nonunique diet; + = food called for in diet at less than maximum available amount; 0 = food not called for in diet; bullet (•) = nonbinding constraint other than food.

Table 5.17 Selected Components of Diets 6–10

Diet	Objective	Energy (MJ/day)	Energy Rate[a] (kJ/feed-minute)	Protein (g/day)	Protein Rate[a] (mg/feed-minute)	Feeding Time (minutes/day)	Milk (g/day)	Nursing Time (minutes/day)
Diet 6	Maximize energy intake	**4.69**	14.2	29.1	74.8	436[b]	668	139
Diet 7	Maximize protein intake	4.35	13.3	**31.0**	75.8	436[b]	630	131
Diet 8	Minimize feeding time	0.871[b]	16.6	3.94[b]	72.5	**50.9**	119	24.8
Diet 9	Maximize energy intake rate	1.38	**20.7**	3.94[b]	32.4	82.4	0	0
Diet 10	Maximize protein intake rate	0.871[b]	0.120	11.8	**113**	114	0	0

Note: Boldface values are for the objective function values of each diet, so they are the highest values in their columns.

[a] Rates for energy and protein yields are mass weighted.

[b] For energy and protein, these values are at their minima; for feeding time, at their maxima.

Table 6.1 Observed Diet: Time Budgets (A_j), Intake Rates (B_j), Unit Masses (C_j), Food Units Eaten per Day (A_jB_j), and Grams of Food Eaten per Day ($A_jB_jC_j$) for the Fifty-two Core Foods

			Unit Intake Rate Samples			Time Budget Samples									Daily Consumption				
				Units per Minute of Feeding		Mean Bout Length (Minutes)		Number of Bouts	Bouts per 100 Minutes		Minutes per Day			Units per Day		Grams per 100 Units	Grams per Day	
Time Budget Rank (j)	Food Code	Units	Minutes in Rate Samples	B_j	SE	L_j	SE	(n_j)	$100 R_j$	SE	A_j	SE		A_jB_j	SE	$(100 C_j)$	$A_jB_jC_j$	SE
1	PCMX	—	—	—	—	1.65	0.17	1350	6.05	0.17	65.93	7.04		—	—	—	316.94	45.22
2	GRCX	524.1 corms	137.85	3.80	0.17	0.47*	0.04	2539	11.39	0.23	35.32	3.09		134.22	13.19	2.35	3.15	0.31
3	GRLU	2458.6 leaves	87.52	28.09ᵃ	0.57	0.36	0.04	1958	8.78	0.20	20.86	2.37		585.96	67.63	6.40	37.50	4.33
4	FTGX	112.2 grams	65.71	1.71	0.16	0.37	0.04	1493	6.70	0.17	16.35	1.82		27.96	4.07	100	27.96	4.07
5	TODU	910.5 seeds without coats	113.52	8.02	0.27	1.27	0.11	385	1.73	0.088	14.47	1.46		116.05	12.34	3.91	4.54	0.48
6	SKCX	338.7 husked corms	43.24	7.83	0.43	0.74	0.06	440	1.97	0.094	9.64	0.91		75.48	8.24	1.37	1.03	0.11
7	SCTX	71.9 lateral meristems	64.29	1.12	0.13	0.53	0.04	561	2.52	0.11	8.80	0.77		9.86	1.43	28.58	2.82	0.41
8	TCFX	485 husked ovaries	64.86	7.48	0.34	0.68	0.07	414	1.86	0.091	8.33	0.95		62.31	7.65	3.30	2.06	0.25
9	CDLU	288.4 tillers	22.82	12.64	0.74	0.45	0.04	546	2.45	0.11	7.27	0.73		91.89	10.68	6.40	5.88	0.68
10	XXXG	434 seeds, etc.	25.47	17.04	0.82	0.32	0.06	664	2.98	0.12	6.29	1.21		107.18	21.25	3.44	3.69	0.73
11	TOBS	801 flowers	48.95	16.36ᵇ	0.58	1.85*	0.32	112	0.50	0.047	6.13	1.20		100.29	19.95	7.27	7.29	1.45
12	FTDG	841.5 seeds	48.90	17.21	0.59	0.26	0.03	795	3.57	0.13	6.12	0.74		105.33	13.24	1.64	1.73	0.22
13	TTFX	1079 fruits	45.43	23.75	0.72	0.46	0.06	417	1.87	0.092	5.68	0.79		134.90	19.20	4.36	5.88	0.84
14	FTDR	332.5 seeds	42.12	7.89	0.43	0.38	0.04	435	1.95	0.094	4.89	0.57		38.58	4.97	1.64	0.633	0.081
15	FTHG	60.6 grams	25.86	2.34	0.30	0.50	0.06	287	1.29	0.076	4.25	0.57		9.95	1.85	100	9.95	1.85
16	SMCX	165.5 corms	20.49	8.08	0.63	0.71	0.07	176	0.79	0.059	3.70	0.46		29.90	4.39	4.61	1.38	0.20
17	ATFR	321 fruits	36.31	8.84	0.49	0.42	0.06	218	0.98	0.066	2.71	0.43		23.96	4.03	26.90	6.44	1.08
18	TODR	94.6 seeds	14.00	6.76	0.70	0.72	0.12	118	0.53	0.049	2.51	0.48		16.97	3.69	3.44	0.584	0.127
19	FTDU	69.9 seeds	6.93	10.09	1.21	1.03	0.18	74	0.33	0.039	2.26	0.47		22.80	5.47	2.42	0.552	0.132
20	FTBS	272 flowers	36.37	7.48	0.45	0.88	0.14	80	0.36	0.040	2.08	0.41		15.56	3.21	14.13	2.20	0.45
21	SPFU	472.4ᵈ fruits	20.27ᵈ	23.31ᵈ	1.07	0.47*	0.07	142	0.64	0.053	1.98	0.34		46.15	8.20	6.45	2.98	0.53
22	SPFS	472.4ᵈ fruits	20.27ᵈ	23.31ᵈ	1.07	1.05	0.19	63	0.28	0.036	1.96	0.43		45.69	10.24	6.45	2.95	0.66
23	YYYW	49 droplets, etc	26.35	1.86	0.27	0.20	0.03	277	1.24	0.075	1.64	0.27		3.05	0.67	5	0.458	0.10

24	SPHK	472.4[a] fruits	20.27[d]	23.31[a]	1.07	0.95	0.11	55	0.25	0.033	1.55	0.35	36.13	8.33	6.45	2.33	0.54
25	GRTU	288 tillers	24.37	11.82	0.70	0.38	0.04	137	0.61	0.052	1.54	0.21	18.20	2.71	3.20	0.582	0.087
26	ATFX	137 fruits	18.45	7.43	0.63	0.23*	0.03	215	0.96	0.066	1.46	0.22	10.84	1.88	26.90	2.92	0.51
27	XXXD	170 seeds, etc.	35.24	4.82	0.37	0.06	0.03	769	3.45	0.12	1.37	0.69	6.60	3.36	3.44	0.227	0.116
28	TFLX	192 leaves	10.17	18.88	1.36	0.36	0.05	97	0.44	0.044	1.03	0.18	19.45	3.68	5.77	1.12	0.21
29	SPLU	78 leaves	20.90	3.73	0.42	0.35	0.04	94	0.42	0.043	0.975	0.15	3.64	0.69	7.65	0.278	0.053
30	ATFU	67.5 fruits	14.77	4.57	0.56	0.35	0.04	90	0.40	0.043	0.932	0.15	4.26	0.86	23.72	1.01	0.20
31	SPLX	29.5 leaves	5.03	5.87	1.08	0.25	0.04	101	0.45	0.045	0.747	0.14	4.39	1.15	7.65	0.34	0.088
32	FTHP	0.6 grams	3.15	0.19	0.25	0.62	0.17	40	0.18	0.028	0.734	0.23	0.14	0.189	(100)	0.139	0.189
33	SMLU	35 blades	7.68	4.56	0.77	0.55	0.08	45	0.20	0.030	0.733	0.15	3.34	0.89	4.27	0.14	0.038
34	FTHW	— grams	—	1.71	0.16	0.46	0.12	53	0.24	0.033	0.722	0.22	1.24	0.39	100	1.24	0.39
35	WSFX	42 fruits	12.02	3.49	0.54	0.48	0.07	47	0.21	0.031	0.668	0.14	2.33	0.61	8.37	0.195	0.051
36	COCX	93.5 hulled corms	13.63	6.86	0.71	0.77	0.16	28	0.13	0.024	0.638	0.18	4.38	1.32	4.03	0.176	0.053
37	INKX	23.5 insects	6.24	3.77	0.78	0.29	0.05	74	0.33	0.039	0.635	0.13	2.39	0.70	12.65	0.303	0.088
38	DUKU	48 larvae	16.80	2.86	0.41	0.17	0.05	115	0.52	0.048	0.578	0.18	1.65	0.57	10	0.165	0.057
39	CDSX	4.5 cm	5.15	0.87	0.87	0.36	0.08	51	0.23	0.032	0.543	0.14	0.472	0.488	4.73	0.022	0.023
40	SPWX	0.3 cm	13.56	0.022	0.04	0.40	0.07	42	0.19	0.029	0.497	0.12	0.011	0.020	100	0.011	0.020
41	SFFX	1395.8 fruits	6.25	223.33	5.98	0.23*	0.05	72	0.32	0.038	0.490	0.12	109.43	26.96	0.688	0.75	0.185
42	TODG	102 seeds	4.35	23.45	2.32	0.20	0.03	79	0.35	0.040	0.467	0.087	10.95	2.31	3.44	0.377	0.079
43	TOBR	178 flowers	13.08	13.61[c]	1.02	0.08	0.02	185	0.83	0.061	0.438	0.12	5.96	1.69	7.27	0.433	0.123
44	XXWX	0 grams	8.63	0	0	0.46	0.16	32	0.14	0.025	0.436	0.17	0	0	100	0	0
45	SULU	2.26 grams	8.12	0.28	0.19	0.25	0.03	56	0.25	0.034	0.414	0.075	0.116	0.081	100	0.116	0.081
46	XXLX	28.5 leaves	6.28	4.54	0.85	0.33*	0.06	41	0.18	0.029	0.400	0.095	1.82	0.55	6.68	0.121	0.037
47	GRDU	31.5 closed seedheads	3.97	7.94	1.41	0.53	0.09	25	0.11	0.022	0.392	0.10	3.11	0.97	9.66	0.30	0.093
48	SPFX	472.4[d] fruits	20.27[d]	23.31[d]	1.07	0.71	0.08	18	0.08	0.019	0.378	0.099	8.81	2.34	6.45	0.568	0.151
49	TFBS	191 flowers	7.48	25.54	1.85	0.41	0.07	30	0.135	0.025	0.364	0.092	9.30	2.44	1.21	0.112	0.030
50	ACDD	515 seeds	21.44	24.02	1.06	0.72	0.25	17	0.076	0.018	0.362	0.15	8.70	3.62	1.89	0.164	0.068
51	CTFX	4 fruits	6.07	0.66	0.33	0.62	0.13	19	0.085	0.020	0.348	0.11	0.230	0.136	10	0.023	0.014
52	SBFX	14 fruits without calyx	8.72	1.61	0.43	0.55	0.12	20	0.090	0.020	0.326	0.10	0.525	0.213	39.17	0.206	0.084

Note: The special method for estimating the intake of milk (PCMX) is described in the text; the intake value given for milk (316.94) is the first approximation (see text). Four-letter food abbreviations as in table 6.2. Time budget data from table 4.1; unit mass values from table 6.4.

[a] 29.59 without Pedro's value (table 7.10).

[b] 16.37 without Pedro's value (table 7.10).

[c] 60.0 without Pedro's value (table 7.10).

[d] Data for all *Salvadora persica* fruit (SPF*) pooled.

*Asterisk indicates that the product-limit estimate for that food may be an underestimate because the longest bout was censored.

Table 6.2 Abbreviations for Fifty-two Core Foods, Arranged Alphabetically

Time Budget Rank	Food Code	Food
50	ACDD	Acacia seeds (fever or umbrella tree) picked from dung
17	ATFR	Ripe fruit of *Azima tetracantha*
30	ATFU	Green fruit of *Azima tetracantha*
26	ATFX	Fruit of *Azima tetracantha*, ripeness unknown
9	CDLU	Green leaves of Bermuda grass, *Cynodon nlemfuensis* = "*dactylon*"
39	CDSX	Stolons of Bermuda grass, condition unknown
36	COCX	Corms of *Cyperus obtusiflorus*, condition unknown
51	CTFX	Fruit of *Capparis tomentosa*, condition unknown
38	DUKU	Dung-beetle larvae
20	FTBS	Fever tree blossoms picked from tree, *Acacia xanthophloea*
12	FTDG	Abscised dry fever tree seeds picked up from ground
14	FTDR	Fever tree seeds from dry pods
19	FTDU	Fever tree seeds from green pods
4	FTGX	Gum (condition unspecified) picked from fever tree
15	FTHG	Fever tree gum picked up from ground
32	FTHP	Fever tree gum picked from other plants
34	FTHW	Fever tree gum picked from logs
2	GRCX	Corms of unidentified grass or sedge, condition unknown
47	GRDU	Green seedheads of unidentified grass or sedge
3	GRLU	Green leaves of unidentified grass or sedge
25	GRTU	Blade bases of green leaves of unidentified grass or sedge
37	INKX	Unidentified arthropods
1	PCMX	Milk
52	SBFX	Fruit of *Solanum* "*dubium*" = *S. coagulans,* condition unknown
7	SCTX	Blade bases of reed grass, *Sporobolus consimilis,* condition unknown
41	SFFX	Fruit of "sticky-fruit plant," *Commicarpus pedunculosus,* condition unknown
6	SKCX	Corms of *Sporobolus* "*kentrophyllus*" = *S. rangei*, condition unknown
16	SMCX	Corms of *Sporobolus cordofanus*, condition unknown
33	SMLU	Green leaves of *Sporobolus cordofanus*
24	SPFR	Ripe fruit of toothbrush bush, *Salvadora persica*
22	SPFS	Semiripe fruit of toothbrush bush
21	SPFU	Green fruit of toothbrush bush
48	SPFX	Fruit of toothbrush bush, ripeness unknown
29	SPLU	Green leaves of toothbrush bush
31	SPLX	Leaves of toothbrush bush, condition unknown
40	SPWX	Wood of toothbrush bush, condition unknown
45	SULU	Green leaves (and cryptic flowers?) of *Suaeda monoica*
8	TCFX	Fruit of *Trianthema ceratosepala*, condition unknown
49	TFBS	"Trumpet-flower" blossoms (*Lycium* "*europaeum*") on plant
28	TFLX	Leaves of "trumpet-flower plant," condition unknown
43	TOBR	Abscised umbrella tree blossoms *Acacia tortilis* picked up from ground
11	TOBS	Umbrella tree blossoms picked from tree
42	TODG	Umbrella tree dry seeds picked up from ground
18	TODR	Umbrella tree seeds from dry pods
5	TODU	Umbrella tree seeds from green pods
13	TTFX	Fruit of devil's thorn, *Tribulus terrestris*, condition unknown
35	WSFX	Fruit of *Withania somnifera*, ripeness unknown
46	XXLX	Unidentified leaves (not grass or sedge), condition unknown
44	XXWX	Wood, unidentified
27	XXXD	Unidentified items (mostly acacia seeds?) picked from dung
10	XXXG	Unidentified items (mostly acacia seeds?) picked up from ground
23	XXXW	Unidentified items (fever tree gum? arthropods?) picked from wood

Table 6.3 Conversion Factors Used to Convert Field Records in Nonstandard Units to Standard Units

Code	Food	Conversion Factors
TODR	*Acacia tortilis,* seeds from dry pods	Assumed same as TODU
TODU	*Acacia tortilis,* seeds from green pods	1 pod = 6.72 seeds eaten (SD 2.45, SE 0.14, $N = 300$)
FTDG	*Acacia xanthophloea,* seeds from dry pods on ground	1 pod = 4.67 seeds*
FTGX⎫ FTHG⎬ FTHP⎭	*Acacia xanthopholea,* gum	1 droplet, bite, HM ≈ 0.05 g
FTDR	*Acacia xanthophloea,* seeds from dry pods	1 pod = 4.67 seeds eaten*
FTDU	*Acacia xanthophloea,* seeds from green pods	1 pod = 4.67 seeds eaten*
SFFX⎫ SFFS⎬	*Commicarpus pedunculosus,* fruit	1 bite = 32/36 = 0.9 stem units
		1 stem unit = 17.19 fruits (SE 0.68, SD 6.97, $N = 105$)
CDLU	*Cynodon nlemfuensis,* green leaves	1 bite ≈ 1/8 HM
		1 HM = 79/25 = 3.165 tillers
		1 tiller = 4.07 leaves (SE 9.28, SD 1.46, $N = 28$)
CDSX	*Cynodon nlemfuensis,* stolons, rhizomes	1 HM ≈ 1 cm
COCX	*Cyperus obtusiflorus,* corms	1 HM = 4/3 corms
TFBS	*Lycium "europaeum,"* blossoms	1 HM = 14/14 = 1.0 blossom
TFLX	*Lycium "europaeum,"* leaves	1 HM = 47/23 = 2.0 leaves
SPFS⎫ SPFR⎬ SPFU⎭	*Salvadora persica,* fruit	1 bite = 59/49 = 1.2 fruits 1 HM ≈ 1 bite
SPLU⎫ SPLX⎬	*Salvadora persica,* leaves	1 cm terminal growth ≈ 6 leaves
SCTX	*Sporobolus consimilis,* blade bases	1 blade base = 2.69 cm (SE 0.13, $N = 132$)
SKCX	*Sporobolus rangei,* corms	1 HM = 39/37 = 1.05 corms
SMCX	*Sporobolus cordofanus,* corms	1 HM = 11/11 = 1.0 corm 1 bite ≈ 1 corm
SMLU	*Sporobolus cordofanus,* green leaves	1 HM = 4/4 = 1 tiller 1 bite = 5/5 = 1 tiller
SULU	*Suaeda monoica,* green leaves	1 HM ≈ 0.02 g 1 bite ≈ 0.05 g
TTFX	*Tribulus terrestris,* fruit	1 HM = 7/7 = 1.0 fruit
GRCX	Grass (species unknown), corms	1 HM = 37/35 = 1.06 corms
GRLU	Grass (species unknown), green leaves	1 HM = 163.5/102 = 1.60 leaves (bites, tillers assumed same as CDLU)

Note: Nonstandard units are shown on the left of the equations. The equals sign is used for empirically determined conversion factors, for which the sample sizes are shown as fractions. For example, for CDLU, 1 HM = 79/25 indicates that in a sample of 25 hand-to-mouth movements (HM), 79 tillers were eaten. The approximately equals sign is used for "guesstimates," where empirical data are lacking. Only the amount of the portion eaten is shown.

*On average, 82.6% (= 6.72 seeds) of the 8.14 ± 0.14 seeds per green *Acacia tortilis* pod were eaten ($N = 300$). On average, each dry *A. xanthophloea* pod contained 5.65 ± 0.14 seeds (mean ± SE). Assuming the same percentage in both species, whether dry or green, 0.826 × 5.65 = 4.67 seeds were eaten from each *A. xanthophloea* pod.

Table 6.4 Unit Mass of Core Foods

Time Budget Rank	Food Code	Samples	Number (n)	Units	Total Mass of N Units (g)	Mass per 100 Units ($100 \text{ g}/n = 100\,C_i$)
1	PCMX	(Not needed)				
2	GRCX	(Taken as weighted mean of SKCX, SMCX, COCX, nos. 6, 16, and 36 below)		Corms		2.35
3	GRLU	(Assumed same as CDLU, no. 9 below)		Tillers		6.40
4	FTGX	(Not needed)				(100)
5	TODU	14, 38A	2549	Seeds without coats	99.63	3.91*
6	SKCX	100, SAs 16A	1341	Husked corms	18.35	1.37
7	SCTX	97, 98, SAs 12A	412	Blade bases	117.75	28.58
8	TCFX	SAs 6A, 114	1525	Husked ovaries (fruits)	50.35	3.30
9	CDLU	50, 51, SAs 8A	975	Tillers	62.40	6.40
10	XXXG	(Assumed same as TODR, no. 18 below)		Seeds, etc.	3.44	
11	TOBS	2, 3, 4, 5, SAs 13A	2224	Flowers	161.7	7.27
12	FTDG	(Assumed same as FTDR, no. 14 below)		Seeds		1.64
13	TTFX	117, 118, SAs 24A	2812	Fruit	122.58	4.36
14	FTDR	31–34	5131	Seeds	84.3	1.64
15	FTHG	(Not needed)				(100)
16	SMCX	104, SAs 3A	344	Corms	15.85	4.61
17	ATFR	41, 42, 43, 45, SAs 9A, 9AS	2170	Fruit	583.76	26.90
18	TODR	16N	125	Seeds	4.30	3.44
19	FTDU	36	551	Seeds	13.35	2.42
20	FTBS	19, 20, SAs 5A	573	Flowers	80.95	14.13
21	SPFU	(Assumed same as SPFR, no. 24 below)				6.45
22	SPFS	(Assumed same as SPFR, no. 24 below)				6.45
23	XXXW	(Assumed primarily droplets of FTGX, no. 4 above)				5
24	SPFR	SAs 4A	759	Fruit	48.95	6.45**
25	GRTU	(Assumed 1/2 of CDLU, no. 9 above)				3.20
26	ATFX	(Assumed same as ATFR, no. 17 above)				26.90
27	XXXD	(Assumed same as TODR, no. 18 above)		Seeds, etc.		3.44

Table 6.4 (*Continued*)

Time Budget Rank	Food Code	Samples	Number (n)	Units	Total Mass of N Units (g)	Mass per 100 Units (100 g/n = 100 C_i)
28	TFLX	75–77, SAs 11A	4280	Leaves	247.0	5.77
29	SPLU	83, 84, SAs 1A	1779	Terminal leaves	136.0	7.65
30	ATFU	44, SAs 34A	473	Fruit	112.2	23.72
31	SPLX	(Assumed same as SPLU, no. 29 above)		Leaves		7.65
32	FTHP	(Not needed)				(100)
33	SMLU	(Assumed 2/3 of CDLU, no. 9 above)		Leaves		4.27
34	FTHW	(Not needed)				(100)
35	WSFX	122–24, SAs 35A, 36B	1518	Fruit	127.0	8.37
36	COCX	63, 64	629	Hulled corms	36.23	4.03
37	INKX	125N (based on grasshoppers)	51	Insects	6.45	12.65
38	DUKU	(No data available)	?	Larvae	?	10
39	CDSX	56	400	Stolons	18.93	4.73
40	SPWX	(No data available)		g/cm		100
41	SFFX	48 and SAs 30B for SFFR	9042	Fruit	62.23	0.688
42	TODG	(Assumed same as TODR, no. 18 above)				3.44
43	TOBR	(Assumed same as TOBS, no. 11 above)				7.27
44	XXWX	(Not needed)				(100)
45	SULU	(Not needed)				(100)
46	XXLX	(Taken as weighed mean of TFLX and SPLU, nos. 28 and 29, above)				6.68
47	GRDU	54 for CDDU, 110 for SSDU	770	Closed seedheads	74.4	9.66
48	SPFX	(Assumed same as SPFR, no. 24 above)				6.45
49	TFBS	73, SAs 10A	2014	Flowers	24.40	1.21
50	ACDD	(Taken as weighed mean of FTDR and TODR, nos. 14 and 19, above)				1.89
51	CTFX	(No data available)		Fruit		10
52	SBFX	92, SAs 33A	316	Fruit without calyx	123.78	39.17

Note: Food codes as in table 6.2. Means, where given, are weighted by time budgets (table 6.1).

*Another sample (87N) gave 0.95.

**Including specimen M331/75, which was inadvertently omitted, would add 497 seeds and 38.1 g, changing the mean mass per 100 units to 4.52 g.

Table 6.5 Nutrient Intake from Seasonally Adjusted, Pooled Mean Diet

Time Budget Rank	Food Code	Mass (g/day)	Water (g/day)	Ash (mg/day)	Lipid (mg/day)	Protein (mg/day)	Fiber (mg/day)	Carbo-hydrate (mg/day)	Energy (kJ/day)	Time (minutes/day)
					A. By Foods					
1	PCMX-1	316.9	272.6	951	15530	5071	0	22503	1022	65.93
	PCMX-2	260.9	224.4	783	12784	4174	0	18524	841	65.93
2	GRCX	2.69	0.74	197	40	256	436	1020	29	30.15
3	GRLU	35.51	24.75	1314	781	2272	2805	3586	165	19.74
4	FTGX	24.27	5.19	1481	267	291	1772	15267	293	14.20
5	TODU	3.81	2.58	61	27	202	190	746	19	12.14
6	SKCX	0.56	0.17	23	8	58	93	217	6	5.26
7	SCTX	2.07	1.63	60	19	95	97	174	6	6.47
8	TCFX	2.06	1.29	68	56	134	301	212	12	8.35
9	CDLU	4.86	3.40	175	107	306	384	486	22	6.00
10	XXXG	3.69	0.39	140	144	705	760	1553	53	6.30
11	TOBS	3.33	2.67	40	67	137	123	296	11	2.80
12	FTDG	1.54	0.15	60	62	288	328	655	22	5.45
13	TTFX	3.71	3.05	82	22	145	126	285	10	3.58
14	FTDR	0.44	0.04	17	17	81	93	185	6	3.36
15	FTHG	8.51	1.82	519	94	102	621	5354	103	3.63
16	SMCX	0.73	0.18	73	9	70	124	280	8	1.97
17	ATFR	5.12	3.85	148	138	251	179	558	20	2.15
18	TODR	0.42	0.06	13	15	87	74	169	6	1.80
19	FTDU	0.40	0.30	8	10	43	11	27	2	1.64
20	FTBS	1.91	1.43	31	19	21	73	339	8	1.81
21	SPFU	1.90	1.40	110	15	114	78	188	7	1.26
22	SPFS	2.10	1.66	74	19	71	80	204	6	1.40
23	XXXW	0.43	0.09	26	5	5	31	271	5	7.56
24	SPFR	1.37	1.15	16	12	10	48	130	3	0.91
25	GRTU	0.33	0.24	10	10	28	22	25	2	0.88
26	ATFX	2.52	1.88	76	71	131	93	267	10	1.26
27	XXXD	0.21	0.02	8	8	40	43	88	3	1.26
28	TFLX	0.63	0.54	22	4	28	12	25	1	0.58
29	SPLU	0.16	0.12	11	1	8	5	21	1	0.57
30	ATFU	0.78	0.56	31	30	55	40	65	4	0.72
31	SPLX	0.18	0.13	12	1	9	5	23	1	0.40
32	FTHP	0.11	0.02	7	1	1	8	69	1	0.57
33	SMLU	0.07	0.04	6	2	6	6	11	0	0.36
34	FTHW	0.63	0.14	38	7	8	46	397	8	0.37
35	WSFX	0.10	0.07	1	4	6	10	11	1	0.35
36	COCX	0.05	0.02	2	2	2	5	16	0	0.18
37	INKX	0.20	0.13	2	8	54	5	6	1	0.42
38	DUKU	0.07	0.05	1	2	7	2	1	0	0.23
39	CDSX	0.01	0.01	0	0	0	0	1	0	0.29
40	SPWX	0.01	0.00	0	0	0	0	0	0	0.27
41	SFFX	0.52	0.30	22	50	68	25	56	4	0.34
42	TODG	0.27	0.04	9	10	56	48	109	4	0.33
43	TOBR	0.14	0.11	2	3	6	5	12	0	0.14
44	XXWX	0.00	0.00	0	0	0	0	0	0	0.47
45	SULU	0.09	0.08	4	1	3	1	7	0	0.33
46	XXLX	0.06	0.05	3	0	3	1	5	0	0.19
47	GRDU	0.01	0.01	0	0	0	1	1	0	0.01

Table 6.5 (*Continued*)

Time Budget Rank	Food Code	Mass (g/day)	Water (g/day)	Ash (mg/day)	Lipid (mg/day)	Protein (mg/day)	Fiber (mg/day)	Carbo-hydrate (mg/day)	Energy (kJ/day)	Time (minutes/day)
48	SPFX	0.45	0.36	17	4	16	17	44	1	0.30
49	TFBS	0.04	0.03	1	0	2	2	3	0	0.13
50	ACDD	0.06	0.01	2	2	12	13	27	1	0.14
51	CTFX	0.01	0.01	0	0	1	0	1	0	0.17
52	SBFX	0.12	0.08	1	2	6	6	18	1	0.18

B. By Food Categories

	Mass (g/day)	Water (g/day)	Ash (mg/day)	Lipid (mg/day)	Protein (mg/day)	Fiber (mg/day)	Carbo-hydrate (mg/day)	Energy (kJ/day)	Time (minutes/day)
Core foods	380.2	287.4	5808	14959	10475	9250	52038	1708	225.3
All foods	391.5	293.4	6285	15166	11073	10129	55222	1791	240.5
Nonmilk core	119.3	63.0	5025	2175	6301	9250	33514	867	159.4
Seeds	10.8	3.6	319	295	1514	1560	3561	116	32.4
Corms	4.0	1.1	295	59	386	658	1534	44	37.6
Grass leaves	40.4	28.2	1495	890	2584	3195	4083	187	26.1
Other leaves	1.1	0.9	52	8	51	25	81	3	2.1
Gum	34.0	7.3	2071	374	407	2479	21358	410	26.3
Fruits	20.8	15.7	647	422	1007	1004	2040	79	21.0
Grass meristems, etc.	2.4	1.9	70	29	123	120	199	8	7.4
Flowers	5.4	4.2	73	89	166	203	650	19	4.9
Wood, etc.	0	0	0	0	0	0	0	0	0.7
Runners	0	0	0	0	0	0	1	0	0.3
Milk	260.9	224.4	783	12784	4174	0	18524	841	65.9
Meat	0.3	0.2	3	9	61	7	7	2	0.7

Note: Feeding time (minutes per day) on any nonmilk food *j* calculated from the seasonally adjusted mass intake times $B_j^{-1}C_j^{-1}$ (values and definitions in table 6.1). Food codes in panel A as in table 6.2. Food types in panel B as defined in table 4.4. Summary totals in panel B that are marked "all foods" (1.095 times core food totals) include estimated contribution of noncore foods; other summary values do not. Summary values (panel B) use second approximation of milk intake (PCMX-2; see endnote 6). See table A8.1, column 17, for standard errors of mass values.

Table 6.6 Characteristics of Suckling Bouts and Suckling Time Budgets, and Estimates (First Approximations) of Daily Milk Intakes

Infant	Age (weeks)	Mean Bout Length (minutes)	SE	Number of Bouts	Bout Rate[a]	SE	Percentage of Daytime[b]	SE	Mean Minutes per Daytime	SE	Milk Intake (ml/24 hours)	SE
Alice	30–40	1.46	0.56	25	4.44	0.89	6.49	2.81	42.82	18.52	166.9	72.2
	40–50	2.32	0.64	37	3.40	0.56	7.88	2.53	52.03	16.71	276.0	88.6
	50–60	1.39	0.53	25	5.68	1.14	7.89	3.40	52.07	22.42	252.5	108.7
	60–70	2.75	1.05	15	1.34	0.35	3.69	1.70	24.37	11.23	163.4	75.3
Bristle	30–40	1.34	0.31	100	11.06	1.11	14.82	3.73	97.80	24.65	381.2	96.1
	40–50	1.45	0.34	84	7.27	0.79	10.54	2.73	69.60	18.00	369.1	95.5
Dotty	50–60	5.54	2.89	10	2.21	0.70	12.24	7.47	80.81	49.30	391.8	239.1
Eno	30–40	1.80	0.47	89	13.70	1.45	24.66	6.95	162.76	45.87	634.5	178.8
	40–50	1.49	0.33	74	8.35	0.97	12.44	3.11	82.09	20.53	435.4	108.9
	50–60	1.34	0.39	50	3.69	0.52	4.95	1.60	32.67	10.57	158.4	51.3
Fred	30–40	1.90	0.66	12	1.64	0.47	3.12	1.41	20.62	9.31	80.4	36.3
	40–50	0.92	0.20	49	4.33	0.62	3.99	1.04	26.32	6.85	139.6	36.3
	50–60	2.10	1.42	11	4.84	1.46	10.16	7.52	67.06	49.65	325.2	240.8
	60–70	1.14	0.42	53	4.92	0.68	5.61	2.21	37.04	14.56	248.3	97.6
Hans	30–40	1.92	0.39	163	12.03	0.94	23.10	5.03	152.44	33.19	594.2	129.4
Ozzie	30–40	1.53	0.51	56	7.22	0.97	11.05	3.97	72.93	26.19	284.3	102.1
	40–50	1.58	0.45	82	7.38	0.82	11.67	3.56	76.99	23.52	408.4	124.7
	50–60	1.41	0.56	23	10.20	2.13	14.38	6.45	94.93	42.58	460.4	206.5
	60–70	3.42	0.82	33	2.87	0.50	9.82	2.91	64.80	19.20	434.4	128.7
Pedro	30–40	3.09	0.64	21	2.91	0.63	8.99	2.70	59.32	17.85	231.2	69.6
Pooh	30–40	—	—	0	0	0	0	0	0	0	0	0
	40–50	1.28	0.44	37	3.95	0.65	5.05	1.93	33.35	12.71	176.9	67.4
	50–60	1.15	0.47	29	3.62	0.67	4.16	1.87	27.44	12.32	133.1	59.7
	60–70	0.26	0.08	11	2.48	0.75	0.64	0.28	4.26	1.83	28.5	12.3
Striper	50–60	1.83	0.74	12	2.20	0.64	4.03	2.00	26.60	13.22	129.0	64.1
Summer	30–40	1.23	0.34	119	13.33	1.22	16.39	4.77	108.20	31.51	421.8	122.8
	40–50	1.63	0.40	80	9.07	1.01	14.78	3.99	97.57	26.31	517.4	139.5
	50–60	1.90	0.96	50	7.52	1.06	14.29	7.50	94.29	49.47	457.2	239.9

Means[c]

30–40	1.66 (1.78[d])	0.20 (0.18[d])	585	8.85 (8.29[d])	0.37 (0.35[d])	14.70 (14.76[d])	1.87 (1.50[d])	97.00 (89.61[d])	12.36 (17.86[d])	(349.3[d])	69.6 (38.49[d])
30–40	1.66 (1.78[d])	0.20 (0.18[d])	585	8.85 (8.29[d])	0.37 (0.35[d])	14.70 (14.76[d])	1.87 (1.50[d])	97.00 (89.61[d])	12.36 (17.86[d])	(349.3[d])	69.6 (38.49[d])
40–50	1.51 (1.52)	0.19 (0.16)	443	6.16 (6.25)	0.29 (0.30)	9.30 (9.50)	1.25 (1.08)	61.40 (62.28)	8.26 (9.93)	(331.8)	52.7 (37.71)
50–60	1.85 (2.08)	0.40 (0.45)	210	4.46 (5.00)	0.31 (0.41)	8.25 (10.40)	1.87 (1.90)	54.45 (59.48)	12.37 (10.23)	(288.4)	41.6 (60.92)
60–70	1.94 (1.89)	0.38 (0.35)	112	2.96 (2.90)	0.28 (0.30)	5.74 (5.48)	1.06 (1.01)	37.90 (32.62)	8.24 (12.67)	(218.7)	85.0 (44.66)

Note: See text for methods of estimation. Omitted lines (e.g., Bristle at 50–60 weeks) indicate lack of samples. In the last four rows the mean of mean bout length and bout rates were used to calculate the parenthetical values for percentage of daytime and the method 2 time budgets in table 6.8. The parenthetical minutes per daytime, which were used to calculate individual milk intakes, are the means of the corresponding values in the table.

[a] Bouts/(100 minutes of in-sight sample time × k), that is, n_{ij}(suckle)/$s_{ij}k$ of appendix 9.

[b] Eleven hours, 0700–1800 h.

[c] First value of each pair is pooled mean. Parenthetical value is mean of individual values. Values for number of bouts are totals.

[d] Calculated without Pooh's very small sample (14.2 minutes) at 30–40 weeks.

Table 6.7 Mean Nursing Time (Minutes per Day) at Various Ages

Age in Weeks	Males	Females
30–40	80.6 ($n = 4$)	104.6 ($n = 3$)
40–50	57.6 ($n = 3$)	66.3 ($n = 4$)
50–60	81.0 ($n = 2$)	52.3 ($n = 6$)
60–70	50.0 ($n = 2$)	14.3 ($n = 2$)

Note: Values are means of individual means given in table 6.6.

Table 6.8 Age Differences in Nursing Time Budgets (Mean Number of Minutes on the Nipple per Day), Calculated by Four Methods

Method	Parameter type	Age (weeks)			
		30–40	40–50	50–60	60–70
1	Individual × age-specific values	89.6	62.6	59.5	32.6
2	Age-specific means of values above	97.4[a]	62.7	68.6	36.2
3	Age-specific pooled values	97.0	61.4	54.4	37.9
4	Ditto if different, otherwise overall				
	pooled values	96.4	67.1	48.5	32.2
	Spread (range: median)	4.2%	4.4%	17.2%	8.1%

Note: See text for details.

[a]Calculated without Pooh's very small sample (14.2 minutes) at 30–40 weeks.

Table 6.9 Numbers of Nursing Bouts Terminated by the Infant and by Its Mother

Name	30–40 Weeks		40–50 Weeks		50–60 Weeks		60–70 Weeks		Totals		
	Vol.	Invol.	Vol.	Invol.	Vol.	Invol.	Vol.	Invol.	Vol.	Invol.	
Alice	1	0	14	7	18	3	8	1	41	11	
Bristle	24	9	31	13	—	—	—	—	55	22	
Eno	21	9	22	23	20	12	—	—	63	44	
Fred	—	—	12	10	0	3	22	20	34	33	
Hans	52	12	—	—	—	—	—	—	52	12	
Ozzie	4	20	22	19	8	9	10	5	44	53	
Pedro	3	4	—	—	—	—	—	—	3	4	
Pooh	—	—	4	4	2	9	10	1	16	14	
Summer	19	13	33	8	9	11	—	—	61	32	
Totals	124	67	138	84	57	47	50	27	369	225	594

Note: Vol. = voluntary; invol. = involuntary.

Table 6.10 Nutrient Intake Rates (Mass or Energy per Minute of Feeding on Specified Food) of Seasonally Adjusted, Pooled Mean Diet

Time Budget Rank	Food Code	Mass (mg/min)	Water (mg/min)	Ash (mg/min)	Lipid (mg/min)	Protein (mg/min)	Fiber (mg/min)	Carbo- hydrate (mg/min)	Energy (kJ/min)
				A. By Foods					
1	PCMX-1	4807	4134	14.4	235.6	76.9	0	341	15.5
	PCMX-2	3957	3403	11.9	193.9	63.3	0	281	12.8
2	GRCX	89	25	6.5	1.3	8.5	14.5	34	1.0
3	GRLU	1799	1254	66.5	39.6	115.1	142.1	182	8.3
4	FTGX	1709	366	104.3	18.8	20.5	124.8	1075	20.6
5	TODU	314	213	5.0	2.2	16.6	15.7	61	1.6
6	SKCX	107	31	4.4	1.6	11.0	17.7	41	1.2
7	SCTX	320	252	9.3	2.9	14.7	15.0	27	1.0
8	TCFX	247	155	8.1	6.7	16.0	36.0	25	1.5
9	CDLU	809	566	29.1	17.8	51.0	63.9	81	3.7
10	XXXG	586	62	22.3	22.9	112.0	120.8	247	8.4
11	TOBS	1189	954	14.3	23.8	48.8	44.0	106	4.0
12	FTDG	282	27	11.0	11.3	52.8	60.1	120	4.1
13	TTFX	1035	852	22.8	6.2	40.4	35.2	80	2.7
14	FTDR	129	13	5.0	5.2	24.2	27.6	55	1.9
15	FTHG	2342	501	142.9	25.8	28.1	171.0	1473	28.3
16	SMCX	372	89	37.2	4.5	35.8	62.9	142	4.0
17	ATFR	2375	1786	68.9	64.1	116.4	83.1	259	9.5
18	TODR	233	33	7.4	8.4	48.4	41.2	94	3.2
19	FTDU	244	184	4.9	6.3	26.1	6.6	16	1.0
20	FTBS	1057	789	16.9	10.6	11.6	40.2	187	4.3
21	SPFU	1504	1104	87.2	12.0	90.2	61.7	149	5.2
22	SPFS	1504	1186	52.6	13.5	51.1	57.1	146	4.5
23	XXXW	57	12	3.5	0.6	0.7	4.2	36	0.7
24	SPFR	1504	1268	18.0	13.5	10.5	52.6	143	3.8
25	GRTU	378	270	11.7	11.3	31.4	25.3	29	1.7
26	ATFX	2000	1494	60.0	56.0	104.0	74.0	212	8.1
27	XXXD	166	17	6.3	6.5	31.7	34.2	70	2.4
28	TFLX	1089	930	38.1	7.6	49.0	20.7	42	2.0
29	SPLU	285	206	18.5	2.0	14.0	8.6	37	1.0
30	ATFU	1083	777	43.3	41.2	76.9	55.3	91	4.9
31	SPLX	449	324	29.2	3.1	22.0	13.5	58	1.6
32	FTHP	190	41	11.6	2.1	2.3	13.9	120	2.3
33	SMLU	195	109	16.9	4.3	16.5	16.7	31	1.2
34	FTHW	1709	366	104.3	18.8	20.5	124.8	1075	20.6
35	WSFX	292	201	3.2	10.2	16.9	30.1	31	1.6
36	COCX	276	129	12.2	9.4	10.5	26.3	90	2.4
37	INKX	477	299	5.7	18.1	127.8	11.4	15	3.4
38	DUKU	286	232	4.0	7.7	30.3	8.0	4	1.0

Table 6.10 (*Continued*)

Time Budget Rank	Mass (mg/min)	Water (mg/min)	Ash (mg/min)	Lipid (mg/min)	Protein (mg/min)	Fiber (mg/min)	Carbo- hydrate (mg/min)	Energy (kJ/min)
			B. By Food Categories					
Seeds	334.2	111.0	9.8	9.1	46.7	48.1	109.8	3.573
Corms	107.6	29.5	7.9	1.6	10.3	17.5	40.8	1.159
Grass leaves	1548.9	1079.8	57.3	34.1	99.0	122.4	156.4	7.174
Other leaves	541.1	436.6	24.9	3.8	24.6	12.1	39.0	1.322
Gum	1289.5	276.0	78.7	14.2	15.5	94.1	811.1	15.563
Fruits	989.9	746.5	30.9	20.1	48.0	47.9	97.3	3.769
Grass meristems, etc.	327.2	253.8	9.6	3.9	16.7	16.3	27.1	1.086
Flowers	1111.0	869.1	14.9	18.2	33.9	41.6	133.2	3.987
Wood, etc.	8.1	0	0	0	0	0	0	0
Runners	41.2	35.8	0.3	0.1	1.6	1.2	2.1	0.081
Milk	3957.2	3403.2	11.9	193.9	63.3	0	281.0	12.755
Meat	408.9	275.1	5.1	14.4	93.1	10.2	10.9	2.546

Note: Calculated from values in table 6.5. Summary means are pooled rates: total mass divided by total time. Milk values calculated without Pooh's sample at 30–40 weeks. Food abbreviations as in table 6.2. Food types in panel B as defined in table 4.4.

Table 6.11 Cation Elements of Foods

Food	Units	P	K	Ca	Mg	Na	Units	Al	Ba	Fe	Sr	B	Cu	Zn	Mn	Cr
Salvadora persica, green leaves (SPLU, SPLX)	%	0.12	0.84	1.45	0.26	0.03	ppm	59.6	12.4	48.4	206	28.9	2.5	4.8	26.2	4.1
Intake (% × 3.42)	mg/day	0.41	2.87	4.96	0.89	0.10	μg/day	20.4	4.24	16.6	70.5	9.88	0.86	1.64	8.96	1.40
Acacia tortilis, green seeds (TODU)	%	0.10	0.37	0.15	0.10	0.01	ppm	5.3	3.0	11.8	18.4	6.8	2.4	5.6	6.2	1.3
Intake (% × 38.07)	mg/day	3.81	14.1	5.71	3.81	0.38	μg/day	20.2	11.4	44.9	70.1	25.9	9.14	21.3	23.6	5.0
Sporobolus cordofanus, corms (SMCX)	%	0.40	1.38	1.27	0.69	0.90	ppm	4290	55.4	4050	192	37.0	10.7	18.1	216	14.2
Intake (% × 7.34)	mg/day	2.94	10.1	9.3	5.06	6.61	μg/day	3149	40.7	2973	141	27.2	7.89	13.3	159	10.4
Total, above three foods	mg/day	7.16	27.1	20.0	9.76	7.09	μg/day	3190	56.3	3035	282	63.0	17.9	36.2	192	16.8
Intake, all foods (total/0.01247)	mg/day	574	2173	1604	783	569	mg/day	256	4.5	243	22.6	5.1	1.4	2.9	15.4	1.4
Requirement	mg/day	1175	939.6	234.9	391.5	861.3	mg/day	?	?	39.2	?	?	0.783	7.83	7.83	0.196
Intake/requirement	%	49	231	683	200	66	%	—	—	621	—	—	179	37	197	714
Tolerance	mg/day	5873	11745	7829	1958	13702	mg/day	392	7.9	1175	1175	59	313	392	392	392
Intake/tolerance	%	11	19	19	37	4	%	65	57	19.1	2	8	0.4	0.7	4	0.3

Note: Elemental analysis by emission spectroscopy (Altmann, Post, and Klein 1986). Requirements and tolerances from table A1.2.

Table 6.12 Nutritive Value of Actual and Optimal Diets versus Requirements and Tolerances

Diet	Mass Intake (g/day)	Mass Safety Margin	Water Intake (g/day)	Water Surplus	Mineral Intake (g/day)	Mineral Surplus	Mineral Safety Margin	Lipid Intake (g/day)	Lipid Surplus	Lipid Safety Margin
Actual	391.5	60.9%	293.4	13.8%	6.3	-7.2%	87.7%	15.2	924.7%	98.5%
Diet 6	1002.4	0	744.0	5.6%	17.3	0	86.8%	39.5	2565.5%	96.1%
Diet 7	1002.4	0	763.4	7.0%	17.3	0	86.8%	38.2	2478.4%	96.2%
Diet 8	174.1	82.6%	126.2	-47.2%	3.0		86.8%	7.1	377.0%	99.3%
Diet 9	114.0	88.6%	24.6	-89.7%	6.6	235.3%	55.7%	1.8	23.3%	99.8%
Diet 10	161.7	83.9%	105.9	-56.3%	5.7	104.2%	73.0%	4.0	171.4%	99.6%

Diet	Protein Intake (g/day)	Protein Surplus	Protein Safety Margin	Protein Rate (mg/min)	Fiber Intake (g/day)	Fiber Surplus	Fiber Safety Margin	Carbohydrate Intake (g/day)	Carbohydrate Safety Margin	Energy Intake (kJ/day)	Energy Surplus	Time Rate (kJ/min)	Time Minutes per Day	Time Safety Margin
Actual	11.1	181.1%	98.9%	63.5	10.1	2849.7%	99.0%	55.2	94.5%	1790.8	105.6%	12.10	240.5	44.8%
Diet 6	29.1	639.3%	97.1%	66.8	27.1	2892.7%	97.3%	145.4	85.5%	4694.6	439.0%	10.77	436.0	0%
Diet 7	*31.0*	686.3%	96.9%	*71.1*	26.7	3190.1%	97.3%	126.0	87.4%	4346.8	399.1%	9.97	436.0	0%
Diet 8	3.94	0	99.6%	77.5	3.4	1999.6%	99.7%	30.5	97.0%	871.0	0	17.13	*50.9*	88.3%
Diet 9	3.94	0	99.6%	47.8	9.6	2962.8%	99.0%	67.4	93.3%	1383.2	58.8%	20.65	*82.4*	81.1%
Diet 10	11.8	199.5%	98.8%	*112.6*	13.0	6554.3%	98.7%	21.4	97.9%	871.0	0	7.64	114.1	73.8%

Note: "Actual" diet is the seasonally adjusted mean (from tables 6.5 and 6.10). For each diet component, "surplus" and "safety margin" indicate relative slack values, that is, the percentage deviation of the diet from the baboons' requirement (minimum) and tolerance (maximum), respectively, for that food component (table 5.3). For water, the larger requirement (smaller excess) was used for each diet. Time requirement (minutes/day) for milk in optimal diets calculated from 0.20833 minutes/g, based on first approximation of milk intake without Pooh's 30–40 week sample. (Among optimal diets, only diets 6–8 include milk and are thus affected.) For each optimal diet, the value of the objective function is in italics. Objective function rates (diets 9, 10) and those of the actual diet are mass-weighted means; other rates are pooled means, that is, total energy or protein from all foods divided by total feeding time.

Table 7.1 Aggregate Primary Food Categories, Their Component Foods, and Sample Sizes

A. Component Foods of Aggregate Primary Categories

Code	Aggregate Primary Category	Component Foods	PLE Sample Size	Total Sample
A	*Lycium "europaeum"* blossoms	TFBS, TFBX	31	31
B	*Sporobolus rangei* corms	SKCS, SKCU, SKCX	430	444
C	*Acacia* seeds from dung	ACDD, TODD	22	23
D	Green grass seedheads	CDDU, GRDU, SMDU, SVDU	44	45
E	*Azima tetracantha* fruit	ATFR, ATFS, ATFU, ATFX	470	530
F	*Capparis tomentosa* fruit	CTFR, CTFS, CTFU, CTFX	30	37
G	*Salvadora persica* fruit	SPFR, SPFS, SPFU, SPFX	228	278
H	*Lycium "europaeum"* fruit	TFFR, TFFX	32	32
I	*Withania somnifera* fruit	WSFR, WSFU, WSFX	48	58
J	*Commicarpus pedunculosus* fruit	SFFS, SFFX	72	73
K	*Solanum dubium* fruit	SBFU, SBFX	20	21
L	*Trianthema ceratosepala* fruit	TCFU, TCFX	388	422
M	Fever tree gum from the tree	FTGR, FTGX, FTGU	1334	1504
N	Fever tree gum on other objects	FTHG, FTHP, FTHW, FTHX	380	417
O	*Azima tetracantha* leaves	ATLR, ATLU, ATLX	24	25
Q	*Lycium "europaeum"* leaves	TFLU, TFLX	109	113
R	Fever tree leaves	FTLU, FTLX	25	25
S	Arthropoda except grasshoppers	INKX, DUKU, BEKX, HYKX, DUZX, INKU	230	232
T	Animal material except milk and arthropods	BIKX, BINX, CGNX, GAKX, GAZX, GGKX, GSKX, HRKX, PCJX, POQX, SNKX, VTKX, WIAX, XXAX	37	219
U	*Cynodon nlemfuensis* stolons	CDSU, CDSR, CDSS, CDSX	88	88
V	*Sporobolus consimilis* blade bases	SCTU, SCTX	521	574
W	Wood except *Salvadora persica*	**W* exept SPW*	114	125
X	*Salvadora persica* wood	SPWR, SPWU, SPWX	59	64
Y	Petioles, pedicels, blade bases except *Sporobolus consimilis*	**T* except SCT*	173	174
BB	Blossoms	**B*	421	450
CC	Corms, grass or sedge	**C* except CBCX, ODCX, XXCX	3052	3232
DD	Seeds	**D*, XXXD, XXXG	3349	3530
FF	Fruits	**F*	1721	1895
GG	Green leaves, grass or sedge	CDLU, CPLU, GRLU, PJLU, SALU, SCLU, SKLU, SMLU, SSLU, SVLU	2495	2595
NN	Nongreen leaves, grass or sedge	CDLR, CDLS, CDLX, GRLX	8	8
OO	Leaves, not grass or sedge	**L* except those in categories GG or NN	486	536
PD	Pods, rinds	**P*	21	21
PT	Other plant parts (bulbs, roots, etc.)	ATEU, CBCX, FTQU, FTRU, FUEX, GREU, ODCX, SPEU, TOEU, XXCX, XXRX	29	31
RS	Rhizomes, stolons	**S*	120	120
UU	Gum, sap	**G* AND **H*	1716	1923
XX	Miscellaneous	**P*, **W*, **S*, EAEX and category PT	345	363
ZZ	Food or food taxon unknown	**XX, XXX* except XXXD and XXXG	961	1036

Table 7.1 (*Continued*)

B. Aggregate Primary Categories of Foods

Rank	Food Code	Food Number	Aggregate Primary Category	Frequency	Rank	Food Code	Food Number	Aggregate Primary Category	Frequency
1	GRCX	117		2539	26	TOPR	16		118
2	GRLU	121		1958	27	DUKU	271		115
3	FTGX	29	M	1493	28	TOBS	8		112
4	PCMX	270		1350	29	SPLX	168		101
5	FTDG	24		795	30	TFLX	140	Q	97
6	XXXD	246		769	31	SPLU	167		94
7	XXXG	247	DD	664	32	ATFU	56		90
8	SCTX	191		561	33	FTBS	23		80
9	CDLU	90		546	34	TODG	10		79
10	*XXXX*	250	ZZ	529	35	FTPU	39		74
11	SKCX	202	B	440	36	INKX	252		74
12	FTPR	38		435	37	SFFX	82	J	72
13	TTFX	216		417	38	SPFS	163		63
14	TCFX	214	L	414	39	SULU	209		56
15	TOPU	18		385	40	SPFR	162		55
16	*WAVG*	276		315	41	FTHW	34		53
17	FTHG	30		287	42	CDSX	95		51
18	XXXW	249	ZZ	277	43	WSFX	222		47
19	ATFR	54		218	44	SMLU	197		45
20	ATFX	57		215	45	SPWX	171		42
21	TOBR	7		185	46	XXLX	239		41
22	*GHKX*	267		180	47	FTHP	31		40
23	SMCX	194		176	48	*RMXX*	160		40
24	SPFU	164		142	49	*CDSU*	94		35
25	GRTU	126		137	50	*WAVP*	276		35

Note: Food codes as in table 6.2. An asterisk is a "wild card" that indicates every possible letter. PLE = product-limit estimate of bout length. Foods in panel B are rank-ordered by frequency, and only foods that were eaten at least thirty-four times in my samples are listed. Foods without aggregate primary categories were analyzed individually. In panel B, italics indicate noncore foods, and the number to the right of each food code is its row in table 4.1.

Table 7.2 Age Difference in Feeding Bout Lengths

Food or Food Code	30–40 Weeks		40–50 Weeks		50–60 Weeks		60–70 Weeks	
	Mean	SE	Mean	SE	Mean	SE	Mean	SE
All seeds	0.39	0.068	0.28	0.058	0.38	0.064	0.18	0.046
Arthropods	0.31	0.098	0.24	0.056	0.36	0.069	0.13	0.041
ATF	0.34	0.053	0.28*	0.036	0.29	0.033	0.57	0.105
ATFX	0.32*	0.066	0.18	0.023	0.23	0.034	0.36	0.029
Blossoms	0.23	0.096	0.81	0.117	1.41	0.394	1.22	0.330
DUKU	0.02	0.000	0.17	0.044	0.50	0.359	—	—
Fruit	0.56	0.058	0.48	0.052	0.45	0.056	0.59	0.088
FTDG	0.30	0.064	0.30	0.041	0.24	0.030	0.20	0.036
FTDR	0.41	0.052	0.33	0.045	0.47	0.064	0.31	0.041
FTG	0.31	0.033	0.39	0.044	0.32	0.048	0.43	0.051
FTGX	Run in category FTG							
GRCX	0.41	0.042	0.49	0.041	0.50*	0.062	0.40	0.053
GRLU	0.32	0.032	0.43	0.055	0.35	0.040	0.31	0.041
GRTU	0.27	0.094	0.32	0.041	0.39	0.063	0.53	0.075
INKX	0.42	0.123	0.29	0.055	0.30	0.092	0.07	0.049
Leftover	0.21	0.024	0.17	0.043	0.19	0.035	0.23	0.053
Mixed	0.37	0.067	0.31	0.041	0.46	0.074	0.72	0.184
Petioles	0.27	0.054	0.38	0.056	0.38	0.061	0.56	0.074
Seeds	0.39	0.068	0.28	0.058	0.38	0.064	0.18	0.046
SKC	0.82	0.097	0.91	0.110	0.59	0.091	0.62	0.060
SKCX	Run in category SKC							
SPF	0.76	0.095	0.67	0.159	0.57	0.117	0.19	0.044
SPFR	0.81*	0.132	1.62	0.540	—	—	0.07	0.000
SPW	0.24	0.068	0.34*	0.084	0.45	0.134	0.68	0.105
TOBR	0.08	0.014	0.15	0.118	—	—	0.04	0.021
TODR	0.75	0.185	0.92	0.244	0.46	0.082	0.22	0.049
TODU	1.71	0.353	1.33	0.207	1.16	0.105	0.13	0.100
WAVG	0.12	0.017	0.15	0.018	0.17	0.027	0.18	0.026
XXXD	0.10*	0.045	0.05	0.032	0.05	0.029	0.04	0.020
XXXG	Run in category "all seeds"							
XXXW	Run in category "leftover"							

Note: Table gives product-limit estimates of the mean duration and the standard error of the mean for each identified food and each aggregate primary food class that was eaten at least forty times in the overall sample of all infants (table 6.1) and that showed significant age differences ($p \leq 0.05$) on the Mantel/Cox test. An asterisk indicates a mean value that may be an underestimate because the longest bout was censored. A dash indicates that in my samples no infant of that age ate that food. Four-letter food abbreviations as in table 6.2. Seven foods that occurred at least forty times and whose primary class is an aggregate (table 4.1, col. d) were analyzed with their aggregate; a cross-reference is given for those for which the aggregate deviated significantly.

Table 7.3 Individual Feeding Bout Lengths within Ten-Week Age Blocks

Age 30–40 Weeks

Food or Food Code	Alice		Bristle		Eno		Fred		Hans		Ozzie		Pedro		Pooh		Summer		
	Mean	SE	Mean	SE	Mean	SE	Mean	SE	Mean	SE	Mean	SE	Mean	SE	Mean	SE	Mean	SE	CV(%)
All seeds	◆0.55	0.159	◆0.18	0.038	0.27	0.119	◆0.69	0.226	0.39	0.128	◆0.55	0.172	0.35	0.071	0.50	0	◆0.26	0.037	40.3
ATF	0.46	0.108	0.25	0.062	0.41	0.127	0.28	0.041	0.33*	0.075	0.24	0.055	0.22	0.150	0	0	0.02	0	48.4
Blossoms	0.16	0	0.29	0.031	0.03	0.010	0.65	0	0.58	0.117	0.44	0.132	0.08	0.014	0	0.065	1.36	0.572	96.3
Corms	0.32	0.050	0.18	0.030	0.50*	0.074	0.58	0.087	0.55	0.063	0.43	0.081	0.30	0.195	0.24	0	0.54	0.093	36.6
Fruit	0.55*	0.107	0.39	0.066	0.55	0.141	0.67	0.116	0.44	0.063	0.64*	0.087	0.39	0.202	0	0	0.67	0.173	22.1
FTG	◆0.44	0.072	0.19	0.038	0.40	0.168	◆0.35	0.052	0.21	0.039	◆0.45	0.100	0.25	0	0	0	0.23	0.043	34.0
FTGX	Run in category FTG																		
PCMX	◆1.46	0.558	◆1.34	0.306	1.80	0.474	◆1.90	0.659	1.92	0.391	◆1.53	0.509	3.09	0.639	0	0	◆1.23	0.340	32.9
SKC	0	0	0	0	0.89	0.452	◆0.75	0.136	0.83	0.160	◆1.34	0.558	0.07	0	0	0	0.92	0.169	51.6
SKCX	Run in category SKC																		
SPF	◆0.67*	0.166	0.44	0.325	0.05	0	0.78	0.142	0.13	0	◆0.72*	0.099	0.97	0	0	0	0	0	64.0
SPFU	◆0.81*	0.230	0.44	0.325	0.05	0	◆0.43	0.097	0	0	◆0.32	0.049	0.97	0	0	0	0	0	66.5
TTFX	0		0.49	0.094	0		0		0.29	0.072	0	0	0		0		0		25
XXXD	◆0.01	0.001	0.02	0	◆0.11*	0.079	◆0.22	0.080	0.05	0.026	◆0.17	0.044	0.20	0.183	0		0.08	0.060	85.7
XXXG	Run in category "all seeds"																		

Age 40–50 Weeks

Food or Food Code	Alice		Bristle		Eno		Fred		Ozzie		Pooh		Summer		
	Mean	SE	Mean	SE	Mean	SE	Mean	SE	Mean	SE	Mean	SE	Mean	SE	CV(%)
All seeds	◆0.31	0.090	◆0.26	0.097	0.39	0.072	◆0.15	0.035	◆0.22	0.054	0.46	0.139	0.29	0.103	33.5
Arthropods	0.33	0.069	0.32	0.124	0	0	0.18	0.125	0.26	0.45	0.41		0.19	0.047	31.5
Blossoms	0.63*	0.134	0.22	0	1.37	0.526	1.23	0.297	0.54	0.081	1.46	0.764	◆0.26	0.115	64.6
DUKU	◆0.23	0.133	0	0	0	0	◆0.02	0.007	◆0.24	0.046	0	0	0	0	76.1
Fruit	0.47	0.095	0.36	0.047	1.00	0.224	0.54*	0.101	0.39	0.073	0.72	0.146	0.40*	0.055	41.8
FTG	◆0.57	0.088	◆0.24	0.044	0.41	0.090	◆0.54	0.093	◆0.51	0.083	0.31	0.107	◆0.29	0.041	32.4
FTGX	Run in category FTG														

Food or Food Code	Alice Mean	Alice SE	Dotty Mean	Dotty SE	Eno Mean	Eno SE	Fred Mean	Fred SE	Ozzie Mean	Ozzie SE	Pooh Mean	Pooh SE	Striper Mean	Striper SE	Summer Mean	Summer SE	CV (%)
FTDU	◆1.93	0.780	0	0.062	1.14	0.277	◆0.18	0.093	◆0.96*	0.273	0	0	0	0			68.2
Green leaves	0.62	0.121	0.43	0.045	0.39	0.060	0.43	0.075	0.37	0.054	0.33	0.060	0.26	0.029			27.8
GRTU	0	0	◆0.31	0.059	0.07	0.082	0	0.082	0	0.087	0	0	◆0.34	0.073			61.7
Gums	0.53	0.075	0.27	0.073	0.41	0.117	0.51	0.082	0.53	0.037	0.37	0.085	0.31	0.046			25.7
Leftovers	◆0.10	0.029	◆0.20	0.055	0.74	0.381	◆0.12	0.026	◆0.08	0.044	0.26	0.045	◆0.17	0.039			96.3
Meat	0.26	0.088	0.13	0.062	0.07*	0.058	0.19	0.120	0.24	0	0.27	0.107	0.12	0.034			42.6
Petioles	0	0	0.38	0	0.07	0	0	0	0	0	0	0	0.40	0.85			65.3
XXXG	Run in category "all seeds"																
XXXW	Run in category "leftovers"																

Age 50–60 Weeks

Food or Food Code	Alice Mean	Alice SE	Dotty Mean	Dotty SE	Eno Mean	Eno SE	Fred Mean	Fred SE	Ozzie Mean	Ozzie SE	Pooh Mean	Pooh SE	Striper Mean	Striper SE	Summer Mean	Summer SE	CV (%)
All seeds	0.15	0.050	1.10	0.161	0.25	0.048	0.36*	0.160	0.17	0.064	0.16	0.040	0.77	0.132	0.32	0.092	84.0
CDLU	◆0.59	0.165	0.17	0.063	0.53	0.083	◆0.27	0.150	◆0.10	0.000	0.67	0.226	0.43	0.185	0	0	55.4
Corms	0.42	0.159	0.76*	0.138	0.79	0.117	0.74	0.159	0.38	0.156	0.62	0.147	0.48	0.070	0.37	0.036	31.3
FTG	0.59	0.141	0.20	0.059	0.38	0.077	1.08	0.533	0.48	0	0.36	0.106	0.15	0.032	0.23	0.051	69.1
FTGX	Run in category FTG																
FTH	0	0	0.29	0.095	0.57	0.097	0	0	0.82	0	0.61	0.126	0.17	0.062	0.18	0.045	60.4
GRCX	◆0.34	0.164	◆0.78*	0.149	0.58	0.117	◆0.58	0.183	◆0.38	0.156	0.64	0.159	◆0.44	0.069	0.34	0.039	31.4
Gum	0.59	0.141	0.25	0.066	0.45	0.065	1.08	0.533	0.65	0.167	0.54	0.099	0.15	0.038	0.22	0.044	61.3
Leftovers	0.28	0.067	0.21	0.054	0.24	0.070	0.09*	0.028	0	0	0.10	0.048	0.23	0.059	0.10	0.029	44.5
Mixed	0.42	0.139	0.37	0.199	0.44	0.090	0.21	0.058	1.14	0.582	0.17*	0.045	0.19	0.067	0.82	0.217	72.9
Other leaves	0.48	0.109	0.26	0.042	0.65	0.242	0	0	0.36	0.143	0.09	0.033	0.18	0.063	0.28	0.102	57.4
SCT	0.82*	0.246	0	0	0.58	0.058	0.80	0.176	0	0	0.30	0.042	0	0	0.36	0.055	42.1
SCTX	◆0.82*	0.246	0	0	0.58	0.058	◆0.80	0.176	0	0	0.30	0.042	0	0	0.36	0.055	42.1
SKC	0	0	0	0	0.58	0.212	0.88	0.221	0	0	0	0	0	0	0.43	0.056	36.4
SKCX	Run in category SKC																
TCF	0.18	0.024	0	0	0.76	0.152	0	0	0	0	0.33	0.141	0	0	0.69	0.225	57.1
XXXD	0.05	0.025	◆0.13	0.107	0.05	0.029	0	0	0.06	0.043	0.03	0.015	◆0.14	0.058	0.03	0.011	82.2

Table 7.3 (*Continued*)

| | Age 50–60 Weeks | | | | | | | | | | | | | | | |
| | Alice | | Dotty | | Eno | | Fred | | Ozzie | | Pooh | | Striper | | Summer | | |
Food or Food Code	Mean	SE	Mean	SE	Mean	SE	Mean	SE	Mean	SE	Mean	SE	Mean	SE	Mean	SE	CV (%)
XXXG	Run in category "all seeds"																
XXXW	Run in category "leftovers"																

| | Age 60–70 Weeks | | | | | | | | | | |
| | Alice | | Fred | | Ozzie | | Pooh | | |
Food or Food Code	Mean	SE	Mean	SE	Mean	SE	Mean	SE	CV (%)
All seeds	◆0.23	0.045	◆0.20	0.086	◆0.16	0.039	0.10	0.046	32.6
GRCX	◆0.32	0.052	◆0.47	0.081	◆0.49	0.078	0.30	0.089	25.0
Other leaves	0.35	0.049	0.22	0.054	0.29	0.084	0.63	0.183	48.2
PCMX	◆2.75	1.054	◆1.14	0.418	◆3.42	0.821	0.26	0.081	76.6
SCT	0.63	0.077	0.44	0.054	0.60	0.074	0.38	0.108	23.7
SCTX	◆0.63	0.077	◆0.44	0.054	◆0.60	0.074	0.38	0.108	23.7
SKC	◆0.68	0.068	0	0	◆0.33	0.068	0.64	0.140	34.9
SKCX	Run in category SKC								
XXXG	Run in category "all seeds"								

Note: The table lists only those foods and aggregate food classes for which sample size was adequate (see text) and on which individuals differed significantly ($p \le 0.05$, Mantel/Cox test) at the indicated age. Values are product-limit estimates of the mean feeding bout duration (minutes) and the standard error of the mean. Asterisk as in table 7.2. At each age, omitted infants were not sampled at that age. A mean of zero indicates that in my samples on the infant during this age, it did not feed on this food. (The statistical tests are based only on those that did.) Food codes are explained in table 6.2. CV is the coefficient of variation ($100 \, \sigma/\mu$). Any individual food (distinguished by a four-letter abbreviation) that had samples large enough to use for table 7.5 is indicated by diamonds (◆) placed in the compared yearlings' columns and next to that food's mean bout lengths. A food's means are listed in its own row if it was tested separately for this table, otherwise in the row of its aggregate class.

Table 7.4 Feeding Bout Lengths for the Six "Well-Sampled" Infants at 30–70 Weeks of Age

Food Code	Alice		Eno		Fred		Ozzie		Pooh		Summer		
	Mean	SE	Mean	SE	Mean	SE	Mean	SE	Mean	SE	Mean	SE	CV (%)
ATFR	0.53*	0.146	0.34	0.087	0.42	0.169	0.46	0.108	0.66	0.142	0.24	0.065	33.2
ATFX	0.21	0.045	0.33	0.081	0.19	0.033	0.16	0.030	0.30	0.059	0.14*	0.022	34.6
DUKU	0.21	0.080	0.02	0	0.02	0.007	0.25	0.048	0.12	0	0	0	85.5
FTBS	0.65*	0.146	0	0	1.26	0.322	0.57	0.110	1.55	0.562	0	0	47.2
FTGX	0.48	0.054	0.39	0.063	0.48	0.056	0.45	0.053	0.32	0.067	0.26	0.034	22.9
FTHP	0.30	0.175	1.14	0.287	0.02	0	0.82	0.120	0.18	0.113	0	0	95.6
FTDR	0.32	0.048	0.43	0.047	0.36	0.081	0.26	0.040	0.29	0.072	0.40	0.074	19.0
FTDU	1.93	0.780	1.07	0.241	0.32*	0.153	0.96*	0.273	0.72	0.276	0	0	59.4
GRCX	0.36	0.046	0.55	0.059	0.49	0.062	0.44	0.045	0.51	0.066	0.37	0.043	17.0
GRLU	0.46	0.075	0.37	0.044	0.39	0.059	0.31	0.040	0.36	0.037	0.27	0.030	18.3
GRTU	0.53	0.075	0.43	0.083	0	0	0	0	0	0	0.31	0.056	26.0
SCTX	0.67	0.072	0.53	0.067	0.49	0.051	0.62	0.085	0.34	0.059	0.47	0.066	22.5
SPLU	0.23	0.041	0.37	0.096	0.29	0.165	0.59	0.140	0.05	0	0.52	0.152	57.7
SPWX	0.66	0.123	0.33*	0.127	0.12	0.018	0.18	0.043	0.21	0.180	0.35	0.033	62.8
SULU	0.34	0.115	0.31	0.092	0.06	0.035	0.22	0.059	0.06	0	0.32	0.200	59.3
TOBR	0	0	0.15	0.118	0.32	0	0.02	0.005	0	0	0	0	92.1
XXXG	0.09	0.032	0.08	0.049	0.09	0.025	0.11	0.036	0.14	0.038	0.05	0.022	32.3

Note: The table lists only those core foods that showed significant differences. Values are product-limit estimates of mean bout durations (minutes) and their standard errors. Asterisks as in table 7.2.

Table 7.5 Individual Differences in Feeding Bout Lengths for Infants Born at the Same Time of Year

Food Code	Age Block (weeks)	Group I Sample Size AC	FR	OZ	Group I Significance	Group I Order (increasing)	Group II Sample Size BS	SR	Group II Significance	Group II Order (increasing)	Group III Sample Size DT	ST	Group III Significance	Group III Order (increasing)
ATFR	40–50	7	11	9	NS		19	6	NS		—	—		
ATFX	30–40	7	6	13	NS		—	—			—	—		
CDLU	40–50	22	30	27	NS		30	21	NS		—	—		
	30–40	62	47	40	NS		—	—			—	—		
	40–50	18	29	21	NS		—	—			—	—		
	♦ 50–60	4	12	59	NS		—	—			—	—		
DUKU	♦ 40–50	2	36	70	**	FR-AC-OZ	—	—			—	—		
FTBS	40–50	23	13	14	NS		—	—			—	—		
FTDG	30–40	16	7	21	*	AC-OZ-FR	—	—			—	—		
	40–50	58	50	108	*	FR-OZ-AC	—	—			—	—		
	60–70	11	20	72	NS		—	—			—	—		
FTDR	30–40	—	—	—			15	46	NS		—	—		
	40–50	—	—	—			12	14	NS		—	—		
	60–70	43	33	49	NS		—	—			—	—		
FTGX	30–40	46	51	43	NS		136	79	NS		—	—		
	♦ 40–50	68	47	56	NS		—	—			36	35	NS	
	50–60	124	73	124	NS		—	—			—	—		
FTHG	40–50	9	6	16	NS		5	10	NS		—	—		
	60–70	11	5	17	NS		—	—			—	—		
GRCX	30–40	62	72	73	NS		—	—			—	—		
	40–50	40	77	129	*	AC-FR-OZ	9	57	NS		—	—		
	50–60	12	16	4	NS		185	5	NS		209	528	**	ST-DT
	60–70	118	39	104	NS		—	—			—	—		
GRLU	30–40	—	—	—			142	26	NS		—	—		
	40–50	109	150	48	NS		50	121	**	SR-BS	—	—		
	50–60	33	4	3	NS		—	—			—	—		
	60–70	136	165	107	NS		—	—			—	—		
GRTU	♦ 40–50	—	—	—			38	19	NS		—	—		
PCMX	♦ 30–40	14	11	50	NS		87	112	NS		—	—		
	40–50	32	43	74	NS		77	71	NS		—	—		
	50–60	22	11	22	NS		—	—			—	—		

Food	Bout length	AC	FR	OZ	order	Grp I sig	BR	SR	II sig (a)	II sig (b)	DT	ST (ST‑DT)	III sig
SCTX	♦60–70	14	49	31	FR-AC-OZ	*	14	—	—	—	—	—	—
	40–50	11	25	29		NS	15	37	NS	NS	—	—	—
	50–60	6	21	0		NS	—	12	NS	—	—	—	—
	♦60–70	61	114	45		*	—	—	—	—	—	—	—
SKCX	♦30–40	0	79	7	FR-OZ-AC	NS	—	—	—	—	—	—	—
	40–50	0	22	25		NS	—	—	—	—	—	—	—
	♦60–70	104	0	22		**	—	—	—	—	—	—	—
SMCX	50–60	—	—	10	OZ-AC	—	—	—	—	—	17	60	NS
SPFS	30–40	5	23	17		NS	—	—	—	—	—	—	—
SPFR	30–40	0	21	33		NS	—	—	—	—	—	—	—
SPFU	♦30–40	18	25	—	OZ-FR-AC	*	—	—	—	—	37	10	NS
SPLX	50–60	—	—	—		—	—	—	—	—	—	—	—
TCFX	30–40	5	3	20		NS	5	—	NS	—	—	—	—
	40–50	27	38	16		NS	—	39	NS	—	—	—	—
	60–70	2	20	34		NS	—	—	—	—	—	—	—
TOBS	60–70	8	22	12		NS	—	—	—	—	—	—	—
TODU	30–40	19	8	16		NS	—	—	—	—	88	104	*
	50–60	—	—	—		—	—	—	—	—	—	—	—
TTFX	40–50	—	—	—		—	38	153	NS	NS	—	—	—
WAVG	30–40	—	—	—		—	26	9	NS	NS	—	—	—
	40–50	7	21	14		NS	—	—	—	—	—	—	—
	60–70	19	9	27		*	—	—	—	—	—	—	—
XXXD	30–40	7	7	22	OZ-AC-FR	**	17	26	NS	—	—	—	—
	40–50	42	22	98		NS	—	—	—	—	—	—	—
	50–60	—	—	—		—	—	—	—	—	—	—	—
XXXG	60–70	23	34	69		NS	—	—	—	—	10	22	NS
	♦30–40	8	6	19		NS	19	6	NS	NS	—	—	—
	♦40–50	47	60	61		NS	43	20	NS	NS	—	—	—
	♦60–70	18	25	28		NS	—	—	—	—	—	—	—
XXXW	♦40–50	17	11	43		NS	12	25	NS	—	—	—	—

Note: There are three birth groups: I: Alice (AC), Fred (FR), and Ozzie (OZ), II: Bristle (BR) and Summer (SR), and III: Dotty (DT) and Striper (ST). The table lists only those foods for which sample size was adequate (see text). Foods marked with a diamond (♦) were significantly different when all available infants were compared (table 7.3). A dash indicates that the sample was too small. NS = not significant at 0.05 level; * = significant at 0.05 level; ** = significant at 0.01 level. For significant cases, "order" gives the ordering of the infants by increasing bout length. (For the bout length values, see table 7.2.) Food acronyms are decoded in table 6.2. Foods whose 'primary category' (table 4.1, col. d) is an aggregate were separated from the other foods in their aggregate before analysis.

Table 7.6 Feeding Bout Rates, by Individual and Age

A. Age Block 30–40 Weeks

Food Code	AC	BS	DT	EN	FR	HN	OZ	PD	PH	SR	ST	Total	All	CV
PCMX	44.4	110.5	—	137.0	16.4	120.2	72.2	29.0	0	133.2	—	88.5	v	71
GRCX	119.0	9.9	—	83.1	102.7	107.0	109.6	4.1	504.4	67.2	—	76.5	v	121
GRLU	1.7	182.4	—	109.2	2.7	45.0	7.7	37.3	0.0	143.3	—	69.7	v	117
FTGX	94.2	48.6	—	12.3	82.2	57.5	69.6	1.3	0	35.8	—	49.9	v	77
TODU	60.4	0	—	0	13.7	10.3	25.7	0	0	0	—	11.8	v	164
SKCX	0	0	—	4.6	113.7	18.4	9.0	1.3	0	21.2	—	20.8	v	195
SCTX	3.5	17.6	—	15.3	13.7	15.4	2.5	26.3	0	50.4	—	18.9	v	95
TCFX	8.8	5.5	—	44.6	5.4	27.3	27.0	1.3	0	48.1	—	21.9	v	99
CDLU	117.3	34.2	—	3.0	69.8	2.2	51.5	6.9	0	24.6	—	33.2	v	114
XXXG	14.2	21.0	—	33.8	9.5	23.6	24.5	20.7	0	6.7	—	19.3	s	60
TOBS	0	0	—	0	0	0	0	8.3	63.0	19.0	—	3.4	v	217
FTDG	30.2	6.6	—	23.0	10.9	5.1	28.3	2.7	0	10.0	—	13.1	v	95
TTFX	0	42.0	—	0	0	30.9	0	0	0	0	—	12.1	v	201
FTDR	5.3	18.7	—	0	0	14.0	1.2	52.6	0	52.6	—	18.9	v	135
FTHG	3.5	8.8	—	20.0	2.7	25.0	11.6	1.3	0	11.2	—	11.9	v	91
SMCX	1.7	0	—	0	1.3	5.9	9.0	0	0	0	—	2.5	v	162
ATFR	5.3	1.1	—	32.3	0	14.7	1.2	1.3	0	0	—	7.1	v	173
TODR	24.8	0	—	20.0	1.3	4.4	7.7	4.1	0	1.1	—	6.6	v	129
FTDU	0	0	—	6.1	0	0	0	0	0	0	—	0.6	v	299
FTBS	1.7	0	—	0	0	0	16.7	1.3	0	0	—	2.1	v	269
SPFU	44.4	2.2	—	1.5	42.4	0	43.8	1.3	0	0	—	14.2	v	141
SPFS	8.8	0	—	0	45.2	0	23.2	0	0	0	—	8.4	v	183
XXXW	5.3	3.3	—	9.2	1.3	9.5	5.1	11.0	0	2.2	—	6.0	n	
SPFR	1.7	0	—	0	31.5	0	27.0	0	0	0	—	6.8	v	191
GRTU	0	1.1	—	0	0	11.0	0	0	0	0	—	2.4	v	270
ATFX	17.7	0	—	23.0	9.5	8.1	19.3	0	0	1.1	—	8.9	v	105
XXXD	12.4	3.3	—	35.4	9.5	26.5	28.3	5.5	0	8.9	—	16.6	v	86
TFLX	0	7.7	—	9.2	1.3	0	3.8	1.3	0	0	—	2.7	v	136
SPLU	0	11.0	—	3.0	1.3	5.9	2.5	4.1	0	8.9	—	5.1	n	
ATFU	8.8	22.1	—	0	10.9	2.2	6.4	1.3	0	0	—	6.3	v	127
SPLX	5.3	0	—	3.0	0	0.7	2.5	1.3	0	2.2	—	1.6	n	
FTHP	3.5	3.3	—	3.0	0	3.6	3.8	0	0	0	—	2.2	n	
SMLU	0	0	—	0	0	4.4	0	0	0	0	—	0.9	s	300

Food Code	AC	BS	DT	EN	FR	HN	OZ	PD	PH	SR	ST	Total	All	CV
FTHW	3.5	0	—	0	0	0.7	0	0	0	0	—	0.4	n	
WSFX	0	0	—	0	0	6.6	2.5	0	0	2.2	—	1.9	n	
COCX	0	0	—	0	0	0	0	0	0	0	—	0	n	
INKX	1.7	0	—	6.1	1.3	2.9	2.5	0	0	2.2	—	2.1	s	299
DUKU	0	0	—	4.6	0	0	0	0	0	0	—	0.4	n	
CDSX	5.3	7.7	—	0	1.3	8.1	0	0	0	5.6	—	4.0	n	
SPWX	0	0	—	1.5	5.4	1.4	1.2	0	0	0	—	1.2	v	299
SFFX	0	14.3	—	0	0	15.4	0	0	0	0	—	3.1	v	129
TODG	7.1	0	—	15.3	1.3	0	5.1	0	0	0	—	4.8	n	
TOBR	0	0	—	0	0	0	0	217.4	0	0	—	23.7	n	
XXWX	1.7	3.3	—	1.5	0	1.4	1.2	1.3	0	0	—	1.2	s	299
SULU	3.5	3.3	—	7.6	1.3	9.5	7.7	1.3	0	2.2	—	4.9	n	
XXLX	1.7	9.9	—	1.5	0	1.4	1.2	0	0	1.1	—	2.4	v	146
GRDU	0	0	—	0	0	2.9	9.0	0	0	0	—	0.6	n	
SPFX	5.3	0	—	0	1.3	0.7	0	0	0	0	—	1.8	n	
TFBS	0	4.4	—	4.6	0	0	0	0	0	0	—	1.0	n	
ACOD	1.7	0	—	0	0	0	1.2	0	0	0	—	0.3	v	175
CTFX	0	0	—	0	0	0	0	0	0	2.2	—	0.3		
SBFX	0	0	—	0	0	5.9	0	0	0	0	—	1.2		
Total	677.2	604.8	—	675.8	612.6	657.5	674.4	446.0	567.4	664.1	—	628.4		300

B. Age Block 40–50 Weeks

Food Code	AC	BS	DT	EN	FR	HN	OZ	PD	PH	SR	ST	Total	All	CV
PCMX	33.9	72.7	—	83.4	43.3	—	73.8	—	39.4	90.6	—	61.6	v	36
GRCX	39.4	167.0	—	108.2	70.7	—	120.6	—	135.4	5.6	—	94.2	v	61
GRLU	111.1	44.1	—	141.0	134.4	—	43.2	—	6.4	141.7	—	87.3	v	63
FTGX	69.7	122.0	—	86.8	53.0	—	58.5	—	17.0	102.0	—	73.0	v	47
TODU	0	6.0	—	2.2	0	—	0.9	—	21.3	11.3	—	5.5	v	132
SKCX	0	29.4	—	10.1	21.2	—	23.4	—	3.2	0	—	13.3	v	97
SCTX	11.9	12.9	—	13.5	23.0	—	29.7	—	1.0	18.1	—	16.1	v	57
TCFX	26.6	6.9	—	28.2	36.2	—	14.4	—	14.9	12.4	—	20.0	v	52
CDLU	17.4	2.5	—	25.9	25.6	—	18.9	—	32.0	2.2	—	17.6	v	64
XXXG	43.1	38.0	—	22.5	53.9	—	55.8	—	12.8	22.6	—	36.9	v	46
TOBS	0	0	—	9.0	0	—	0	—	0	0	—	1.1	v	264
FTDG	55.0	11.2	—	34.9	46.8	—	98.1	—	38.4	5.6	—	42.6	v	73
TTFX	0	32.8	—	0	0	—	0	—	0	176.8	—	26.9	v	220

Table 7.6 (*Continued*)

Food Code	AC	BS	DT	EN	FR	HN	OZ	PD	PH	SR	ST	Total	All	CV
FTDR	0	10.3	—	46.2	0	—	4.5	—	0	17.0	—	10.1	v	149
FTHG	11.0	6.9	—	12.4	5.3	—	14.4	—	10.6	13.6	—	10.4	n	100
SMCX	0	7.7	—	5.6	4.4	—	0.9	—	0	2.2	—	3.0	v	81
ATFR	7.3	19.0	—	0	11.5	—	10.8	—	29.8	9.0	—	8.7	v	157
TODR	0.9	5.1	—	4.5	0	—	6.3	—	0	0	—	6.3	v	147
FTDU	8.2	0	—	31.5	5.3	—	8.1	—	4.2	0	—	7.2	v	121
FTBS	25.7	0.8	—	0	12.3	—	13.5	—	19.2	0	—	8.4	v	150
SPFU	2.7	0	—	0	0.8	—	0	—	7.4	10.2	—	4.4	v	264
SPFS	0	0	—	0	0	—	0	—	3.2	0	—	0.9	v	81
XXXW	15.6	10.3	—	5.6	9.7	—	38.7	—	8.5	28.3	—	16.1	v	236
SPFR	0	0	—	0	0.8	—	0	—	0	0	—	1.2	v	171
GRTU	0	33.7	—	1.1	0	—	0	—	5.3	21.5	—	8.2	v	74
ATFX	22.0	7.7	—	1.1	29.1	—	26.1	—	25.6	11.3	—	15.4	v	76
XXXD	39.4	14.7	—	16.9	19.4	—	88.2	—	1.0	29.4	—	34.0	v	109
TFLX	0.9	6.0	—	2.2	0.8	—	15.3	—	1.0	13.6	—	5.7	v	103
SPLU	13.7	10.3	—	4.5	0	—	2.7	—	3.2	2.2	—	5.1	v	
ATFU	1.8	3.4	—	0	0	—	2.7	—	3.2	4.5	—	2.2	n	
SPLX	2.7	3.4	—	3.3	0	—	1.8	—	11.7	1.1	—	2.2	n	
FTHP	0	0.8	—	0	0.8	—	0	—	0	0	—	1.8	v	225
SMLU	0	14.7	—	0	0	—	0	—	8.5	2.2	—	2.6	v	226
FTHW	5.5	0	—	0	0	—	2.7	—	0	0	—	2.3	v	143
WSFX	0	7.7	—	2.2	0	—	0	—	0	1.1	—	1.6	v	179
COCX	0	4.3	—	0	0	—	0	—	1.0	0	—	0.6	v	264
INKX	4.5	3.4	—	0	11.5	—	5.4	—	0	6.8	—	4.8	s	81
DUKU	1.8	0	—	0	31.8	—	63.9	—	1.0	0	—	15.1	v	179
CDSX	1.8	2.5	—	4.5	0	—	0.9	—	1.0	1.1	—	1.6	n	
SPWX	0.9	1.7	—	3.3	0	—	1.8	—	1.0	2.2	—	1.5	n	
SFFX	0	13.8	—	0	0	—	0	—	1.0	3.4	—	2.6	v	210
TODG	4.5	0	—	4.5	0.8	—	15.3	—	7.4	0	—	4.7	v	116
TOBR	0	0	—	7.8	0	—	0	—	0	0	—	0.9	v	264
XXWX	4.5	1.7	—	2.2	1.7	—	5.4	—	0	0	—	2.3	n	
SULU	1.8	5.1	—	2.2	0.8	—	0.9	—	0	3.4	—	1.6	n	
XXLX	0.9	1.7	—	0	1.7	—	2.7	—	0	5.6	—	1.5	n	
GRDU	0	0	—	0	0	—	0	—	0	0	—	0.6	v	264
SPFX	0.9	0	—	0	0	—	0.9	—	1.0	0	—	0.4	n	

	AC	BS	DT	EN	FR	HN	OZ	PD	PH	SR	ST	Total	All	CV
TFBS	0.9	0	—	0	0	—	19.8	—	0	0	—	3.1	v	251
ACDD	0	0	—	2.2	0	—	0.9	—	0	0	—	0.4	n	
CTFX	1.8	0	—	0	0	—	0.9	—	0	0	—	0.4	n	
SBFX	0	0.8	—	0	0	—	0	—	0	3.4	—	0.5	s	208
Total	591.3	745.4	—	731.0	658.1	—	893.2	—	476.9	783.3	—	699.0		

C. Age Block 50–60 Weeks

Food Code	AC	BS	DT	EN	FR	HN	OZ	PD	PH	SR	ST	Total	All	CV
PCMX	56.7	—	22.1	36.9	48.3	—	102.0	—	36.1	75.1	22.0	44.5	v	55
GRCX	29.5	—	541.4	44.3	74.7	—	17.7	—	34.9	184.9	1025.9	222.6	v	147
GRLU	74.9	—	8.8	147.7	21.9	—	17.7	—	208.2	21.0	5.5	91.2	v	119
FTGX	22.7	—	83.9	70.1	35.1	—	4.4	—	18.7	109.7	69.7	59.0	v	71
TODU	0	—	254.1	5.1	87.9	—	0	—	0	25.5	231.2	56.2	v	171
SKCX	0	—	0	5.1	101.1	—	0	—	0	49.6	0	12.7	v	185
SCTX	18.1	—	0	8.8	0	—	0	—	11.2	27.0	0	14.8	v	163
TCFX	18.1	—	0	19.2	0	—	0	—	6.2	16.5	1.8	10.8	v	113
CDLU	15.8	—	24.3	37.6	8.7	—	4.4	—	37.4	0	5.5	22.2	v	88
XXXG	36.3	—	8.8	47.2	4.3	—	44.3	—	82.2	36.0	9.1	40.3	v	77
TOBS	13.6	—	0	0	13.1	—	79.8	—	0	0	0	5.7	v	206
FTDG	34.0	—	2.2	62.0	0	—	17.7	—	205.7	18.0	14.6	61.3	v	153
TTFX	0	—	0	93.0	0	—	0	—	0	3.0	0	27.1	v	272
FTDR	20.4	—	2.2	45.8	8.7	—	17.7	—	6.2	10.5	14.6	20.8	v	85
FTHG	0	—	17.6	34.7	0	—	4.4	—	24.9	19.5	14.6	20.5	v	85
SMCX	2.2	—	37.5	29.5	0	—	0	—	2.4	15.0	111.9	27.8	v	153
ATFR	36.3	—	0	34.7	0	—	4.4	—	7.4	13.5	0	16.7	v	126
TODR	0	—	0	0.7	4.3	—	0	—	2.4	19.5	11.0	4.8	v	147
FTDU	0	—	0	0	39.5	—	0	—	8.7	0	0	3.3	v	230
FTBS	0	—	0	0	0	—	0	—	6.2	4.5	0	1.0	v	282
SPFU	0	—	0	5.1	0	—	0	—	0	0	0	2.1	n	
SPFS	0	—	0	0	0	—	0	—	0	0	0	0	n	
XXXW	4.5	—	6.6	16.2	8.7	—	0	—	41.1	27.0	11.0	18.2	v	94
SPFR	0	—	0	0	0	—	0	—	0	0	0	0	n	
GRTU	0	—	6.6	13.2	0	—	0	—	18.7	12.0	5.5	5.5	v	185
ATFX	4.5	—	6.6	1.4	0	—	0	—	0	24.0	0	8.7	v	117
XXXD	34.0	—	24.3	42.1	4.3	—	22.1	—	167.0	30.0	40.3	56.2	v	111
TFLX	4.5	—	11.0	1.4	0	—	35.4	—	0	1.5	9.1	4.8	v	150
SPLU	2.2	—	4.4	2.2	0	—	0	—	0	1.5	1.8	1.6	n	

Table 7.6 (*Continued*)

Food Code	AC	BS	DT	EN	FR	HN	OZ	PD	PH	SR	ST	Total	All	CV
ATFU	2.2	—	8.8	5.9	0	—	0	—	3.7	1.5	7.3	4.4	n	
SPLX	0	—	108.2	1.4	0	—	4.4	—	3.7	0	22.0	14.2	v	214
FTHP	0	—	2.2	3.6	0	—	0	—	1.2	0	9.1	2.5	n	
SMLU	0	—	8.8	11.8	0	—	0	—	0	0	0	4.2	v	187
FTHW	0	—	4.4	0.7	4.3	—	0	—	31.1	0	9.1	7.0	v	190
WSFX	9.0	—	0	4.4	0	—	0	—	0	15.0	0	4.4	v	133
COCX	0	—	0	6.6	0	—	0	—	0	3.0	0	2.3	n	
INKX	2.2	—	2.2	0.7	0	—	0	—	8.7	7.5	3.6	3.6	n	
DUKU	2.2	—	0	0	0	—	4.4	—	1.2	0	0	0.6	n	
COSX	9.0	—	0	0.7	0	—	8.8	—	0	0	0	1.4	v	175
SPWX	0	—	0	4.4	0	—	0	—	0	3.0	1.8	1.9	n	
SFFX	0	—	0	18.4	0	—	0	—	0	10.5	0	6.7	v	194
TODG	2.2	—	0	0	0	—	0	—	8.7	0	1.8	1.9	v	188
TOBR	0	—	0	0	0	—	0	—	0	0	0	0	n	
XXWX	0	—	0	0	0	—	0	—	0	6.0	5.5	1.4	n	
SULU	0	—	2.2	2.9	0	—	0	—	1.2	0	0	1.2	n	
XXLX	0	—	8.8	2.2	0	—	0	—	0	3.0	1.8	2.1	n	
GROU	0	—	0	11.8	0	—	0	—	0	0	3.3	3.3	v	282
SPFX	0	—	0	0	0	—	0	—	1.2	0	3.6	0.6	n	
TFBS	0	—	0	0	0	—	0	—	0	0	0	0	n	
ACDD	0	—	0	0	0	—	4.4	—	4.9	0	0	1.0	n	
CTFX	4.5	—	0	0.7	0	—	8.8	—	0	0	0	1.0	n	
SBFX	0	—	0	5.9	0	—	0	—	0	0	0	1.6	s	282
Total	460.9	—	1202.3	888.0	466.2	—	403.6	—	992.3	795.4	1672.0	930.1		

D. Age Block 60–70 Weeks

Food Code	AC	BS	DT	EN	FR	HN	OZ	PD	PH	SR	ST	Total	All	CV
PCMX	13.4	—	—	—	49.2	—	28.7	—	24.8	—	—	29.5	v	51
GRCX	112.7	—	—	—	36.2	—	94.8	—	72.1	—	—	80.8	v	42
GRLU	125.3	—	—	—	156.0	—	96.5	—	45.0	—	—	115.9	v	45
FTGX	119.9	—	—	—	77.0	—	117.4	—	18.0	—	—	95.0	v	57
TODU	1.7	—	—	—	0	—	0	—	0	—	—	0.5	n	

SKCX	94.8	0	20.0	38.3	38.5	v	106
SCTX	58.1	110.5	44.3	33.8	66.0	v	55
TCFX	1.7	19.5	32.1	31.5	19.5	v	67
CDLU	3.5	13.0	53.9	31.5	24.8	v	87
XXXG	16.1	24.1	24.3	18.0	21.1	v	21
TOBS	9.8	25.0	13.9	0	14.2	v	85
FTOG	9.8	18.5	62.6	20.2	29.5	v	85
TTFX	13.4	0	0	0	3.9	v	200
FTDR	42.0	30.6	46.1	13.5	36.7	s	44
FTHG	11.6	4.6	15.6	0	9.5	n	87
SMCX	4.4	0.9	0	0	1.5	s	
ATFR	1.7	1.8	1.7	51.8	7.6	n	175
TODR	0.8	1.8	0.8	2.2	1.3	s	47
FTDU	0	0	0	4.5	0.5	v	200
FTBS	0	0	0	0	0	n	
SPFU	0.8	0	4.3	0	1.5	n	
SPFS	0	0	0	0	0	n	
XXXW	1.7	2.7	20.0	15.7	9.2	v	91
SPFR	0	0.9	0	0	0.2	n	
GRTU	32.2	0	0	0	9.5	v	200
ATFX	0	0	0	9.0	1.0	v	200
XXXD	20.5	31.5	61.7	47.3	39.3	v	45
TFLX	1.7	2.7	6.0	6.7	3.9	v	56
SPLU	6.2	4.6	2.6	0	3.9	n	
ATFU	5.3	1.8	2.6	0	2.9	n	
SPLX	0.8	2.7	2.6	0	1.8	n	
FTHP	0	0	0	0	0	n	
SMLU	0	0	0	0	0	n	
FTHW	0	0	0	0	0	n	
WSFX	0.8	0	0	0	0.2	v	
COCX	0	11.1	5.2	0	3.1	n	200
INKX	0.8	0.9	0	0	2.1	n	
DUKU	0	0	0	0	0	n	
CDSX	0	0.9	3.4	0	1.3	n	
SPWX	10.7	0	0	4.5	3.6	v	133
SFFX	0	0	0	0	0	n	
TODG	0.8	2.7	17.3	0	1.0	v	
TOBR	0	0.9	0	0	5.5	n	186
XXWX	0	0	0	0	0		
SULU	0.8	0.9	2.6	0	1.3		

Table 7.6 (Continued)

Food Code	AC	BS	DT	EN	FR	HN	OZ	PD	PH	SR	ST	Total	All	CV
XXLX	1.7	—	—	—	0.9	—	0.8	—	0	—	—	1.0	n	
GRDU	0	—	—	—	0	—	0	—	0	—	—	0	n	
SPFX	0	—	—	—	0	—	0	—	0	—	—	0	n	
TFBS	0	—	—	—	0	—	0	—	0	—	—	0	n	
ACDD	0	—	—	—	4.6	—	1.7	—	0	—	—	1.8	n	
CTFX	0	—	—	—	1.8	—	6.0	—	0	—	—	2.3	n	
SBFX	0	—	—	—	0	—	0	—	0	—	—	0	n	
Total	727.7	—	—	—	641.7	—	790.7	—	489.2	—	—	694.4		

E. Rates for All Ages Combined

Food Code	AC	BS	DT	EN	FR	HN	OZ	PD	PH	SR	ST	Total	All	CV	WSI	By Age
PCMX	31.7	89.3	22.1	73.7	39.5	**120.2**	59.4	29.0	35.0	**102.0**	22.0	60.5	v	61	v	v
	3.1	6.5	6.9	5.0	3.5	9.4	4.2	6.3	3.9	6.4	6.3	1.6				
GRCX	77.5	98.0	**541.4**	72.6	66.6	107.0	101.8	4.1	88.6	77.0	**1025.9**	113.8	v	149	v	v
	4.9	6.9	34.5	5.0	4.5	8.8	5.5	2.3	6.3	5.6	43.3	2.2				
GRLU	91.9	104.8	8.8	**137.0**	103.3	45.0	51.8	37.3	87.7	**109.4**	5.5	87.8	v	61	v	v
	5.3	7.1	4.4	6.8	5.7	5.7	3.9	7.1	6.3	6.6	3.1	1.9				
FTGX	**85.0**	**89.8**	83.9	62.2	66.6	57.5	78.1	1.3	17.7	79.9	69.7	66.9	v	45	v	v
	5.1	6.6	13.6	4.6	4.5	6.5	4.8	1.3	2.8	5.7	11.3	1.7				
TODU	11.2	3.3	**254.1**	3.1	3.1	10.3	6.4	0	9.0	11.0	**231.2**	17.2	v	193	v	v
	1.8	1.2	23.7	1.0	0.9	2.7	1.4	1.3	2.0	2.1	20.6	0.8				
SKCX	**33.0**	16.5	0	6.5	**40.1**	18.4	17.1	1.3	9.0	21.3	0	19.7	v	89	v	v
	3.2	2.8	0	1.5	3.5	3.6	2.2	1.3	2.0	2.9	0	0.9				
SCTX	27.4	15.0	0	11.7	**56.2**	15.4	26.3	26.3	11.3	**32.3**	0	25.1	v	79	v	v
	2.9	2.7	0	2.0	4.2	3.3	2.8	6.0	2.2	3.6	0	1.0				
TCFX	13.7	6.3	0	**27.6**	20.8	**27.3**	22.6	1.3	15.0	26.6	1.8	18.5	v	73	v	v
	2.0	1.7	0	3.0	2.5	4.4	2.6	1.3	2.6	3.3	1.8	0.9				
CDLU	29.9	16.5	24.3	26.3	30.3	2.2	**38.0**	6.9	**33.6**	9.8	5.5	24.4	v	62	v	v
	3.0	2.8	7.3	3.0	3.0	1.2	3.4	3.0	3.9	2.0	3.1	1.0				
XXXG	27.7	30.5	8.8	**36.6**	30.0	23.6	36.4	20.7	**39.1**	20.4	9.1	29.7	v	40	v	v
	2.9	3.8	4.4	3.5	3.0	4.1	3.3	5.3	4.2	2.8	4.1	1.1				
TOBS	5.2	0	0	2.7	9.4	0	**10.4**	**8.3**	0	6.9	0	5.0	v	108	v	v

FTDG	1.2	0	0	0.9	1.7	0	1.7	3.3	0	1.6	0	0.4	v	106	v	v
TTFX	32.0	9.2	2.2	44.9	25.5	5.1	4.1	2.7	6.6	10.6	14.6	35.6	v	142	v	v
FTDR	4.6	36.9	0	43.6	30.9	1.9	0	1.9	0	64.7	5.1	18.7	v	74	v	v
FTHG	1.2	4.2	0	3.8	4.7	4.7	0	0	0	5.1	0	0.9	v	57	v	v
SMCX	18.3	14.0	2.2	35.6	11.0	14.0	19.3	52.6	5.0	28.2	14.6	19.5	v	194	v	v
ATFR	2.3	2.6	2.2	3.5	1.8	3.2	2.4	8.5	1.5	3.4	5.1	0.9	v	88	v	v
TODR	8.4	7.7	17.6	24.5	25.0	4.1	13.4	1.3	13.6	14.3	14.6	12.8	v	75	v	v
FTDU	1.6	1.9	6.2	2.9	4.3	1.1	2.0	1.3	2.4	2.4	5.1	0.7	v	148	v	v
FTBS	2.1	4.3	37.5	15.5	5.9	2.2	2.4	1.3	0.9	4.9	111.9	7.8	v	151	v	v
SPFU	0.8	1.4	9.1	2.3	2.0	0.8	0.8	0	0.6	1.4	74.3	0.5	v	99	n	v
SPFS	9.0	11.1	0	23.5	14.7	4.7	4.9	1.3	13.1	6.9	0	9.7	v	178	v	v
XXXW	1.6	2.3	0	2.8	3.3	1.2	1.2	1.3	2.4	1.6	0	0.6	v	47	v	v
SPFR	4.9	2.9	0	6.2	4.4	1.2	4.2	4.1	14.0	5.7	11.0	5.2	v	177	v	v
GRTU	1.2	1.1	0	1.4	1.8	0.6	1.1	2.3	2.5	1.5	4.4	0.4	v	127	v	v
ATFX	2.8	0	0	11.0	4.7	2.7	2.7	0	4.0	0	0	3.3	v	49	v	v
XXXD	0.9	0	0	1.9	1.2	1.2	0.9	0	1.3	0	0	0.3	v	69	v	v
TFLX	9.0	0	0	0	4.4	4.4	8.5	0	4.0	0	0	3.5	v	84	v	v
SPLU	1.6	0	0	0	1.1	1.1	1.6	0	1.3	0	0	0.4	v	68	n	n
ATFU	1.5	0	0	0	0.7	2.8	1.9	0	1.9	1.4	0	6.3	v	72	n	n
SPLX	2.1	1.9	108.2	2.4	0.9	0.7	2.4	1.3	2.7	1.2	22.0	4.5	v	241	v	v

Table 7.6 (*Continued*)

Food Code	AC	BS	DT	EN	FR	HN	OZ	PD	PH	SR	ST	Total	All	CV	WSI	By Age
FTHP	0.8	0.9	15.4	0.9	0.5	0.7	0.8	1.3	1.1	0.7	6.3	0.4	v	115	v	n
	0.6	1.9	2.2	2.4	0.3	3.6	0.9	0	**5.4**	0	**9.1**	1.7				
SMLU	0.4	0.9	2.2	0.9	0.3	1.6	0.5	0	1.5	0	4.1	0.2	v	140	v	v
	0	**8.2**	**8.8**	5.5	0	4.4	0	0	0	0	0	2.0				
FTHW	2.4	2.0	4.4	1.3	0	1.8	0.9	0	**15.0**	0.8	**9.1**	0.3	v	161	v	v
	0.8	0	4.4	0.3	0	0.7	0.5	0	2.6	0.5	4.1	2.3				
WSFX	1.5	4.3	3.1	0.3	0.3	**6.6**	0.6	0	0	0	0	0.3	v	124	s	v
	0.6	1.4	0	2.7	0.3	2.2	0.4	0	0	0	0	2.1				
COCX	0	2.4	0	0.9	**3.7**	0	0	0	0	**5.3**	0	0.3	v	158	v	v
	0	1.0	0	3.1	1.0	0	0	0	0	1.4	0	1.2				
INKX	2.4	1.9	2.2	1.0	**4.7**	2.9	4.2	0	3.6	**5.3**	3.6	0.2	v	51	n	n
	0.8	0.9	2.2	1.7	1.2	1.4	1.1	0	1.2	1.4	2.5	3.3				
DUKU	0.9	0	0	0.7	**11.3**	0	**22.0**	0	0.4	0	0	0.3	v	217	v	v
	0.5	0	0	1.0	1.8	0	2.6	0	0.4	0	0	5.1				
CDSX	2.8	**4.8**	0	0.5	0.6	**8.1**	2.1	0	0.4	2.4	0	0.4	s	118	n	s
	0.9	1.5	0	1.7	0.4	2.4	0.8	0	0.4	1.0	0	2.2				
SPWX	**4.0**	0.9	0	0.7	1.2	1.4	0.9	0	1.3	1.6	1.8	0.3	s	81	n	n
	1.1	0.6	0	3.4	0.6	1.0	0.5	0	0.7	0.8	1.8	1.8				
SFFX	0	**7.7**	0	1.0	0	**15.4**	0	0	0	4.0	1.8	0.2	v	159	v	v
	0	1.9	0	**8.6**	0	3.3	0	0	0	1.2	1.8	3.2				
TODG	3.4	6.3	0	1.7	1.5	0	**6.4**	0	**6.3**	0	1.8	0.3	v	98	n	s
	1.0	1.7	0	4.8	0.7	0	1.4	0	1.7	0	1.8	3.5				
TOBR	0	0	0	1.2	0.3	0	**6.1**	**217.4**	1.7	0	0	0.3	v	317	v	v
	0	0	0	2.4	0.3	0	1.3	17.3	0	0	0	8.2				
XXWX	1.8	**2.4**	0	0.9	0.6	1.4	2.1	0	0	1.6	**5.5**	0.6	n		n	n
	0.7	1.0	0	1.0	0.4	1.0	0.8	0	0	0.8	3.1	1.4				
SULU	1.5	**4.3**	2.2	0.5	0.9	**9.5**	3.0	1.3	0.4	0.8	0	0.2	v	105	n	v
	0.6	1.4	2.2	3.8	0.5	2.6	0.9	1.3	0.4	0.5	0	2.5				
XXLX	1.2	**5.3**	**8.8**	1.1	0.9	1.4	1.5	1.3	0	2.4	1.8	0.3	n		s	s
	0.6	1.6	4.4	1.3	0.5	1.0	0.6	1.3	0	1.0	1.8	1.8				
GRDU	0	0	0	**5.5**	0	**2.9**	0	0	0	2.0	0	1.1	v	191	v	v
	0	0	0	1.3	0	1.4	0	0	0	0.9	0	0.2				
SPFX	1.2	0	0	1.3	0.3	0.7	**2.4**	0	0.9	0	**3.6**	0.8	n		n	n
	0.6	0	0	0	0.3	0.7	0.8	0	0.6	0	2.5	0.1				

Food															
TFBS	0.3	**1.9**	0	1.0	1.5	0	**6.7**	0	0	9.0	0	1.3	v	222	v
	0.3	0.9	0	0.5	0.7	0	1.4	0	0	0	0	0.2			
ACDD	0.3	0	0	0.6	0	0	**1.5**	**1.8**	0	0	0	0.7	v	137	n
	0.3	0	0	0.4	0	0	0.6	0.9	0	0	0	0.1			
CTFX	**1.2**	0	0	0.3	0.6	0	**3.0**	0	0	0.8	0	0.8	n		n
	0.6	0	0	0.3	0.4	0	0.9	0	0	0.5	0	0.1			
SBFX	0	0.4	0	2.7	0	**5.9**	0	0	0	1.2	0	0.8	v	196	s
	0	0.4	0	0.9	0	2.0	0	0	0	0.7	0	0.2			
Total	635.9	683.7	**1202.3**	792.1	628.2	657.5	771.2	668.0	446.0	743.0	**1672.0**	726.1			
	14.0	18.2	51.5	16.5	14.0	22.0	15.3	17.4	24.8	17.4	55.3	5.7			

Note: Feeding bout rates (bouts per thousand minutes of daytime) on the fifty-two core foods are given for each infant during each of the four ten-week age blocks in which it was sampled (panels A–D) and pooled over all sampled ages, 30–70 weeks (panel E). For conciseness, rate values are truncated after the first decimal place and standard errors (calculated as in chapter 3) have been omitted, except in panel E, where they are given immediately below the individual rates.

Each rate in panels A–D was obtained by dividing the number of feeding bouts of that yearling during that age block by the number of minutes of in-sight sample time on that yearling at that age (table 3.1), then dividing by the normalizing factor $k = 1.116$ (see appendix A9). These rates per potential foraging minute were then multiplied by one thousand for ease of tabulation. Marginal values (total in each panel) were calculated from pooled data, not by averaging. The overall rates ("total" column of panel E) were obtained by dividing the total bout frequency of each food by the product-limit sample time of 22,299.84 minutes, which is the sum of all activity-specific times (appendix 9).

The numbers of bouts are not presented but can be obtained by back-calculation: multiply each rate per thousand minutes in table 7.6 by 1.116×10^{-3} and then by the number of minutes of sample time (table 3.1) for that yearling at that age.

Statistical tests (eq. 3.4) were used to detect individual differences (across each row) and also age changes (across corresponding marginal cells in the first four panels). In the right margin of each panel A–D, rate differences between individuals of that age are summarized as follows: n = not significant ($p > 0.05$), s = significant ($0.05 \geq p \geq 0.01$), v = very significant ($p \leq 0.01$). Coefficients of variation (CV) in each panel are given only for those foods on which the sampled infants' feeding bout rates differed significantly (s or v in column "all").

In panel E, for each core food, rate differences were tested statistically across the pooled 30–70-week values of all eleven infants ("all"), across these same values for just the six well-sampled infants (WSI), whose names are given in boldface at the top of panel E, and across the four age values, one for each age block, given in the right margins of panels A–D (by age). In panel E, the two highest rates in each row are printed in boldface.

Food codes as in table 6.2. Names of infants abbreviated as in table 3.2.

Table 7.7 Time Budgets by Age for Core Foods

Time Budget Rank	Food Code	Sig. Var.	30–40 Weeks		40–50 Weeks		50–60 Weeks		50–60 Weeks (1975–76 only)		60–70 Weeks	
			Minutes	SE	Minutes	SE	Minutes	SE	Minutes	SE	Minutes	SE
1	PCMX	LR	96.4	10.4	67.1	7.4	48.5	5.9	55.1	6.8	32.2	4.4
2	GRCX	LR	20.7	2.3	30.5	2.8	73.5	9.4	21.8	3.0	21.3	3.1
3	GRLU	LR	14.7	1.6	24.8	3.3	21.1	2.6	26.3	3.3	23.7	3.3
4	FTGX	LR	10.2	1.2	18.8	2.3	12.5	2.0	11.5	1.9	27.0	3.5
5	TODU	LR	13.3	3.1	4.9	1.1	43.1	4.7	5.0	1.1	0	0.1
6	SKCX	LR	11.3	1.6	8.0	1.3	5.0	1.0	6.3	1.3	15.8	2.0
7	SCTX	R	6.6	0.8	5.6	0.7	5.2	0.7	6.6	0.9	23.1	2.3
8	TCFX	R	9.9	1.3	9.0	1.2	4.7	0.8	6.0	1.1	8.8	1.4
9	CDLU	R	9.9	1.2	5.3	0.7	6.6	0.9	7.3	1.0	7.4	1.0
10	XXXG	LR	5.0	1.0	6.8	1.5	10.1	1.9	12.2	2.3	2.5	0.7
11	TOBS	R	4.3	1.1	1.4	0.5	7.0	1.8	8.9	2.3	17.4	3.8
12	FTDG	R	2.6	0.6	8.5	1.3	9.7	1.3	12.0	1.7	3.9	0.8
13	TTFX	R	3.7	0.6	8.2	1.2	8.3	1.3	10.5	1.6	1.2	0.4
14	FTDR	LR	5.1	0.8	2.2	0.4	6.5	1.1	7.4	1.3	7.5	1.2
15	FTHG	R	3.9	0.6	3.4	0.6	6.8	1.1	7.2	1.2	3.1	0.6
16	SMCX	R	1.2	0.3	1.4	0.3	13.0	1.8	6.7	1.2	0.7	0.3
17	ATFR	R	2.0	0.4	2.4	0.5	4.7	0.8	5.9	1.1	2.1	0.5
18	TODR	LR	3.3	1.0	3.9	1.2	1.5	0.4	1.4	0.4	0.2	0.1
19	FTDU	R	0.4	0.2	4.9	1.1	2.3	0.7	2.9	0.9	0.4	0.3
20	FTBS	R	1.2	0.4	4.9	1.0	0.6	0.3	0.8	0.4	0	0
21	SPFU	R	4.4	0.8	1.4	0.3	0.7	0.2	0.8	0.3	0.5	0.2
22	SPFS	R	5.9	1.3	0.7	0.3	0	0	0	0	0	0
23	XXXW	LR	0.8	0.2	1.8	0.5	2.3	0.5	2.6	0.6	1.4	0.4
24	SPFR	LR	3.6	0.8	1.3	0.6	0	0	0	0	0	0
25	GRTU	LR	0.4	0.2	1.7	0.3	1.4	0.4	1.8	0.5	3.3	0.7
26	ATFX	LR	1.9	0.5	1.8	0.3	1.3	0.3	1.4	0.3	0.3	0.2
27	XXXD	LR	1.1	0.5	1.1	0.7	1.9	1.1	2.1	1.2	1.0	0.5
28	TFLX	—	1.0	0.2	1.0	0.2	1.0	0.2	0.8	0.3	1.0	0.2
29	SPLU	—	1.0	0.1	1.0	0.1	1.0	0.1	0.3	0.1	1.0	0.1
30	ATFU	R	1.5	0.3	0.5	0.1	1.0	0.3	0.8	0.2	0.7	0.2
31	SPLX	R	0.3	0.1	0.4	0.1	2.4	0.5	0.3	0.1	0.3	0.1

No.	Food	Sig. var.	1	2	3	4	5	6	7	8
32	FTHP	—	0.7	0.2	0.7	0.7	0.3	0.7	0.7	0.2
33	SMLU	R	0.3	0.1	1.0	1.5	0.5	1.6	0	0
34	FTHW	R	0.1	0.1	0.7	2.1	0.7	2.1	0	0
35	WSFX	R	0.6	0.2	0.5	1.4	0.5	1.8	0.1	0.1
36	COCX	R	0	0	0.4	1.2	0.6	1.5	1.6	0.6
37	INKX	L	0.9	0.3	0.6	0.7	0.3	0.8	0.2	0.1
38	DUKU	LR	0	0	1.7	0.2	0.3	0.3	0	0
39	CDSX	R	1.0	0.3	0.4	0.4	0.2	0.5	0.3	0.2
40	SPWX	—	0.5	0.1	0.5	0.5	0.2	0.6	0.5	0.1
41	SFFX	R	0.5	0.2	0.4	1.0	0.4	1.3	0	0
42	TODG	R	0.6	0.2	0.6	0.3	0.1	0.3	0.1	0.1
43	TOBR	LR	1.3	0.2	0.1	0	0	0	0.2	0.1
44	XXWX	—	0.4	0.2	0.4	0.4	0.2	0.3	0.4	0.2
45	SULU	R	0.8	0.2	0.3	0.2	0.1	0.2	0.2	0.1
46	XXLX	R	0.5	0.2	0.3	0.5	0.1	0.3	0.2	0.1
47	GRDU	R	0.2	0.1	0.2	1.2	0.5	1.5	0	0.1
48	SPFX	—	0.4	0.1	0.4	0.4	0.1	0.1	0.4	0
49	TFBS	R	0.3	0.1	0.9	0	0	0	0	0.1
50	ACDD	—	0.4	0.2	0.4	0.4	0.4	0.6	0.4	0.2
51	CTFX	—	0.4	0.1	0.4	0.4	0.3	0.6	0.4	0.1
52	SBFX	R	0.4	0.2	0.2	0.6	0.3	0.8	0	0
Total, core			258.03 ± 12.13		245.78 ± 9.83	317.49 ± 13.42	248.43 ± 10.17		213.43 ± 9.17	
Total, nonmilk core			161.63 ± 6.24		178.68 ± 6.47	268.99 ± 12.05	193.33 ± 7.56		181.23 ± 8.05	
Total, nonmilk			176.98 ± 6.83		195.65 ± 7.09	294.54 ± 13.20	211.70 ± 8.28		198.45 ± 8.81	
Total, all foods			273.38 ± 12.44		262.75 ± 10.25	343.04 ± 14.46	266.80 ± 10.71		230.65 ± 9.85	

Note: Values are the average number of minutes per day spent feeding on each food (pooled values, all yearlings, all seasons, observed diet). "Sig. var." (significant variable) indicates whether time budget differences with age were due to significant age changes in bout length (L), bout rate (R), or both.

Table 7.8 Individual Intakes of Macronutrients, Observed (Core plus Noncore) versus Optimal Diets

		Water				Minerals					
Infant	Age (weeks)	Optimal Intake (g/day)	Observed Intake (g/day)	SE	Deviation (%)	Optimal Intake (g/day)	Observed Intake (g/day)	SE	Deviation (%)	Min (g/day)	Max (g/day)
Alice	30–40	720.8	254.5	64.3	−64.7	18.16	9.35	0.67	−48.5	6.36	48.2
	40–50	715.3	334.9	78.4	−53.2	19.25	8.56	0.48	−55.5	8.02	60.7
	50–60	579.2	302.2	96.4	−47.8	12.84	5.04	0.54	−60.7	6.64	50.3
	60–70	727.9	228.6	66.7	−68.6	18.13	8.33	0.45	−54.1	5.93	44.9
Mean (Alice)					−58.6				−54.7		
Bristle	30–40	723.5	424.9	85.0	−41.3	18.68	6.59	0.41	−64.7	9.19	69.6
	40–50	722.1	390.2	84.4	−46.0	19.85	7.09	0.38	−64.3	8.82	66.8
Mean (Bristle)					−43.7				−64.5		
Dotty	50–60	743.4	427.9*	211.2	−42.4	17.04	8.75	0.86	−48.7	9.90	74.9
Eno	30–40	659.9	634.5*	158.0	−3.8	18.17	7.05	0.67	−61.2	13.50	102.2
	40–50	736.5	481.3	96.4	−34.7	17.96	9.07	0.51	−49.5	10.90	82.5
	50–60	720.8	272.0	45.6	−62.3	19.12	10.54	0.41	−44.9	6.96	52.7
Mean (Eno)					−33.6				−51.9		
Fred	30–40	725.7	188.0	33.2	−74.1	17.83	7.75	0.48	−56.5	4.70	35.6
	40–50	715.1	219.0	32.4	−69.4	19.28	7.70	0.37	−57.2	5.49	41.6
	50–60	754.9	348.2*	212.9	−53.9	16.88	5.94	1.21	−64.8	7.86	59.5
	60–70	729.0	323.0	86.5	−55.7	17.99	7.40	0.45	−58.9	7.47	56.6
Mean (Fred)					−63.3				−59.4		
Hans	30–40	721.5	577.1*	114.3	−20.0	19.64	6.51	0.48	−66.9	12.40	93.9
Ozzie	30–40	732.0	361.9	90.5	−50.6	17.63	8.54	0.56	−51.6	8.39	63.5
	40–50	714.7	426.7	110.2	−40.3	19.32	7.23	0.49	−62.6	9.66	73.2
	50–60	754.9	530.6*	183.8	−29.7	16.88	4.53	0.79	−73.2	11.03	83.6
	60–70	729.2	468.7	113.8	−35.7	17.92	9.26	0.55	−48.3	10.79	81.7
Mean (Ozzie)					−39.1				−58.9		
Pedro	30–40	732.5	244.4	61.7	−77.8	17.46	2.33	0.27	−86.7	5.14	38.9
Pooh	30–40	743.4	5.41	1.25	−99.3	17.04	1.22	0.35	−92.8	0.34	2.56
	40–50	732.6	208.7	59.8	−71.5	17.52	4.36	0.32	−75.1	4.81	36.4
	50–60	715.0	223.7	53.0	−68.7	19.70	8.87	0.49	−55.0	5.83	44.1
	60–70	654.0	87.9	12.7	−86.6	14.57	3.64	0.34	−75.1	2.19	16.6
Mean (Pooh)					−75.6				−68.4		
Striper	50–60	743.4	186.4	−74.9	−74.9	17.4	8.35	0.46	−51.0	5.11	38.7

	Age (weeks)	Optimal Intake (g/day)	Observed Intake (g/day)	SE	Deviation (%)	Min (g/day)	PDR ("Surplus")	NMILKP	NMILK-PDO	NMILK-PDR
Summer	30–40	730.7	458.7	108.7	−37.2	18.11	6.11	−66.3	9.87	74.8
	40–50	719.6	598.1*	123.4	−16.9	18.91	10.04	−46.9	13.09	99.1
	50–60	722.6	452.6*	211.9	−37.4	19.72	6.80	−65.5	10.99	86.4
Mean (Summer)					−30.5			−59.6		
Mean Deviation ± SD (all infants, n = 27 blocks)					−50.2 ± 19.9			±59.8 ± 10.2		

							Proteins			
Infant	Age (weeks)	Optimal Intake (g/day)	Observed Intake (g/day)	SE	Deviation (%)	Min (g/day)	PDR ("Surplus")	NMILKP	NMILK-PDO	NMILK-PDR
Alice	30–40	27.85	12.78	1.35	−54.1	3.56	.721	10.04	−.639	1.820
	40–50	29.82	13.28	1.51	−55.5	3.86	.709	8.74	−.707	1.264
	50–60	25.50	11.01	1.89	−56.8	4.15	.623	6.86	−.731	.653
	60–70	28.31	10.20	1.28	−64.0	4.43	.566	7.52	−.734	.698
Mean (Alice)					−57.6		.655	8.290	−.703	1.109
Bristle	30–40	28.18	14.45	1.63	−48.7	3.56	.754	8.19	−.709	1.301
	40–50	31.95	12.94	1.60	−59.5	3.86	.702	6.87	−.785	.780
Mean (Bristle)					−54.1		.728	7.530	−.747	1.040
Dotty	50–60	27.35	16.12	3.97	−41.1	4.15	.743	9.68	−.646	1.333
Eno	30–40	28.83	18.51	2.98	−35.8	3.56	.808	8.11	−.719	1.278
	40–50	29.35	17.25	1.86	−41.2	3.86	.776	10.10	−.656	1.617
	50–60	29.72	15.32	0.94	−48.5	4.15	.729	12.72	−.572	2.065
Mean (Eno)					−41.8		.771	10.310	−.649	1.653
Fred	30–40	27.57	9.02	0.75	−67.3	3.56	.605	7.70	−.721	1.163
	40–50	29.82	11.56	0.73	−61.2	3.86	.666	9.27	−.689	1.402
	50–60	30.46	10.66	4.01	−65.0	4.15	.611	5.32	−.825	.282
	60–70	28.66	12.45	1.66	−56.6	4.43	.644	8.37	−.708	.889
Mean (Fred)					−62.5		.632	7.665	−.736	.934
Hans	30–40	31.29	15.47	2.14	−50.6	3.56	.770	5.71	−.818	.604
Ozzie	30–40	28.18	12.75	1.74	−54.8	3.56	.721	8.08	−.713	1.270
	40–50	29.82	14.69	2.08	−50.7	3.86	.737	7.98	−.732	1.067
	50–60	30.46	15.35	3.60	−49.6	4.15	.730	7.78	−.745	.875
	60–70	28.66	14.75	2.14	−48.5	4.43	.700	7.61	−.734	.718
Mean (Ozzie)					−50.9		.722	7.863	−.731	.982

Table 7.8 (*Continued*)

Infant	Age (weeks)	Optimal Intake (g/day)	Observed Intake (g/day)	SE	Deviation (%)	Min (g/day)	PDR ("Surplus")	NMILKP	NMILK-PDO	NMILK-PDR
Pedro	30–40	29.11	7.70	1.19	−73.5	3.56	.538	3.90	−.866	.096
Pooh	30–40	27.35	2.22	0.78	−91.9	3.56	−.604	2.22	−.919	−.376
	40–50	28.37	8.16	1.15	−71.2	3.86	.527	5.25	−.815	.360
	50–60	29.97	14.54	1.11	−51.5	4.15	.715	12.35	−.588	1.976
	60–70	26.56	5.94	0.52	−77.6	4.43	.254	5.47	−.794	.235
Mean (Pooh)					−66.8		.499	7.691	−.732	.857
Striper	50–60	27.35	12.62	1.15	−53.9	4.15	.671	10.50	−.616	1.530
Summer	30–40	28.92	14.56	2.07	−49.7	3.56	.755	7.63	−.736	1.143
	40–50	28.99	19.58	2.35	−32.5	3.86	.803	11.08	−.618	1.870
	50–60	31.78	13.49	3.96	−57.6	4.15	.692	5.98	−.812	.441
Mean (Summer)					−46.6		.750	8.230	−.722	1.152
Mean Deviation ± SD (all infants, *n* = 27 blocks)					−54.7 ± 10.8					

Infant	Age (weeks)	Fiber					Other Carbohydrate			
		Optimal Intake (g/day)	Observed Intake (g/day)	SE	Deviation (%)	Min (g/day)	Optimal Intake (g/day)	Observed Intake (g/day)	SE	Deviation (%)
Alice	30–40	27.28	13.15	0.76	−51.8	0.40	167.93	67.03	6.90	−60.1
	40–50	29.79	13.88	0.62	−53.4	0.45	169.75	75.45	7.35	−55.5
	50–60	22.01	8.25	0.72	−62.5	0.28	87.00	41.78	8.57	−52.0
	60–70	28.62	13.11	0.56	−54.2	0.40	159.27	71.25	6.57	−55.3
Mean (Alice)					−55.5					−55.7
Bristle	30–40	29.11	10.31	0.52	−64.6	0.37	163.36	53.01	7.27	−67.6
	40–50	30.12	10.03	0.40	−66.7	0.42	157.37	68.58	7.23	−56.4
Mean (Bristle)					−65.7					−62.0
Dotty	50–60	25.79	13.95	0.77	−45.9	0.50	148.72	83.70	17.91	−43.7
Eno	30–40	28.72	10.02	0.62	−65.1	0.51	156.65	75.65	13.41	−51.7
	40–50	28.31	14.77	0.69	−47.8	0.52	147.88	82.16	8.52	−44.4
	50–60	29.48	17.72	0.58	−39.9	0.46	163.33	74.31	4.96	−54.5
Mean (Eno)					−50.9					−50.2

Infant	Age (weeks)									
Fred	30–40	26.13	11.41	0.61	−56.3	0.29	164.90	49.26	4.05	−70.1
	40–50	29.82	13.70	0.59	−54.1	0.35	169.94	55.22	3.87	−67.5
	50–60	26.48	8.19	1.27	−69.1	0.37	124.03	63.00	20.04	−49.2
	60–70	28.54	12.33	0.57	−56.8	0.38	158.15	61.23	7.69	−61.3
Mean (Fred)					−59.1					−62.0
Hans	30–40	29.93	8.19	0.39	−72.6	0.49	159.30	77.06	9.73	−51.6
Ozzie	30–40	26.08	12.51	0.65	−52.0	0.43	158.24	72.22	8.36	−54.4
	40–50	29.88	11.38	0.48	−61.9	0.46	170.23	75.27	9.53	−55.8
	50–60	26.48	7.49	1.13	−71.7	0.37	124.03	53.43	15.50	−56.9
	60–70	28.56	13.56	0.56	−52.5	0.54	157.91	92.39	10.07	−41.5
Mean (Ozzie)					−59.5					−52.2
Pedro	30–40	28.25	4.12	0.36	−85.4	0.18	154.54	25.57	5.17	−83.5
Pooh	30–40	25.79	3.13	1.03	−87.9	0.049	148.72	7.23	2.23	−95.1
	40–50	26.06	7.37	0.41	−71.7	0.24	157.69	39.12	5.24	−75.2
	50–60	30.62	17.12	0.75	−44.1	0.40	168.74	61.68	5.74	−63.4
	60–70	23.24	6.31	0.52	−72.8	0.13	104.64	19.14	2.36	−81.7
Mean (Pooh)					−62.9					−73.4
Striper	50–60	25.79	15.60	0.71	−39.5	0.38	148.72	64.00	5.83	−57.0
Summer	30–40	28.72	10.38	0.55	−63.9	0.39	155.76	56.98	9.22	−63.4
	40–50	29.31	14.41	0.63	−50.8	0.55	166.00	85.03	10.47	−48.8
	50–60	30.01	9.28	0.54	−69.1	0.46	157.28	75.90	17.83	−51.7
Mean (Summer)					−61.3					−54.6
Mean Deviation ± SD (all infants, n = 27 blocks)					59.1 ± 11.3					−58.3 ± 10.7

Energy

Infant	Age (weeks)	Optimal Intake (kJ/day)	Observed Intake (kJ/day)	SE	Deviation ("Shortfall") (%)	Min (kJ/day)	JDR	MILKJ	NMILKJ	NMILK-JDO	NMILK-JDR
Alice	30–40	5066	1911	256.6	−62.3	672	1.8438	552	1359	−.732	1.022
	40–50	5097	2288	302.1	−55.1	755	2.0305	914	1374	−.730	.820
	50–60	3239	1547	367.5	−52.2	836	.8505	836	711	−.780	−.150
	60–70	4949	1936	259.6	−60.9	913	1.1205	541	1395	−.718	.528
Mean (Alice)					−57.6		1.4613	711	1209.8	−.740	.555
Bristle	30–40	4998	2056	321.3	−58.9	672	2.0595	1260	796	−.841	.185
	40–50	5019	2257	319.0	−55.0	755	1.9894	1220	1037	−.793	.374
Mean (Bristle)					−57.0		2.0245	1240	916.5	−.817	.279
Dotty	50–60	4713	2637	796.2	−44.0	836	2.1543	1300	1337	−.716	.599
Eno	30–40	4726	2963	595.7	−37.3	672	3.4092	2100	863	−.817	.284

Table 7.8 (Continued)

Infant	Age (weeks)	Optimal Intake (kJ/day)	Observed Intake (kJ/day)	SE	Deviation ("Shortfall") (%)	Min (kJ/day)	JDR	MILKJ	NMILKJ	NMILK-JDO	NMILK-JDR
	40–50	4763	2777	366.9	−41.7	755	2.678	1440	1337	−.719	.771
	50–60	5039	2186	182.9	−56.6	836	1.615	524	1662	−.670	.988
Mean (Eno)					−45.2		2.567	1355	1287.3	−.736	.681
Fred	30–40	4991	1359	137.2	−72.8	672	1.022	266	1093	−.781	.626
	40–50	5100	1677	134.6	−67.1	755	1.221	462	1215	−.762	.609
	50–60	4359	1996	819.5	−54.2	836	1.388	1080	916	−.790	.096
	60–70	4930	1965	329.1	−60.1	913	1.152	822	1143	−.768	.252
Mean (Fred)					−63.6		1.196	658	1091.8	−.775	.396
Hans	30–40	5028	2813	431.1	−44.1	672	3.186	1970	843	−.832	.254
Ozzie	30–40	4892	2200	347.0	−55.0	672	2.274	941	1259	−.743	.874
	40–50	5106	2492	416.9	−51.2	755	2.301	1350	1142	−.776	.513
	50–60	4359	2198	692.1	−49.6	836	1.629	1520	678	−.844	−.189
	60–70	4929	2870	432.3	−41.8	913	2.144	1440	1430	−.710	.566
Mean (Ozzie)					−49.4		2.087	1313	1127.3	−.768	.441
Pedro	30–40	4887	1077	232.2	−78.0	672	.603	765	312	−.936	−.536
Pooh	30–40	4713	216	69.4	−95.4	672	−.679	0	216	−.954	−.679
	40–50	4880	1262	226.8	−74.1	755	.672	586	676	−.861	−.105
	50–60	5092	1896	213.0	−62.8	836	1.268	441	1455	−.714	.740
	60–70	3711	633	66.8	−82.9	913	−.307	94.30	539	−.855	−.410
Mean (Pooh)					−73.3		.544	374	889.8	−.810	.075
Striper	50–60	4713	1797	224.8	−61.9	836	1.150	427	1370	−.709	.639
Summer	30–40	4900	2202	409.7	−55.1	672	2.277	1400	802	−.836	.193
	40–50	5058	3005	465.3	−40.6	755	2.980	1710	1295	−.744	.715
	50–60	5012	2527	797.3	−49.6	836	2.023	1510	1017	−.797	.217
Mean (Summer)					−48.4		2.427	1540	1038.0	−.792	.375
Mean Deviation ± SD (all infants, n = 27 blocks)					−56.5 ± 11.6						

		Lipids					Total Mass			
Infant	Age (weeks)	Optimal Intake (g/day)	Observed Intake (g/day)	SE	Deviation (%)	Min (g/day)	Optimal Intake (g/day)	Observed Intake (g/day)	SE	Deviation (%)
Alice	30–40	39.64	11.23	3.64	−71.7	1.28	1002.4	367.9	64.8	−63.3
	40–50	37.90	17.17	4.46	−54.7	1.44	1002.4	463.3	78.9	−53.8
	50–60	29.49	15.40	5.48	−47.8	1.59	756.4	383.6	97.0	−49.3
	60–70	39.55	11.07	3.79	−72.0	1.74	1002.4	342.6	67.1	−65.8
Mean (Alice)					−61.6					−58.1
Bristle	30–40	38.93	21.89	4.84	−43.8	1.28	1002.4	531.2	85.5	−47.0
	40–50	40.43	21.12	4.81	−47.8	1.44	1002.4	509.9	84.9	−49.1
Mean (Bristle)					−45.8					−48.1
Dottie	50–60	39.48	21.69*	12.03	−45.1	1.59	1002.4	572.0	212.3	−42.9
Eno	30–40	34.58	34.81*	9.00	0.67	1.28	927.4	780.5	158.9	−15.8
	40–50	39.35	25.43	5.48	−35.4	1.44	1000.0	630.1	97.0	−37.0
	50–60	39.40	12.56	2.59	−68.1	1.59	1002.4	402.4	46.0	−59.9
Mean (Eno)					−34.3					−37.6
Fred	30–40	39.57	6.43	1.83	−83.8	1.28	1002.4	271.9	33.5	−72.9
	40–50	37.89	10.41	1.83	−72.5	1.44	1002.4	317.6	32.7	−68.3
	50–60	39.43	18.21*	12.12	−53.8	1.59	993	454.2	214.2	−54.3
	60–70	39.45	15.75	4.91	−60.1	1.74	1002.4	432.0	87.0	−56.9
Mean (Fred)					−67.6					−63.1
Hans	30–40	40.12	32.09*	6.51	−20.0	1.28	1002.4	716.5	114.9	−28.5
Ozzie	30–40	39.62	16.97	5.14	−57.2	1.28	1002.4	484.8	91.1	−51.6
	40–50	37.88	23.17	6.28	−38.8	1.44	1002.4	558.5	110.8	−44.3
	50–60	39.43	26.42*	10.41	−33.0	1.59	992.8	637.7	184.8	−35.8
	60–70	39.52	24.88	6.48	−37.0	1.74	1002.4	623.7	114.5	−37.8
Mean (Ozzie)					−41.5					−42.4
Pedro	30–40	39.85	12.92	3.50	−67.6	1.28	1002.4	297.0	62.0	−70.4
Pooh	30–40	39.48	0.40	0.16	−99.0	1.28	1002.4	19.5	2.9	−98.1
	40–50	39.48	10.36	3.39	−73.8	1.44	1002.4	278.2	60.1	−72.2
	50–60	37.81	10.88	3.01	−71.2	1.59	1002.4	336.8	53.4	−66.4
	60–70	33.46	3.64	0.66	−89.1	1.74	857.0	126.4	13.0	−85.3

Table 7.8 (*Continued*)

Infant	Age (weeks)	Optimal Intake (g/day)	Observed Intake (g/day)	SE	Deviation (%)	Min (g/day)	Optimal Intake (g/day)	Observed Intake (g/day)	SE	Deviation (%)
Mean (Pooh)					−78.0					−74.6
Striper	50–60	39.48	8.46	3.23	−78.6		1002.4	295.5	57.2	−70.5
Summer	30–40	39.53	23.97	6.18	−39.4	1.28	1002.4	570.8	109.3	−43.1
	40–50	39.05	29.51*	7.02	−24.4	1.44	1002.4	756.6	124.1	−24.5
	50–60	40.38	25.24*	12.07	−37.5	1.59	1002.4	583.3	213.0	−41.8
Mean (Summer)					−33.8					−36.5
Mean deviation ± SD (all infants)					−52.7 ± 21.4					−52.2 ± 16.6

Note: Imbibed water is not included. Deviation (%) = 100 (observed intake − optimal intake)/optimal intake. Means are calculated without the values from Pooh's very small sample when 30–40 weeks old. Reqirements (min) and tolerances (max) are calculated from formulas in table 5.3, using mass of observed diet as Φ. Minimum values are tabulated only if nonzero including food-bound water; maximum values are not tabulated if at consumption capacity. Intakes marked with an asterisk are within two standard errors of the value in the corresponding energy-maximizing diet. Abbreviations for energy and protein variables (JDR, etc.) are defined in table 8.4.

Table 7.9 Mean Composition of Observed Diets (g/day; kJ/day), by age

Nutrient	Age (weeks)			
	30–40[a]	40–50	50–60	60–70
Water	393.0 ± 161.2	379.8 ± 139.9	343.0 ± 119.5	277.1 ± 160.2
Mineral	6.8 ± 2.1	7.7 ± 1.8	7.4 ± 2.1	7.2 ± 2.5
Lipid	20.0 ± 10.5	19.6 ± 7.3	17.4 ± 6.7	13.8 ± 8.9
Protein	13.2 ± 3.5	13.9 ± 3.7	13.6 ± 2.1	10.8 ± 3.8
Fiber	10.0 ± 2.8	12.2 ± 2.7	12.2 ± 4.3	11.3 ± 3.4
Other carbohydrates	59.6 ± 17.3	68.7 ± 16.3	64.7 ± 13.3	61.0 ± 30.8
Energy	2072.4 ± 643.8	2251.2 ± 607.1	2098.0 ± 366.1	1850.8 ± 920.6
Total diet mass[b]	502.6 ± 186.1	502.0 ± 168.4	458.2 ± 125.8	381.2 ± 206.4

Note: Values are means ± SD of corresponding individual values in table 7.8.

[a]Calculated without Pooh's small sample at 30–40 weeks.

[b]Overall mean mass ($n = 28$) = 455.4 ± 182.3; mean + 3 SD = 1002.4.

Table 7.10 Unit Intake Rates of Pedro Versus Those of Other Yearlings

Food Code	Food Unit	Intake Rate (units per minute)		
		Pedro	Others (pooled mean)	All[a] (pooled mean)
GRLU	Green grass leaves	4.84	29.59	28.09
TOBS	Umbrella tree flowers from tree	15.00	16.37	16.36
TOBR	Umbrella tree flowers from ground	11.83	60.0	13.61

[a] From table 6.1.

Table 7.11 Corrected Mean Daily Nutrient Intakes of Pedro

Food Code	Correction Factor[a]	Water	Ash	Fat	Protein	Fiber	Other Carbohydrates	Energy
GRLU	0.17230	.171E1	.481E0	.054E0	.157E0	.193E0	.246E0	.133E2
TOBS	0.91687	.888E1	.133E0	.221E0	.454E0	.410E0	.981E0	.370E2
TOBR	0.86921	.835E1	.125E1	.208E0	.217E0	.385E0	.926E0	.349E2
All foods	—	234.6	2.26	12.62	6.65	3.12	24.2	1015

Note: Energy in k/J day, all other food components in g/day. Exponent notation: En indicates a factor of 10^n.

[a] Ratio of Pedro's unit intake rate (table 7.10) to pooled rate of all weanlings, which was used to calculate values in table 7.8.

Table 7.12 Predicted and Observed Nutrient Deficiencies

	Age (weeks)			
	30–40[a]	40–50	50–60	60–70
Energy				
Requirement (kJ/day)	673	755	835	913
Normal deviates below requirement	1.95	2.20	3.13	0.82
Expected percentage deficient	2.56	1.4	0.09	20.6
Expected number of deficients	0.20	0.10	0.007	0.82
Observed deficients	0	0	0	1
Protein				
Requirement (g/day)	3.56	3.86	4.15	4.43
Nornal deviates below requirement	2.52	2.60	3.90	1.35
Expected percentage deficient	0.6	0.5	0.005	8.9
Expected number of deficients	0.05	0.03	0.000	0.35
Observed deficients	0	0	0	0
Fats				
Requirement (g/day)	1.28	1.44	1.59	1.74
Normal deviates below requirement	1.63	2.39	2.20	1.17
Expected percentage deficient	5.2	0.84	1.4	12.1
Expected number of deficients	0.42	0.06	0.01	0.48
Observed deficients	0	0	0	0
Ash				
Requirement (proportion of diet)	0.0173	0.0173	0.0173	0.0173
Normal deviates below requirement	0.48	(0.27 above)	0.05	0.44
Expected percentage deficient	31.6	60.6	48.0	33.0
Expected number deficient	2.5	4.2	3.8	1.3
Observed number deficient	5	5	5	2
Maximum (proportion of diet)	0.131	0.131	0.131	0.131
Normal deviates above maximum	>7	>7	>7	>7
Expected percentage above maximum	<0.01	<0.01	<0.01	<0.01
Expected number above maximum	<0.001	<0.001	<0.001	<0.001
Observed number above maximum	0	0	0	0
Fiber				
Requirement (% of dry matter)	0.35	0.35	0.35	0.35
Normal deviates below requirement	3.12	5.12	2.53	3.28
Expected percentage deficient	0.091	1.53	0.57	0.52
Expected number of deficients	0.007	0.11	0.05	0.02
Observed deficients	0	0	0	0

[a]Calculated without Pooh's very small sample at this age.

Table 7.13 Mineral: Diet Mass Ratios as Percentages of Dietary Mineral Requirement (1.73% of Diet)

0–40 Weeks		40–50 Weeks		50–60 Weeks		60–70 Weeks	
PH	358%						
FR	166%			ST	164%		
AC	146%	FR	140%	PH	153%	PH	165%
OZ	101%	AC	107%	EN	151%	AC	140%
— — — — — — — — — — — — 100% of requirement — — — — — — — — — — — —							
BS	72%	PH	91%	DT	88%	FR	99%
SR	62%	EN	83%	AC	75%	OZ	86%
HN	52%	BS	80%	FR	75%		
EN	52%	SR	76%	SR	67%		
PD	45%	OZ	75%	OZ	41%		

Note: Mineral and diet mass values from table 7.8. Abbreviations of infant names as in table 3.2.

Table 8.1 In-Sight Sample Time by Semimonth and Infant in Each Ten-Week Age Block

Infant	Semimonth (-1, -2) and Sample Time (minutes)												Total
AC1	Aug-2	141.167	Sep-1	309.110	Sep-2	53.850							504.127
AC2	Oct-1	196.720	Oct-2	171.833	Nov-1	197.583	Nov-2	206.767	Dec-1	202.717			975.62
AC3	Jan-2	196.100	Feb-1	198.550									394.65
AC4	Feb-2	200.250	Mar-1	209.183	Mar-2	195.500	Apr-1	206.583	May-1	189.500			1001.016
BS1	Mar-2	397.618	Apr-2	208.750	May-1	203.950							810.318
BS2	Jun-1	407.217	Jun-2	417.450	Jul-1	210.350							1035.017
DT3	Jul-2	405.430											405.43
EN1	Nov-1	198.550	Nov-2	183.350	Dec-1	90.117	Dec-2	110.067					582.084
EN2	Feb-1	198.667	Feb-2	188.117	Mar-2	407.533							794.317
EN3	Apr-2	211.050	May-1	419.650	May-2	182.800	Jun-1	399.333					1212.833
FR1	Aug-1	131.700	Sep-1	369.370	Sep-2	152.740							653.81
FR2	Oct-1	203.320	Oct-2	189.533	Nov-1	206.367	Nov-2	209.300	Dec-1	204.400			1012.92
FR3	Feb-1	203.717											203.717
FR4	Feb-2	201.083	Mar-1	197.183	Mar-2	193.733	Apr-1	168.567	May-1	204.267			964.833
HN1	May-1	198.367	Jun-1	409.050	Jun-2	406.550	Jul-1	200.183					1214.15
OZ1	Aug-1	127.150	Aug-2	195.533	Sep-1	162.080	Sep-2	210.070					694.833
OZ2	Oct-1	192.340	Oct-2	196.133	Nov-1	200.217	Nov-2	208.483	Dec-1	197.983			995.156
OZ3	Feb-1	202.033											202.033
OZ4	Feb-2	195.083	Mar-1	214.283	Mar-2	201.000	Apr-1	210.883	Apr-2	208.827			1030.076
PD1	Feb-2	216.950	Mar-1	429.967									646.917
PH1	Jul-2	14.217											14.217
PH2	Aug-1	124.617	Aug-2	169.933	Sep-1	174.030	Sep-2	192.040	Oct-1	179.260			839.88
PH3	Oct-2	307.347	Nov-1	194.233	Nov-2	217.167							718.747
PH4	Jan-1	188.717	Feb-1	208.717									397.434
SR1	Feb-2	203.570	Mar-1	212.850	Mar-2	383.633							800.053
SR2	Apr-2	202.900	May-1	390.283	Jun-1	197.250							790.433
SR3	Jun-2	405.733	Jul-1	190.150									595.883
ST3	Jul-2	488.210											488.21
Total	Jan-1	188.717	Jan-2	196.100	Feb-1	1011.684	Feb-2	1205.053	Mar-1	1263.466	Mar-2	1779.017	19,978.71
	Apr-1	586.033	Apr-2	831.527	May-1	1606.017	May-2	182.800	Jun-1	1412.850	Jun-2	1229.733	
	Jul-1	600.683	Jul-2	907.857	Aug-1	383.467	Aug-2	506.633	Sep-1	1014.590	Sep-2	608.700	
	Oct-1	771.640	Oct-2	864.846	Nov-1	802.717	Nov-2	1002.133	Dec-1	912.384	Dec-2	110.067	

Note: Abbreviations for infants' names (AC, BS, etc.) and age blocks (1, . . . , 4) as in table 3.1.

Table 8.2 Energy-Maximizing Optimal Diets for Each Infant during Each Age Block in Which It Was Sampled

Time Budget Rank	Food Code	Infant, Age Block								
		AC-1	AC-2	AC-3	AC-4	BS-1	BS-2	DT-3	EN-1	EN-2
1	PCMX	693.83	621.16	477.24	687.89	665.98	639.56	683.59	551.96	674.10
2	GRCX							2.03		
3	GRLU	72.74	111.50	124.00	124.01	124.01	124.01	124.01	124.01	124.01
4	FTGX	83.85	75.14	21.09	83.85	83.85	83.85	83.85	61.51	73.37
5	TODU		0.19						0.15	
7	SCTX			10.37						
8	TCFX		1.35	3.24	0.25	0.27		6.51	2.24	
9	CDLU	3.48	13.33	8.19	7.70	13.67			11.20	11.60
10	XXXG	7.27	7.27	7.27	7.27	7.27	7.27	7.27	7.27	7.27
12	FTDG	9.79	8.77	7.36	9.79	9.79	9.79	9.79	7.18	9.79
14	FTDR	0.14	0.53	0.83	1.59	1.70			0.83	0.47
15	FTHG	30.49	27.32	7.67	30.49	30.49	30.49	30.49	22.37	26.68
16	SMCX	24.99	4.93	14.07	24.12	19.28	30.75			27.15
17	ATFR	20.22	48.90	48.90	5.97	18.15	43.93		48.90	30.57
18	TODR	5.84	6.23		0.99		0.77		6.85	1.79
19	FTDU								4.41	
20	FTBS		3.24							
21	SPFU	6.21	6.57					24.55	2.21	
23	XXXW		0.08	0.20	0.09	0.10		0.40	0.14	5.65
26	ATFX	15.33	15.16	15.33	1.81	3.91	3.12	15.33	15.33	0.26
27	XXXD	0.42	0.78	0.54	0.82	0.55		1.08	0.79	2.05
28	TFLX			4.11					3.09	
30	ATFU	1.88	2.18	0.81	5.45	2.77		7.21	1.23	2.61
31	SPLX			3.36						1.67
32	FTHP	0.21	0.61		0.24	0.26		1.99	0.65	
34	FTHW	20.96	41.46		8.09	17.41			50.15	
35	WSFX							4.31		
36	COCX					1.78				
37	INKX	0.45	1.31	0.45	0.22	0.46	0.11		1.33	0.22
41	SFFX						28.76			
42	TODG	4.04	4.21	1.17	1.78	0.61			3.45	0.56
50	ACDD	0.27	0.17	0.22		0.11			0.14	0.22

Infant, Age Block

Time Budget Rank	Food Code	EN-3	FR-1	FR-2	FR-3	FR-4	HN-1	OZ-1	OZ-2	OZ-3
1	PCMX	643.74	690.88	621.03	667.19	678.46	640.04	689.77	620.66	667.19
2	GRCX		0.33					0.30		
3	GRLU	124.01	60.01	111.56	124.01	124.01	124.01	72.05	112.02	124.01
4	FTGX	83.85	83.85	75.39	41.93	83.85	83.85	83.85	75.51	41.93
5	TODU			0.19					0.19	
7	SCTX									
8	TCFX	0.18	1.31	1.31				1.19	1.30	
9	CDLU	12.93	10.88	13.10	16.28	11.07	3.43	12.82	12.77	16.28
10	XXXG	7.27	7.27	7.27	7.27	7.27	7.27	7.27	7.27	7.27
12	FTDG	9.79	9.79	8.80	9.79	9.79	9.79	9.79	8.82	9.79
14	FTDR	1.82	0.31	0.52		1.68	0.55	0.40	0.52	
15	FTHG	30.49	30.49	27.42	15.25	30.49	30.49	30.49	27.46	15.25
16	SMCX	14.57	15.80	5.24	27.98	18.49	25.74	14.40	5.51	27.98
17	ATFR	31.93	25.24	48.90	48.90	15.18	40.88	20.49	48.90	48.90
18	TODR		5.42	6.15		2.35	0.62	4.86	6.07	
19	FTDU									
20	FTBS			3.22					3.10	
21	SPFU		16.70	6.47				20.41	6.29	
23	XXXW	0.07	0.04	0.08				0.04	0.08	
26	ATFX	3.99	15.33	15.15	15.33	3.22	3.78	15.33	15.13	15.33
27	XXXD	0.73	0.47	0.77		0.87	0.18	0.83	0.76	
28	TFLX				8.18					8.18
30	ATFU	2.50	3.14	2.18	3.61	5.00	0.59	4.61	2.12	3.61
31	SPLX				6.67					6.67
32	FTHP	0.32	0.87	0.61				0.97	0.60	
34	FTHW	17.78	19.31	41.35		7.24	5.59	7.97	41.62	
35	WSFX		0.43					0.39		
36	COCX	2.44	0.42	1.29		1.49	1.15	0.81	1.28	
37	INKX	0.47				0.38	0.23			
41	SFFX	12.95					24.14			
42	TODG	0.41	3.75	4.23		1.38		3.18	4.23	
50	ACDD	0.15	0.34	0.17	0.44	0.19	0.07	0.18	0.18	0.44

Table 8.2 (Continued)

Time Budget Rank	Food Code	Infant, Age Block									
		OZ-4	PD-1	PH-1	PH-2	PH-3	PH-4	SR-1	SR-2	SR-3	ST-3
1	PCMX	682.42	689.82	683.59	682.37	619.55	553.15	683.27	644.19	642.38	683.59
2	GRCX			2.03	0.24					2.03	2.03
3	GRLU	124.01	124.01	124.01	69.57	124.01	124.01	124.01	124.01	124.01	124.01
4	FTGX	83.85	83.85	83.85	83.85	71.18	41.93	83.85	83.85	83.85	83.85
5	TODU					0.28	4.96				
7	SCTX						3.09				
8	TCFX	0.21		6.51	0.97	1.97			0.27	6.51	6.51
9	CDLU	12.62	10.65		16.81	9.84	16.28	8.08	18.72		
10	XXXG	7.27	7.27	7.27	7.27	7.27	7.27	7.27	7.27	7.27	7.27
12	FTDG	9.79	9.79	9.79	9.79	8.31	5.14	9.79	9.79	9.79	9.79
14	FTDR	1.69	1.01		0.31	0.37		0.64	2.50		
15	FTHG	30.49	30.49	30.49	30.49	25.89	8.01	30.49	30.49	30.49	30.49
16	SMCX	19.21	18.59		11.46	11.96	14.69	26.28	7.85	30.35	
17	ATFR	9.40	8.20		26.69	48.90	48.90	17.95	24.28	41.10	
18	TODR	2.22	5.05		5.45	4.33		2.93		1.21	
19	FTDU										
20	FTBS				3.43	4.30					
21	SPFU			24.55	19.41	0.12	0.19				24.55
23	XXXW	0.08		0.40	0.03	14.90	15.33		0.10		0.40
26	ATFX	3.01	2.57	15.33	15.33	0.46	0.26	1.95	5.75	4.89	15.33
27	XXXD	0.87	1.08	1.08	0.86			0.56	0.81		1.08
28	TFLX						8.18				
30	ATFU	5.81	7.21	7.21	5.06		1.89	3.75	3.63	7.21	7.21
31	SPLX						3.50				
32	FTHP	0.20		1.99	0.96	0.27			0.26	1.99	1.99
34	FTHW	6.93			7.08	43.42			25.66		
35	WSFX			4.31	0.32						4.31
36	COCX								3.48		
37	INKX	0.36	0.31		0.99	0.77		0.23	0.68	0.17	
41	SFFX								7.99	26.90	
42	TODG	1.89	2.35		3.53	3.99		1.22	0.60		
50	ACDD	0.08	0.15		0.16	0.31	0.23	0.11	0.22		

Note: Abbreviations for foods as in table 6.2. Abbreviations for infants and age blocks as in table 3.1. Diet values in grams per day; a blank indicates zero. Core foods that are not listed have a measured value of zero for all infants, all ages.

Table 8.3 Indices of Consistency Across Age in Pairwise Rankings of Deviations from Content of Energy-Maximizing Optimal Diets

Nutrient	Like-Aged Infants			All Infants		
	Inconsistencies		Consistency Index	Inconsistencies		Consistency Index
	Actual	Possible		Actual	Possible	
Water	1	10	90%	7	42	83%
Mineral	4	10	60%	19	41	54%
Lipid	1	10	90%	7	42	83%
Protein	1	8	88%	10	39	74%
Fiber	3	7	57%	18	39	54%
Other carbohydrate	6	10	40%	11	40	73%
Energy	2	10	80%	4	40	90%
Total	18	65	72%	76	283	73%

Note: See text for details.

Table 8.4 Individual Values for Fitness Indicators and Predictor Variables

		Fitness Indicators									Relational Variables					
Name	Sex	Reproductive Period (days)[1]	Age at Menarche (days)[2]	Age at First Conception (days)[2]	Live Births[3]	Live Births per 1000 Days[4]	Yearlings[5]	Yearlings per 1000 Days[6]	Survival to Six Years[7]	Mother's Rank[8]	Play Bout Length[9]	Play Bouts per Day[9]	Play Time per Day[9]	Involuntary Sucking Terminations[10]	Number of Food Firsts[11]	Number of Food Seconds[12]
(a)	(b)	(c)	(d)	(e)	(f)	(g)	(h)	(i)	(j)	(k)	(l)	(m)	(n)	(o)	(p)	(q)
NAME	SEX	REPD	PUBAGE	CONAGE	INF	INFRATE	YRLG	YRLGRATE	SURVIVE6	BRANK	PLAYDUR	PLAYFREQ	PLAYTIME	REBUFF	FOOD1	FOOD2
Alice	f	2550	1862	2353	2	.784	2	.784	yes	2	.10	12.62	1.262	21.6	2	5
Bristle	m	—	—	—	—	—	—	—	yes	17	.12	28.26	3.391	28.6	4	6
Dotty	f	4035	1637	2073	5	1.239	4	.991	yes	2	—	—	—	—	5	3
Eno	f	3158	1844	2284	4	1.267	3	.950	yes	17	.06	12.49	0.749	40.6	6	7
Fred	m	—	—	—	—	—	—	—	yes	10	.07	25.37	1.776	50.0	5	7
Hans	m	—	—	—	—	—	—	—	yes	12	.09	35.88	3.229	34.4	2	3
Ozzie	m	—	—	—	—	—	—	—	yes	9	.18	22.81	4.106	54.6	10	7
Pedro	m	—	—	—	—	—	—	—	no	9	.08	6.120	0.490	57.1	2	1
Pooh	f	0	—	—	0	—	0	—	no	15	0	0	0	46.7	5	3
Striper	f	1532	1657	2191	2	1.305	0	—	yes	3	—	—	—	—	5	5
Summer	f	2677	1628	2117	4	1.494	2	.747	yes	8	.10	32.30	3.230	34.4	2	5

		Energy Variables				Nonmilk Energy			Protein Variables			Nonmilk Protein		
Name	Sex	Observed (kJ/day)	Deviation from Optimum[13]	Deviation from Requirement[14]	Milk (kJ/day)[15]	(kJ/day)[16]	Deviation from Optimum[17]	Deviation from Requirement[18]	Observed (g/day)	Deviation from Optimum[13]	Deviation from Requirement[14]	(g/day)[16]	Deviation from Optimum[17]	Deviation from Requirement[18]
		(r)	(s)	(t)	(u)	(v)	(w)	(x)	(y)	(z)	(a')	(b')	(c')	(d')
		J	JDO	JDR	MILKJ	NMILKJ	NMILK-JDO	NMILK-JDR	P	PDO	PDR	NMILKP	NMILK-PDO	NMILK-PDR
Alice	f	1921	−.5763	1.461	710.8	1210	−.7402	.5551	11.818	−.5759	.6549	8.290	−.7029	1.109
Bristle	m	2157	−.5695	2.024	1240	916.5	−.8171	.2790	13.695	−.5411	.7277	7.530	−.7472	1.040
Dotty	f	2637	−.4405	2.154	1300	1337	−.7163	.5993	16.12	−.4106	.7426	9.680	−.6461	1.333
Eno	f	2642	−.4521	2.567	1355	1287	−.7356	.6810	17.027	−.4182	.7710	10.310	−.6489	1.653
Fred	m	1749	−.6356	1.196	657.5	1092	−.7752	.3958	10.923	−.6252	.6316	7.665	−.7358	.9339
Hans	m	2813	−.4405	3.186	1970	843.0	−.8323	.2545	15.47	−.5056	.7699	5.710	−.8175	.6039
Ozzie	m	2440	−.4939	2.087	1313	1127	−.7683	.4408	14.385	−.5091	.7218	7.863	−.7312	.9824
Pedro	m	1077	−.7796	0.6027	765.0	312.0	−.9362	−.5357	7.70	−.7355	.5377	3.900	−.8660	.0955
Pooh	f	1264	−.7328	0.544	373.8	889.8	−.8102	.0752	9.547	−.6679	.4986	7.691	−.7323	.8571
Striper	f	1797	−.6187	1.150	427.0	1370	−.7093	.6388	12.62	−.5386	.6712	10.500	−.6161	1.530
Summer	f	2578	−.4841	2.427	1540	1038	−.7925	.3751	15.877	−.4656	.7502	8.230	−.7219	1.152

Note: For each yearling, diet values are unweighted means of its ten-week values (table 7.8) except that Pooh's small sample at 30–40 weeks was omitted.

[1] Days from puberty to death or December 31, 1988, whichever is earlier.

[2] PUBAGE = age at first persistent sexual skin swelling that was followed by menstruation (Altmann, Altmann, and Hausfater 1981); CONAGE = age at onset of deturgescence of sexual skin in first conception cycle.

[3] Through December 31, 1988.

[4] INFRATE = 1000 INF/REPD.

[5] Number of offspring surviving through twelve months, as of December 31, 1988.

[6] YRLGRATE = 1000 YRLG/REPD.

Table 8.4 (*Continued*)

[7] Alice died on January 30, 1987. Bristle was last seen November 21, 1984. Dotty was still alive on August 7, 1997. Eno died on July 25, 1990, Hans on April 12, 1992, Ozzie on February 1, 1991. Fred was last seen February 10, 1984. Pedro died on January 10, 1976. Pooh on October 19, 1976. Striper on May 29, 1982. Summer on April 20, 1987. Dates of disappearances are taken to be dates of death.

[8] Dominance rank of mother at birth of subject.

[9] Mean bout length is product-limit estimate of mean duration; mean bouts of play per day calculated from product of mean bout length and mean bouts per day, assuming independence.

[10] Involuntary suckling terminations: percentage of nursing bouts terminated by the mother.

[11] Number of core foods for which this infant's bout rate was highest.

[12] Number of core foods for which this infant's bout rate was second highest.

[13] Mean shortfall: mean across ten-week age blocks of relative deviations of average daily intakes from intake in energy-maximizing optimal diets: JDO = J/energy optimum − 1 and PDO = P/protein optimum − 1, from table 7.8.

[14] Mean surplus: mean across ten-week age blocks of relative deviation of average daily intake, all foods, from minima: JDR = J/energy requirement − 1 and PDR = P/protein requirement − 1.

[15] Milk kJ/day: ml/day (table 5.8) × milk density (1.027 g/ml) × milk energy (3.2233 kJ/g, table 5.1).

[16] NMILKJ = J − milk energy and NMILKP = P − milk protein; g milk protein/day = ml milk/day (table 6.6) × milk density (1.027 g/ml) × milk protein concentration (0.016 g protein/g milk, table 5.1).

[17] Mean across ten-week age blocks of relative deviations of average daily intake from intake in energy-maximizing optimal diets: NMILKJDO = NMILKJ/energy optimum − 1 and NMILKPDO = NMILKP/protein optimum − 1, from table 7.8.

[18] Mean across ten-week age blocks of relative deviations of mean daily intake from requirements at that age: NMILKJDR = NMILKJ/energy requirement − 1, and NMILKPDR = NMILKP/protein requirement − 1.

Table 8.5 Descriptive Statistics for Potential Predictors of Fitness

Variable	N	Mean	SD	Minimum	Maximum
REPD	6	2325	1402	0	4035
PUBAGE	5	1726	116.95	1628	1862
CONAGE	5	2204	115.65	2073	2353
INF	6	2.83	1.83	0	5
INFRATE	5	1.22	0.26	0.78	1.49
YRLG	6	1.83	1.6	0	4
YRLGRATE	5	0.69	0.4	0	0.99
BRANK	11	9.45	5.54	2	17
PLAYDUR	9	0.09	0.05	0	0.18
PLAYFREQ	9	19.54	12.3	0	35.88
PLAYTIME	9	2.03	1.49	0	4.11
REBUFF	9	40.89	12.12	21.60	57.10
FOOD1	11	4.36	2.42	2	10
FOOD2	11	4.73	2	1	7
J	11	2098	585.74	1077	2813
JDO	11	-0.57	0.12	-0.78	-0.44
JDR	11	1.76	0.84	0.54	3.19
MILKJ	11	1059	504.16	373.77	1970
NMILKJ	11	1038	300.66	312.00	1370
NMILKJDO	11	-0.78	0.06	-0.94	-0.71
NMILKJDR	11	0.34	0.34	-0.54	0.68
P	11	13.2	2.97	7.70	17.03
PDO	11	-0.54	0.10	-0.74	-0.41
PDR	11	0.68	0.09	0.50	0.77
NMILKP	11	7.94	1.92	3.90	10.50
NMILKPDO	11	-0.72	0.07	-0.87	-0.62
NMILKPDR	11	1.03	0.43	0.10	1.65

Note: Variables as defined in notes to table 8.4. Statistics for the first seven variables (REPD through YRLGRATE) are for females only; the rest are for males and females pooled.

Table 8.6 Product-Moment Correlation Coefficients for Fitness Functions and Predictors

	REPD	PUB-AGE	CON-AGE	INF	INF RATE	YRLG	YRLG RATE	BRANK	PLAY DUR	PLAY FREQ	PLAY TIME	RE-BUFF
REPD	1	−.005	−.339	.945	.007	.934	.892	−.348	.829	.663	.551	−.546
PUBAGE	5/.994	1	.939	−.357	−.720	.113	.312	.373	−.439	−.997	−.965	−.264
CONAGE	5/.577	5/.018	1	−.651	−.731	−.227	−.003	.227	−.233	−.957	−.885	−.467
INF	6/.005	5/.555	5/.234	1	.507	.873	.698	−.213	.700	.794	.673	−.243
INFRATE	5/.991	5/.171	5/.160	5/.383	1	−.054	−.158	.367	−.203	.743	.602	.799
YRLG	6/.006	5/.856	5/.714	6/.023	5/.931	1	.938	−.133	.701	.533	.395	−.418
YRLGRATE	5/.042	5/.609	5/.996	5/.190	5/.800	5/.018	1	.334	−.985	−.646	−.779	.626
BRANK	6/.499	5/.536	5/.714	6/.685	5/.543	6/.802	5/.583	1	−.331	−.028	−.068	.139
PLAYDUR	4/.171	3/.711	3/.850	4/.300	3/.870	4/.299	3/.110	9/.385	1	.547	.820	−.053
PLAYFREQ	4/.337	3/.048	3/.187	4/.206	3/.467	4/.467	3/.553	9/.943	9/.127	1	.863	−.330
PLAYTIME	4/.449	3/.169	3/.309	4/.327	3/.589	4/.605	3/.431	9/.862	9/.007	9/.003	1	−.191
REBUFF	4/.455	3/.830	3/.691	4/.757	3/.411	4/.582	3/.570	9/.722	9/.892	9/.386	9/.622	1
FOOD1	6/.889	5/.997	5/.842	6/.984	5/.666	6/.982	5/.927	11/.691	9/.371	9/.803	9/.548	9/.176
FOOD2	6/.599	5/.259	5/.240	6/.611	5/.953	6/.794	5/.954	11/.535	9/.291	9/.363	9/.286	9/.705
J	6/.009	5/.782	5/.426	6/.001	5/.388	6/.027	5/.130	11/.971	9/.206	9/.023	9/.043	9/.261
JDO	6/.002	5/.883	5/.486	6/.001	5/.540	6/.015	5/.077	11/.863	9/.201	9/.039	9/.060	9/.234
JDR	6/.027	5/.986	5/.643	6/.008	5/.419	6/.048	5/.136	11/.645	9/.251	9/.017	9/.052	9/.247
MILKJ	6/.058	5/.786	5/.474	6/.016	5/.392	6/.051	5/.135	11/.519	9/.194	9/.014	9/.034	9/.475
NMILKJ	6/.175	5/.924	5/.949	6/.265	5/.754	6/.409	5/.630	11/.304	9/.687	9/.580	9/.579	9/.268
NMILKJDO	6/.239	5/.883	5/.883	6/.376	5/.598	6/.479	5/.620	11/.241	9/.755	9/.722	9/.700	9/.273
NMILKJDR	6/.100	5/.491	5/.570	6/.178	5/.642	6/.297	5/.851	11/.464	9/.564	9/.353	9/.406	9/.172
P	6/.028	5/.748	5/.423	6/.003	5/.252	6/.059	5/.277	11/.898	9/.330	9/.070	9/.102	9/.262
PDO	6/.013	5/.658	5/.327	6/.001	5/.300	6/.045	5/.298	11/.795	9/.328	9/.113	9/.138	9/.238
PDR	6/.017	5/.767	5/.454	6/.006	5/.233	6/.092	5/.289	11/.984	9/.122	9/.015	9/.034	9/.224
NMILKP	6/.384	5/.926	5/.894	6/.347	5/.664	6/.631	5/.565	11/.522	9/.940	9/.979	9/.959	9/.357
NMILKPDO	6/.424	5/.830	5/.842	6/.428	5/.829	6/.697	5/.464	11/.349	9/.902	9/.868	9/.917	9/.364
NMILKPDR	6/.272	5/.867	5/.935	6/.238	5/.595	6/.499	5/.717	11/.701	9/.935	9/.761	9/.799	9/.250

FOOD1	FOOD2	J	JDO	JDR	MILKJ	NMILKJ	NMILK-JDO	NMILK-JDR	P	PDO	PDR	NMILKP	NMILK-PDO	NMILK-PDR
−.074	.274	.922	.960	.863	.797	.636	.569	.730	.859	.906	.893	.439	.407	.537
.002	.626	−.172	−.092	−.011	−.169	.060	.092	.412	−.199	−.272	−.184	−.059	−.134	.105
−.125	.645	−.469	−.416	−.284	−.426	.040	.092	.345	−.471	−.559	−.444	−.083	−.125	.051
.011	.265	.980	.982	.926	.896	.544	.446	.632	.957	.978	.937	,470	.403	.570
.266	.037	.503	.370	.475	.500	−.194	−.321	−.285	.633	.585	.653	.268	.135	.324
.012	.138	.862	.899	.815	.810	.418	.363	.514	.794	.822	.741	.251	.205	.348
−.057	−.036	.767	.837	.760	.762	−.295	−.304	−.117	.608	.588	.596	−.349	−.435	−.224
.136	.210	.013	−.059	.157	.218	−.342	−.386	−.247	.044	−.089	−.007	−.217	−.313	−.131
.340	.396	.467	.471	.428	.477	.157	.122	.223	.368	.369	.554	−.030	−.048	.032
−.097	.345	.738	.692	.760	.775	.214	.139	.352	.629	.565	.769	.010	−.065	.119
.232	.400	.681	.647	.662	.705	.215	.150	.317	.579	.535	.704	.020	−.041	.100
.495	−.147	−.419	−.442	−.431	−.275	−.414	−.410	−.498	−.419	−.438	−.450	−.349	−.344	−.428
1	.559	.155	.186	.014	−.071	.422	.414	.367	.210	.250	.122	.415	.395	.368
11/.074	1	.373	.390	.287	.071	.607	.563	.671	.418	.409	.455	.552	.480	.614
11/.648	11/.259	1	.993	.965	.858	.509	.451	.635	.970	.927	.959	.383	.310	.480
11/.583	11/.236	11/.000	1	.936	.807	.582	.531	.695	.970	.946	.949	.449	.384	.536
11/.967	11/.392	11/.000	11/.000	1	.938	.306	.242	.468	.907	.822	.933	.182	.097	.305
11/.836	11/.835	11/.001	11/.003	11/.000	1	−.005	−.070	.156	.766	.653	.809	−.117	−.197	.003
11/.196	11/.048	11/.110	11/.061	11/.359	11/.989	1	.996	.977	.606	.710	.512	.943	.935	.930
11/.206	11/.071	11/.163	11/.093	11/.473	11/.839	11/.000	1	.959	.551	.667	.450	.942	.943	.917
11/.267	11/.024	11/.036	11/.018	11/.146	11/.648	11/.000	11/.000	1	.709	.780	.646	.893	.864	.915
11/.536	11/.201	11/.000	11/.000	11/.000	11/.006	11/.048	11/.079	11/.015	1	.981	.944	.549	.476	.640
11/.458	11/.211	11/.000	11/.000	11/.000	11/.002	11/.029	11/.014	11/.005	11/.000	1	.907	.665	.611	.734
11/.721	11/.160	11/.000	11/.000	11/.000	11/.003	11/.108	11/.165	11/.032	11/.000	11/.000	1	.407	.338	.521
11/.205	11/.079	11/.245	11/.166	11/.592	11/.732	11/.000	11/.000	11/.000	11/.080	11/.026	11/.214	1	.992	.985
11/.230	11/.135	11/.353	11/.243	11/.776	11/.562	11/.000	11/.000	11/.001	11/.139	11/.046	11/.309	11/.000	1	.959
11/.265	11/.045	11/.135	11/.090	11/.362	11/.994	11/.000	11/.000	11/.000	11/.034	11/.010	11/.100	11/.000	11/.000	1

Note: Values on and above the main diagonal are correlation coefficients. In those below the diagonal, the value before the slash is the sample size; the value after the slash is the probability of getting a coefficient that far from zero by chance. Variables as defined in notes to table 8.4.

Table 8.7 Regression Models of Female Fitness

Model Number	Fitness Measure	V	N	p	R^2_{adj}	Intercept	Play Freq	Mother's Rank	Energy Variables J	JDO	JDR	MILKJ	P	PDO	Protein Variables PDR	NMILK-PDO
1	Reproductive life span	1	6	**	.887	8764	—	—	—	11707	—	—	—	—	—	—
2	Reproductive life span	2	6	**	.976	42171	—	—	-6.857	45770	—	—	—	—	—	—
3	Reproductive life span	3	6	**	.996	27701	—	—	—	25252	—	—	-570.5	—	—	5298
4	Age at menarche	1	3	*	.989	1996	-11.39	—	—	—	—	—	—	—	—	—
5	Age at menarche	3	5	*	.994	7691	—	29.01	—	2650	—	—	—	—	-6688	—
6	Age at first conception	3	5	*	.994	7003	—	26.56	—	1714	—	—	—	—	-5710	—
7	Number of infants	1	6	**	.958	11.52	—	—	—	15.80	—	—	—	—	—	—
8	Number of infants	2	6	**	.9995	-6.887	—	—	6.480E-3	—	-2.414	—	—	—	—	—
9	Infants per 1000 days	3	5	.068	.989	23.13	—	—	-7.268E-3	—	—	4.044E-3	—	19.42	—	—
10	Infants per 1000 days	3	5	.064	.990	-13.24	—	-0.03404	—	-6.679	—	—	—	—	15.70	—
11	Number of yearlings	1	6	*	.755	8.751	—	—	—	12.58	—	—	—	—	—	—
12	Number of yearlings	2	6	**	.939	33.22	—	—	—	30.53	—	—	—	—	-21.47	—
13	Number of yearlings	3	6	**	.997	56.87	—	—	—	55.76	—	—	—	—	-18.03	—
14	Yearlings per 1000 days	3	5	*	.999	6.468	—	—	-5.658E-3	17.08	—	—	—	-14.68	—	6.136

Note: For each fitness measure, the table first gives a one-variable linear regression model based on the predictor variable with the highest significant correlation coefficient, then the most significant two-variable model found by stepwise multiple regression, and then, where that fell short of R^2_{adj} = 0.99, the smallest significant model (fewest variables, preferably nutrient based) that did not. The table gives the number of variables in each model (V), the number of subjects available for each regression (N), the probability (p) that all coefficients other than the y-intercept are significantly different from zero (* denotes $0.05 \geq p > 0.01$, ** denotes $p \leq 0.01$), the adjusted R^2 values (R^2_{adj}), and the linear regression coefficients. Blank cells indicate zero coefficients; cells with dashes indicate variables that were not entered in the regression. Variables and scaling of variables as in table 8.4. In the regression coefficients, E-n designates a factor of 10^{-n}.

Table 8.8 Fitness Predicted from Energy Shortfall and Protein Surplus

Model Number	Fitness Measure	R^2_{adj}	p	Intercept	Energy Shortfall	Protein Surplus
15	Reproductive life span	0.887	**	8764	11707	—
	Reproductive life span		NS		—	
16	Reproductive life span	0.859	*	12905	14744	−3633
—	Age at puberty	(no significant model found)				
—	Age at first conception	(no significant model found)				
17	Number of infants	0.958	**	11.520	15.80	—
18	Number of infants	0.848	**	−8.863	—	17.20
19	Number of infants	0.944	**	11.77	15.98	−0.2174
—	Infants per 1000 days	(no significant model found)				
20	Number of yearlings	0.755	*	8.751	12.58	—
	Number of yearlings		NS		—	
21	Number of yearlings	0.939	**	33.22	30.53	−21.47
—	Yearlings per 1000 days	(no significant model found)				
	Survival to age six					
22	Discriminant function		**	90	−62.16	−9.798
23	Logistic regression		**	−24.29	41.69	91.27

Note: Survivorship analysis is based on males (5) and females (6) combined. All other analyses are based only on females. For each fitness indicator except survivorship, coefficients of three linear regression models are listed, based, in order, on the mean relative deviation of each female's energy intake from that of her energy-maximizing optimal diet (energy shortfall), the mean deviation of her protein intakes from her protein requirements (protein surplus), and their linear combination. NS indicates not significant ($p > 0.05$). Other abbreviations, variable scaling, and format as in table 8.7.

Table 8.9 Predicting Fitness from Protein Variables

Model Number	Fitness Measure	V	N	R^2_{adj}	p	Intercept	P	PDO	PDR	NMILKPDR
								Protein Variables		
24	Reproductive life span	1	6	.773	*	8795		12563		
—	Age at puberty	(no significant model found)								
—	Age at first conception	(no significant model found)								
25	Number of infants	1	6	.944	**	11.98		17.76		
26	Number of infants	4	6	1.000	**	18.40	−0.3250	25.10	6.550	−2.048
—	Infants per 1000 days	(no significant model found)								
27	Number of yearlings	1	6	.595	*	8.555		13.05		
—	Yearlings per 1000 days	(no significant model found)								

Note: Symbols and abbreviations as in table 8.7.

Table 8.10 Fitness Predicted from Nonmilk Nutrient Variables (Protein and Energy)

Model	Fitness Measure	V	N	p	R^2_{adj}	Intercept	NMILK-JDO	NMILK-JDR	NMILKJ	NMILKP	NMILK-PDR	NMILK-PDO
							Energy Variables			Protein Variables		
28	Reproductive life span	4	6	0.018	0.999	−325936		−56853	100.13		27059	−298799
29	Number of infants	3	6	0.006	0.991	−415.43	−409.09		0.1057		−11.696	
30	Number of infants	4	6	0.003	1.000	−430.83	−433.60		0.1104		−13.085	10.090
31	Number of yearlings	4	6	0.042	0.996	−582.33			0.04615	23.372	−41.776	−544.52

Note: Variables and scaling of variables as in table 8.4.

Table 8.11 Predicting Play Characteristics

Model Number	Play Variable	V	N	R^2_{adj}	p	Inter-cept	Relational Variables				Energy Variables									Protein Variables		
							REBUFF	FOOD1	FOOD2	J	JDO	MILKJ	N-MILKJ	N-MILK-JDO	N-MILK-JDR	P	PDO	PDR	N-MILK-P	N-MILK-PDO	N-MILK-PDR
32	PLAYDUR	4	9	0.823	*	2.138	—	—	—	—		-5.50E-4					3.047	1.261			-0.6153
33	PLAYDUR	5	9	0.936	*	1.023	—	—	—	—		-1.13E-2				2.250		1.546	-2.140		-0.6306
34	PLAYDUR	6	9	0.991	**	0.4582	—	—	—	—		-8.715E-3				1.690	1.417	1.400	-1.633		-0.6165
35	PLAYFREQ	1	9	0.543	*	-1.133	—	—	—	—		0.01875									
36	PLAYFREQ	4	9	0.900	**	-214.2	—	—	—	0.2714	-1013					-36.14	784.2				
37	PLAYFREQ	5	9	0.998	*	-665.5	—	—	—	0.4763	-991.0			-990.8		-53.60				1302	
38	PLAYFREQ	6	9	0.981	*	163.2	-0.3925	-3.213	13.79			0.07252						-363.0			-18.18
39	PLAYTIME	1	9	0.427	*	-5.063												10.53			
40	PLAYTIME	5	9	0.921	*	116.9	-0.1075		1.864								82.83	-67.92	-3.648		
41	PLAYTIME	6	9	0.963	*	150.6	-0.103	-0.2254	2.422								107.74	-91.02	-4.457		
42	PLAYTIME	7	9	1.000	**	153.6	-0.1176	-0.2299	2.606					-7.285			114.7	-97.77	-4.547		
43	PLAYTIME	7	9	1.000	**	223.0						-0.3037			186.6	-172.7	-172.7	262.4	51.34	542.3	-248.2

Note: PLAYDUR = duration (in minutes) of play bouts, PLAYFREQ = play bouts per day, PLAYTIME = minutes of play per day. Other symbols, abbreviations, and scaling as in table 8.7.

Table 9.1 Umbrella Trees: Composition of Seeds, Seed Coats, and Pods

	Seed without Coat		Discarded Seed Coat		Seed with Coat		Discarded Pod per Seed		Pod, Seed, and Coat	
	%	mg/Seed	%	mg/Seed	%	mg/Seed	%	mg/Seed	%	mg/Seed
Unit mass	100	45.2	100	45.6	100	90.8	100	229	100	319.8
Water	67.9	30.7	56.7	25.9	62.2	56.5	67.90	155.50	66.3	212.0
Ash	1.6	0.70	1.0	0.46	1.3	1.18	2.20	5.04	1.9	6.07
Lipid	0.65	0.29	0.6	0.27	0.6	0.57	0.55	1.26	0.57	1.83
Protein	5.3	2.39	3.3	1.51	4.3	3.90	2.90	6.64	3.3	10.5
Fiber	5.0	2.23	12.6	5.75	8.8	8.00	10.80	24.7	10.2	32.7
Carbohydrate	19.6	8.9	25.9	11.8	22.8	20.7	15.70	36.0	17.7	56.6
Energy										
J/mg	5.03		6.99		6.02		4.95		5.24	
J/seed	227.5		318.9		546.4		1134		1680	
Folic acid										
µg/100 g	2.1		16.8		9.45		—		—	
µg/seed		9.5×10^{-4}		76.6×10^{-4}		85.8×10^{-4}		—		—
Ascorbic acid										
mg/100 g	<3.9		<4.8		<8.7		—		—	
mg/seed		$<17.6 \times 10^{-4}$		$<21.9 \times 10^{-4}$		79.0×10^{-4}		—		—
Riboflavin										
mg/100 g	0.20		1.7		0.95		—		—	
mg/seed		0.90×10^{-4}		7.8×10^{-4}		8.6×10^{-4}		—		—
Trypsin inhibitor										
TIU/g	32.84		52.86		42.89		52.86[a]		50.01	
TIU/seed	1.484		2.410		3.894		12.10		15.99	

Note: Composition data from Altmann, Post, and Klein 1984. Seed proximate composition from mean of samples 38B and M331/75, unit weight from those plus 38A, vitamins from 38A. Seed coat proximate composition from sample 39B, unit weight from that plus 39A, vitamins from 39A. Pod proximate composition from mean of samples 40B and M332/75. Pod weight per seed based on mean pod weight of 1.539 g/pod (table 6.3) and intake of 6.72 seeds eaten per pod (above samples) and intake of 6.72 seeds eaten per pod (table 6.3). Trypsin-inhibitor values from samples 38A (seeds) and 39A (coats). Other values by computation. TIU = trypsin-inhibitor unit.

[a] Assumed value.

Table 9.2 *Sporobolus rangei:* Composition of Corm Core, Corm Sheath, and Whole Corm

	Corm Core		Corm Sheath		Whole Corm	
	%	mg/Unit	%	mg/Unit	%	mg/Unit
Unit mass	100	13.09	100	7.76	100	20.9
Water	33.3	4.36	21.6	1.68	28.9	6.04
Ash	3.2	0.42	7.8	0.61	4.9	1.02
Lipid	1.9	0.25	3.2	0.25	2.4	0.50
Protein	11.0	1.44	8.1	0.63	9.9	2.07
Fiber	14.7	1.92	29.9	2.32	20.4	4.24
Carbohydrate	35.9	4.70	29.5	2.29	33.5	6.99
Energy						
J/mg	10.5		12.0		11.1	
J/unit	138		93.2		231	
Folic acid						
μg/100 g	< 25.1		—		—	
μg/unit	< 32.9 × 10^{-4}		—		—	
Ascorbic acid						
mg/100 g	< 25.1		—		—	
mg/unit	< 32.9 × 10^{-4}		—		—	
Riboflavin						
mg/100 g	< 0.25		—		—	
mg/unit	< 0.33 × 10^{-4}		—		—	

Note: Composition data from Altmann, Post, and Klein 1984. Unit weight and proximate composition of corm cores and corm sheaths from samples 16B and 17B, respectively, and of whole corms by summation. Vitamin content of corm cores from sample 15A.

Table 9.3 *Trianthema ceratosepala:* Composition of Fruit, Fruit Wall, and Flower Bud

	Fruit		Fruit Wall		Fruit + Wall		Whole Flower	
	%	mg/Fruit	%	mg/Calyx	%	mg/Unit	%	mg/Flower
Unit mass	100	26.1	100	90.0	100	116	100	145
Water	62.7	16.4	79.1	71.2	75.5	87.6	81.0	117
Ash	3.3	0.86	5.3	4.77	4.9	5.63	2.6	3.77
Lipid	2.7	0.70	1.0	0.90	1.4	1.60	1.0	1.45
Protein	6.5	1.70	2.2	1.98	3.2	3.68	2.7	3.92
Fiber	14.6	3.81	4.9	4.41	7.1	8.22	5.0	7.25
Carbohydrate	10.3	2.69	7.5	6.75	7.9	9.19	7.8	11.3
Energy								
J/mg	5.95		2.70		3.39		2.83	
J/unit	155.4		243		398		411	
Ascorbic acid								
mg/100 g	14.1		4.5		9.3		9.3	
mg/100 units	0.37		0.98		1.35		1.35	
Riboflavin								
mg/100 g	0.87		−0.63[a]		0.24		0.24	
mg/100 units	0.023		0.012		0.035		0.035	
Folic acid								
μg/100 g	14.1		15.7		14.9		14.9	
μg/unit	36.8×10^{-4}		141.3×10^{-4}		172.8×10^{-4}			

Note: Composition data from Altmann, Post, and Klein 1984. Fruit proximate composition and unit mass from mean of samples 1896/74 and 776/75. Proximate composition and unit mass calyx from 1900/74. Proximate composition and unit mass of floral buds from sample 7B. Vitamin content of fruit and floral buds from samples 6A and 7A, respectively. All other values by computation.

[a] This negative computed value for the fruit wall indicates an inconsistency between the riboflavin values for fruit and flower.

Table 9.4 *Withania somnifera:* Composition of Unripe and Ripe Fruit

	Unripe Fruit		Ripe Fruit	
	%	mg/Fruit	%	mg/Fruit
Unit mass	100	83.4	100	78.6
Water	74.6	62.2	61.1	48.0
Ash	1.3	1.1	0.9	0.7
Lipid	4.3	3.6	2.8	2.2
Protein	5.1	4.3	7.3	5.7
Fiber	8.2	6.8	12.9	10.1
Carbohydrate	6.4	5.34	15.1	11.9
Energy				
J/mg	4.63		6.65	
J/fruit	386		523	
Ascorbic acid				
mg/100 g	—		61.8	
mg/100 fruits	—		4.86	
Riboflavin				
mg/100 g	—		0.43	
mg/100 fruits	—		0.034	
Folic acid				
μg/100 g	—		8.9	
μg/100 fruits	—		0.70	

Note: Data from Altmann, Post, and Klein 1984. Unripe fruit values from sample 36B, ripe fruit from 645/76.

Table 9.5 *Azima tetracantha*: Composition of Unripe, Semiripe, and Ripe Fruit

	Unripe Fruit		Semiripe Fruit		Ripe Fruit	
	%	mg/Fruit	%	mg/Fruit	%	mg/Fruit
Unit mass	100	218	100	259	100	270
Water	71.7	156	76.5	198	74.6	201
Ash	4.0	8.7	3.2	8.3	2.7	7.3
Lipid	3.8	8.3	2.8	7.3	2.6	7.0
Protein	7.05	15.4	5.3	13.7	4.75	12.8
Fiber	5.1	11.1	3.1	8.0	3.75	10.1
Carbohydrate	8.35	18.2	9.1	23.6	11.8	31.9
Energy						
J/mg	4.54		3.74		4.14	
J/fruit	990		967		1117	
Ascorbic acid						
mg/100 g	41.5		—		46.0	
mg/100 fruits	9.05		—		12.4	
Riboflavin						
mg/100 g	0.19		—		0.145	
mg/100 fruits	0.041		—		0.039	
Folic acid						
μg/100 g	5.6		—		9.2	
μg/fruit	122×10^{-4}		—		201×10^{-4}	

Note: Composition data from Altmann, Post, and Klein 1984. For unripe fruit, proximate composition and unit mass from mean of samples 38B and M324/75; vitamins from 34A. For semiripe fruit, proximate composition and unit mass from 1895/74; vitamins from average of ripe and unripe fruit values. For ripe fruit, proximate composition and unit mass from mean of 9B and 777/75; vitamins from mean of 9A and 9AS.

Table 9.6 Foods of Amboseli Vervet Monkeys Compared with Those of Yearling Baboons

| Vervets | | | Baboons | | |
| | | Major Food? | Eaten by Baboons? | Core Food? | Reference Numbers in Table 4.1 |
Species	Food				
Acacia xanthophloea	1. Exudate	yes	yes	*yes	27–35
	2. Thorns	yes	once	no	41
	3. Young leaves	yes	yes	no	36–37
	4. Flowers	yes	yes	*yes	22–23
	5. Young seedpods	yes	no	—	—
	6. Mature seeds	yes	yes	yes	25–26, 39–40
	7. Dry seeds	yes	yes	yes	24, 38
	8. Cotyledons	no	no	—	—
A. tortilis	9. Young leaves ⎫ 10. Terminal leaves⎭	yes	once	no	15
	11. Flowers	yes	yes	*yes	7–8
	12. Young seedpods	no	(adults)	—	—
	13. Mature seeds	yes	yes	*yes	17–19
	14. Dry seeds	no	yes	*yes	10, 16
	15. Exudate	no	twice	—	14
	16. Young thorns	no	no	—	—
Lycium "europaeum"	17. Leaves ⎫ 18. Terminal growth⎭	yes	yes	yes	138–40
	19. Fruit	yes	yes	no[a]	136–37
	20. Flowers	no	yes	yes	134–35
Azima tetracantha	21. Fruit	yes	yes	*yes	54–57
	22. Young leaves	no	yes	no	58–60
Salvadora persica	23. Fruit	yes	yes	*yes	162–65
	24. Young leaves	no	yes	yes	167–68
Withania somnifera	25. Fruit	no	yes	yes	220–22
Tribulus terrestris	26. Fruit	yes	yes	*yes	216
	27. Dry seeds	yes	no	—	—
Amaranthus graecizans	28. Leaves and inflorescences	yes	no	—	—
Viscum hildebrandtii	29. Fruit	no	no	—	—
	30. Flowers	no	no	—	—
Commicarpus pedunculosus	31. Leaves	no	no	—	—
	32. Fruit	no	yes	yes	81–82
Achyranthes aspera	33. Leaves	no	once	no	174[b]
Trianthema ceratosepala	34. Flower buds	no	yes	*yes	212–14
Dicliptera "albicularis" [*albicaulis*]	35. Flowers	no	no	—	—
Setaria verticillata	36. Young shoots	yes	yes	no	178–79
Cynodon "dactylon" [*nlemfuensis*]	37. Succulent inner portions of stolons	yes	yes[c]	no	92–95
	38. Seedhead bases	yes	yes[d]	no	87
	39. Terminal leaves	no	yes	no	90–91
	40. Unopened seedheads	no	yes	no	86
Sporobolus cordofanus	41. Unopened seedheads	no	no	—	—
	42. Succulent stem bases	yes	once	no	198
Sporobolus consimilis	43. Bases of stems	no	yes	*yes	190–91
Suaeda monoica	44. Terminal growth (shoots + flowers)	yes	yes	yes	209–10

Table 9.6 (*Continued*)

| Species | Food | Vervets | | | Baboons | |
		Major Food?	Eaten by Baboons?	Core Food?	Reference Numbers in Table 4.1
Insects (mostly grasshoppers, also termites, beetles, etc.)	45. Whole bodies	no	yes	yes	251, 252, 258, 264, 267, 271, 272
Justicia uncinulata	46. Young leaves	no	no	—	—
Solanum setaceum	47. Fruit	no	no	—	—
Solanum incanum	48. Flowers	no	no	—	—
Capparis tomentosa	49. Fruit[e]	no	yes	*yes	68–71, 73–74
Coccinia grandis	50. Leaves	no	no	—	—
	51. Young fruit	no	no	—	—
Pentarrhinum insipidum	52. Leaves	no	no	—	—
Cyphostemma sp.	53. Leaves	no	no	—	—
Chlorophytum sp. nr. *bakeri*	54. Young leaves	no	yes	no	80
Dactyloctenium bogdanii	55. Succulent stem bases	no	yes	no	113

Note: Vervet data from Klein 1978, table 2.2. Of the fifty-five recorded foods of vervets, twenty-two "major" foods (ibid., fig. 2.3) are listed. Similarly, baboon foods that ranked among the top twenty-two by time expended (table 6.1 above) are marked with an asterisk. Foods eaten by baboons but not vervets, such as corms, are not listed.

[a] Probably a sampling error: I took few samples when it was most abundant.

[b] Probably misidentified as *Sericocomopsis hildebrandtii* (footnote, table 4.3).

[c] Baboons ate inner and outer portion of stolons.

[d] Baboons ate entire seedhead.

[e] "Rarely eaten" (Klein 1978, 83).

Table 9.7 Percentage of Feeding Time Devoted to Various Plant Parts by Wild Baboons

Species	Location	Leaves	Flowers	Fruit and Seeds	Corms, Etc.	Animal Matter	Other
Papio anubis	Bole, Ethiopia[1]	32.9	7.4	54.9	1.6	2.7	0.7
	Gilgil, Kenya[2]	28.8	2.6	34.2	27.0	1.9	5.7
	Gombe, Tanzania[3]	13.9	2.2	48.6	6.6	13.1	15.6
	Laikipia, Kenya[4]	26.3	29.5	24.3	16.5	0	3.4
	Shai Hill, Ghana[5]	7.5	4.6	58.8	16.5	0	12.4
Papio cynocephalus	Amboseli, Kenya[6]	14.6	5.1	28.0	30.9	1.5	19.9
	Amboseli, Kenya[7]	23	9	18	34	0	16
	Amboseli, Kenya[7]	17	4	17	50	0	12
	Amboseli, Kenya[8]	22.3	3.1	33.5	23.6	0.4	17.1
	Amboseli, Kenya[9]	20.5	2.0	44.8	20.3	0.12	12.3
	Amboseli, Kenya[9]	19.3	1.7	48.8	21.0	0.76	8.4
	Amboseli, Kenya[9]	21.0	5.7	43.1	22.8	0.13	7.3
	Mikumi, Tanzania[10]	13.1	2.8	51.4	21.8	4.6	6.4
	Ruaha, Tanzania[11]	18.7	1.4	15.5	51.6	8.7	4.4
Papio hamadryas	Erer-Gota, Ethiopia[12]	10.0	21.5	66.5	2	0	0
Papio papio	Mount Assirik, Senegal[13]	7.8	8.6	73.5	4.0	2.5	3.9
Papio ursinus	Cape Reserve, S.A.[14]	16.3	12.8	45.2	24.7	1.3	0
	Drakensberg, S.A.[15]	26	14	3	53	4	0
	Kuiseb, Namibia[16]	26.5		4.3		trace	0
	Okavango, Botswana[16]	10.2		9.4		9.5	1
	Suikerbosrand, S.A.[17]	8.0	7.3	43.3	39.0	2.6	0
Mean[18]		18.3	7.3	39.6	24.6	2.3	7.7
Standard deviation		7.3	7.4	18.6	15.8	3.4	6.5
Minimum		7.5	1.4	3.0	1.6	0	0
Maximum		32.9	29.5	73.5	53.0	13.1	19.9

Note: "Fruit" includes seeds and ovaries as well as fleshy fruits. "Corms" includes all underground plant foods: roots, bulbs, tubers, corms, and rhizomes.

[1] Dunbar and Dunbar 1974, table 6. Subjects: all visible animals. Dunbar 1988, table 13.2, gives very different values for this study.

[2] Harding 1976, tables 3 and 4, and pers. comm.). Time spent feeding on species from which two parts were eaten (e.g., fruit and leaves from *Opuntia*) was apportioned equally between them. Grass seedheads assigned to "fruits and seeds." Subjects: adult males. Different values for this study are given by Dunbar 1988, table 13.2, and by Whiten, Byrne, Barton, Waterman, and Henzi 1991, table 1. In this same population, Johnson 1989, fig. 1 (subjects: eight juvenile males, eight juvenile females), recorded 35% of feeding time on underground storage organs and 50% on "grass," meaning grass blades and seeds and also ground-level herbs.

[3] J. Oliver, pers. comm. to Dunbar 1988, table 13.2. Value in "other" adjusted to achieve 100% as total.

[4] Barton, Whiten, Byrne, and English 1993, table 1. *Sansevieria* base included in "corm" category. Value in "other" adjusted to achieve 100% as total. Subjects: nineteen adult females, seven adult males. Different values for this study are given by Whiten, Byrne, Barton, Waterman, and Henzi 1991, table 1, based on Barton 1989.

[5] Depew 1983 from Dunbar 1988, table 13.2.

[6] Post 1978, table 21, assigning Post's gum, other, nothing, and unknown to "other." Subjects: two adult males, two adult females. Slightly different values for this study are given by Whiten, Byrne, Barton, Waterman, and Henzi 1991, table 1, and by Dunbar 1988, table 13.2.

[7] Silk 1987, fig. 2, and pers. comm. Subjects: pregnant females in, respectively, Hook's and Alto's groups, assigning "sap" (gum)—respectively 11% and 7% in the two groups—to "other."

Table 9.7 (*Continued*)

---

[8] This study (table 6.5), excluding milk, assigning petioles, pedicels, and grass blade bases (meristems) to "leaves," and assigning gum (16.5%) to "other." Subjects: six female yearlings, five male yearlings.

[9] Stacey 1986, table 2, Limp's, Hook's, and Alto's groups, respectively, assigning time on grass to "leaves" and to "corms" in the ratio 1:0.8911, as in this study. Subjects: one adult male and two adult females in each group.

[10] Rhine, Norton, Wynn, and Wynn 1989, table 2, instantaneous means, normalized. Grass assigned to "leaves," pod to "fruits and seeds," bud to "flowers," unknown and bark to "other." Subjects: eighteen adult males, forty-six adult females. Different values for this population are given by Whiten, Byrne, Barton, Waterman, and Henzi 1991, table 1, based on table 7 in Norton, Rhine, Wynn, and Wynn 1987, which is of species frequencies, not feeding time.

[11] Rasmussen 1978, from Dunbar 1988, table 13.2.

[12] Kurt in Kummer 1968, table 14. Kurt's "other foods picked from the ground" assigned to "fruits and seeds," assuming these were primarily grass and acacia seeds. Subjects: all visible animals. Different values for this study are given by Dunbar 1988, table 13.2.

[13] Sharman 1981, from Dunbar 1988, table 13.2.

[14] Davidge 1978, table 9, averaging across months, with *Carpobrutus* time assigned equally to leaves and blossoms. Other "foods" (clay, dirt) described in text and many plant foods in appendix 1 are not included in Davidge's table. Values shown here differ from those in Dunbar 1988, table 13.2.

[15] Whiten, Byrne, and Henzi 1987, from Whiten, Byrne, Barton, Waterman, and Henzi 1991, table 1.

[16] Hamilton, Buskirk, and Buskirk 1978, table 1, averaging across four group-by-date columns. Subjects: all visible animals.

[17] C. Anderson, from Dunbar 1988, table 13.2.

[18] Values from *Papio ursinus* at Kuiseb and Okavango omitted from summary statistics.

Table A1.1 Growth Rates of Baboons (*Papio* spp.) during the First Year of Life

Species	Age	Locale	Growth Rate (g/day)	Reference
Papio cynocephalus	0–1 year	Field	5–6	Altmann 1980
Papio cynocephalus	0–1 year	Field	4.5	Altmann and Alberts 1986
"*Papio cynocephalus*"	0–1/2 year	Lab	8.4	Berchelman, Vice and Kalter 1971
Papio anubis	0–1 year	Enclosure	Males 8.44, females 8.00	Coelho 1985
"*Papio cynocephalus*"	0–1 year	Lab	Males 9.15, females 7.67	Glassman et al. 1984
Papio sp.	0–16 weeks	Lab	Males 12.67, females 10.7	Lewis et al. 1984
Papio sp.	0–16 weeks	Lab	Males 7.54, females 4.9	Lewis et al. 1984
"*Papio cynocephalus*"	0–104 days	Lab	Males 8.4, females 8.6	McMahan, Wigodsky, and Moore 1976
Papio anubis	0–1 year	Field	Males 5.5, females 4.5	Nicolson 1982
Papio sp. and *Papio anubis*	0–1 year	Lab	Males 8.6, females 7.5	Snow 1967, fig. 41
Papio cynocephalus and *Papio papio*	0–60 days	Lab	5.04	Vice et al. 1966
Papio cynocephalus and *Papio anubis*	0–10 weeks	Lab	8 [a]	Roberts, Cole, and Coward 1985
Papio cynocephalus and *Papio anubis*	0–10 weeks	Lab	7.42 [b]	Roberts, Cole, and Coward 1985
Papio cynocephalus and *Papio anubis*	0–10 weeks	Lab	5.40 [c]	Roberts, Cole, and Coward 1985

Note: All lab infants raised on formulas except those in Berchelman, Vice, and Kalter (1971) and Roberts, Cole, and Coward (1985), which were suckled by their mothers. Quotation marks around *Papio cynocephalus* indicate publications that apparently do not restrict this name to yellow baboons.

[a] Mother fed ad libitum.

[b] Mother fed 80% of ad libitum after second week.

[c] Mother fed 60% of ad libitum after second week.

Table A1.2 Cation Requirements and Tolerance

Element	RDA[a]	MTL[b]	Limits Adopted Here (g/day)[i]
Sodium, %	0.22–0.44	3.5[c]	$22 \times 10^{-4} \Phi \leq Na \leq .035 \Phi$
Potassium, %	0.24–1.09	3	$24 \times 10^{-4} \Phi \leq K \leq .03 \Phi$
Calcium, %	0.60–0.80	1–2[d]	$.006 \Phi \leq Ca \leq .02 \Phi$
Phosphorus, %	0.30–0.40	0.6–1.5[d]	$.003 \Phi \leq P \leq .015 \Phi$
Magnesium, %	0.10–0.15	0.5	$.001 \Phi \leq Mg \leq .005 \Phi$
Iron, ppm	100	500–3000	$10^{-4} \Phi \leq Fe \leq .003 \Phi$
Zinc, ppm	20	300–1000	$2 \times 10^{-5} \Phi \leq Zn \leq .001 \Phi$
Iodine, ppm	2	5–400	$2 \times 10^{-6} \Phi \leq I \leq 4 \times 10^{-4} \Phi$
Chromium, ppm	0.5	1000[e]	$5 \times 10^{-7} \Phi \leq Cr \leq .001 \Phi$
Manganese, ppm	~20	400–1000	$2 \times 10^{-5} \Phi \leq Mn \leq .001 \Phi$
Copper, ppm	~2	25–800	$2 \times 10^{-6} \Phi \leq Cu \leq 8 \times 10^{-4} \Phi$
Selenium, ppm	~0.1	2[f]	$10^{-7} \Phi \leq Se \leq 2 \times 10^{-6} \Phi$
Fluoride, ppm	~0.5	40–150[g]	$5 \times 10^{-7} \Phi \leq F \leq 15 \times 10^{-6} \Phi$
Aluminum, ppm	—	1000	$Al \leq .001 \Phi$
Barium, ppm	—	20	$Ba \leq 2 \times 10^{-5} \Phi$
Strontium, ppm	—	2000–3000	$Sr \leq .003 \Phi$
Boron, ppm	—	150[h]	$B \leq 15 \times 10^{-5} \Phi$

[a] Recommended daily allowances, from Nicolosi and Hunt 1979, table 4. Given as percentage of the diet (%) or as parts per million (ppm) in the diet.

[b] Maximum tolerable levels, from NRC 1980, table 1, based on five species of domestic mammals.

[c] Value for cattle only, given as 9 g NaCl.

[d] Ratio of calcium to phosphorus is important.

[e] Value for poultry only; expressed as CrCl.

[f] Value for swine only.

[g] As sodium fluoride or fluorides of similar toxicity.

[h] Value for cattle only.

[i] The notation Na, K, etc. refers here to the consumed amount (g/day) of sodium, potassium, etc., and Φ represents the total mass of the diet (g/day).

Table A8.1 Seasonal Characteristics of Core Foods, 1975–76

Time Budget Food Rank (j) (1)	Code (1)	Mean (2)	SD (3)	Mean + 3 SD (4)	Max (5)	Max (4,5) (s) (6)	Semi-months Available p (7)	h (8)	Fraction of Year (f) (9)	Sample-Minutes (T) (10)	Fraction (t) (11)	Correction Factor (f/t) (12)	Seasonality Constraint (s) (13)	Observed Diet E (14)	SE (15)	Seasonally Adjusted Diet Ê (16)	SE (17)	In-Season Intake (V) (18)	Peak Intake (W) (19)
1	PCMX	304.318	171.842	819.844	652.00	819.844	24	0	1.0000	19,085.07	1.0000	1.0000	819.844	316.94	45.22	260.90	—	260.90	260.90
2	GRCX	4.079	6.529	23.666	26.60	26.600	18	2	0.7917	17,677.79	0.9263	0.8547	22.735	3.15	0.310	2.692	0.265	3.580	30.228
3	GRLU	31.583	27.749	114.830	86.50	114.830	21	2	0.9167	18,476.36	0.9681	0.9469	108.729	37.50	4.33	35.507	4.100	40.496	124.005
4	FTGX	24.867	17.338	76.881	58.20	76.881	19	3	0.8542	18,778.89	0.9840	0.8681	66.740	27.96	4.07	24.272	3.533	30.494	83.848
5	TODU	6.232	15.370	52.342	61.00	61.000	8	4	0.4167	9,483.74	0.4969	0.8385	51.148	4.54	0.480	3.807	0.402	10.964	147.313
6	SKCX	1.046	1.661	6.029	6.04	6.040	5	8	0.3750	13,063.15	0.6845	0.5479	3.309	1.03	0.110	0.564	0.060	2.174	12.746
7	SCTX	2.825	3.706	13.943	17.10	17.100	14	6	0.7083	18,394.19	0.9638	0.7349	12.567	2.82	0.410	2.073	0.301	3.442	20.873
8	TCFX	1.802	1.570	6.512	5.34	6.512	24	0	1.0000	19,085.06	1.0000	1.0000	6.512	2.06	0.250	2.060	0.250	2.060	6.512
9	CDLU	5.826	6.459	25.203	28.20	28.200	17	4	0.7917	18,288.28	0.9583	0.8262	23.298	5.88	0.680	4.858	0.562	6.782	32.525
10	XXXG	2.847	1.474	7.269	5.21	7.269	24	0	1.0000	19,085.06	1.0000	1.0000	7.269	3.69	0.730	3.690	0.730	3.690	7.269
11	TOBS	9.957	23.145	79.392	116.00	116.000	1	2	0.0833	3,480.20	0.1824	0.4570	53.011	7.29	1.45	3.331	0.663	59.951	953.947
12	FTDG	1.653	1.944	7.485	9.20	9.200	20	2	0.8750	18,786.27	0.9843	0.8889	8.178	1.73	0.220	1.538	0.196	1.841	9.792
13	TTFX	4.307	11.850	39.857	55.60	55.600	3	2	0.1667	5,046.29	0.2644	0.6303	35.047	5.88	0.840	3.706	0.529	27.799	262.859
14	FTDR	0.566	0.588	2.330	1.84	2.330	11	6	0.5833	16,212.16	0.8495	0.6867	1.600	0.633	0.081	0.435	0.056	0.905	3.331
15	FTHG	8.426	6.620	28.286	26.80	28.286	19	2	0.8333	18,590.17	0.9741	0.8555	24.199	9.95	1.85	8.512	1.583	10.725	30.490
16	SMCX	1.522	3.849	13.069	19.50	19.500	3	8	0.2917	10,466.54	0.5484	0.5318	10.371	1.38	0.200	0.734	0.106	3.954	55.877
17	ATFR	6.303	8.941	33.126	34.20	34.200	13	5	0.6458	15,513.93	0.8129	0.7945	27.172	6.44	1.08	5.117	0.858	9.200	48.857
18	TODR	0.601	0.998	3.595	4.22	4.220	9	6	0.5000	13,331.38	0.6985	0.7158	3.021	0.584	0.127	0.418	0.091	1.045	7.552
19	FTDU	0.699	1.667	5.700	6.57	6.570	2	3	0.1458	3,838.98	0.2012	0.7250	4.763	0.552	0.132	0.400	0.096	3.919	46.649
20	FTBS	1.769	3.931	13.562	15.80	15.800	4	2	0.2083	4,569.13	0.2394	0.8702	13.749	2.20	0.450	1.914	0.392	11.028	79.198
21	SPFU	3.456	7.646	26.394	35.70	35.700	7	7	0.4375	13,079.00	0.6853	0.6384	22.791	2.98	0.530	1.902	0.338	5.798	69.459
22	SPFS	3.155	9.947	32.996	47.10	47.100	1	7	0.1042	2,792.61	0.1463	0.7119	33.530	2.95	0.660	2.100	0.470	32.262	515.106
23	XXXW	0.121	0.083	0.370	0.296	0.370	21	2	0.9167	18,578.43	0.9735	0.9417	0.348	0.458	0.100	0.431	0.094	0.492	0.397
24	SPFR	2.276	6.560	21.956	25.30	25.300	1	4	0.1250	4,056.08	0.2125	0.5882	14.881	2.33	0.540	1.370	0.318	18.275	198.431
25	GRTU	0.429	0.996	3.417	4.26	4.260	3	3	0.1875	6,251.34	0.3276	0.5724	2.439	0.582	0.087	0.333	0.050	2.369	17.338
26	ATFX	2.877	3.064	12.069	9.75	12.069	16	5	0.7708	17,052.76	0.8935	0.8627	10.412	2.92	0.510	2.519	0.440	3.710	15.333

Table A8.1 (*Continued*)

Time Rank (*j*) (1)	Budget Food Code (1)	Mean (2)	SD (3)	Mean + 3 SD (4)	Max (5)	Max (4, 5) (*s*) (6)	Semi-months Available *p* (7)	*h* (8)	Fraction of Year (*f*) (9)	Sample-Minutes (*T*) (10)	Fraction (*t*) (11)	Correction Factor (*f/t*) (12)	Seasonality Constraint (*ŝ*) (13)	Observed Diet *E* (14)	SE (15)	Seasonally Adjusted Diet *Ê* (16)	SE (17)	In-Season Intake (*V*) (18)	Peak Intake (*W*) (19)
27	XXXD	0.235	0.232	0.931	1.03	1.030	21	2	0.9167	18,974.38	0.9942	0.9220	0.950	0.227	0.116	0.209	0.107	0.239	1.083
28	TFLX	1.334	1.908	7.058	9.18	9.180	6	10	0.4583	15,579.14	0.8163	0.5615	5.154	1.12	0.210	0.629	0.118	1.996	16.358
29	SPLU	0.290	0.162	0.776	0.908	0.908	7	8	0.4583	14,848.45	0.7780	0.5891	0.535	0.278	0.053	0.164	0.031	0.487	1.591
30	ATFU	1.029	1.071	4.242	5.54	5.540	14	4	0.6667	16,492.86	0.8642	0.7714	4.274	1.01	0.200	0.779	0.154	1.315	7.212
31	SPLX	0.471	1.508	4.995	8.02	8.020	5	8	0.3750	13,675.35	0.7165	0.5233	4.197	0.340	0.088	0.178	0.046	0.685	16.168
32	FTHP	0.142	0.159	0.619	0.912	0.912	8	6	0.4583	11,140.70	0.5837	0.7852	0.716	0.139	0.189	0.109	0.148	0.303	1.989
33	SMLU	0.106	0.271	0.919	1.04	1.040	1	2	0.0833	3,201.67	0.1678	0.4967	0.517	0.140	0.038	0.070	0.019	1.251	9.297
34	FTHW	1.230	3.211	10.863	16.20	16.200	2	5	0.1875	7,034.22	0.3686	0.5087	8.241	1.24	0.390	0.631	0.198	5.233	68.367
35	WSFX	0.208	0.313	1.147	1.39	1.390	3	6	0.2500	9,242.71	0.4843	0.5162	0.718	0.195	0.051	0.101	0.026	0.604	4.305
36	COCX	0.172	0.326	1.150	1.57	1.570	0	3	0.0625	4,248.60	0.2226	0.2808	0.441	0.176	0.053	0.049	0.015	1.581	14.106
37	INKX	0.348	0.208	0.972	1.05	1.050	7	9	0.4792	13,739.34	0.7199	0.6656	0.699	0.303	0.088	0.202	0.059	0.586	2.029
38	DUKU	0.128	0.544	1.760	2.90	2.900	0	3	0.0625	2,926.19	0.1533	0.4076	1.182	0.165	0.057	0.067	0.023	2.153	37.834
39	CDSX	0.025	0.023	0.094	0.089	0.094	6	7	0.3958	13,889.12	0.7277	0.5439	0.051	0.022	0.023	0.012	0.013	0.041	0.177
40	SPWX	0.011	0.010	0.041	0.057	0.057	3	10	0.3333	11,043.38	0.5786	0.5761	0.033	0.011	0.020	0.006	0.012	0.031	0.160
41	SFFX	0.514	1.237	4.225	4.31	4.310	2	2	0.1250	3,440.27	0.1803	0.6934	2.989	0.750	0.185	0.520	0.128	5.546	31.873
42	TODG	0.354	0.516	1.902	1.64	1.902	6	7	0.3958	10,565.82	0.5536	0.7150	1.360	0.377	0.079	0.270	0.056	0.932	4.701
43	TOBR	0.452	2.152	6.908	11.40	11.400	0	2	0.0417	2,468.52	0.1293	0.3221	3.672	0.433	0.123	0.139	0.040	6.698	176.334
44	XXWX	0	0	0	0	0	1	9	0.2292	10,990.57	0.5759	0.3979	0	0	0	0	0	0	0
45	SULU	0.119	0.079	0.356	0.231	0.356	14	4	0.6667	15,789.68	0.8273	0.8058	0.287	0.116	0.081	0.093	0.065	0.158	0.484
46	XXLX	0.119	0.112	0.455	.657	0.657	4	9	0.3542	13,890.53	0.7278	0.4866	0.320	0.121	0.037	0.059	0.018	0.254	1.381
47	GRDU	0.220	0.645	2.155	3.17	3.170	0	2	0.0417	3,018.87	0.1582	0.2634	0.835	0.030	0.093	0.008	0.024	0.379	40.076
48	SPFX	0.801	1.276	4.629	6.36	6.360	1	5	0.1458	3,486.50	0.1827	0.7983	5.077	0.568	0.151	0.453	0.121	5.330	59.676
49	TFBS	0.091	0.310	1.021	1.66	1.660	0	3	0.0625	3,349.93	0.1755	0.3561	0.591	0.112	0.030	0.040	0.011	1.276	18.917
50	ACDD	0.164	0	0.164	0.164	0.164	0	7	0.1458	7,102.00	0.3721	0.3919	0.064	0.164	0.068	0.064	0.027	0.881	0.881
51	CTFX	0.023	0	0.023	0.023	0.023	2	6	0.2083	8,078.40	0.4233	0.4922	0.011	0.023	0.014	0.011	0.007	0.087	0.087
52	SBFX	0.131	0.375	1.256	1.35	1.350	1	2	0.0833	2,825.38	0.1480	0.5629	0.760	0.206	0.084	0.116	0.047	2.088	13.682

Note:

Column 1: Core foods, abbreviated as in table 6.2.

Column 2: Mean of twenty-eight individual × age block intakes (g/day).

Column 3: Standard deviation of these twenty-eight values.

Column 4: Mean intake (g/day) plus three standard deviations = col. 2 + col. 3.

Column 5: Highest of these twenty-eight values.

Column 6: Greater of columns 4 and 5.

Columns 7 and 8: Number of semimonths that the food was "present" (p) or "half-present" (h) (see text). From table A8.2.

Column 9: Fraction of year that this food was available $f = (2p + h)/48$.

Columns 10 and 11: Respectively, the number (T) and proportion (t) of sample-minutes in 1975–76 during which this food was available. T calculated from values in tables A8.2 and A8.3; $t = T/19085.07$.

Column 12: Correction factor = f/t = col. 9/col. 11.

Column 13: Seasonality constraint (used for optimal diets 6–10) = col. 6 × col. 12.

Column 14: Observed diet (annual pooled mean intake, g/day), from table 6.1.

Column 15: Standard error of column 14.

Columns 16 and 17: Seasonally adjusted diet (annual pooled mean intake, g/day, adjusted for seasonal differences in observation time) and its standard errors, where col. 16 = col. 12 × col. 14 and col. 17 = col. 12 × col. 15. Milk intake (food 1, PCMX) in col. 16 from endnote 6.

Column 18: Mean daily intake (g/day) during portion of year each food was "in season" (present or half-present) = (col. 14 [col. 7 + col. 8])/(col. 11 [col. 8/2 + col. 7]). For milk (food 1, PCMX), mean daily intake assumed to be independent of season.

Column 19: Mean daily intake (g/day) on days when food was "present" = (col. 6 [col. 8 + col. 7])/(col. 11 [col. 8/2 + col. 7]). Milk value as in columns 16 and 18.

Table A8.2 Seasonal Availability of Core Foods

Time Budget Rank (j)	Food Code	JAN		FEB		MAR		APR		MAY		JUN		JUL		AUG		SEP		OCT		NOV		DEC		Totals	
		1	2	1	2	1	2	1	2	1	2	1	2	1	2	1	2	1	2	1	2	1	2	1	2	p_j	h_j
1	PCMX	p	p	p	p	p	p	p	p	p	p̂	p	p	p	p	p	p	p	p	p	p	p	p	p	p	24	0
2	GRCX			h				p	p	p	p	p	p	p	p̂	p	p	p	p	p	p	p	h	p	p	18	2
3	GRLU	p	p	p	p	p	p	p	p	p	p	p	p	p	p̂	p	p	h		p	h	p	p	p	p	21	2
4	FTGX	h		h	p	p	p	p	p	h	p	p	p	p	p̂	p	p	p	h	p	p	p	p	h		19	3
5	TODU			h		p		p		p		p	p	p	p̂	p	h	p	h	h	p			h		8	4
6	SKCX							h				h	p	h							p					5	8
7	SCTX	h		p	p	p	p	p	p	h		p	p	p		p	p	p	p	h	p		p	h		14	6
8	TCFX	p	p	p	p	p	p	p	p	p	p	p	p	p	p̂	p	p	p	p	p	p	p	p	p	p	24	0
9	CDLU	h		p	p	p	p	p	p	p	p	p	p	p		p	p	p	p	p	p	p	p	p	p	17	4
10	XXXG	p	p	p	p	p	h	p	p	p	p	p	p	p	p̂	p	p	p	p	p	p	p	p	p	p	24	0
11	TOBS			p	p	h																		h		1	2
12	FTDG	h		p	p	p	p	p		p	p	p	p	p	p	p	p	p	p	p	p	p	p	h		20	2
13	TTFX											h	p	h				h		h	h					3	2
14	FTDR	h		p	h	p	p	p	p	p		p	p	p		p	h	h		p	p		h			11	6
15	FTHG			p	p	p	p	p	p	p	h	p	p	p	p̂	p	h	p	h	h	p	p	p	h		19	2
16	SMCX					p	h	h				p	p	p							h					3	8
17	ATFR						h			h		p	p	p		p		p		p	p	p	p	p	h	13	5
18	TODR	p	p	p	p	p		p				p	p	p	p̂	h		p		p	p	h	h	h	h	9	6
19	FTDU			h								p	h	h				h		h		h				2	3
20	FTBS					h												p								4	2
21	SPFU				h					h		p	p	p	p̂		h	p	h	p	p	h	p	h		7	7
22	SPFS													p				h		h		h				1	3
23	XXXW	p	p	p	p	p	h	p	p	p	p	p	p	p	p̂	p	h	p	h	p	p	p	p	p	p	21	2
24	SPFR									p		p	p	p		h	h	h		h	p		p			1	4
25	GRTU				h	h		h		h		h	h	p		h		h		p	h					3	3
26	ATFX	p	p	p	p	p	p	p		h	p	p	p	p	p̂	p	p	p	h	p	p	p	p	p	h	16	5
27	XXXD	h	p	p	p	p	p	p	p	h	p	p	p	p	p̂	p	p	p	h	p	p	p	p	p	h	21	2
28	TFLX	h		h	h	p	p	p	h	h	h	h	h	h		p		h	h	h	h	h	p	h	h	6	10
29	SPLU					p	p					p	h	p		p		p	h	p	p	h	p			7	8
30	ATFU					p	p		h	p		p	p	p		p	p	h		h	p	h	p	h		14	4
31	SPLX								h		h	p	h	p		p		p		p	p	p	p			5	8
32	FTHP										h	p	p	p	p̂	p		h		p	h	p	p		p	8	6
33	SMLU									h	p	p	h	p		p		p		h		h		h		1	2

Code	Food																									
34	FTHW					h			h		h		h	h	h		p	h		2	5
35	WSFX		h	h		h			h		h	h	h	p	h	p̂	p	h		3	6
36	COCX	h							h	h	h	h	h		p		p			0	3
37	INKX			h		h			h	h	h	p	h	h	p	h	p	h	h	7	9
38	DUKU	h												p		p		p	h	0	3
39	CDSX			h	p	h		h	p	h	h	p	p	h		h	p		h	6	7
40	SPWX	h		h					h	p	p	p	p	p	h		h	h		3	10
41	SFFX						h			h	p	h			h					2	2
42	TODG		h	h		h			h	p	p	h	p		p	h	p	h		6	7
43	TOBR			h	h				h	h			h							0	2
44	XXWX			h		h			h	h	p	p	h	p	p	p	p	h		1	9
45	SULU			h	p	p			h	p	p	h	h	p	p	p	p	h		14	4
46	XXLX			h	p	p			h	p	p	h	p	h	p	p	p	h		4	9
47	GRDU			h		h			h			h	h							0	2
48	SPFX									h	h	h	h	p̂	h	h				1	5
49	TFBS		h	h		h	h		h			h		h						0	3
50	ACDD		h	h		h			h		h			h		p		h		0	7
51	CTFX	h	p	p		p		h	h			h	p		h		p	h	h	2	6
52	SBFX						h	p	h											1	2

Note: The first and second semimonth of each month are indicated by 1 and 2. Foods were eaten in all semimonths marked p (present) or h (half-present) and probably also where presence is interpolated (p̂). Food codes as in table 6.2.

Table A8.3 Semimonthly in-Sight Sample Time, Bout Frequency, and Bout Rate:
Core Foods, 1975–76 Data Only

Semi-month	In-Sight Sample Time		Bout Frequency	Bout Rate (bouts/1000 minutes)
	(minutes/day)	(%)		
Jan-1	188.72	1.0	58	307.3
Jan-2	196.10	1.0	69	351.9
Feb-1	1011.68	5.3	640	632.6
Feb-2	1205.05	6.3	729	605.0
Mar-1	1263.47	6.6	834	660.1
Mar-2	1779.02	9.3	1441	810.0
Apr-1	586.03	3.1	418	713.3
Apr-2	831.53	4.4	786	945.3
May-1	1606.02	8.4	1567	975.7
May-2	182.80	1.0	187	1023.0
Jun-1	1412.85	7.4	1107	783.5
Jun-2	1229.73	6.4	966	785.5
Jul-1	600.68	3.1	512	852.4
Jul-2	14.22	0.1	9	632.9
Aug-1	383.47	2.0	322	839.7
Aug-2	506.63	2.7	294	580.3
Sep-1	1014.59	5.3	708	697.8
Sep-2	608.70	3.2	385	632.5
Oct-1	771.64	4.0	477	618.2
Oct-2	864.85	4.5	713	824.4
Nov-1	802.72	4.2	631	786.1
Nov-2	1002.13	5.3	930	928.0
Dec-1	912.38	4.8	897	983.1
Dec-2	110.07	0.6	56	508.8
Total	19085.07	100	14,736	772.1

Notes

1. This section, which occupies the rest of this chapter, can be skipped by readers impatient to get to the results, which begin in chapter 4.

2. The search for theories is not unlike the search for an optimal diet. I hope that the approach taken in this work will be judged not by whether it achieves the unlikely goal indicated above, but by whether it is moving in the right direction and is, for the questions I pose, preferable to any other available approach . . . or perhaps I should say (to pursue the analogy) preferable to any other approach for which I have the competence.

Readers familiar with theories of grammar will know the distinction in linguistics (Chomsky 1968) between competence (the ability to produce and distinguish grammatical sentences from non-sentence strings of words) and performance (what one actually says). Many linguists take as their task the development of a theory of competence and would leave to others, such as psychologists, the task of predicting what an individual will actually say in given circumstances. Just so, I consider that the goal of foraging theory is to specify the animals' foraging competence and its consequence. The task of predicting what animals actually do in particular circumstances is a closely related task but a different one.

3. Post's data, like mine, are from a one-year study of Alto's Group. Bronikowski and J. Altmann (1996) have recently published long-term time-budget data for this group and two other Amboseli baboon groups. For adult females in Alto's Group over nine years, 1982–90, feeding time plus moving time averaged 69.9% of nine daylight hours (SD 5.49) (from ibid., table 2). The mean ratio of move time to feed time during that period averaged 0.594 (SD 0.19). Feeding itself averaged 44.7% of the time, which, at 660 minutes per day of potential foraging time, comes to 295 minutes per day.

4. In the descriptions of diets 6–10, an asterisk indicates a food with zero slack; that is, a food that would be prescribed in larger quantities if it were not for its limited seasonal availability.

5. The estimate in chapter 4 that, on average, the yearlings spent 23.74% of their *observed* feeding time on milk was obtained from mean daily nursing time (65.93 minutes) divided by mean daily foraging time (277.74 minutes). The latter value, referred to in chapter 6, was estimated from 660 times the ratio (9,384.23/22,299.84) of total product-limit feeding time ($\Sigma L_i n_i$ for the 277 foods in table 4.1) over that plus the product-limit estimate of in-sight-not-feeding time (12,915.61 minutes). The seasonally adjusted estimate of *actual* feeding time is 250.5 minutes per day (table 6.5). Cf. 295 minutes for adult females, note 3 above.

551

6. With the newer growth rate value, 4.5 g/day (appendix 1), second-approximation estimates of milk intakes for the four age classes are, respectively, 292.0, 273.1, 234.5, and 176.1 ml of milk per day; the sample-time weighted mean is 254.0 ml/day, equivalent to 260.9 g/day.

7. A value of $8.95 \times 10^5 M^{0.762}$ J/d for the total metabolism of eutherian mammals is from kJ/$d = 4.63$ g$^{0.762}$ given by Nagy (1994, 44), which is based on a larger sample than the value of 4.57 g$^{0.765}$ given on p. 46 of that article (Nagy, pers. comm.). It differs significantly from that in a previous compilation (Nagy 1987) as a result of a larger sample size and using species means. In the new data set, unlike the previous one, no significant effects of diet on total metabolism were detected. The current value is also larger than the value used in S. Altmann (1987), namely, $8 \times 10^5 M^{0.71}$ J d^{-1} (from Garland 1983), which was based on a still smaller sample size. All these estimates are based on the doubly labeled water method (references in Nagy 1987).

Hayssen and Lacy (1985) give regressions of mammalian basal metabolic rates on body mass in the form log BMR (ml O_2/g-hr) $= a + b$ log body mass (g), and for eutherian mammals, they estimate $a = 0.649$, $b = -0.304$. At 20.08 J per ml O_2, this becomes BMR (J/24 hr) $= 2.72 \times 10^5 M^{0.696}$. Thus, for eutherian mammals of one kilogram of body mass, total metabolism is, on average, about 3.3 times basal metabolism.

8. Note that in calculating the energy value of fever tree gum, I assumed that this polymerized sugar is digestible and utilizable as an energy source. However, gums are widely assumed not to be directly digestible by vertebrates (Simmen 1994); digesting them may require gut fermentation (Chivers and Hladik 1980; Nash 1986). That such fermentation can go on even in animals without specialized fermentation chambers is evident from the ability of humans and chimpanzees to digest large percentages of the cellulose, hemicellulose, and neutral-detergent fiber in their diets (Milton and Demment 1988; Van Soest 1994). The sugars in fever tree gum are primarily D-galactose and D-arabinose. Fresh fever tree gum is virtually tasteless to those humans in our research group who have tried it, yet the baboons are highly attracted to it and presumably do taste it. If the sugar-digesting enzymes are isomer-specific, we are faced with the following interesting question, to which I have not found an answer in the literature. Can animals digest all and only the sugar isomers that they can taste?

9. For the reader's convenience, let me repeat the objectives of the five optimal diets.

Diet 6: maximize energy intake
Diet 7: maximize protein intake
Diet 8: minimize feeding time
Diet 9: maximize energy intake rate
Diet 10: maximize protein intake rate

10. Other estimated feeding time budgets for Amboseli baboons have been published by J. Altmann (1980), Post (1978), Post, Hausfater, and McCuskey (1980), Post (1981), Slatkin (1975), Slatkin and Hausfater (1976), Silk (1987), Pereira (1984), Stacey (1986). Variability in these estimates may be due not only to age differences but also to differences in sex, individual subjects, methods, groups, and years (and thus habitat conditions). Recently some attempts have been made at looking for sources

of this variability. Bronikowski and Altmann (1996) looked at effects of year-to-year changes in climate. Their study and those by J. Altmann and Muruthi (1988) and Muruthi (1997) demonstrate the impact of food availability. Alberts, J. Altmann, and Wilson (1996) and Rasmussen (1985) examined the effects of sexual consortships on foraging time. At Mikumi National Park, Tanzania, Wasser and Wasser (1995) demonstrated that over the first three months of age, yellow baboons decreased nipple contact with age more rapidly if they were born later in the birth year (December–November).

11. Elsewhere in this book I have used the term "yearling" to mean an individual roughly one year of age—in particular, an individual in the age range of the subjects of this study, 30 to 70 weeks of age. In the rest of this chapter I use "yearling" in a more restricted sense, to mean those that have survived one year but are less than two years old at the time.

12. The following algorithm was used to locate trios of well-connected variables in the table of correlations (table 9.7). (In what follows, values on or below the main diagonal are to be ignored.) In the upper right triangle of the matrix of pairwise correlations, with variables ordered the same way in rows and columns, mark the cells of all highly correlated pairs. Check each marked cell in turn, moving from left to right in each row. See whether the first marked cell has at least one other marked cell to its right in the same row. If not—that is, if it is the rightmost marked cell in its row—go to the next marked cell (leftmost marked cell in the next row). Again check and move on if there is no marked cell to the right, continuing until you are in a marked cell for which there is one. Call the marked cell you are in A_1 and the next marked cell in the row A_2. Go to the transpose of A_1. Move to the right in that row until you encounter a marked cell. If in that column there is a marked cell in the same row as A_1 (other than A_1 itself), the variables of these three cells are a well-connected trio, as defined above. Repeat this procedure for every other marked cell that is in the same row as the transpose of A_1. Now go on to the next marked cell, A_2, and repeat the procedure above. Continue until all marked cells have been done.

13. Inadvertently, sample 1895/74 was included with semiripe fruit in compiling table 10.5, but with ripe fruit in compiling table 6.1. The sample contained a mixture of white and greenish white fruit, and its inclusion with samples of completely ripe fruit in table 6.1 probably gives a value for that category that is closer to what was actually consumed when "ripe fruit" was scored.

14. This value is based on whole body composition of *Homo sapiens,* the only primate for which I found data. By one year of age, the infant's protein concentration is nearly asymptotic at 3% of fat-free mass (Moulton 1923). Assuming a fat content of 10% and a nitrogen:protein conversion factor of 6.25 for animal tissue (Merrill and Watt 1973), the protein content of a yearling baboon is approximately 17%. For comparison, whole body protein content of six species of nonprimate mammals, tabulated by Robbins (1983, table 10.4), ranged from 10% to 20%.

15. A more recent estimate of day-journey length in Amboseli baboons (J. Altmann and Samuels 1992) uses a method—pace counts on individuals of known mean pace length—that we believe to be considerably more accurate than the various map-based methods used in our previous estimates, particularly because it includes local travel that does not show up on the scale of our day-journey maps. During 1983–84, a sample of adult females walked eight to ten kilometers per day.

16. The literature on this and other survival functions is not consistent on inclusion of interval endpoints. The inclusion of t shown here is consistent with Gross and Clark (1975), whereas Kaufman (1977) uses $\mathrm{pr}\{T > t\}$.

17. The calculations of milk intake noted in this section that are indicated as a first approximation and that are used at various places elsewhere in this book (e.g., "PCMX-1" in various tables) were carried out before more recent evidence (appendix 1) suggesting that in Alto's Group, average growth rate of yearlings has been closer to 4.5 g per day, which, at a birth weight of 0.775 kg, leads to the following formula for body mass M) in kilograms.

$$(5.2')\qquad M = 0.775 + 0.0315w \quad \text{(second approximation)}.$$

At the ten-week age interval midpoints, this formula gives 1.88, 2.19, 2.51, and 2.82 kg, respectively. At the mean sample age, 47.5 weeks, average body mass would be 2.27 kg.

If this preliminary estimate of 4.5 g/day holds up, then the milk intake values given in table 6.6 are correspondingly too large. At the four indicated ages, the milk intakes, calculated from the ratios of the respective body-mass values, would be, respectively, 83.6%, 82.3%, 81.3%, and 80.5% of the tabulated values. (See also note 6 above.) Wherever feasible, I have recalculated values, such as nutrient intake levels, that depend on the quantity of milk. In each section, I indicate whether first- or second-approximation milk intakes were used.

Literature Cited

Acheson, K. J., I. T. Campbell, O. G. Edholm, D. S. Miller, and M. J. Stock. 1980. The measurement of food and energy intake in man—an evaluation of some techniques. *Am. J. Clin. Nutr.* 33:1147–54.

Adolph, E. F. 1949. Quantitative relations in the physiological constituents of animals. *Science* 109:579–85.

al-Bagieh, N. H., A. Idowu, and N. O. Salako. 1994. Effects of aqueous extract of miswak on the in vitro growth of *Candida albicans. Microbios* 80:107–13.

Albanese, A. A., and L. L. Orto. 1973. The proteins and amino acids. In *Modern nutrition in health and disease,* 5th ed., ed. R. S. Goodhart and M. E. Shils. Philadelphia: Lea and Febiger.

Alberts, S. C., J. Altmann, and M. L. Wilson. 1996. Mate guarding constraints foraging activity of male baboons. *Anim. Behav.* 51:1269–77.

Alexander, R. D. 1979. *Darwinism and human affairs.* Seattle: University of Washington Press.

Alison, J. B. 1951. Interpretation of nitrogen balance data. *Fed. Proc.* 10:676–83.

Altmann, J. 1974. Observational study of behavior: Sampling methods. *Behaviour* 49:227–67.

———. 1980. *Baboon mothers and infants.* Cambridge: Harvard University Press.

———. 1983. Costs of reproduction in baboons. In *Behavioral energetics: Vertebrate costs of survival,* ed. W. P. Aspey and S. I. Lustick, 67–88. Columbus: Ohio State University Press.

Altmann, J., and S. Alberts. 1987. Body size and growth rates in a wild primate population. *Oecologia* 72:15–20.

Altmann, J., S. A. Altmann, G. Hausfater, and S. McCuskey. 1977. Life history of yellow baboons: Physical development, reproductive parameters and infant mortality. *Primates* 18:315–30.

Altmann, J., G. Hausfater, and S. A. Altmann. 1985. Demography of Amboseli baboons, 1963–1983. *Am. J. Primatol.* 8:113–25.

———. 1988. Determinants of reproductive success in savannah baboons (*Papio cynocephalus*). In *Reproductive success,* ed. T. H. Clutton-Brock, 403–18. Chicago: University of Chicago Press.

Altmann, J., and P. Muruthi. 1988. Differences in daily life between semiprovisioned and wild-feeding baboons. *Am. J. Primatol.* 15:213–21.

Altmann, J., and A. Samuels. 1992. Costs of parental care in a non-human primate: Infant carrying in baboons. *Behav. Ecol. Sociobiol.* 29:391–98.

555

Altmann, J., D. Schoeller, S. A. Altmann, P. Muruthi, and R. M. Sapolsky. 1993. Body size and fatness of free-living baboons reflect food availability and activity levels. *Am. J. Primatol.* 30:149–61.

Altmann, S. A. 1969. Changes in the acacia woodland of the Masai-Amboseli Game Reserve, 1963–1969. Report to Director, Kenya National Parks, Nairobi.

———. 1974. Baboons, space, time, and energy. *Am. Zool.* 14:221–48.

———. 1977. The acacia woodland of Amboseli National Park: Current status and future developments. Report to Director, Kenya National Parks, Nairobi.

———. 1979a. Altruistic behaviour: The fallacy of kin deployment. *Anim. Behav.* 27:958–59.

———. 1979b. Field preservation of food samples. Manuscript.

———. 1984. What is the dual of the energy-maximization problem? *Am. Nat.* 123:433–41.

———. 1985. More on hominid diet before fire. *Curr. Anthropol.* 26:661–62.

———. 1987. The impact of locomotor energetics on mammalian foraging. *J. Zool. Lond.* 211:215–25.

———. 1991. Diets of yearling female primates (*Papio cynocephalus*) predict lifetime fitness. *Proc. Natl. Acad. Sci. USA* 88:420–23.

Altmann, S. A., and J. Altmann. 1970. *Baboon ecology: African field research.* Chicago: University of Chicago Press.

———. 1979. Demographic constraints on behavior and social organization. In *Primate ecology and human origins,* ed. I. S. Bernstein and E. O. Smith, 47–63. New York: Garland STPM.

Altmann, S. A., D. G. Post, and D. F. Klein. 1987. Nutrients and toxins of plants in Amboseli, Kenya. *Afr. J. Ecol.* 25:279–93.

Altmann, S. A., and J. M. Shopland. 1997. Do baboons time feeding interruptions optimally? Manuscript.

Altmann, S. A., and S. S. Wagner. 1978. A general model of optimal diet. In *Recent advances in primatology.* ed. D. J. Chivers and J. Herbert, 1:407–14. London: Academic.

Ambe, K. S., and K. Sohonie. 1956. Trypsin inhibition in plant metabolism. *Experientia* 12:302.

Anand, C. R., and H. M. Linds-Wiler. 1974. Effect of protein intakes on calcium balance on young men given 500 mg. of calcium daily. *J. Nutr.* 104:695–700.

Anderson, G. H. 1988. Metabolic regulation of food intake. In *Modern nutrition in health and disease.* 7th ed., ed. M. E. Shils and V. R. Young. Philadelphia: Lea and Febiger.

Anderson, B., L. G. Leskell, and M. Rungren. 1982. Regulation of water intake. *Ann. Rev.* 2:73–89.

Andrewartha, H. G. 1971. *Introduction to the study of animal populations,* 2d ed. Chicago: University of Chicago Press.

Andrewartha, H. G., and L. C. Birch. 1954. *The distribution and abundance of animals.* Chicago: University of Chicago Press.

Andrews, H. P., R. D. Snee, and M. H. Sarner. 1980. Graphical display of means. *Am. Stat.* 34 (4): 195–99.

Ankel-Simon, F. 1983. *A survey of living primates and their anatomy.* New York: Macmillan.

Apgar, J. 1978. Nutrient deficiencies in animals: Zinc. In *Effects of nutrient deficiencies in animals,* ed. J. M. Rechcigl. West Palm Beach, Fla.: CRC.

Arkin, H., and R. R. Colton. 1955. *Statistical methods.* New York: Barnes and Noble.

Armanious, M. W., W. M. Britton, and H. L. Fuller. 1973. Effect of methionine and choline on tannic acid and tannin toxicity in the laying hen. *Poultr. Sci.* 52:2160–68.

Armstrong, E. 1982. A look at relative brain size in mammals. *Neurosci. Lett.* 34 (2): 101–4.

Arnold, S. J. 1983. Morphology, performance and fitness. *Am. Zool.* 23:347–61.

Atwater, W. O. 1902. On the digestibility and availability of food materials. *Conn. Agric. Exp. Stn. Ann. Rep.* 16:180–209.

Atwater, W. O., and A. P. Bryant. 1900a. Composition of common food materials— available nutrients and fuel value. *Conn. Agric. Exp. Stn. Ann. Rep.* 12:111–23.

———. 1900b. The availability and fuel value of food materials. *Conn. Agric. Exp. Stn. Ann. Rep.* 12:73–110.

Ausman, L. M., and D. L. Gallina. 1979. Liquid formulas and protein requirements of nonhuman primates. In *Primates in nutritional research,* ed. K. C. Hayes, 39–57. New York: Academic.

Barlow, R. E., and F. Proschan. 1975. *Statistical theory of reliability and life testing: Probability models.* New York: Holt, Rinehart and Winston.

Bartholomew, G. A. 1972. Energy metabolism. In *Animal physiology: Principles and adaptations,* ed. M. S. Gordon, 44–72. New York: Macmillan.

Barton, R. A. 1989. Foraging strategies, diet and competition in olive baboons. Ph.D. thesis, University of St. Andrews.

———. 1990. Feeding, reproduction and social organisation in female olive baboons (*Papio anubis*). In *Baboons: Selected Proceedings XII Congress IPS,* ed. M. Thiago de Mello, A. Whiten, and R. W. Byrne, 29–37. N.p.

———. 1993. Sociospatial mechanisms of feeding competition in female olive baboons, *Papio anubis. Anim. Behav.* 46:791–802.

Barton, R. A., and A. Whiten. 1993. Feeding competition among female olive baboons, *Papio anubis. Anim. Behav.* 46:777–89.

———. 1994. Reducing complex diets to simple rules: Food selection by olive baboons. *Behav. Ecol. Sociobiol.* 35:283–93.

Barton, R. A., A. Whiten, R. W. Byrne, and M. English. 1993. Chemical composition of plant foods: Implications for the interpretation of intra- and interspecific differences in diet. *Folia Primatol.* 61:1–20.

Barton, R. A., A. Whiten, S. C. Strum, R. W. Byrne, and A. J. Simpson, 1992. Habitat use and resource availability in baboons. *Anim. Behav.* 43:831–44.

Basu, T. K. 1981. Nutrition and metabolism of drugs. In *CRC handbook of nutritional requirements in a functional context,* ed. M. Rechcigl, 307–24. Boca Raton, Fla.: CRC.

Beach, F. A. 1945. Current concepts of play in animals. *Am. Nat.* 79:523–41.

Beaton, G. H. 1974. Epidemiology of iron deficiency. In *Iron in biochemistry and medicine,* ed. A. Jacobs and M. Worwood, 323. New York: Academic.

Beaton, G. H., J. Milner, P. Corey, V. McGuire, M. Cousins, E. Stewart, M. de Ramos, D. Hewitt, P. V. Grambsch, N. Kassim, and J. A. Little. 1979. Sources of variance in 24-hour dietary recall data: Implications for nutrition study design and interpretation. *Am. J. Clin. Nutr.* 32:2546–59.

Beck, B. B. 1982. Chimpocentrism: Bias in cognitive ethology. *J. Hum. Evol.* 11:3–17.

Begon, M., J. L. Harper, and C. R. Townsend. 1986. *Ecology: Individuals, populations, and communities.* Sunderland, Mass.: Sinauer.

Behrensmeyer, A. K. 1981. Vertebrate paleoecology in a recent East African ecosystem. In *Communities of the past,* ed. J. Gray, 591–615. Stroudsburg, Pa.: Hutchinson Ross.

———. 1993. The bones of Amboseli. *Natl. Geogr. Res. Explor.* 9:402–21.

Behrensmeyer, A. K., and D. E. D. Boaz. 1980. The recent bones of Amboseli National Park, Kenya, in relation to East African paleoecology. In *Fossils in the making; Vertebrate taphonomy and paleoecology,* ed. A. K. Behrensmeyer, 72–92. Chicago: University of Chicago Press.

Bell, R. H. V. 1970. The use of the herb layer by grazing ungulates in the Serengeti. In *Animal populations in relation to their food resources,* ed. A. Watson, 111–24. Oxford: Blackwell.

———. 1971. A grazing ecosystem in the Serengeti. *Sci. Am.* 225 (1): 86–93.

Belovsky, G. E. 1978a. Diet optimization in a generalist herbivore: The moose. *Theor. Popul. Biol.* 14:105–34.

———. 1978b. The time-energy budget of a moose. *Theor. Popul. Biol.* 14:76–104.

———. 1986. Hunter-gatherer foraging: A linear programming approach. *J. Anthropol. Archeol.* 6:29–76.

Belovsky, G. E., and O. J. Schmitz. 1994. Plant defenses and optimal foraging by mammalian herbivores. *J. Mammal.* 75:816–32.

Benedetti, J., and K. Yuen. 1977. Life tables and survival functions. In *BMDP-77,* ed. M. B. Brown, 743–70. Berkeley: University of California Press.

Benjamin, L. S. 1961. The effect of frustration on the nonnutritive sucking of the infant rhesus monkey. *J. Comp. Physiol. Psychol.* 54:700–703.

———. 1967. The beginning of thumbsucking. *Child Dev.* 38:1065–78.

Bennett, A. F., and J. A. Ruben. 1979. Endothermy and activity in vertebrates. *Science* 206:649–54.

Benton, A. W., R. A. Morse, and A. F. Gunnison. 1964. Bee venom tolerance in white mice in relation to diet. *Science* 145:1690–94.

Berchelmann, M. L., T. E. Vice, and S. S. Kalter, 1971. The hemogram of the maternally-reared neonatal and infant baboon. *Lab. Anim. Sci.* 21:564–71.

Berg, B. N. 1960. Nutrition and longevity in the rat. I. Food intake in relation to size, health, and fertility. *J. Nutr.* 71:242–54.

Berg, B. N., and H. S. Simms. 1960. Nutrition and longevity in the rat. II. Longevity and onset of disease with different levels of food intake. *J. Nutr.* 71:255.

Bishop, A. 1962. Control of the hand in lower primates. *Ann. N.Y. Acad. Sci.* 102:316–37.

Bitran, G. R., and A. G. Novaes. 1973. Linear programming with a fractional objective function. *Operations Res.* 21:22–29.

Blackman, F. F. 1905. Optima and limiting factors. *Ann. Bot.* 19:281–95.

Blaxter, K. L. 1961. Lactation and the growth of the young. In *Milk,* ed. S. S. Kon and A. F. Cowie. New York: academic.

———. 1971. The comparative biology of lactation. In *Lactation,* ed. J. R. Falconer, 51–69. In London: Butterworth.

Blaxter, K. L., and J. C. Waterlow, eds. 1985. *Nutritional adaptation in man.* London: Libbey.

Block, R. J., and H. H. Mitchell. 1946. The correlation of the amino-acid composition of proteins with their nutritive value. *Nutr. Abstr. Rev.* 16:249–78.

Boesch, C., and H. Boesch. 1983. Optimization of nut-cracking with natural hammers by wild chimpanzees. *Behaviour* 83:265–86.

———. 1984. Mental map in wild chimpanzees: An analysis of hammer transports for nut cracking. *Primates* 25:160–70.

Boltzmann, L. 1889. *Der zweite Hauptsatz der mechanischen Wärmetheorie.* Vienna: Gerpod. Reprinted as The second law of thermodynamics, in *Ludwig Boltzmann's Theoretical Physics and Philosophical Problems,* ed. B. McGuinness, 13–32. Dordrecht, Holland: D. Reidel.

Bonner, J. T. 1965. *Size and cycle.* Princeton: Princeton University Press.

Boyd, C. E., and C. P. Goodyear. 1971. Nutritive quality of food in ecological systems. *Arch. Hydrobiol.* 69:256–70.

Breslow, N. 1970. A generalized Kruskal-Wallis test for comparing K samples subject to unequal pattern of censorship. *Biometrika* 57:379–94.

Bressers, M., E. Meelis, P. Haccou, and M. Kruk. 1991. When did it really start or stop: The impact of censored observations on the analysis of duration. *Behav. Processes* 23:1–20.

Brody, S., R. C. Proctor, and U. S. Ashworth. 1934. Growth and development with special reference to domestic animals. *Res. Bull. Mo. Agric. Exp. Stn.* 220:32.

Bronikowski, A. M., and J. Altmann. 1996. Foraging in a variable environment: Weather patterns and the behavioral ecology of baboons. *Behav. Ecol. Sociobiol.* 39:11–25.

Brower, L. P., W. N. Ryerson, L. L. Coppinger, and S. C. Glasier. 1968. Ecological chemistry and the palatability spectrum. *Science* 161:1349–51.

Brown, J. L. 1975. *The evolution of behavior.* New York: Norton.

Brown, M. B. 1977. *BMDP-77.* Berkeley: University of California Press.

Buddenbreck, W. 1934. Über die kinetische und statische Leistung grosser und kleiner Tiere und ihre Bedeutung für den Gesamtstoffwechsel. *Naturwissenschaft* 22:675–80.

Burkitt, D. P. 1982. Dietary fiber as a protection against disease. In *Adverse effects of foods,* ed. E. F. P. Jelliffe and D. P. Jelliffe, 483–95. New York: Plenum.

Buss, D. H. 1968. Gross composition and variation of the components of baboon milk during natural lactation. *J. Nutr.* 96 (4):421–26.

———. 1971. Mammary glands and lactation. In *Comparative reproduction of non-human primates,* ed. E. S. E. Hafez. Springfield, Ill.: Thomas.

Buss, D. H., and O. M. Reed. 1970. Lactation of baboons fed a low protein maintenance diet. *Lab. Anim. Care* 20:709–12.

Buss, D. H., and W. R. Voss. 1971. Evaluation of four methods for estimating the milk yield of baboons. *J. Nutr.* 101:901–9.

Byrne, R. W. 1994. The evolution of intelligence. In *The evolution of behavior,* ed. P. J. B. Slater and T. R. Halliday, 223–65. Cambridge: Cambridge University Press.

————. 1996. Relating brain size to intelligence. In *Modelling the early human mind,* ed. P. A. Mellors and K. R. Gibson, 1–8. Cambridge: Macdonald Institute for Archaeological Research.

Byrne, R. W., and A. Whiten. 1988. *Machiavellian intelligence.* Oxford: Oxford University Press.

Calder, W. 1983. Body size, mortality, and longevity. *J. Theor. Biol.* 102:135–44.

Calloway, D. H., and S. Margen. 1971. Variation in endogenous nitrogen excretion and dietary nitrogen utilization as determinants of human protein requirements. *J. Nutr.* 101:205–16.

Cant, J. G. H. 1980. What limits primates? *Primates* 21:538–44.

Caraco, T. 1980. On foraging time allocation in a stochastic environment. *Ecology* 6:119–28.

————. 1981. Risk-sensitivity and foraging groups. *Ecology* 62 (3): 527–31.

Caraco, T., S. Martindale, and T. S. Whittam. 1980. An empirical demonstration of risk-sensitive foraging preferences. *Anim. Behav.* 28:820–30.

Carmel, J. 1976. The prediction of diets of high energy and protein value by linear programming. *Ecol. Food Nutr.* 5:161–77.

Carpenter, K. J. 1994. *Protein and energy: A study of changing ideas in nutrition.* Cambridge: Cambridge University Press.

Casimir, M. J. 1975. Feeding ecology and nutrition of an eastern gorilla group in the Mt. Kahuzi region (République de Zaire). *Folia Primatol.* 24:81–136.

Castracane, V. D., K. C. Copeland, P. Reyes, and T. J. Kuehl. 1986. Pubertal endocrinology of yellow baboons (*Papio cynocepnalus*). *Am. J. Primatol.* 11:263–70.

Chamberlin, J. G., and R. E. Stickney. 1973. Improvement of children's diet in developing countries: An analytical approach to evaluation of alternative strategies. *Nutr. Rep. Int.* 7:71–84.

Chapman, C. A., and L. J. Chapman. 1990. Dietary variability in primate populations. *Primates* 31:121–28.

Charnes, A., and W. W. Cooper. 1962. Programming with linear fractional functionals. *Logist. Quart.* 9:181–86.

Charnov, E. L. 1976. Optimal foraging: The marginal value theorem. *Theor. Popul. Biol.* 9:129–35.

Charnov, E. L., G. H. Orians, and K. Hyatt. 1976. Ecological implications of resource depression. *Am. Nat.* 110:247–59.

Chavon, J. K., and S. S. Kadam. 1989. Protease inhibitors. In *CRC handbook of world food legumes,* 123–33. Boca Raton, Fla.: CRC.

Cheeke, P. R., and G. R. Garman. 1974. Influence of dietary protein and sulfur amino acid levels on the toxicity of *Senecio jacobaea* (tansy ragwort) to rats. *Nutr. Rep. Int* 9:197–207.

Cheney, D. L., and R. M. Seyfarth. 1990. *How monkeys see the world.* Chicago: University of Chicago Press.

Cheney, D. L., R. M. Seyfarth, S. J. Andelman, and P. C. Lee. 1988. Reproductive success in vervet monkeys. In *Reproductive success,* ed. T. H. Clutton-Brock, 384–402. Chicago: University of Chicago Press.

Chivers, D. J., and C. M. Hladik. 1978. Primate feeding behaviour in relation to food availability and composition. In *Recent advances in primatology.* ed. D. J. Chivers and J. Herbert, 1:209–10. London: Academic.

———. 1984. Diet and gut morphology in primates. In *Food acquisition and processing in primates,* ed. D. J. Chivers, B. A. Wood, and A. Bilsborough, 213–30. New York: Plenum.

Chomsky, N. 1968. *Language and mind.* New York: Harcourt, Brace and World.

Cleland, J. B., and R. V. Southcott. 1969. Illness following the eating of seal liver in Australian waters. *Med. J. Aust.* 1:760.

Clutton-Brock, T. H., and P. H. Harvey. 1977. Species differences in feeding and ranging behavior in primates. In *Primate ecology: Studies of feeding and ranging behaviour in lemurs, monkeys, and apes,* 539–56. London: Academic.

———. 1980. Primates, brains, and ecology. *J. Zool. Lond.* 190:309–23.

———. 1983. The functional significance of variation in body size among mammals. In *Advances in the study of mammalian behavior,* ed. J. F. Eisenberg and D. G. Kleiman, 632–63. Shippensburg, Pa.: American Society of Mammalogists.

Coelho, A. M., C. A. Bramblett, and L. B. Quick. 1977. Social organization and food resource availability in primates: A socio-bioenergetic analysis of diet and disease hypotheses. *Am. J. Phys. Anthropol.* 46:253–64.

Coelho, A. M., L. S. Coelho, C. A. Bramblett, S. S. Bramblett, and L. Quick. 1976. Ecology, population characteristics and sympatric association in primates: A socio-bioenergetic analysis of howler and spider monkeys in Tikal, Guatemala. *Yearb. Phys. Anthropol.* 20:96–135.

Cole. L. C. 1954. The population consequences of life history phenomena. *Quart Rev. Biol.* 29:103–37.

Coley, P. D., J. P. Bryant, and F. S. Chapin III. 1985. Resource availability and plant anti-herbivore defense. *Science* 230:895–99.

Collier, G. H. 1980. An ecological analysis of motivation. In *Analysis of motivational processes,* ed. F. Toates and T. Halliday, 125–51. London: Academic.

———. 1982. Determinants of choice. In *Nebraska Symposium on Motivation,* 69–127. Lincoln: University of Nebraska Press.

Coon, J. M. 1976. Natural food toxicants—a perspective. In *Present knowledge in nutrition,* ed. D. M. Hegsted, C. O. Chichester, W. J. Darby, K. W. McNutt, R. M. Stalvey, and E. H. Stotz, 528–46. Washington, D.C.: Nutrition Foundation.

Cooper-Driver, G. A., and F. Swain. 1976. Cyanogenic polymorphism in relation to herbivore predation. *Nature* (London) 260:604.

Cork, S. J. 1994. Digestive constraints on dietary scope in small and moderately-small mammals: How much do we really understand? In *The digestive system in mammals: Food, form and function,* ed. D. J. Chivers and P. Langer, 337–69. Cambridge: Cambridge University Press.

Corkhill, L. 1952. Cyanogenesis in white clover (*Trifolium repens* L.). VI. Experiments with high-glycoside and glycoside-free strains. *N.Z.J. Sci. Technol.* A34:1–16.

Cornforth, J. W., and A. J. Henry. 1952. The isolation of L-stachydrine from the fruit of *Capparis tomentosa. J. Chem. Soc.,* 601–3.

Coward, W. A., et al. 1980. Measurement of milk intake in suckling mice. *J. Nutr.* 110 (2):371–72.

Cowlishaw, G. 1997. Trade-offs between foraging and predation risk in habitat use in a desert baboon population. *Anim. Behav.* 53:667–86.

Cowper, W. 1785. The time-piece. In *The poetical works of William Cowper.* 3d ed., ed. H. S. Milford. London: Oxford University Press.

Cox, D. R., and P. A. W. Lewis. 1966. *The statistical analysis of series of events.* London: Methuen.

Cummings, J. H. 1984. Microbial digestion of complex carbohydrates in man. *Proc. Nutr. Soc.* 43:35–44.

Cummins, K. W., and J. C. Wuycheck. 1971. Caloric equivalents for investigations in ecological energetics. *Mitt. Int. Verein. Limnol.* (Communications Int. Assoc. Appl. Limnol. 18:1–158.

Dantzig, G. B. 1963. *Linear programming and extensions.* Princeton: Princeton University Press.

Dart, R. A. 1953. The predatory transition from ape to man. *Int. Anthropol. Linguistic Rev.* 1:201–18.

Darwin, C. 1896. *The expression of the emotions in man and animals.* New York: Appleton.

Daulatabad, C. D., V. A. Desai, K. M. Hosamani, and A. M. Jamkhandi. 1991. Novel fatty acids in *Azima tetracantha* seed oil. *J. Amer. Oil Chemists Soc.* 68:978–79.

Dave, Y. S., N. D. Patel, and A. R. S. Menon. 1985. The pericarpic structure of *Withania somnifera. Proc. Indian Acad. Sci.: Plant Sci.* 94:677–82.

David, H. A. 1974. Parametric approaches to the theory of competing risks. In *Reliability and biometry,* ed. F. Proschan and R. J. Serfling, 275–90. Philadelphia: SIAM.

David, H. A., and M. L. Moeschberger. 1978. *The theory of competing risks.* London: Griffin.

Davidson, S., R. Passmore, and J. F. Brock. 1972. *Human nutrition and dietetics.* Baltimore: Williams and Wilkins.

Davies, F. G., B. Clausen, and L. F. Lund. 1972. The pathogenicity of Rift Valley fever virus for the baboon. *Trans. R. Soc. Trop. Med. Hyg.* 66:363.

Dawkins, R. 1982. *The extended phenotype.* San Francisco: Freeman.

Day, P. L. 1944. The nutritional requirements of primates other than man. *Vitam. Horm.* 2:71–105.

———. 1962. *Nutrient requirements of domestic animals.* 10. *Nutrient requirements of the young monkey.* Publ. no. 990. Washington, D.C.: National Research Council.

Delius, J. D. 1969. A stochastic analysis of the maintenance behaviour of skylarks. *Behaviour* 33:137–78.

Demment, M. W. 1983. Feeding ecology and the evolution of body size of baboons. *Afr. J. Ecol.* 21:219–33.

Dent, J. B. 1966. The evaluation of economically optimal rations for bacon pigs formulated by curve fitting and linear programming techniques. *Anim. Prod.* 8:213–20.

Depew, L. A. 1997. Ecology and behavior of baboons *(Papio anubis)* in the Shai Hills Game Production Reserve, Ghana. M.Sc. thesis, Cape Coast University, Ghana.

DeVore, I., and K. R. L. Hall. 1965. Baboon ecology. In *Primate behavior: Field studies of monkeys and apes,* ed. I. DeVore. New York: Holt, Rinehart, and Winston.

Dubos, R. 1979. Nutritional ambiguities. *Nat. Hist.* 89 (7): 14–21.

Dunbar, R. I. M., and P. Dunbar. 1974. Ecological relations and niche separation between sympatric terrestrial primates in Ethiopia. *Folia Primatol.* 21:36–60.

———. 1975. *Social dynamics of gelada baboons.* Basel: Karger.

Durnin, J. V. G. A., et al. 1976. Symposium on "Sex differences in response to nutritional variables." *Proc. Nutr. Soc.* 35:145–89.

Dykes, O. 1709. *English proverbs.* 3d ed. London: G. Sawbridge.

Ehle, F. R., J. B. Robertson, and P. J. Van Soest. 1982. Influence of dietary fibers on fermentation in the human large intestine. *J. Nutr.* 112:158–66.

Ehrlich, P. R., and L. C. Birch. 1967. The "balance of nature" and "population control." *Am. Nat.* 101:97–107.

Eisenberg, J. F. 1981. *The mammalian radiations: An analysis of trends in evolution, adaptation, and behavior.* Chicago: University of Chicago Press.

Eisenberg, J. F., and D. E. Wilson. 1978. Relative brain size and feeding strategies in the Chiroptera. *Evolution* 32:740–51.

Eisenhart, C. 1968. Expression of the uncertainties of final results. *Science* 160: 1201–4.

Elman, R. 1939. Time factor in retention of nitrogen after intravenous injection of a mixture of animo-acids. *Proc. Soc. Exp. Biol. Med.* 40:484–87.

Elton, C. 1927. *Animal ecology.* London: Methuen.

Elvin-Lewis, M. 1982. The therapeutic potential of plants used in dental folk medicine. *Odontostomatol. Trop.* 3:107–15.

Emlen, J. M. 1966. The role of time and energy in food preference. *Am. Nat.* 100: 611–17.

———. 1987. Evolutionary ecology and the optimality assumption. In *The latest on the best,* ed. J. Dupré. Cambridge: MIT Press.

Emlen, J. M., and M. G. R. Emlen. 1975. Optimal choice in diet: Test of a hypothesis. *Am. Nat.* 109:427–35.

Eppen, G. D., and F. J. Gould. 1979. *Quantitative concepts for management.* Englewood Cliffs, N.J.: Prentice-Hall.

Estabrook, F., and A. E. Dunham. 1976. Optimal diet as a function of absolute abundance, relative abundance, and relative value of available prey. *Am. Nat.* 110: 401–13.

Everitt, A. V., and B. Porter. 1976. Nutrition and aging. In *Hypothalamus, pituitary, and aging,* ed. A. V. Everitt and J. A. Burgess, 570–613. Springfield, Ill.: C. C. Thomas.

Ezmirly, S. T., J. C. Cheng, and S. R. Wilson. 1979. Saudi Arabian medicinal plants: *Salvadora persica. Planta Medica* 35:191–92.

Fa, J. E., and C. H. Southwick, eds. 1988. *Ecology and behavior of food-enhanced primate groups.* New York: Liss.

Fabry, P. 1967. Metabolic consequences of the pattern of food intake. *Handb. Physiol.* 1 (6): 31–49.

Fagen, R. M., and D. Y. Young. 1978. Temporal patterns of behaviors: Durations, intervals, latencies, and sequences. In *Quantitative ethology,* ed. P. W. Colgan, 79–114. New York: John Wiley.

Fairbanks, L. 1975. Communication of food quality in captive *Macaca nemestrina* and free-ranging *Ateles geoffroyi. Primates* 16 (2): 181–90.

FAO (Food and Agriculture Organization). 1975. Population, food supply and agricultural development. In *The state of food and agriculture, 1974,* 150. Rome: FAO.

FAO/WHO. 1962. Calcium requirements. *WHO Tech. Rep. Ser.* 230:1–54.

———. 1965. Protein requirements. *WHO Tech. Rep. Ser.* 301:1–71.

———. 1973. Energy and protein requirements. *WHO Tech. Rep. Ser.* 522:1–118.

Faulk, W. P., and R. K. Chandra. 1981. Nutrition and resistance. In *CRC handbook of nutritional requirements in a functional context,* ed. M. Rechcigl, 2:555–71. Boca Raton, Fla.: CRC.

Fedigan, L. M., L. Fedigan, S. Gouzoules, and H. Gouzoules. 1986. Lifetime reproductive success in female Japanese macaques. *Folia Primatol.* 47:142–57.

Feller, W. 1957. *An introduction to probability theory and its applications.* New York: John Wiley.

Fenchel, T. 1974. Intrinsic rate of natural increase: The relationship with body size. *Oecologia* 14: 317–26.

Fleming, T. H. 1975. The role of small mammals in tropical ecosystems. In *Small mammals: Their productivity and population dynamics,* ed. F. B. Golley and K. Petrusewicz, 269–98. Cambridge: Cambridge University Press.

Fletemeyer, J. R. 1978. Communication about potentially harmful foods in free-ranging chacma baboons. *Primates* 19:223–26.

FNB (Food and Nutrition Board). 1980. *Recommended dietary allowances,* 9th ed. Washington, D.C.: National Academy.

———. 1986. *Nutrient adequacy.* Washington, D.C.: National Academy.

———. 1989. *Recommended dietary allowances,* 10th ed. Washington, D.C.: National Academy.

Foley, W. J., and C. McArthur. 1994. The effects and costs of allelochemicals for mammalian herbivores: An ecological perspective. In *The digestive system in mammals: Food, Form and Function,* ed. D. J. Chivers and P. Langer, 370–91. Cambridge: Cambridge University Press.

Forthman Quick, D. L. 1997. Activity budgets and the consumption of human food in two troops of baboons, *Papio anubis.* In *Primate ecology and conservation,* ed. J. G. Else and P. C. Lee. Cambridge: Cambridge University Press.

Forthman Quick, D. L., and M. W. Demment. 1988. Dynamics of exploitation: Differential energetic adaptations of two troops of baboons to recent human contact. In *Ecology and behavior of food-enhanced primate groups,* ed. J. E. Fa and C. H. Southwick, 25–51. New York: Liss.

Fox, L. R., and P. A. Morrow. 1981. Specialization: Species property or local phenomenon? *Science* 211:887–93.

Freeland, W. J., and D. H. Janzen. 1974. Strategies in herbivory by mammals: The role of secondary plant compounds. *Am. Nat.* 108:269–89.

Friedman, M. 1975. *There's no such thing as a free lunch.* LaSalle, Ill.: Open Court.

Frisancho, A. R., J. Sanchez, D. Pallardel, and L. Yanez. 1973. Adaptive significance of small body size under poor socioeconomic conditions in southern Peru. *Am. J. Phys. Anthropol.* 39:255–61.

Frisch, R. E. 1980. Fatness, puberty, and fertility. *Nat. Hist.* 89:16–27.

Fritz, R. S., and E. L. Simms, eds. 1992. *Plant resistance to herbivores and pathogens.* Chicago: University of Chicago Press.

Fukuda, F. 1988. Influence of artificial food supply on population parameters and dispersal in the Hakone T troop of Japanese macaques. *Primates* 29:477–92.

Fuller, H. L., S. I. Chang, and D. K. Potter. 1967. Detoxication of dietary tannic acid by chicks. *J. Nutr.* 91:477–81.

Futuyma, D. J. 1986. *Evolutionary Biology,* 2d ed. Sunderland, Mass.: Sinauer.

Gabriel, K. R. 1978. A simple method of multiple comparison of means. *SASA* 73: 724–29.

Galdikas, B. M. F. 1982. An unusual instance of tool-use among wild orangutans (*Pongo pygmaeus*) in Tanjung Puting Reserve, Indonesian Borneo. *Primates* 23: 138–39.

Galef, B. G., Jr. 1976. Social transmission of acquired behavior: A discussion of traditional and social learning in vertebrates. In *Advances in the study of behavior,* ed. J. S. Rosenblatt, R. A. Hinde, E. Shaw, and C. Beer, 77–100. New York: Academic.

———. 1992. The question of animal culture. *Hum. Nat.* 3:157–78.

Galef, B. G., Jr., and M. Beck. 1990. Diet selection and poison avoidance by mammals individually and in social groups. In *Handbook of neurobiology,* vol. 11, ed. E. M. Stricker, 329–49. New York: Plenum.

Galef, B. G., Jr., and P. W. Henderson. 1972. Mother's milk: A determinant of the feeding preferences of weaning rat pups. *J. Comp. Physiol. Psychol.* 78:213–19.

Galef, B. G., Jr., and M. Stein. 1985. Demonstrator influence on observer diet preference: Analysis of critical social interactions and olfactory signals. *Anim. Learn. Behav.* 13:31–38.

Galton, F. 1871. Gregariousness in cattle and in men. *Macmillan's Mag. Lond.* 23: 353–57.

Garland, T. 1983. Scaling the ecological cost of transport to body mass in terrestrial mammals. *Am. Nat.* 121:571–87.

Gaulin, S. J. C. 1979. A Jarman/Bell model of primate feeding niches. *Hum. Ecol.* 7:1–20.

Gazi, M. I., A. Lambourne, and A. H. Chagla. 1987. The antiplaque effect of toothpaste containing *Salvadora persica* compared with chlorhexidine gluconate. *Clin. Prev. Dentistry* 9 (6): 3–8.

Geiger, E. 1950. The role of the time factor in protein synthesis. *Science* 111:594–99.

Gershoff, S. N., S. B. Andrus, D. M. Hegstead, and E. A. Lentin. 1957. Vitamin A deficiency in cats. *Lab. Invest.* 6:227–40.

Gillet, H. 1964. Agrostologie et zoocynégétique en République Centrafricaine. *J. Agric. Trop. Bot. Appl.* 11:267–330.

Gittleman, J. L., and O. T. Oftedal. 1987. Comparative growth and lactation energetics in carnivores. *Symp. Zool. Soc. Lond.* 57:41–77.

Glander, K. E. 1978. Howling monkey feeding behavior and plant secondary compounds: A study of stategies. In *The ecology of arboreal folivores,* ed. G. G. Montgomery. Washington, D.C.: Smithsonian Institution Press.

———. 1981. Feeding patterns in mantled howling monkeys. In *Foraging behavior: Ecological, ethological, and psychological approaches,* ed. A. Kamil and T. D. Sargent, 231–59. New York: Garland.

————. 1982. The impact of plant secondary compounds on primate feeding behavior. *Yearb. Phys. Anthropol.* 25:1–18.

————. 1994. Nonhuman primate self-medication with wild plant foods. In *Eating on the wild side,* ed. N. L. Etkin, 227–39. Tucson: University of Arizona Press.

Glassman, D. M., A. M. Coelho, K. D. Carey, and C. A. Bramblett. 1984. Weight growth in savannah baboons: A longitudinal study from birth to adulthood. *Growth* 48: 425–33.

Golley, F. B. 1961. Energy values of ecological materials. *Ecology* 42:581–84.

Goodall, J. 1963. Feeding behaviour of wild chimpanzees. *Symp. Zool. Soc. Lond.* 10:39–47.

Goodman, L. A. 1960. On the exact variance of products. *J. Am. Stat. Assoc.* 55: 708–13.

Gopalan, C., and B. Narasinga Rao. 1974. *Dietary allowances for Indians.* ICMR Special Report Series 60. N.p.: National Institute of Nutrition [India].

Graf, E., and J. W. Eaton. 1985. Dietary phytate and calcium bioavailability. In *Nutritional bioavailability of calcium,* ed. C. Kies. Washington, D.C.: American Chemical Society.

Greenberg, L. D. 1970. Nutritional requirements for macaque monkeys. In *Feeding and nutrition of nonhuman primates,* ed. R. S. Harris, 117–57. New York: Academic.

Gross, A. J., and V. A. Clark. 1975. *Survival distributions: reliability applications in the biomedical sciences.* New York: John Wiley.

Gruber, S. H. 1978. Mechanisms of color vision: An ethologist's primer. In *The behavioral significance of color,* ed. J. E. H. Burtt. New York: Garland.

Guigoz, Y., and H. N. Munro. 1985. Nutrition and aging. In *Handbook of the biology of aging,* 2d ed., ed. C. E. Finch and E. L. Schneider. New York: Van Nostrand.

Guilbert, H. R., C. E. Howell, and G. H. Hart. 1940. Minimum vitamin A requirements of mammalian species. *J. Nutr.* 19:91.

Hailman, J. P. 1974. A stochastic model of leaf-scratching bouts in two emberizine species. *Wilson Bull.* 86:296–98.

Hairston, N. G., F. E. Smith, and L. B. Slodbokin. 1960. Community structure, population control and competition. *Am. Nat.* 94:421–25.

Hall, K. R. L. 1960. Social vigilance behaviour of the chacma baboon. *Papio ursinus. Behaviour* 16:261–94.

————. 1963. Variations in the ecology of the chacma baboon, *Papio ursinus. Symp. Zool. Soc. Lond.* 10:1–28.

————. 1965. Ecology and behavior of baboons, patas, and vervet monkeys in Uganda. In *The baboon in medical research,* ed. H. Vagtborg, 43–61. Austin: University of Texas Press.

Hallfrisch, J., A. Powell, C. Carafelli, S. Reiser, and E. S. Prather. 1987. Mineral balances of men and women consuming high fiber diets with complex or simple carbohydrates. *J. Nutr.* 117:48–55.

Hamilton, W. D. 1971. Geometry for the selfish herd. *J.Theor. Biol.* 31:295–311.

Hamilton, W. J., III, R. E. Buskirk, and W. H. Buskirk. 1978a. Environmental developmental determinants of object manipulation by chacma baboons in two southern African environments. *J. Hum. Evol.* 7:206–16.

————. 1978b. Omnivory and utilization of food resources by chacma baboons, *Papio ursinus. Am. Nat.* 112:911–24.

Hamilton, W. J., III, and C. Busse. 1982. Social dominance and predatory behavior of chacma baboons. *J. Hum. Evol.* 11 (7): 567–73.

Hanwell, A., and M. Peaker. 1977. Physiological effects of lactation on the mother. *Symp. Zool. Soc. Lond.* 41:297–312.

Harborne. J. B. 1978. *Biochemical aspects of plant and animal coevolution.* New York: Academic.

Harding, R. S. O. 1976. Ranging patterns of a troop of baboons (*Papio anubis*) in Kenya. *Folia Primatol.* 25:143–85.

Harding, R. S. O., and G. Teleki, eds. 1981. *Omnivorous primates.* New York: Columbia University Press.

Harestad, A. S., and F. L. Bunnell. 1979. Home range and body weight—a reevaluation. *Ecology* 60 (2): 389–402.

Harlow, H. F. 1951. Primate learning. In *Comparative psychology,* ed. C. P. Stone. New York: Prentice-Hall.

Harris, R. S., ed. 1970. *Feeding and nutrition of nonhuman primates.* New York: Academic.

Haukioja, E., and P. Niemala. 1976. Does birch defend itself against herbivores? *Rep. Kevo Subarct. Res. Stn.* 13:44–47.

Hausfater, G. 1975. *Dominance and reproduction in baboons (*Papio cynocephalus*): A quantitative analysis.* Basel: Karger.

———. 1976. Predatory behavior of yellow baboons. *Behaviour* 56:44–68.

Hausfater, G., and H. Bearce. 1976. Acacia tree exudates: Their composition and use as a food source by baboons. *E. Afr. Wildl. J.* 14:241–43.

Hayes, K. C., ed. 1979. *Primates in nutritional research.* New York: Academic.

Hayssen, V., and R. C. Lacy. 1985. Basal metabolic rates in mammals: Taxonomic differences in the allometry of BMR and body mass. *Comp. Biochem. Physiol.* 81:741–54.

Hegsted, D. M. 1972. Problems in the use and interpretation of the recommended dietary allowances. *Ecol. Food Nutr.* 1:255–65.

Hegsted, D. M., S. A. Schuette, M. B. Zemel, and H. M. Links-Wiler. 1981. Urinary calcium and calcium balance in young men as affected by level of protein and phosphorus intake. *J. Nutr.* 111:553–62.

Hemmingsen, A. M. 1960. Energy metabolism as related to body size and respiratory surfaces, and its evolution. *Rep. Steno Mem. Hosp. Nordinsk Insulin Lab.* 9:6–110.

Hernandez-Camacho, J., and R. W. Cooper. 1976. The non-human primates of Colombia. In *Neotropical primates: Field studies and conservation,* ed. R. W. Thorington Jr. and P. G. Heltne, 35–69. Washington, D.C.: National Academy of Sciences.

Hershkovitz, P. 1977. *Living New World monkeys (Platyrrhini).* Chicago: University of Chicago Press.

Hill, W. C. O. 1966. *Primates: Comparative anatomy and taxonomy,* Vol. 6. Edinburgh: Edinburgh University Press.

Hillier, F. S., and G. J. Lieberman. 1974. *Operations research.* San Francisco: Holden-Day.

Hippocrates. 1939. *The genuine works of Hippocrates.* Trans. Francis Adams. Baltimore: Williams and Wilkins.

Hixon, M. A. 1982. Energy maximizers and time minimizers: Theory and reality. *Am. Nat.* 119:596–99.

Hladik, C. M. 1978. Adaptive strategies of primates in relation to leaf-eating. In *The ecology of arboreal folivores,* ed. G. G. Montgomery, 373–95. Washington, D.C.: Smithsonian Institution Press.

Hladik, C. M., and D. J. Chivers. 1994. Food and the digestive system. In *The digestive system of mammals,* ed. D. J. Chivers and P. Langer, 65–73. Cambridge: Cambridge University Press.

Hoage, R. J., and L. Goldman, eds. 1986. *Animal intelligence.* Washington D.C.: Smithsonian Institution Press.

Hockett, C. F. 1964. Scheduling. In *Cross-cultural understanding: Epistemology in anthropology,* ed. F. S. C. Northrop and H. H. Livingston, 125–44. New York: Harper and Row.

Hodges, C. M., and L. L. Wolf. 1981. Optimal foraging in bumblebees: Why is nectar left behind in flowers? *Behav. Ecol. Sociobiol.* 9 (1): 41–44.

Holt, L. E., E. Halac Jr., and C. N. Kajdi. 1962. The concept of protein stores and its implications in diet. *J. Am. Med. Assoc.* 181:699–705.

Holt, L. E., and S. E. Snyderman. 1965. Protein and amino acid requirements of infants and children. *Nutr. Abstr. Rev.* 35:1–13.

Homewood, K. M. 1978. Feeding strategy of Tana mangabey (*Cercocebus galeritus galeritus*). *J. Zool. Lond.* 186:375–91.

Howell, N. 1990. *Surviving fieldwork.* Washington, D.C.: American Anthropological Association.

Huges, D. W., and K. Genest. 1973. Alkaloids. *Phytochemistry* 2:118–70.

Hughes, R. N. 1979. Optimal diets under the energy maximization premise: The effects of recognition time and learning. *Am. Nat.* 113:209–21.

Hummer, R. L. 1970. Observations of the feeding of baboons. In *Feeding and nutrition of nonhuman primates,* ed. R. S. Harris, 183–203. New York: Academic.

Humphrey, N. K. 1976. The social function of intellect. In *Growing points in ethology,* ed. P. P. G. Bateson and R. A. Hinde, 303–17. Cambridge: Cambridge University Press.

Hunter, J. S. 1980. The national system of scientific measurement. *Science* 210: 869–74.

ICNND (United States Interdepartmental Committee on Nutrition for National Defense). 1963. *Manual for nutrition surveys,* 2d ed. Bethesda, Md.: ICNND.

Inge, W. R. 1927. *Confessio fidei: Outspoken essays.* 2d ser. New York: Longmans, Green.

Isaac, G. Ll., and D. C. Crader. 1981. To what extent were early hominids carnivorous? An archeological perspective. In *Omnivorous primates,* ed. R. S. O. Harding and G. Teleki, 37–103. New York: Columbia University Press.

Isbell, L. A. 1990. Sudden short-term increase in mortality of vervet monkeys (*Cercopithecus aethiops*) due to leopard predation in Amboseli National Park, Kenya. *Am. J. Primatol.* 21:41–52.

Iwamoto, T. 1979. Feeding ecology. In *Ecological and sociological studies of gelada baboons,* ed. M. Kawai, 279–330. Basel: Karger.

Jacobs, G. H. 1996. Primate photopigments and primate color vision. *Proc. Natl. Acad. Sci. USA* 93:577–81.

Janson, C. H., and S. Boinski. 1992. Morphological and behavioral adaptations for foraging in generalist primates: The case of the Cebines. *Am. J. Phys. Anthropol.* 88:483–98.

Janzen, D. H. 1977. How southern cowpea weevil larvae (Bruchidae: *Callosobruchus maculatus*) die on nonhost seeds. *Ecology* 58:921–27.

Jarman, P. J. 1972. The use of drinking sites, wallows and salt licks by herbivores in the flooded middle Zambezi valley. *E. Afr. Wildl. J.* 9:158–61.

———. 1974. The social organisation of antelope in relation to their ecology. *Behavior* 48:215–67.

———. 1982. Prospects for interspecific comparison in sociobiology. In *Current problems in sociobiology,* ed. King's College Sociobiology Group, 323–42. Cambridge: Cambridge University Press.

Jarman, P. J., and A. R. E. Sinclair. 1979. Feeding strategy and the pattern of resource-partitioning in ungulates. In *Serengeti Dynamics of an ecosystem,* ed. A. R. E. Sinclair and M. Norton-Griffiths, 130–63. Chicago: University of Chicago Press.

Jerison, H. J. 1973. *The evolution of the brain and intelligence.* New York: Academic.

Johnson, F. E., ed. 1987. *Nutritional anthropology.* New York: Liss.

Johnson, J. A. 1989. Supplanting by olive baboons: Dominance rank differences and resource value. *Behav. Ecol. Sociobiol.* 24:277–83.

Johnson, N. J., and S. Kotz. 1969. *Discrete distributions.* Boston: Houghton Mifflin.

Jolly, A. 1966. Lemur social intelligence and primate intelligence. *Science* 153:501–6.

Jolly, C. J. 1970. The seed-eaters: A new model of hominid differentiation based on a baboon analogy. *Man* 5:5–26.

———. 1993. Species, subspecies, and baboon systematics. In *species, species concepts, and primate evolution,* ed. W. H. Kimbel and L. B. Martin, 67–107. New York: Plenum.

Jones, C. 1970. Stomach contents and gastrointestinal relationships of monkeys collected in Rio Muni, West Africa. *Mammalia* 34:107–17.

Jones, D. A. 1966. On the polymorphism of cyanogenesis in *Lotus corniculatus* L. selection by animals. *Can. J. Genet. Cytol.* 8:556–67.

Jones, D. B. 1941. Factors for converting percentages of nitrogen in foods and feeds into percentages of protein. *U.S. Dept. Agric. Circ.* 183:1–22.

Kahumbu, P., and R. M. Eley. 1991. Teeth emergence in wild olive baboons in Kenya and formulation of a dental schedule for aging wild baboon populations. *Am. J. Primatol.* 23:1–9.

Kalter, S. S., ed. 1983. *Viral and immunological diseases in nonhuman primates.* New York: Liss.

Kamil, A. 1994. A synthetic approach to the study of animal intelligence. In *Behavioral mechanisms in evolutionary ecology,* ed. L. A. Real, 11–45. Chicago: University of Chicago Press.

Kaplan, A. 1964. *The conduct of inquiry.* San Francisco: Chandler

Kaplan, E. L., and P. Meier. 1958. Non-parametric estimation from incomplete observations. *J. Am. Stat. Assoc.* 53:457–81.

Kaufmann, A. 1977. *Mathematical models for the study of the reliability of systems.* New York: Academic.

Kawai, M., ed. 1979. *Ecological and sociological studies of gelada baboons.* Basel: Karger.

Keay, R. W. J. 1959. *Vegetation map of Africa south of the Tropic of Cancer.* Oxford: Oxford University Press.

Kerlinger, F. N., and E. J. Pedhazur. 1973. *Multiple regression in behavioral research.* New York: Holt, Rinehart and Winston.

Kerr, G. R. 1972. Nutritional requirements of subhuman primates. *Physiol. Rev.* 52:415–67.

Kerr, G. R., J. R. Allen, G. Scheffler, and H. A. Waisman. 1970. Malnutrition studies in the rhesus monkey. I. Effect on physical growth. *Am. J. Clin. Nutr.* 23:239–48.

Kiester, R. 1979. Conspecifics as cues: A mechanism for habitat selection in the Panamanian grass anole (*Anolis auratus*). *Behav. Ecol. Sociobiol.* 5:323–30.

Kincer, J. B. 1941. Climate and weather data for the United States. In *Climate and man: Yearbook of agriculture,* 685–99. Washington, D.C.: USDA

Kleiber, M. 1945. Dietary deficiencies and energy metabolism. *Nutr. Abstr. Rev.* 15:207–20.

———. 1947. Body size and metabolic rate. *Physiol. Rev.* 27:511–41.

———. 1961. *The fire of life: An introduction to animal energetics.* New York: John Wiley.

Klein, D. F. 1978. The diet and reproductive cycle of vervet monkeys (*Cercopithecus aethiops*). Ph.D. diss., New York University.

Kluckhohn, C., and H. A. Murray. 1953. Personality formation: The determinants. In *Personality in nature, society, and culture,* chap. 2. New York: Knopf.

Knapka, J. J., and M. L. Morin. 1979. Open formula natural ingredient diets for nonhuman primates. In *Primates in nutrition research,* ed. K. C. Hayes, 121–38. New York: Academic.

Kon, S. K., and A. T. Cowie, eds. 1961. *Milk: The mammary gland and its secretion.* New York: Academic.

Kopaczewski, W. 1948. Caractères physiques et biologiques du suc des fruits de "*Withania somnifera.*" *Therapie* 3:98–103.

Kortlandt, A. 1984. Habitat richness, foraging range and diet in chimpanzees and some other primates. In *Food acquisition and processing in primates,* ed. D. J. Chivers, B. A. Wood, and A. Bilsborough. New York: Plenum.

Kortlandt, A., and M. Kooij. 1963. Protohominid behaviour in primates. *Symp. Zool. Soc. Lond.* 10:61–88.

Krijnen, C. J., and E. M. Boyd. 1970. Susceptibility to captan pesticide of albino rats fed from weaning on diets containing various levels of protein. *Food Cosmet. Toxicol.* 8:35–42.

Kuhn, H. J. 1964. Zür Kenntnis von Bau and Funktion des Magens der Schlankaffen (Colobidae). *Folia Primatol.* 2:193–221.

Kummer, H. 1968. *Social organization of Hamadryas baboons: A field study.* Chicago: University of Chicago Press.

———. 1982. Social knowledge in free-ranging primates. In *Animal mind—human mind,* ed., D. R. Griffin, 113–30. Berlin: Springer.

Kurland, J. A., and S. J. C. Gaulin. 1987. Comparability among measures of primate diets. *Primates* 28:71–77.

Larson, B. L., and V. R. Smith, eds. 1974. *Lactation A comprehensive treatise.* New York: Academic.

Lawless, J. F. 1982. *Statistical models and methods for lifetime data.* New York: John Wiley.

Lee, E. T. 1980. *Statistical methods for survival data analysis.* Belmont, Mass.: Lifetime Learning.

Lee, R. B. 1979. *The !Kung San.* Cambridge: Cambridge University Press.

Leete, E. 1973. Esophageal cancer. *Science* 179:228.

Leitch, I., ed. 1964. Calcium and phosphorous. In *Nutrition: A comprehensive treatise,* ed. I. Leitch. New York: Academic.

Leung, W. W. 1968. *Food composition tables for use in Africa.* Bethesda, Md.: FAO/HEW, U.S. Public Health Service.

Levin, D. A. 1976. The chemical defences of plants to pathogens and herbivores. *Annu. Rev. Ecol. Syst.* 7:121–59.

Levins, R. 1968. *Evolution in changing environments.* Princeton: Princeton University Press.

Levins, R., and R. Lewontin. 1985. *The dialectical biologist.* Cambridge: Harvard University Press.

Lewis, D. S., H. A. Bertrand, E. J. Masoro, H. C. J. McGill, K. D. Carey, and C. A. McMahan. 1983. Preweaning nutrition and fat development in baboons. *J. Nutr.* 113:2253–59.

Lewontin, R. C. 1979. Sociobiology as an adaptionist program. *Behav. Sci.* 24:5–14.

Liebig, J. 1840. *Chemistry in its application to agriculture and physiology.* London: Taylor and Walton.

Liener, I. E. 1982. Toxic constituents of legumes. In *Chemistry and biochemistry of legumes,* ed. S. K. Arora, 217–57. New Delhi: Oxford and IBH.

Linzell, J. L. 1972. Milk yield, energy loss in milk, and mammary gland weight in different species. *Dairy Sci. Abstr.* 34:351–60.

Liu, J., J. Stamler, A. Dyer, J. McKeever, and P. McKeever. 1978. Statistical methods to assess and minimize the role of intra-individual variability in obscuring the relationship between dietary lipids and serum cholesterol. *J. Chronic Dis.* 31:399–418.

Lotka, A. J. 1913. Evolution from the standpoint of physics, *Scientific American,* suppl., 75:345.

———. 1925. *Elements of physical biology.* Baltimore: Williams and Wilkins. Reprinted 1956, with revisions, as *Elements of mathematical biology.* New York: Dover.

Loy, J. 1988. Effects of supplementary feeding on maturation and fertility in primate groups. In *Ecology and behavior of food-enhanced primate groups,* ed. J. E. Fa and C. H. Southwick, 153–66. New York: Liss.

Lyles, A. M., and A. P. Dobson. 1988. The population dynamics of provisioned and unprovisioned wild primates. In *Ecology and behavior of food-enhanced primate groups,* ed. J. E. Fa and C. H. Southwick, 167–98. New York: Liss.

MacArthur, R. H., and E. R. Pianka. 1966. On the optimal use of a patchy environment. *Am. Nat.* 100:603–9.

Mace, G. M., and P. H. Harvey. 1983. Energetic constraints on home range size. *Am. Nat.* 121:120–32.

Mace, G. H., P. H. Harvey, and T. H. Clutton-Brock. 1981. Brain size and ecology in small mammals. *J. Zool. Lond.* 193:333–54.

MacFarlane, W. V., B. Howard, and B. D. Siebert. 1969. Tritiated water in the measurement of milk intake and tissue growth of ruminants in the field. *Nature* (London) 221:578–79.

Machlis, L. 1977. An analysis of the temporal patterning of pecking in chicks. *Behaviour* 63:1–70.

Machlis, L., P. W. D. Dodd, and J. C. Fentress. 1985. The pooling fallacy: Problems arising when individuals contribute more than one observation to the data set. *Z. Tierpsychol.* 68:201–14.

MacKinnon, J. R., and K. S. MacKinnon. 1978. Comparative feeding ecology of six sympatric primates in West Malasia. In *Recent advances in primatology*. ed. D. J. Chivers and J. Herbert, 305–21. London: Academic.

MacPhail, E. M. 1982. *Brain and intelligence in vertebrates.* Oxford: Clarendon.

Macy, I. G., H. A. Hunscher, E. Donelson, and B. Nims. 1930. Human milk flow. *Am. J. Dis. Child.* 39:1186–1204.

Majumdar, D. N. 1952. Alkaloid constituents of *Withania somnifera. Curr Sci.* 21:46.

Mann. A. E. 1981. Diet and human evolution. In *Omnivorous primates,* ed. R. S. O. Harding and G. Teleki, 10–36. New York: Columbia University Press.

Mann, N. R., R. E. Schafer, and N. D. Singpurwalla. 1974. *Methods for statistical analysis of reliability and life data.* New York: John Wiley.

Mantel, N. 1966. Evaluation of survival data and two new rank order statistics arising in its consideration. *Cancer Chemother. Rep.* 50:163–70.

Marr, J. W. 1971. Individual dietary surveys: Purposes and methods. *World Rev. Nutr. Diet.* 13:105–64.

Martin, P. 1984. The meaning of weaning. *Anim. Behav.* 32:1257–59.

Martz, H. F. 1987. Reliability theory. In *Encyclopedia of physical science and technology,* New York: Academic.

Marwine, A., and G. Collier. 1979. The rat at the waterhole. *J. Comp. Physiol. Psychol.* 93 (2):391–402.

Massiot, G., and C. Deloude. 1986. Pyrrolidine alkaloids. *Alkaloids* 27:269–322.

Master, A. M., and E. T. Oppenheimer. 1929. A simple exercise tolerance test for circulatory efficiency with standard tables for normal individuals. *Am. J. Med. Sci.* 177:223–43.

Maxwell, L. C., D. S. Carlson, J. A. McNamara Jr., and J. A. Faulkner. 1979. Histochemical characteristics of the masseter and temporalis muscles of the rhesus monkey *(Macaca mulatta). Anat. Rec.* 193:389–401.

Maynard, L. A. 1944. The Atwater system of calculating the caloric value of diets. *J. Nutr.* 28:443–52.

Maynard Smith, J. 1978. Optimization theory in evolution. *Annu. Rev. Ecol. Syst.* 9:31–56.

Mayr, E. 1963. *Animal species and evolution.* Cambridge: Harvard University Press.

McCance, R. A. 1976. Critical periods of growth. *Proc. Nutr. Soc.* 35:309–13.

McCay, C. M., L. A. Maynard, G. Sperling, and L. L. Barnes. 1939. Retarded growth, life span, ultimate body size and age changes in albino rat after feeding diets restricted in calories. *J. Nutr.* 18:1–13.

McConnell, E. E., P. A. Basson, and V. deVos. 1971. Nasal ascariasis in the chacma baboon, *Papio ursinus* (Kerr). 1972. *Onderstepoort J. Vet. Res.* 38:207–14.

McGrew, W. C. 1981. Social and cognitive capabilities of nonhuman primates: Lessons from the wild to captivity. *Int. J. Study Anim. Probl.* 2:138–49.

McGrew, W. C., and M. E. Rogers. 1983. Chimpanzees, tools, and termites: New record from Gabon. *Am. J. Primatol.* 5 (2): 171–74.

McKey, D. 1975. The ecology of coevolved seed dispersal systems. In *Coevolution of animals and plants,* ed. L. E. Gilbert and P. H. Raven, 159–91. Austin: University of Texas Press

McMahan, C. A., H. S. Wigodsky, and G. T. Moore. 1976. Weight of the infant baboon (*Papio cynocephalus*) from birth to fifteen weeks. *Lab. Anim. Sci.* 26:928–31.

McNab, B. K. 1983. Ecological and behavioral consequences of adoption to various food resources. In *Advances in the study of mammalian behavior,* ed. J. F. Eisenberg and D. G. Kleiman. 664–97. Shippensburg, Pa.: American Society of Mammalogists.

———. 1986. The influence of food habits on the energetics of eutherian mammals. *Ecol. Monogr.* 56:1–19.

McNair, J. N. 1979. A generalized model of optimal diets. *Theor. Popul Biol.* 15:159–70.

Meier, P. 1975. Estimation of a distribution function from incomplete observations. In *Perspectives in probability and statistics,* ed. J. Gani, 67–87. Sheffield: Applied Probability Trust.

Mellanby, E. 1950. *A story of nutritional research.* Baltimore: Williams and Wilkins.

Menzel, E. W., and E. J. Wyers. 1981. Cognitive aspects of foraging behavior. In *Foraging behavior,* ed. A. C. Kamil and T. D. Sargent. New York: Gartlan STMP.

Merrill, A. L., and B. K. Watt. 1973. Energy value of foods . . . basis and derivation. *U.S. Dept. Agric. Handb.* 74:1–105.

Miller, D. S., and P. R. Payne. 1963. A theory of protein metabolism, *J. Theor. Biol.* 5:398–411.

———. 1968. Longevity and protein intake. *Exp. Geront.* 3:231–34.

Miller, G. A., E. Galanter, and K. H. Pribram. 1960. *Plans and the structure of behavior.* New York: Holt, Rinehart and Winston.

Miller, G. R., A. Watson, and D. Jenkins. 1970. Responses of red grouse populations to experimental improvement of their food. In *Animal populations in relation to their food resources,* ed. A. Watson, 323–35. Oxford: Blackwell.

Miller, J. H. 1960. *Papio doguera* (dog-faced baboon), a primate reservoir host of *Schistosoma mansoni* in East Africa. *Trans. R. Soc. Trop. Med. Hyg.* 54:44–46.

Miller, R. H. 1967. Crotalaria seed morphology, anatomy and identification. *U.S. Dept. Agric. Tech. Bull.* 1373:1–73.

Miller, W. E., and S. A. Altmann. 1958. Ecological observations on the Virginia pitch-nodule moth, *Petrova wenzeli* (Kearfott) including a note on its nomenclature. *Ohio J. Sci.* 58 (5): 273–81.

Milton, K. 1980. *The foraging strategy of howler monkeys: A study in primate economics.* New York: Columbia University Press.

———. 1981. Distribution patterns of tropical plant foods as an evolutionary stimulus to primate mental development. *Am. Anthropol.* 83:534–48.

———. 1984. The role of food-processing factors in primate food choice. In *Adaptations for foraging in nonhuman primates,* ed. P. S. Rodman and J. G. H. Cant, 249–69. New York: Columbia University Press.

Milton, K., and M. W. Demment. 1988. Digestion and passage kinetics of chimpanzees fed high and low fiber diets and comparison with human data. *J. Nutr.* 118:1082–88.

Milton, K., and F. R. Dintzis. 1981. Nitrogen-to-protein conversion factors for tropical plant samples. *Biotropica* 13:177–81.

Mitchell, H. H. 1964. *Comparative nutrition of man and domestic animals.* New York: Academic.

Montgomery, G. G., ed. 1978. *The ecology of arboreal folivores.* Washington, D.C.: Smithsonian Institution Press.

Moreno-Black, G. 1978. The use of scat samples in primate diet analysis. *Primates* 19:215–21.

Mori, A. 1979. An experiment on the relation between the feeding speed and the caloric intake through leaf eating in Japanese monkeys. *Primates* 20:185–95.

Morrison, S. D. 1952. Human milk: Yield, proximate principles and inorganic constituents. *Farnham Royal, Slough, Bucks, U.K., Commonwealth Agr. Bureau.* No. 18, 91.

Morrison, S. D., F. C. Russell, and J. Stevenson. 1949. Estimating food intake by questioning and weighing: A one-day survey of eight subjects. *Brit. J. Nutr.* 3:v.

Morse, D. H. 1971. The insectivorous bird as an adaptive strategy. *Ann. Rev. Ecol. Syst.* 2:177–200.

———. 1980. *Behavioral mechanisms in ecology.* Cambridge: Harvard University Press.

Mortorell, R. 1985. Child growth retardation: A discussion of its causes and its relationship to health. In *Nutritional adaptation in man,* ed. K. Blaxter and J. C. Waterlow. London: Libbey.

Moses, L. E., L. C. Gale, and J. Altmann. 1992. Methods for analysis of unbalanced, longitudinal growth data. *Am. J. Primatol.* 28:49–59.

Moulton, C. R. 1923. Age and chemical composition in mammals. *J. Biol. Chem.* 57:79–97.

Munro, H. N. 1976. Regulation of body protein metabolism in relation to diet. *Proc. Nutr. Soc.* 35:297–308.

———. 1985. Historical perspective on protein requirements: Objectives for the future. In *Nutritional adaptation in Man,* ed. K. Blaxter and J. C. Waterlow. London: Libbey.

———. 1986. Back to basics: An evolutionary odyssey with reflections on the nutrition research of tomorrow. *Ann. Rev. Nutr.* 6:1–12.

Munro, H. N., and M. C. Crim. 1988. The proteins and amino acids. In *Modern nutrition in health and disease,* 9th ed., ed. M. E. Shils and V. R. Young. Philadelphia: Lea and Febiger.

Murdoch, W. W. 1966. Community structure, population control, and competition—a critique. *Am. Nat.* 100:219–26.

Murphy, M. E., and S. D. Pearcy. 1993. Dietary amino acid complementation as a foraging strategy for wild birds. *Physiol. Behav.* 53:689–98.

Murray, J., and A. Murray. 1977. Suppression of infection by famine and its activation by refeeding—a paradox? *Perspect. Biol. Med.* 20:471–84.

Murray, P. F. 1970. The evolutionary biology of cheek pouches in the Cercopithecinae. *Am. J. Phys. Anthropol.,* n.s., 33:139.

Muruthi, P. 1980. Food intake and energy expenditure among adult female baboons (*Pa-*

pio cynocephalus) in Amboseli National Park, Kenya. M.Sc. thesis, University of Nairobi, Nairobi.

———. 1997. Socioecological correlates of parental care and demography in savanna baboons. Ph.D. diss., Princeton University.

Myers, B. J., and R. E. Kuntz. 1965. A checklist of parasites reported from the baboon. *Primates* 6:137–94.

Nagel, U. 1973. A comparison of anubis baboons, hamadryas baboons, and their hybrids at a species border in Ethiopia. *Folia Primatol.* 19:104–65.

Nagy, K. A. 1987. Field metabolic rate and food requirement scaling in mammals and birds. *Ecol. Monogr.* 57:111–28.

———. 1994. Field bioenergetics of mammals: What determines field metabolic rate? *Aust. J. Zool.* 42:43–53.

Napier, J. R., and P. H. Napier, 1967. *A handbook of living primates.* London: Academic.

NAS (National Academy of Sciences). 1975. *The effect of genetic variance on nutritional requirements of animals.* Washington, D.C.: National Academy of Sciences.

———. 1978. *Nutrient requirements of nonhuman primates.* Washington, D.C.: National Academy of Sciences.

Nash, L. T. 1986. Dietary, behavioral, and morphological aspects of gumivory in primates. *Yearb. Phys. Anthropol.* 29:113–37.

Nelson, K. 1964. The temporal patterning of courtship behaviour in the glandulocaudine fishes (Ostariophysi, Characidae). *Behaviour* 24:90–146.

Nelson, R. C., C. J. Dillman, P. Lagasse, and P. Bickett. 1972. Biomechanics of overground versus treadmill running. *Medicine and Science in Sports* 4:233–40.

Nelson, W. 1972. Theory and application of hazard plotting for censored failure data. *Technometrics* 14:945–65.

Newton, M., and G. E. Egli. 1958. The effect of intranasal administration of oxytocin on the let-down of milk in lactating women. *Am. J. Obstet. Gynecol.* 76:103–7.

Nicol, B. M., and P. G. Phillips. 1976a. The utilization of dietary protein by Nigerian men. *Br. J. Nutr.* 36:337–51.

———. 1976b. Endogenous nitrogen excretion and utilization of dietary protein. *Br. J. Nutr.* 35:181–93.

———. 1978. The utilization of proteins and amino acids in diets based on cassava (*Manihot utilissima*), rice or sorghum (*Sorghum sativa*) by young Nigerian men of low income. *Br. J. Nutr.* 39:271–87.

Nicol, S. C. 1978. Rates of water turnover in marsupials and eutherians: A comparative review with new data on the Tasmanian devil. *Aust. J. Zool.* 26:465–73.

Nicolosi, R. J., and R. D. Hunt. 1979. Dietary allowances for nutrients in nonhuman primates. In *Primates in nutritional research,* ed. K. C. Hayes, 11–37. New York: Academic.

Nicolson, N. 1982. Weaning and the development of independence in olive baboons. Ph.D. diss., Harvard University.

Nishida, T., and M. Hiraiwa. 1982. Natural history of a tool-using behavior by wild chimpanzees in feeding upon wood-boring ants. *J. Hum. Evol.* 11:73–99.

Norton, G. W., R. J. Rhine, G. W. Wynn, and R. D. Wynn. 1987. Baboon diet: A five-year study of stability and variability in the plant feeding and habitat of the yellow baboons (*Papio cynocephalus*) of Mikumi National Park, Tanzania. *Folia Primatol.* 48:78–120.

NRC (National Research Council). 1980. *Mineral tolerance of domestic animals.* Washington, D.C.: National Academy of Sciences.

Odani, S., and T. Ikenaka. 1973. Studies on soybean trypsin inhibitor. VIII. Disulfide bridges in soybean Bowman Birk proteinase inhibitor. *J. Biochem.* 74:697.

Oftedal, O. T. 1980. Milk and mammalian evolution. In *Comparative physiology: Primitive mammals,* ed. K. Schmidt-Nielson, L. Bolis, and C. R. Taylor, 31–42. Cambridge: Cambridge University Press.

———. 1984a. Milk composition, milk yield and energy output at peak lactation: A comparative review. *Symp. Zool. Soc. Lond.* 51:33–85.

———. 1984b. Body size and reproductive strategy as correlates of milk energy yield in lactating mammals. *Acta Zool. Fenn.* 171:183–86.

———. 1985. Pregnancy and lactation. In *Bioenergetics of wild herbivores,* ed. R. J. Hudson and R. G. White, 215–38. Boca Raton, Fla.: CRC.

Oltersdorf, U., R. Miltenberger, and H. D. Cremer. 1977. Interactions of non-nutrients with nutrients. *World Rev. Nutr. Diet.* 26:41–134.

Orwell, G. 1954. *Animal farm.* New York: Harcourt.

Pace, N., D. F. Hanson, N. J. Rahlmann, N. J. Barnstein, and M. D. Cannon. 1964. Preliminary observations on some physiological conditions of the pig-tailed monkey *Macaca nemestrina. Aerospace Med.* 35:118–21.

Parker, S. T., and K. R. Gibson. 1977. Object manipulation, tool use and sensorimotor intelligence as feeding adaptations in cebus monkeys and great apes. *J. Hum. Evol.* 6:623–41.

———. 1979. A developmental model for the evolution of language and intelligence in early hominids. *Behav. Brain Sci.* 2:367–408.

Payne, P. R., and J. C. Waterlow. 1971. Relative energy requirements for maintenance, growth, and physical activity. *Lancet* 1971 (2): 210–11.

Peaker, M. 1977. *Comparative aspects of lactation.* London: Academic.

Pekkarinen, M. 1970. Methodology in the collection of food consumption data. *World Rev. Nutr. Diet.* 12:145.

Pennycuick, C. J. 1979. Energy cost of locomotion and the concept of "foraging radius." In *Serengeti: Dynamics of an ecosystem,* ed. A. R. E. Sinclair and M. Norton-Griffiths, 389. Chicago: University of Chicago Press.

Pereira, M. E. 1984. Age changes and sex differences in the social behavior of juvenile yellow baboons (*Papio cynocephalus*). Ph.D. diss., University of Chicago.

Peters, R. 1983. *The ecological implications of body size.* Cambridge: Cambridge University Press.

Petter, J. J., and A. Peter. 1967. The aye-aye of Madagascar. In *Social communication among primates,* ed. S. A. Altmann, chap. 11. Chicago: University of Chicago Press.

Phillips-Conroy, J. E. 1986. Baboons, diet, and disease: Food plant selection and schistosomiasis. In *Current perspectives in primate social dynamics,* ed. D. Taub and F. King, 287–304. New York: Van Nostrand Reinhold.

Phillips-Conroy, J. E., C. F. Hildebolt, J. Altmann, C. J. Jolly, and P. Muruthi. 1993. Periodontal health in free-ranging baboons in Ethiopia and Kenya. *Am. J. Phys. Anthropol.* 90:359–71.

Pike, R. L. 1978. Nutrient deficiencies in animals: Sodium. In *Effects of nutrient deficiencies in animals,* ed. J. M. Rechcigl. West Palm Beach, Fla.: CRC.

Pinder, J. E., III, J. G. Wiener, and M. H. Smith. 1978. The Weibull distribution: A new method of summarizing survivorship data. *Ecology* 51:175–79.

Pirlot, P., and J. Pottier. 1977. Encephalization and quantitative brain composition in bats in relation to their life-habits. *Rev. Can. Biol.* 36 (4):321–36.

Poincaré, H. 1913. *Science and hypotheses.* Trans. G. B. Halstead, Reprinted in *The Foundation of Science.* Lancaster, Pa.: Science Press. Originally published as *La science et l'hypothèse,* 1903.

Pollard, A. J. 1992. The importance of deterrence: Responses of grazing animals to plant variation. In *Plant resistance to herbivores and pathogens,* ed. R. S. Fritz and E. L. Simms, 216–39. Chicago: University of Chicago Press.

Polyak, S. 1957. *The vertebrate visual system.* Chicago: University of Chicago Press.

Pond, C. M. 1977. The significance of lactation in the evolution of mammals. *Evolution* 31:177–99.

Post, D. G. 1978. Feeding and ranging behavior of the yellow baboon, *(Papio cynocephalus).* Ph.D. diss., Yale University.

———. 1981. Activity patterns of yellow baboons *(Papio cynocephalus)* in the Amboseli National Park, Kenya. *Anim. Behav.* 29:357–74.

———. 1982. Feeding behavior of yellow baboons *(Papio cynocephalus)* in the Amboseli National Park, Kenya. *Int. J. Primatol.* 3:403–30.

———. 1984. Is optimization the optimal approach to primate foraging? In *Adaptations for foraging in nonhuman primates,* ed. P. Rodman and J. G. H. Cant, 280–303. New York: Columbia University Press.

Post, D. G., G. Hausfater, and S. A. McCuskey. 1980. Feeding behavior of yellow baboons *(Papio cynocephalus):* Relationship to age, gender, and dominance rank. *Folia Primatol.* 34:170–95.

Powell, G. V. N. 1974. Experimental analysis of the social value of flocking by starlings *(Sturnus vulgaris)* in relation to predation and foraging. *Anim. Behav.* 22:501–5.

Pulliam, H. R. 1975. Diet optimization with nutrient constraints. *Am. Nat.* 109:765–68.

Pyke, G. H., H. R. Pulliam, and E. L. Charnov. 1977. Optimal foraging: A selective review of theory and tests. *Quart. Rev. Biol.* 52 (2):137–54.

Quinlan, R. G. Robson, and A. R. Pack. 1994. A study comparing the efficacy of a toothpaste containing extract of *Salvadora persica* with a standard fluoride toothpaste. *J. New Zealand Soc. Periodont.* 77:7–14.

Rand, W. M., N. S. Seriskaw, and V. R. Young. 1977. Determination of protein allowances in human adults from nitrogen balance data. *Am. J. Clin. Nutr.* 30:1129–34.

Rapport, D. J. 1980. Optimal foraging for complementary resources. *Am. Nat.* 116:324–46.

Rasmussen, K. L. R. 1985. Changes in activity budgets of yellow baboons *(Papio cynocephalus)* during sexual consortships. *Behav. Ecol. Sociobiol.* 17:161–70.

Rath, E. A., and S. W. Thenen. 1979. Use of tritiated water for measurement of 24-hour milk intake in suckling lean and genetically obese (ob/ob) mice. *J. Nutr.* 109: 840–47.

Rayudu, G. V., R. Kardivel, P. Vohra, and F. H. Kratzer. 1970. Effect of various agents in alleviating the toxicity of tannic acid for chickens. *Poultry. Sci.* 49:1323–26.

Rea, J. L., and J. C. Drummond. 1932. On the formation of vitamin A from carotene in the animal organism. *Z. Vitaminforsch.* 1:177–83.

Reese, J. C. 1979. Interactions of allelochemicals with nutrients in herbivore food. In *Herbivores: Their interaction with secondary plant metabolites,* ed. G. A. Rosenthal and D. H. Janzen, 309–30. New York: Academic.

Reidhead, V. A. 1980. The economics of subsistence change: Test of an optimization model. In *Modeling change in prehistoric subsistence economies,* ed. T. K. Earle and A. L. Christenson, 141–56. New York: Academic.

Reiss, M. J. 1986. Belovsky's model of optimal moose size. *J. Theor. Biol.* 122: 237–42.

Rhine, R. J., G. W. Norton, and G. M. Wynn. 1989. Plant feeding of yellow baboons (*Papio cynocephalus*) in Mikumi National Park, Tanzania, and the relationship between seasonal feeding and immature survival. *Int. J. Primatol.* 10:319–42.

Rhine, R. J., G. W. Norton, G. M. Wynn, and R. D. Wynn. 1985. Weaning of free-ranging infant baboons (*Papio cynocephalus*) as indicated by one-zero and instantaneous sampling of feeding. *Int. J. Primatol.* 6 (5): 491–99.

Rhine, R. J., G. W. Norton, G. M. Wynn, R. D. Wynn, and H. B. Rhine. 1986. Insect and meat eating among infant and adult baboons (*Papio cynocephalus*) of Mikumi National Park, Tanzania. *Am. J. Phys. Anthropol.* 70:105–18.

Richmond, C. R., W. H. Langham, and T. T. Trujillo. 1962. Comparative metabolism of tritiated water by mammals. *J. Cell. Comp. Physiol.* 59:45–53.

Richter, C. P. 1942. Total self-regulatory function in animals and human beings. *Harv. Lect. Ser.* 38:63–103.

Rijksen, H. D. 1978. *A field study on Sumatran orang utans: Ecology, behaviour and conservation.* Netherlands: Wageningen.

Riopelle, A. J., C. W. Hill, S. C. Li, R. H. Wolf, H. R. Seibold, and J. L. Smith. 1974. Protein deprivation in primates. I. Nonpregnant adult rhesus monkeys. *Am. J. Clin. Nutr.* 27:13–21.

Robbins, C. T. 1983. *Wildlife feeding and nutrition.* New York: Academic.

Robbins, R. C., and J. A. Gavan. 1966. Utilization of energy and protein of a commercial diet by rhesus monkeys (*Macaca mulatta*). *Lab. Anim. Care* 16:286–91.

Roberts, S. B., T. J. Cole, and W. A. Coward. 1985. Lactational performance in relation to energy intake in the baboon. *Am. J. Clin. Nutr.* 41:1270–76.

Robinson, J. G., and C. H. Janson. 1987. Capuchins, squirrel monkeys, and atelines: Socioecological convergence with Old World primates. In *Primate societies,* ed. B. B. Smuts, D. L. Cheney, R. M. Seyfarth, R. W. Wrangham, and T. T. Struhsaker, 69–82. Chicago: University of Chicago Press.

Rodahl, K., and T. Moore. 1943. The vitamin A content and toxicity of bear and seal liver. *Biochem. J.* 37:166–68.

Rodman, P. S. 1979. Individual activity patterns and the solitary nature of orangutans. In *The great apes,* ed. D. A. Hamburg and E. R. McCown, 235–55. Menlo Park, Calif.: Benjamin/Cummings.

Rodriguez, M. S., and M. J. Irwin. 1972. A conspectus of research on vitamin A requirements in man. *J. Nutr.* 102:909–68.

Roland, D. A. 1978. Effect of nutrient deficiencies in animals: Calcium. In *Effects of nutrient deficiencies in animals,* ed. J. M. Rechcigl. West Palm Beach, Fla.: CRC.

Rose, W. C., and R. L. Wixon. 1955. The amino acid requirements of man. XVI. The role of the nitrogen intake. *J. Biol. Chem.* 217:997–1004.

Rosenthal, G. A., and D. H. Janzen, eds. 1979. *Herbivores: Their interaction with secondary plant metabolites.* New York: Academic.

Ross, M. H. 1961. Length of life and nutrition in the rat. *J. Nutr.* 75:197.

———. 1969. Aging, nutrition and hepatic enzyme activity patterns in the rat. *J. Nutr.* 97:563–601.

Ross, M. H., and G. Grass. 1971. Lasting influence of dietary change on neoplasms in the rat. *J. Nat. Cancer Inst.* 47:1095–1113.

Rowell, T. E. 1966. Forest living baboons in Uganda. *J. Zool. Lond.* 147:344–64.

Rozin, P. 1976. The selection of foods by rats, humans, and other animals. *Advances in the Study of Behavior* 6:21–76.

Rutenberg, G. W., and A. J. Coelho. 1988. Neonatal nutrition and longitudinal growth from birth to adolescence in baboons. *Am. J. Phys. Anthropol.* 75:529–39.

Ryan, C. A., and T. R. Green. 1974. Proteinase inhibitors in natural plant protection. In *Metabolism and regulation of secondary plant products,* ed. V. C. Ryan and E. E. Conn, 123–39. New York: Academic.

Samonds, K. W., and D. M. Hegsted. 1973. Protein requirements of young cebus monkeys (*Cebus albifrons* and *apella*). *Am. J. Clin. Nutr.* 26:30–40.

Saunders, C., and G. Hausfater. 1978. Sexual selection in baboons: A computer simulation of differential reproduction with respect to dominance rank in males. In *Recent advances in primatology,* ed. J. Herbert and D. J. Chivers, 567–76. London: Academic.

Saunders, R. M. 1978. Wheat bran: Composition and digestibility. In *Topics in dietary fibre research,* ed. G. A. Spiller, 43–58. New York: Plenum.

Saxton, J. A., Jr., M. C. Boon, and J. Furth. 1944. Observations on the inhibition of development of spontaneous leukemia in mice by underfeeding. *J. Cancer Res.* 4:401–9.

Schaeffer, A. A. 1910. Selection of food in *Stentor coeruleus. J. Exp. Zool.* 8:75–132.

Schluter, D. 1981. Does the theory of optimal diets apply in complex environments? *Am. Nat.* 118:139–47.

Schmidt, G. H. 1971. *Biology of lactation.* San Francisco: Freeman.

Schmidt-Nielsen, K. 1972. Locomotion: Energy cost of swimming, flying, and running. *Science* 177:222–28.

———. 1979. *Animal physiology: Adaptation and environment.* Cambridge: Cambridge University Press.

Schoener, T. W. 1971. Theory of foraging strategies. *Annu. Rev. Ecol. Syst.* 2:379–404.

Schrage, L. 1981. *Linear programming with LINDO.* Palo Alto, Calif.: Scientific.

Schryver, H. F., and H. F. Hintz. 1978. Effect of nutrient deficiencies in animals: Phosphorus. In *Effects of nutrient deficinecies in animals,* ed. J. M. Rechcigl. West Palm Beach, Fla.: CRC.

Schuette, S. A., and H. M. Links-Wiler. 1982. Effects on Ca and P metabolism in humans by adding meat, meat plus milk, or purified protein plus Ca and P to a low protein diet. *J. Nutr.* 112:338–49.

Schultz, A. H. 1969. *The life of primates.* New York: Universe Books.

Scott, J. 1963. Factors determining skull form in primates. *Symp. Zool. Soc. Lond.* 10:127–43.

Sharman, M., and R. I. M. Dunbar. 1982. Observer bias in selection of study group in baboon field studies. *Primates* 23 (4): 567–73.

Shelford, V. E. 1911. Physiological animal geography. *J. Morph.* 22:551–618.

Shopland, J. M. 1987. Food quality, spatial deployment and the intensity of feeding interference in yellow baboons (*Papio cynocephalus*). *Behav. Ecol. Sociobiol.* 21: 149–56.

Sigg, H., and A. Stolba. 1981. Home range and daily march in a hamadryas baboon troop. *Folia Primatol.* 36:40–75.

Sih, A. 1980a. Optimal foraging: Partial consumption of prey. *Am. Nat.* 116:281–90.

———. 1980b. Optimal behavior: Can foragers balance two conflicting demands? *Science* 210:1041–43.

Silberberg, M., and R. Silberberg. 1955. Diet and life span. *Physiol. Rev.* 35:347–62.

Silk, J. B. 1987. Activities and feeding behavior of free-ranging pregnant baboons. *Int. J. Primatol.* 8:593–613.

Simmen, B. 1994. Taste discrimination and diet differentiation among New World primates. In *The digestive system of mammals,* ed. D. J. Chivers and P. Langer, 150–65. Cambridge: Cambridge University Press.

Simmonds, P. L. 1891. The medicinal and other useful plants of Algeria. *Am. J. Pharm.* 83:76–80.

Simpson, G. G. 1982. *The book of Darwin.* New York: Pocket Books.

Simpson, M. J. A. 1983. Effect of the sex of an infant on the mother-infant relationship and the mother's subsequent reproduction. In *Primate social relationships: An integrated approach,* ed. R. A. Hinde, 53–57. Oxford: Blackwell.

Slater, P. J. B. 1974. Bouts and gaps in the behavior of zebra finches, with special reference to preening. *Rev. Comp. Animal* 8:47–61.

———. 1978. Data collection. In *Quantitative ethology,* ed. P. W. Colgan, 7–24. New York: John Wiley.

Slatkin, M. 1975. A report on the feeding behavior of two East African baboon species. In *Contemporary primatology,* ed. S. Kondo, M. Kawai, A. Ehara, and S. Kawamura, 418–22. Basel: Karger.

Slatkin, M., and G. Hausfater. 1976. A note on the activities of a solitary male baboon. *Primates* 17:311–22.

Slobodkin, L. B., and H. L. Saunders. 1969. On the contribution of environmental predictability to species diversity. In *Diversity and stability in ecological systems,* ed. G. M. Woodwell and H. H. Smith, 82–95. Upton, N.Y.: Brookhaven National Laboratories.

Slonaker, J. R. 1931a. The effects of different percents of protein in the diet. I. Growth. *Am. J. Physiol.* 96:547–56.

———. 1931b. The effects of different percents of protein in the diet. VII. Life span and cause of death. *Am. J. Physiol.* 98:266–75.

Sly, M. R., W. A. du Bruyn, D. J. de Klerk, D. J. Robbins, N. v. d. W. Liebenberg, and W. H. van der Walt. 1980. Evaluation, using baboons, of a commercial cereal diet supplement, designed to prevent/combat malnutrition. *Nutr. Rep. Int.* 22:223–34.

Smith, A. P. 1984. Diet of Leadbeater's possum. *Gymnobelideus leadbeateri* (Marsupialia). *Aust. Wildl. Res.* 11:265–73.

———. 1985. Adaptations to gumivory in marsupial possums and gliders (abstract). *Am. J. Phys. Anthropol.* 66:229.

Smith, C. H., and K. K. Merritt. 1922. The rate of secretion of breast milk. *Am. J. Dis. Child.* 24:413–26.

Smith, D. L. 1986. *Amboseli—nothing short of a miracle.* Nairobi: East African Publishing House.

Smith, E. A. 1992. *Inujjuamiut foraging strategies.* Hawthorne, N.Y.: Aldine de Gruyter.

Smotherman, W. P. 1982. Odor aversion learning by the rat fetus. *Physiol. Behav.* 29:769.

Snodderly, D. M. 1978. Color discriminations during food foraging by a New World monkey. In *Recent advances in primatology,* ed. D. J. Chivers and J. Herbert, 369–71. London: Academic.

Snow, C. C. 1967a. Some observations on the growth and development of the baboon. In *The baboon in medical research,* ed. H. Vagtborg, 187–99. Austin: University of Texas Press.

———. 1967b. The physical growth and development of the open-land baboon. *Papio doguera.* Ph.D. diss., University of Arizona.

Sohonie, K., and P. M. Honawar. 1955. Trypsin inhibitors in Indian foodstuffs. II. Inhibitors in pulses. *J. Sci. Ind. Res. India* 14c:100–104.

Sokal, R. R., and F. J. Rohlf. 1081. *Biometry.* San Francisco: W. H. Freeman.

Southgate, D. A. F., and I. M. Barrett. 1966. The intake and excretion of caloric constituents of milk by babies. *Br. J. Nutr.* 20:363–72.

Southgate, D. A. F., and J. G. V. A. Durnin. 1970. Caloric conversion factors: An experimental reassessment of the factors used in the calculation of energy value of human diets. *Brit. J. Nutr.* 24:517–35.

Spotter, P. D., and A. E. Harper. 1961. Utilization of ingested and orally administered amino acids by the rat. *Proc. Soc. Exp. Biol. Med.* 106:184–88.

Stacey, P. B. 1986. Group size and foraging efficiency in yellow baboons. *Behav. Ecol. Sociobiol.* 18:175–87.

Stahl, A. B. 1984. Hominid dietary selection before fire. *Curr. Anthropol.* 25:151–68.

Stare, F. J. 1958. General summary. *Ann. N.Y. Acad. Sci.* 69:1064–66.

Stein, D. M. 1984. *The sociobiology of infant and adult male baboons.* Norwood N.J.: Ablex.

Stelzner, J. K. 1988. Thermal effects on movement patterns of yellow baboons. *Primates* 29:91–105.

Stenseth, N. C., and L. Hansson. 1979. Optimal food selection: A graphic model. *Am. Nat.* 113:373–89.

Stephens, D. W. 1981. The logic of risk-sensitive foraging preferences. *Anim. Behav.* 29:628–29.

Stephens, D. W., and E. L. Charnov. 1982. Optimal foraging: Some simple stochastic models. *Behav. Ecol. Sociobiol.* 10 (4): 251–63.

Stephens, D. W., and J. R. Krebs. 1986. *Foraging theory.* Princeton: Princeton University Press.

Stigler, G. J. 1945. The cost of subsistence. *J. Farm Econ.* 27:303–14.

Stini, W. A. 1975. Adaptive strategies of human populations under nutritional stress. In *Biosocial interrelations in population adaptation,* ed. E. S. Watts, F. E. Johnston, and G. W. Laskers, 412. Paris: Mouton.

Stock, A. L., and E. F. Wheeler. 1972. Evaluation of meals cooked by large-scale methods: A comparison of chemical analysis and calculation from food tables. *Br. J. Nutr.* 27:439–48.

Strong, R. M. 1926. The order, time and rate of ossification of the albino rat skeleton. *Am. J. Anat.* 36:313–55.

Struhsaker, T. T. 1967a. Ecology of vervet monkeys (*Cercopithecus aethiops*) in the Masai-Amboseli Game Reserve, Kenya. *Ecology* 48:891–904.

———. 1967b. Behavior of vervet monkeys (*Cercopithecus aethiops*). *Univ. Calif. Publ. Zool.* 82:1–64.

———. 1967c. Auditory communication among vervet monkeys. *Cercopithecus aethiops.* In *Social communication among primates,* ed. S. A. Altmann, 281–324. Chicago: University of Chicago Press.

———. 1967d. Social structure among vervet monkeys (*Cercopithecus aethiops*). *Behavior* 29:82–121.

———. 1971. Social behaviour of mother and infant vervet monkeys (*Cercopithecus aethiops*). *Anim. Behav.* 19:233–50.

———. 1973. A recensus of vervet monkeys in the Masai-Amboseli Game Reserve, Kenya. *Ecology* 54:930–32.

———. 1975. *The red colobus monkey.* Chicago: University of Chicago Press.

———. 1976. A further decline in numbers of Amboseli vervet monkeys. *Biotropica* 8 (3): 211–14.

———. 1978. Food habits of five monkey species in the Kibale forest, Uganda. In *Recent advances in primatology,* ed. D. J. Chivers and J. Herbert, 225–48. London: Academic.

Strum, S. C. 1975. Primate predation: Interim report on the development of a tradition in a group of olive baboons. *Science* 187:755–57.

———. 1983. Baboon cues for eating meat. *J. Hum. Evol.* 12 (4): 327–36.

Stunkard, A. J. 1983. Nutrition, aging and obesity: A critical review of a complex relationship. *Int. J. Obesity* 7:201–20.

Sugiyama, Y., and H. Ohsawa. 1982. Population dynamics of Japanese macaques at Ryozenyama. III. Female desertion of the troop. *Primates* 23:31–44.

Sukhatme, P. V. 1974. The protein problem, its size and nature. *J.R. Statist. Soc.,* ser. A, 137:166–99.

———. 1975. Human protein needs and the relative role of energy and protein in meeting them. In *The man-food equation,* ed. F. Steele and A. Bourne, 53–75. London: Academic.

Taggart, N. 1962. Diet, activity and body-weight: A study of variations in a woman. *Br. J. Nutr.* 16:223–35.

Taylor, C. R., N. C. Heglund, and G. M. O. Maloiy. 1982. Energetics and mechanics of terrestrial locomotion. I. Metabolic energy consumption as a function of speed and body size in birds and mammals. *J. Exp. Biol.* 97:1–21.

Temerin, L. A., B. P. Wheatley, and P. S. Rodman. 1984. Body size and foraging in primates. In *Adaptations for foraging in nonhuman primates,* ed. P. S. Rodman and J. G. H. Cant, 217–48. New York: Columbia University Press.

Terborgh, J. 1983. *Five New World primates: A study in comparative ecology.* Princeton: Princeton University Press.

Thiemann, S., and H. C. Kraemer. 1984. Sources of behavioral variance: Implications for sample size decisions. *Am. J. Primatol.* 7:367–76.

Thomas, R. K. 1980. Evolution of intelligence: An approach to its assessment. *Brain Behav. Evol.* 17:454–72.

———. 1982. The assessment of primate intelligence. *J. Hum. Evol.* 11:247–55.

Townsend, C. R., and P. Calow. 1981. *Physiological ecology: An evolutionary approach to resource use.* Sunderland, Mass.: Sinauer.

Treviño, S. 1991. *Graincollection: Humans' natural ecological niche.* New York: Vantage.

Trivers, R. L. 1972. Parental investment and sexual selection. In *Sexual selection and the descent of man, 1871–1971,* ed. B. Campbell, 136–79. Chicago: Aldine.

Truswell, A. S. 1976. A comparative look at recommended nutrient intakes. *Proc. Nutr. Soc.* 35:1–14.

Truswell, A. S., and J. D. L. Hansen. 1976. Medical research among the !Kung. In *Kalahari hunter-gatherers,* ed. R. B. Lee and I. DeVore, 166–94. Cambridge: Harvard University Press.

Uehara, S. 1982. Seasonal changes in the techniques employed by wild chimpanzees in the Mahale Mountains, Tanzania, to feed on termites (*Pseudacanthotermes spiniger*). *Folia Primatol.* 37 (1–2):44–76.

Vagtborg, H., ed. 1965. *The baboon in medical research.* Austin: University of Texas Press.

van der Pijl, L. 1969. *Principles of dispersal in higher plants.* New York: Springer.

Van Soest, P. J. 1994. *Nutritional ecology of the ruminant,* 2d ed. Ithaca, N.Y.: Comstock.

Van Valen, L. 1960. Nonadaptive aspects of evolution. *Am. Nat.* 94:305–8.

———. 1976. Energy and evolution. *Evol. Theory* 1:179–229.

Vice, T. E., H. A. Britton, I. A. Ratner, and S. S. Kalter. 1966. Care and raising of newborn baboons. *Lab. Anim, Care* 16:12–22.

Vorherr, H. 1974. *The breast.* New York: Academic.

Wagner, H. M. 1969. *Principles of operation research.* Englewood Cliffs, N.J.: Prentice-Hall.

Waisel, Y. 1972. *Biology of halophytes.* New York: Academic.

Walker, A. R. P. 1951. Cereals, phytic acid, and calcification. *Lancet* 261:244–48.

Walker, A. R. P., B. F. Walker, and B. D. Richardson. 1976. Whither optimal nutrition? *Am. Heart J.* 92:403–4.

Washburn, S. L., and I. DeVore. 1962. Ecologie et comportement des babouins. *Terre Vie* 133–49.

Wasser, L. M., and S. K. Wasser. 1995. Environmental variation and development rate among free-ranging yellow baboons (*Papio cynocephalus*). *Am. J. Primatol.* 35: 15–30.

Waterlow, J. C. 1964. Protein metabolism in human protein malnutrition. *Proc. R. Soc.* 156:345.

———. 1981. Crisis for nutrition. *Proc. Nutr. Soc.* 40:195–207.

———. 1986. Metabolic adaptation to low intakes of energy and protein. *Annu. Rev. Nutr.* 6:495–526.

Watt, B. K., and A. L. Merrill. 1963. *Composition of foods . . . Raw, processed, prepared.* Handbook 8. Washington, D.C.: USDA.

Watt, H. J. 1925. *The sensory basis and structure of knowledge.* London: Methuen.

Watt, J. M., and M. G. Breyer-Brandwijk. 1962. *The medicinal and poisonous plants of southern and eastern Africa,* 2d ed. Edinburgh: Livingston.

Weibull, W. 1951. A statistical distribution function of wide applicability. *J. Appl. Mech.* 18:292–97.

Weis, A. E. 1992. Plant variation and the evolution of phenotypic plasticity in herbivore performance. In *Plant resistance to herbivores and pathogens,* ed. R. S. Fritz and E. L. Simms, 140–71. Chicago: University of Chicago Press.

Weiskrantz, L. 1985. Introduction [to symposium]: Categorization, cleverness and consciousness. *Philos. Trans. R. Soc. Lond.,* ser. B, 308:3–19.

Welsh, A. H., A. T. Peterson, and S. A. Altmann. 1987. The fallacy of averages. *Am. Nat.* 132:277–88.

Werner, E. E., and D. J. Hall. 1974. Optimal foraging and size selection of prey by the bluegill sunfish (*Lepomis mochrochirus*). *Ecology* 55:1042–52.

Western, D. 1973a. The changing face of Amboseli. *Africana* 5(3): 23–29.

———. 1973b. *The structure, dynamics and changes of the Amboseli ecosystem.* Ph.D. diss., University of Nairobi, Kenya.

———. 1979, Size, life-history, and ecology in mammals. *Afr. J. Ecol.* 17:185–204.

———. 1980. Linking the ecology of past and present mammal communities. In *Fossils in the making,* ed. A. K. Behrensmeyer and A. P. Hill, 41–54. Chicago: University of Chicago Press.

Western, D., and C. Van Praet. 1973. Cyclical changes in the habitat and climate of an East African ecosystem. *Nature* (London) 241:104–6.

Westoby, M. 1978. What are the biological bases of varied diets? *Am. Nat.* 112: 627–31.

Wetzel, M. C., A. E. Atwater, J. V. Wait, and D. C. Stuart. 1975. Neural implications of different profiles between treadmill and overground locomotion timings in cats. *J. Neurophys.* 38:492–501.

White, T. C. R. 1978. The importance of a relative shortage of food in animal ecology. *Oecologia* 33:71–86.

Whitehead, J. M. 1985. Development of feeding selectivity in mantled howling monkeys (*Alouatta palliata*). In *Primate ontogeny, cognition and social behavior,* ed. J. Else and P. Lee, 105–17. Cambridge: Cambridge University Press.

Whiten, A., R. W. Byrne, R. A. Barton, P. G. Waterman, and S. P. Henzi. 1991. Dietary and foraging strategies of baboons. *Philos. Trans. R. Soc. Lond.,* ser. B, 334: 187–97.

Whiten, A., R. W. Byrne, and S. P. Henzi. 1987. The behavioral ecology of mountain baboons. *Int. J. Primatol.* 8:367–87.

Whiten, A., R. W. Byrne, P. G. Waterman, S. P. Henzi, and F. M. McCullough. 1990. Specifying the rules underlying selective foraging in wild mountain baboons, *P. ursinus.* In *XII Congress of the International Primatological Society, Selected Proceedings,* 5–22. Brasilia, Brazil: n.p.

Whitman, C. O. 1896. Evolution and epigenesis. In *Marine Biological Laboratories biological lectures.* Woods Hole, Mass.: Marine Biological Laboratories.

WHO (World Health Organization). 1974. *Handbook on human nutritional requirements.* Geneva: WHO.

Widdowson, E. M. 1976. Changes in the body and its organs during lactation: Nutritional implications. In *Breast-feeding and the mother,* 103–18. Ciba Foundation Symposium 45 (new series). Amsterdam: Elsevier.

Williams, U. V. and S. Nagarajan. 1988. Isorhamnetin 3-O-rutinoside from leaves of *Azima tetracantha* Lam. *Indian J. Chem, B-Org. Chem. Incl. Med. Chem.* 27:397.

Wilson, R. H., and F. de Eds. 1950. Importance of diet in studies of chronic toxicity. *Arch. Industr. Hyg.* 1:73–80.

Wink, M. 1993. Allelochemical properties or the raison d'être of alkaloids. In *The alkaloids: Chemistry and pharmacology,* ed. G. A. Cordell, San Diego, Calif.: Academic.

Wise, A. 1980. Nutrient interrelationships. *Nutr. Abstr. Rev.,* ser. A, 50:319–32.

Wixson, S. K., and J. W. Griffith. 1986. Nutritional deficiency anemias in nonhuman primates. *Lab. Anim. Sci.* 36:231–36.

Woolf, B. 1954. Statistical aspects of dietary surveys. *Proc. Nutr. Soc.* 13:82–94.

Wrangham, R. W., and P. G. Waterman. 1981. Feeding behaviour of vervet monkeys on *Acacia tortilis* and *Acacia xanthophloea:* With special reference to reproductive strategies and tannin production. *J. Anim. Ecol.* 50:715–31.

Wright, S. 1968. *Evolution and the genetics of population,* vol. 1. Chicago: University of Chicago Press.

Wuensch, K. L. 1978. Exposure to onion taste in mother's milk leads to enhanced preference for onion diet among weanling rats. *J. Gen. Psychol.* 99:163–67.

Wynn, V., and C. G. Rob. 1954. Water intoxication. *Lancet* 1:587–94.

Young, C. M., and M. F. Trulson. 1960. Methodology for dietary studies in epidemiological surveys. II. Strengths and weaknesses of existing methods. *Am. J. Public Health* 50:803–14.

Young, T. P., and W. K. Lindsay. 1988. Role of even-age population structure in the disappearance of *Acacia xanthophloea* woodlands. *Afr. J. Ecol.* 26:69–72.

Young, V. R. 1979. Diet as a modulator of aging and longevity. *Fed. Proc.* 1994–2000.

Yuen, D. E., and H. H. Draper. 1983. Long-term effects of excess protein and phosphorus on bone homeostasis in adult mice. *J. Nutr.* 113:1374–80.

Yule, G. U., and M. G. Kendall. 1953. *An introduction to the theory of statistics,* 14th ed., London: Charles Griffin.

Index

Two-letter codes for taxa (e.g., AT = *Azima tetracantha*) are the first two letters of four-letter codes for baboon foods (e.g., ATFR = *Azima tetracantha* ripe fruit), defined in tables 4.1 and 6.2 and on the last page of this book.

Plants are indexed under their scientific names, with cross-references from their vernacular names. Animals are indexed under their vernacular names, with cross-references from their scientific names.

Abbreviations for the Fifty-two Core Foods

ACDD Acacia seeds (fever or umbrella tree) picked from dung
ATFR Ripe fruit of *Azima tetracantha*
ATFU Green fruit of *Azima tetracantha*
ATFX Fruit of *Azima tetracantha,* ripeness unknown
CDLU Green leaves of Bermuda grass, *Cynodon nlemfuensis* = "*dactylon*"
CDSX Stolons of Bermuda grass, condition unknown
COCX Corms of *Cyperus obtusiflorus,* condition unknown
CTFX Fruit of *Capparis tomentosa,* condition unknown
DUKU Dung-beetle larvae
FTBS Fever tree blossoms picked from tree, *Acacia xanthophloea*
FTDG Abscised dry fever tree seeds picked up from ground
FTDR Fever tree seeds from dry pods
FTDU Fever tree seeds from green pods
FTGX Gum (condition unspecified) picked from fever tree
FTHG Fever tree gum picked up from ground
FTHP Fever tree gum picked from other plants
FTHW Fever tree gum picked from logs
GRCX Corms of unidentified grass or sedge, condition unknown
GRDU Green seedheads of unidentified grass or sedge
GRLU Green leaves of unidentified grass or sedge
GRTU Blade bases of green leaves of unidentified grass or sedge
INKX Unidentified arthropods
PCMX Milk
SBFX Fruit of *Solanum "dubium"* = *S. coagulans,* condition unknown
SCTX Blade bases of reed grass, *Sporobolus consimilis,* condition unknown
SFFX Fruit of "sticky-fruit plant." *Commicarpus pedunculosus,* condition unknown
SKCX Corms of *Sporobolus "kentrophyllus"* = *S. rangei,* condition unknown

SMCX Corms of *Sporobolus cordofanus,* condition unknown
SMLU Green leaves of *Sporobolus cordofanus*
SPFR Ripe fruit of toothbrush bush, *Salvadora persica*
SPFS Semiripe fruit of toothbrush bush
SPFU Green fruit of toothbrush bush
SPFX Fruit of toothbrush bush, ripeness unknown
SPLU Green leaves of toothbrush bush
SPLX Leaves of toothbrush bush, condition unknown
SPWX Wood of toothbrush bush, condition unknown
SULU Green leaves (and cryptic flowers?) of *Suaeda monoica*
TCFX Fruit of *Trianthema ceratosepala,* condition unknown
TFBS "Trumpet-flower" blossoms (*Lycium "europaeum"*) on plant
TFLX Leaves of "trumpet-flower plant," condition unknown
TOBR Abscised umbrella tree blossoms *Acacia tortilis* picked up from ground
TOBS Umbrella tree blossoms picked from tree
TODG Umbrella tree dry seeds picked up from ground
TODR Umbrella tree seeds from dry pods
TODU Umbrella tree seeds from green pods
TTFX Fruit of devil's thorn, *Tribulus terrestris,* condition unknown
WSFX Fruit of *Withania somnifera,* ripeness unknown
XXLX Unidentified leaves (not grass or sedge), condition unknown
XXWX Wood, unidentified
XXXD Unidentified items (mostly acacia seeds?) picked from dung
XXXG Unidentified items (mostly acacia seeds?) picked up from ground
XXXW Unidentified items (fever tree gum? arthropods?) picked from wood